THE RISE OF AMPHIBIANS

365 Million Years of Evolution

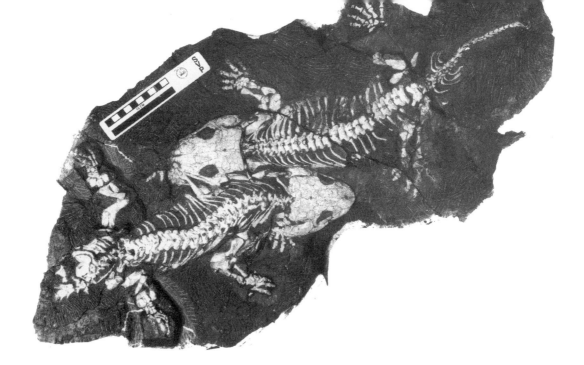

THE RISE OF AMPHIBIANS

365 Million Years of Evolution

ROBERT CARROLL

THE JOHNS HOPKINS UNIVERSITY PRESS

Baltimore

© 2009 The Johns Hopkins University Press
All rights reserved. Published 2009
Printed in the United States of America on acid-free paper
9 8 7 6 5 4 3 2 1

The Johns Hopkins University Press
2715 North Charles Street
Baltimore, Maryland 21218-4363
www.press.jhu.edu

Library of Congress Cataloging-in-Publication Data

Carroll, Robert Lynn, 1938–
    The rise of amphibians : 365 million years of evolution / Robert Carroll.
        p. cm.
    Includes bibliographical references and index.
    ISBN-13: 978-0-8018-9140-3 (hardcover : alk. paper)
    ISBN-10: 0-8018-9140-X (hardcover : alk. paper)
1. Amphibians—Evolution. I. Title.
    QL645.3.C37 2009
    597.8'138—dc22   2008024792

A catalog record for this book is available from the British Library.

*Special discounts are available for bulk purchases of this book. For more
information, please contact Special Sales at 410-516-6936 or specialsales@press
.jhu.edu.*

The Johns Hopkins University Press uses environmentally friendly book
materials, including recycled text paper that is composed of at least 30
percent post-consumer waste, whenever possible. All of our book papers
are acid-free, and our jackets and covers are printed on paper with recycled
content.

To the many students I have supervised during my professional career—undergraduates, graduate students, and postdoctoral fellows—who have assisted me in my research and continue to contribute to my education

CONTENTS

PREFACE

MODERN AMPHIBIANS—the familiar frogs and salamanders of the Northern Hemisphere and the less well-known, elongate, and limbless caecilians of the Southern Hemisphere—appear as relatively minor components of the living fauna, although they actually outnumber the dominant mammals in terms of both numbers of species and individuals. In fact, the three modern amphibian groups are but relics of a much larger radiation that occurred between 370 and 270 million years ago, when their ancestors dominated terrestrial and shallow water environments throughout the world. Of even greater significance, the early amphibians provided an evolutionary link between archaic fish and the ancestors of all more advanced terrestrial vertebrates: the reptiles, birds, and mammals. That is, they were our own ancestors.

We retain many of the skeletal features of the earliest known amphibians, which were able to move about on land more than 365 million years ago. They already exhibited paired limbs, with bones comparable to ours, broadly similar vertebrae, and homologues of our skull bones and sensory organs. But these animals had a much more primitive life history, for they laid eggs in the water and developed via aquatic larvae, as must have been the case for their immediate fish ancestors. This, of course, is the pattern of development observed in the most primitive living amphibians, which makes them potential models for the study of the transition between fish and fully terrestrial vertebrates.

On the other hand, modern frogs, salamanders, and caecilians are far different from one another in both their adult anatomy and ways of life, and none can serve as a model for the ultimate ancestors of terrestrial vertebrates. Fortunately, there is an extremely rich fossil record of earlier amphibians that documents their prior diversity and provides evidence of their origin from primitive fish, as well as relationships to the antecedents of the reptiles, birds, and mammals that subsequently came to dominate the earth.

Despite their importance in the history of life, neither fossil nor living amphibians have ever received the attention given to dinosaurs, birds, or mammals, and no previous book has attempted to document their long and eventful evolution, covering a period of over 365 million years.

Study of the evolution of life is like a detective story, looking for clues from bodies dead for millions or hundreds of millions of years. Like forensic workers, we must understand their anatomy, physiology, and way of life, as well as the cause of their extinction. We also must know when they died and who their relatives were. But as paleontologists or evolutionary biologists, we must also be able to reconstruct the long periods of behavioral, anatomical, and physiological change that linked animals as distinct as fish, amphibians, and reptiles, and to understand the processes and causal factors involved in adaptation to different ways of life and the origin of entirely new structures and body forms.

History, and especially the extremely long history of life, is of interest in itself, but may also help us to understand changes that are occurring in the modern biota and may be expected to change in the future. Human history has been but a brief moment in the evolution of life, but a more complete understanding of the past might help us to prolong our future, or at least that of the few surviving amphibian groups.

This book chronicles the history of amphibians from their origin among archaic fish, through their early dominance, a period of catastrophic decline, and the emergence of their humble living descendants. Our understanding of early amphibian history is based almost entirely on knowledge of the skeletal anatomy as revealed by their fossil record, for the vast majority of amphibian groups are totally extinct, without living survivors. However, the living amphibians do provide much information as to the probable soft anatomy and ways of life of their long-dead ancestors, which allows us to reconstruct their biological diversification during the 365 million years of their evolution.

More broadly, this book is intended not only as a narrative of amphibian history but also as a model, or example, of how evolution has occurred. This involves integration of numerous scientific disciplines, including paleontology (the study of ancient life), herpetology (the study of modern amphibians and reptiles), geology (the study of the earth and the processes that have influenced its history), and astronomy (the history and structure of the universe and the context of the earth within that history), as well as biology (including descriptive and functional anatomy, physiology, and especially evolution and its adaptive, genetic, and molecular basis).

Amphibians are but a single group of vertebrates, but in terms of their basic anatomy, physiology, and ways of life they share major attributes with other animals, including a backbone, an anteriorly placed brain with associated sense organs of smell, sight, and balance, and locomotion based primitively on lateral undulation of the trunk. More broadly, amphibians and other vertebrates share with all other multicellular animals a unique system of genetic control that enables them to evolve complex body parts associated with specialized means of feeding, locomotion, and reproduction.

To understand how these features evolved requires an even wider view, encompassing the very origin of life and the evolution of biochemical and genetic processes that unite all organisms on earth and their ancestors, back to the first appearance of organic compounds that accumulated in the waters of the earth approximately 4 billion years ago. It is only when we place amphibians in this larger context that we can begin to understand their total evolutionary history.

Considering amphibians in this broad context also demonstrates the overall similarities of scientific investigation in all disciplines. The study of evolutionary history depends on procedures shared with astronomy, chemistry, physics, geology, and the health sciences of observing natural phenomena and attempting to find logical and consistent explanations for their occurrence. These explanations take the form of hypotheses that can be tested by further observations or experimentation. Paleontology, like astronomy, is not directly amenable to testing in a laboratory, since one cannot expect to re-create the Big Bang or the origin of life on earth, but one can study the light generated by stars and galaxies millions or even billions of years old, and continue to search for earlier and earlier fossils, in an attempt to establish conditions that existed when they initially formed or first evolved.

ACKNOWLEDGMENTS

I OWE MY INTEREST IN paleontology to my father. Soon after my fifth birthday he showed me fossils of marine invertebrates that he used in teaching a high school science course. I immediately wanted to collect some myself. Collecting began on our farm and in local gravel pits, but later expanded to more distant areas in Michigan and finally to a number of classic vertebrate localities in the western United States, interspersed with visits to many museums. My professional career was most strongly influenced by Alfred Sherwood Romer, who introduced me to the world of Paleozoic amphibians through collecting in the Permian red beds of Texas and a thesis on the armored dissorophids. Romer was a superb supervisor, teaching by example, with the highest research standards in the profession. I also benefited greatly from classes in evolution taught by George Gaylord Simpson and Ernst Mayr and in herpetology by Ernest Williams.

Dr. Romer further broadened my career by arranging for postdoctoral research on the unique fauna of amphibians and reptiles from the upright trees of Joggins, Nova Scotia. This led me first to the Redpath Museum, McGill University, Montreal, where my work was greatly facilitated by Louise Stevenson, and the director, Alice Johannsen, who later offered me positions in the museum and in the Department of Biology, which I have occupied throughout my career. I have profited greatly from the atmosphere of these departments, which have enabled me to supervise a host of undergraduates and graduate students who have greatly increased our understanding of the anatomy and relationships of fossil amphibians and reptiles.

This book could not have been written without the help of colleagues in many other universities and museums throughout the world. These are listed in approximately chronological order: Wann Langston, then at the National Museum of Canada, Ottawa; Errol White and Allan Charig, British Museum (Natural History), London; D. M. S. Watson, University College, London; Alec Panchen, University of Newcastle upon Tyne; Rex Parrington, University Museum of Zoology, Cambridge; Dr. Zăvorka, National Museum of Prague; Hermann Jaeger, Humboldt Museum,

Berlin; Dr. Blüher, Geological Survey of Saxony, Freiberg, Germany; Hermann Prescher, State Museum for Mineralogy and Geology, Dresden; E. C. Olson and Peter Vaughn, University of California, Los Angeles; James Kitching, Bernard Price Institute for Palaeontological Research, Johannesburg, South Africa; Michael Williams and David Dunkle, Cleveland Museum of Natural History; Ian Rolfe and Mahala Andrews, Royal Scottish Museum, Edinburgh; and most recently, collaboration with Hong Li and Qing-Long Shao, director, Inner Mongolia Museum, Hohhot, on the study of Middle Jurassic salamanders.

Special thanks for information on living and fossil caecilians are due to Marvalee Wake for providing a wonderful environment for the study of modern caecilians in her lab at the University of California, Berkeley, and to Farish Jenkins (Harvard) and Denis Walsh (University of Toronto) for inviting me to participate in the study of *Eocaecilia*.

Financial support for the research on which this book was based was provided by the Natural Sciences and Engineering Research Council of Canada.

I thank Vincent Burke, my editor at the Johns Hopkins University Press, for asking me to write this book and for all his support during its genesis. This volume owes its color to Tonino Terenzi, who produced the plates. For assembling, arranging, and labeling the other illustrations, I am greatly indebted to Mary-Ann Lacey, who also attended to other technical aspects of integrating the text. David Green, Redpath Museum, contributed to the final chapter, but also assisted throughout in providing information on the modern amphibian groups. The final form of the text owes much to the proofreading by my wife, Anna Di Turi, to whom I am also grateful for her patience during the many hours I spent bent over a microscope or computer.

THE RISE OF AMPHIBIANS

365 Million Years of Evolution

1

History of the Earth and Life

WHEN WE THINK OF AMPHIBIANS, what immediately comes to mind are the distinctive frogs and salamanders that occupy a seemingly modest role in the ponds and wetlands of our modern environment (Fig. 1.1). However, their biphasic life history, involving the passing from aquatic larvae to terrestrial adults, has long provided biologists with a model for the evolutionary transition between ancestral fish and fully terrestrial vertebrates, including reptiles, birds, and mammals.

Fossils, first discovered nearly 150 years ago, suggested that animals vaguely resembling modern salamanders were among the most primitive land-dwelling vertebrates, indicating a very long history (Dawson and Owen, 1862). We now know that animals with a reproductive physiology broadly similar to that of common frogs and salamanders dominated the earth early in their history and gave rise to all other terrestrial vertebrates (Fig. 1.2).

Most of this book will be devoted to the history of amphibians, from their origin among archaic fish more than 370 million years ago to the present, but to place amphibians within a broader context, this chapter deals with the much longer time scales of the history of the earth and universe, the geological processes of the early earth, and the origin of life leading, eventually, to the appearance of primitive vertebrates.

Nearly everyone has an appreciation of the long span of human history and knows of the prior existence of prehistoric humans, woolly mammoths, and dinosaurs. However, it remains more difficult to imagine the earth prior to the origin of any of the common plants or animals we know today—no flowering plants, no mammals, birds, lizards, turtles, or even insects. Nor can we readily picture the earth without any living organisms—a barren land and sterile sea, beneath an atmosphere with no appreciable oxygen. Yet, this is what has been revealed by geological studies of the history of our world, going back more than 4 billion years.

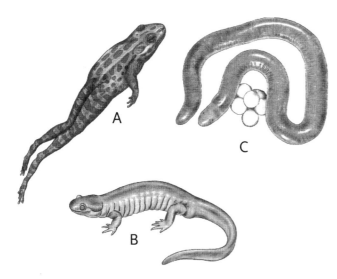

Figure 1.1. The three major groups of living amphibians. *A*, Frogs (Order Anura). *B*, Salamanders (Order Urodela). *C*, Caecilians (Order Gymnophiona).

THE EARTH AND TIME

In order to appreciate this enormous time scale and the nature of changes that occurred on our planet prior to the emergence of vertebrates, it is necessary to have a general familiarity with the geological history of the world, the nature of geological dating, and the changes in the atmosphere, climate, and geography of the earth's surface. We start by considering the context of the earth's history in terms of the largest time scale possible, that of the universe as we know it (Fig. 1.3). Based on the rate of expansion of the universe since the Big Bang, it is currently dated at about 13.7 billion years (Ellis, 2005; Kashlinsky et al., 2005; Stark et al., 2007). This seems an almost immeasurable period of time, but the earth itself is estimated as being approximately 4.6 billion years old, as determined by the greatest age of meteorites that have struck the earth and are assumed to have come from within the solar system (Baker et al., 2005). At that time, and for hundreds of millions of years later, the surface of the planet was molten, and so no evidence of rocks formed on earth is available for dating until well after its initial consolidation. The currently oldest date for rocks that have survived from the early crust is 4.2 billion years (Menneken et al., 2007).

Dating of ancient meteorites and the early crust of the earth is based on the decay of radioactive elements that they contain (Villeneuve, 2004). A number of naturally occurring elements are unstable and decay through time to other isotopes or elements. Such decay proceeds at a constant rate over very long periods of time. Hence, the relative amounts of the original element and its products of decay give a measure of how long ago the rock that contains these elements formed. Elements that decay slowly, over long periods of time, are useful in determining the age of the geological horizons in which they are found. The rate of decay is measured in terms of the half-life of the element—the period over

which one-half of the original mass is lost. For the age of the earth and the periods of time during which amphibians and other vertebrates have lived, the most effective procedures so far developed involve the decay from uranium (U) to lead (Pb) and potassium (K) to argon (Ar). Two modes of decay from uranium are used: ^{235}U (with a half-life of 704 million years) to ^{207}Pb and ^{238}U (with a half-life of 4.5 billion years) to ^{206}Pb. The decay of ^{40}K to ^{39}Ar can be measured directly, but most measurements are now made following irradiation of the potassium to ^{40}Ar. This procedure is referred to as the ^{40}Ar/^{39}Ar method. It is highly accurate and can be accomplished with much smaller samples. The half-life of ^{40}K is 1.25 billion years.

The accuracy of these different means of measuring the age of rocks can be judged by the relative degree of concordance between several samples taken from the same geological horizon, using the same and different isotopes, and from relative differences from the measured age of other geological horizons. The margin of error is considered to be less than 1%, with uncontaminated and unaltered samples. Radiometric dating, sometimes referred to as absolute dating since it provides dates in terms of thousands, millions, or billions of years, provides a reliable temporal framework for the sequence of the important events in the history of the earth and the dating of major geological horizons.

Unfortunately, it is only rarely that rocks that can be used for radiometric dating are found in close association with fossil-bearing horizons. Most radioactive elements are preserved in rocks that were formed under conditions of great heat and pressure, either within the earth or from volcanic eruptions, which would destroy any evidence of plant or animal remains. However, such rocks can provide a guide to the age of fossil-bearing beds if they are in close proximity. If the fossils are found above well-dated igneous layers, their age must be younger, and if below, they must be older. Accurate dating by radiometric means becomes progressively more effective in successively younger beds.

The problem is that long periods within the Paleozoic and Mesozoic lack appropriate horizons for radiometric dating. For example, the Carboniferous, during which the initial divergence of all major groups of amphibians occurred, has 21 radiometrically determined sites, but few, if any, can be associated with sedimentary horizons containing fossil vertebrates (Davydov et al., 2004). In fact, the Paleozoic and Mesozoic beds that contain fossils are only rarely dated directly by radiometric means, but rather by their position relative to other fossiliferous horizons. In many areas in which fossils are found, they occur within a sequence of deposits, one above the other. Some or all of these beds can be consistently recognized by distinctive fossils that are also present in other geographic areas.

While large-scale events in the history of amphibians, as well as those of other animals and plants, can be fitted into a well-established radiometric time scale, details of species succession and the longevity of particular groups are still

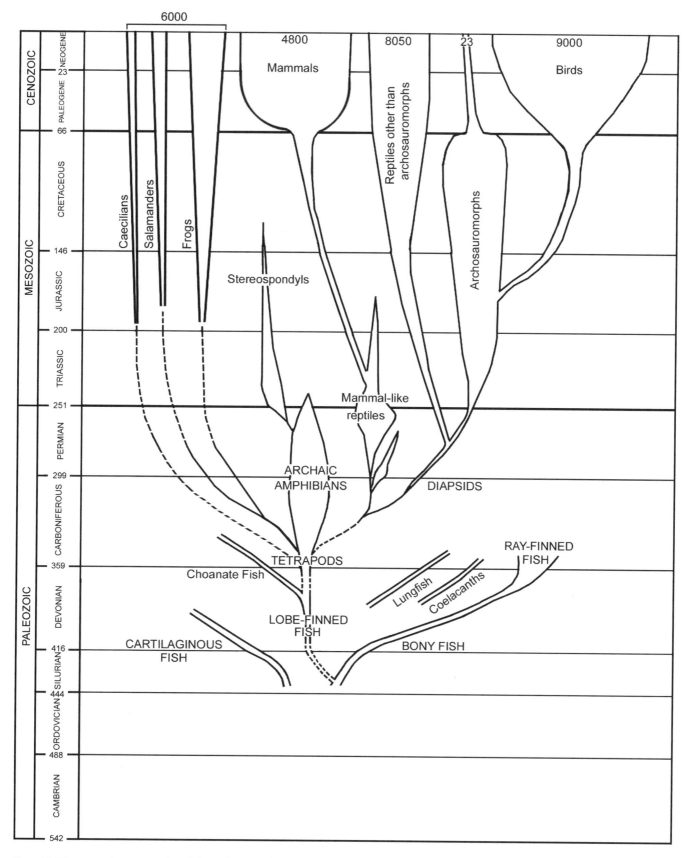

Figure 1.2. Diagrammatic representation of the evolutionary history of terrestrial vertebrates, indicating the times of their divergence and relative diversity. The numbers at the top indicate the approximate number of living species. The width of the lineages indicates the approximate number of families during each of the geological periods, with each mm representing approximately three families. Because of the incompleteness of the fossil record, the number of individual species discovered is extremely small. However, the chance of finding at least one species of a family is consider-ably larger, and so provides a plausible proxy for comparing their diversity throughout evolutionary history. Thicker horizontal lines indicate the times of major extinctions at the end of the Permian, when a great many groups of marine invertebrates died out, and at the end of the Cretaceous, when the dinosaurs and many marine reptiles became extinct. Figures in the left column are in millions of years. Based on data from R. L. Carroll, 1988a, 2001b; Frost et al., 2006; Minelli, 1993; Pough et al., 2004.

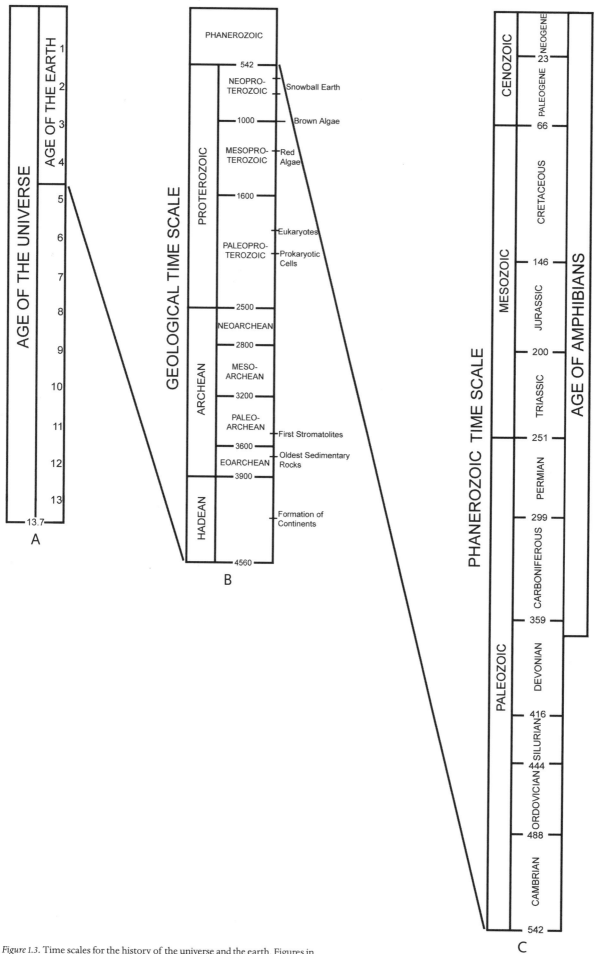

Figure 1.3. Time scales for the history of the universe and the earth. Figures in column A are in billions of years. Based on data from Gradstein et al., 2004.

measured by their presence or absence in particular horizons, distinguished by other species. For example, the sequence of horizons in the Carboniferous that bear fossil amphibians is determined primarily by the presence of particular species of terrestrial plants, which are much more common and whose stratigraphic ranges are much better known. This is referred to as relative dating.

As early as the beginning of the 19th century, long before the development of radiometric dating, geologists recognized a series of stages in the history of life on earth. For sedimentary rocks, three eras were identified, based on the relative similarity of their fossils to modern plants and animals. These were designated the Paleozoic Era (a time of very primitive forms of life), the Mesozoic Era (middle life), and the Cenozoic Era (modern life). Collectively, these three eras were termed the Phanerozoic Eon (a time of conspicuous life). The Phanerozoic is divided into 11 geological periods. The Paleozoic consists of the Cambrian, Ordovician, Silurian, Devonian, Carboniferous (in North America divided into the Mississippian and Pennsylvanian), and Permian. The Mesozoic includes the Triassic, Jurassic, and Cretaceous, and the Cenozoic has only two periods, the Paleogene and Neogene. Alternatively, the Cenozoic has been divided into seven epochs: Paleocene, Eocene, Oligocene, Miocene, Pliocene, Pleistocene, and Holocene (also termed Recent).

Other rock units, generally located beneath these sediments and lacking readily identifiable fossils, were later recognized as representing yet earlier stages in the earth's history: the Proterozoic Eon (before life) and the Archean Eon. Together, these are referred to as the Precambrian. The oldest Archean rocks that have been dated are approximately 3900 million years old. The heavy meteorite bombardment of the earth that peaked at about that time presumably destroyed most of the pre-existing terrestrial crust (Cohen et al., 2000). The earliest period in the history of the earth, going back 4.56 billion years ago, is referred to as the Hadean, in reference to the great heat of the earth's molten crust.

PLATE TECTONICS

The evolution of the earth did not stop with the initial solidification of the crust, but has continued throughout its history in a manner that has greatly influenced the evolution of life. Nineteenth- and early 20th-century geologists recognized that many features of the earth's surface have changed throughout its history—mountains rose and were eroded, oceans expanded over the margins of continents and subsequently retreated, and major portions of continents were uplifted and folded. However, until the 1960s, it was assumed that the general position of the oceans and continents had remained essentially as we see them today, since at least the beginning of the Paleozoic.

Beginning in the 1950s, a great deal of new information was gained about the structure and history of the ocean basins as a result of advanced techniques for mapping the

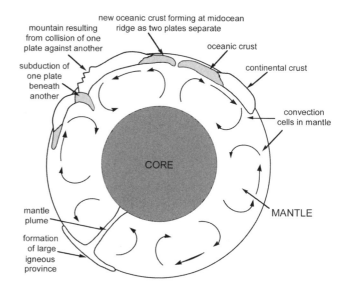

Figure 1.4. The mechanism of continental drift. Heat is generated in the earth's core, which results in circulation within the mantle. Movement in the upper mantle transports the crustal plates. They move away from one another where heat from the mantle is rising, and converge where the currents are sinking. Mantle plumes, which form large igneous provinces, arise from the base of the mantle independent of the position of the crustal plates. From R. L. Carroll, 1997.

seafloor and determining the changes in magnetic polarity of lava flows that run parallel to the mid-oceanic ridges. Ridges, such as that which extends from north to south along the middle of the Atlantic Ocean, were recognized as foci of expansion of the ocean bed that resulted in the progressive separation of North and South America from Europe and Africa, a process termed continental drift (Fig. 1.4). At the same time, evidence was accumulating that other areas of the earth's crust, such as the western edge of North and South America and the plates beneath the Pacific Ocean, were moving toward one another, forcing the oceanic crust beneath the continental crust (subduction) and pushing up the Andes and the coastal mountain ranges in North America.

As knowledge of crustal movements accumulated, a model of the forces of the earth was developed that explained the drifting of the continental plates (Hess, 1962; T. J. Wilson, 1965). It has long been known that the core of the earth not only retained some of the original heat resulting from the formation of the solar system, but has maintained a high temperature as a result of radioactive decay of elements such as uranium. The heat from the iron-nickel core of the earth is transferred to the lighter rocks of the mantle in somewhat the same way as heating up water in a saucepan. The heated, semi-fluid rocks of the mantle move upward in the areas that form the mid-oceanic ridges. From these ridges, the molten rock spreads out laterally, pushing the rigid oceanic and continental plates before it. As the crust and underlying mantle slowly cool, a great distance from where they emerged, they begin to sink back toward the core.

The first evidence for the formation of continental plates dates from approximately 4.2 billion years ago (Knoll, 2003).

Initially, they may have been small, but movements in the mantle resulted in accretion into larger continents and their subsequent break-up. The configuration and position of the continents has profoundly affected the history of amphibians and other forms of life in relationship to changing climates, the potential for dispersal, and periods of mass extinction. A very readable account of the nature of continental drift and its significance to the history of life is provided by Rogers and Santosh (2004) in their book *Continents and Supercontinents*.

As a result of continuous subduction of oceanic crust beneath the continents, no areas of the oceanic basins have survived for more than 200 million years. Continental rocks that have not been subducted may retain their original structure much longer, but all are subject to erosion. Unless sediments are covered by overlying deposits, they are eroded and redeposited elsewhere. The older the rocks, the less likely they are to be preserved in their original condition. Hence, the less likely any fossils they contain will be preserved.

An example of the incompleteness of the fossil record is provided by the state of Michigan, where I grew up. The lower peninsula has a very simple geological structure. It was a huge basin. The central portion exposes little but rocks deposited during the Pennsylvanian Period (also called Upper Carboniferous), dating from about 300 to 310 million years ago. Within these beds are found a great many fossils of plants that formed the great coal deposits of that time. Around the edges of the basin are exposed limestones and shales of the underlying Devonian Period, popularly referred to as the age of fishes, dating from 360 million years ago to about 375 million years ago. In it were preserved marine fossils, including those of fish and many invertebrates, but no terrestrial plants or animals. But all the sediments that might have been deposited on land or in the sea over the past 300 million years were eroded away or covered by the glacial debris that was repeatedly spread across the state over the past one million years. Hence, we have no fossil evidence of any organisms that lived in lower Michigan since the end of the Carboniferous. Now I live in Canada, where thousands of square miles of dinosaur beds are exposed in Alberta, sadly covering up all the deposits where Paleozoic amphibians may lay buried, but inaccessible. The farther back in time you go, the less complete the sedimentary and fossil record, and so the more speculative is the interpretation of the history of life (Fig. 1.5). This problem is especially critical in the Proterozoic and Archean, but remains highly significant, especially for amphibians, into the Cenozoic.

THE BEGINNING OF LIFE

A period of more than 4 billion years passed between the formation of the earth's crust and the origin of amphibians, during which time there were slow, successive changes in the physical and biological aspects of the world that made it possible for the eventual emergence of vertebrates on land. The

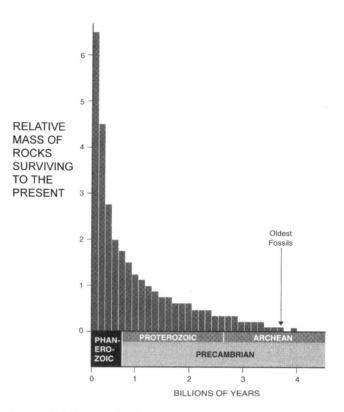

Figure 1.5. Relative mass of rocks surviving to the present. The older the rock, the greater the chance that it no longer exists. From Schopf, 1999.

early stages of these processes were discussed in a highly informative manner by Andrew Knoll in his 2003 book, *Life on a Young Planet*.

What is most striking is the extremely long period between the cooling of the earth's crust to the temperature at which water could persist in the liquid state and the appearance of the most primitive vertebrates. More than 3 billion years were necessary for changes in the chemistry of the atmosphere and oceans to permit the accumulation of oxygen to a level that would support the metabolism of complex multicellular animals.

There is no fossil evidence for the earliest stages in the origin of life. However, a reasonable guess of how it began can be made on the basis of the probable constituents of the ancestral atmosphere and the nature of the rocks of the early earth. Based on the relative proportions of elements in the solar system, the early atmosphere of the earth was probably composed primarily of nitrogen, hydrogen, carbon dioxide, and water vapor. Geologists have concluded that the early atmosphere contained only a very small percentage of oxygen because of the presence of unoxidized iron, sulfur, and uranium in the rocks of the early crust that would have been fully oxidized if a significant amount of oxygen had been available. What is particularly interesting is that early forms of life themselves must have played a vital role in providing the oxygen that was necessary for the origin of advanced plants and animals.

Once water was available in a liquid state on the surface of the earth, a host of chemical reactions could have taken place in this medium. The diversity of these reactions was increased as the action of falling rain and the currents of streams and rivers eroded the earth's surface and washed a variety of rocks and their contained elements into lakes and the primordial ocean. Important constituents included carbon, sulfur, and phosphorus, which make up the tissues of all organisms. Early in the 20th century, scientists hypothesized that life might have begun spontaneously, but this was not demonstrated until 1953, when Miller passed heated water through a mixture of methane (CH_4), ammonia (NH_3), and hydrogen (H_2), and further activated the system with electric discharges, such as would have been produced by lightning on the early earth. After a week, analysis of the resulting molecules indicated the presence of numerous sugars and amino acids common to living systems and other molecules that are required in the synthesis of nucleic acids (Fig. 1.6). A comparable experiment, but including carbon monoxide, resulted in the formation of purine and pyrimidine nucleotides—bases that are constituents of ribonucleic acid (RNA) and deoxyribonucleic acid (DNA). Many similar molecules have also been extracted from stony meteorites (carbonaceous chondrites), proving they can be formed naturally. However, the left- and right-handed isomeres of meteoritic

amino acids occur in about the same frequency, in contrast with living organisms, in which only the left-handed forms are incorporated (Miller, 1992).

Most, if not all, of the basic molecules common to living cells could have been formed in the atmosphere and in pools and larger water bodies of the early earth. The larger and/or more tightly bound molecules might have accumulated preferentially, while smaller ones were probably broken down or integrated into other combinations. One may think of this as a chemical analogue of natural selection, or survival of the "fittest" molecules.

The key to the origin of consistently self-replicating molecules is thought to lie with RNA. RNA is the only nucleic acid present in some viruses, where it acts as the sole genetic material and also serves as a messenger for protein synthesis (messenger RNA). RNA was probably the only nucleic acid in the most primitive cells. Under certain conditions it can act as a catalyst for its own replication (Zubay, 2000; Gesteland et al., 1999). In the early stages leading to the origin of life, the replication of short sequences of nucleic acid would have had little significance beyond the accumulation of large quantities of RNA with different base sequences. On the other hand, if the specific sequence that occurred by chance in some RNAs facilitated their capacity to polymerize or to replicate themselves using the advantageous sequence as a template, they would be perpetuated at the expense of other RNAs with less advantageous sequences. Such a consistently replicating molecule may be considered the first gene.

Life may have been initiated via the accumulation of such isolated molecules, but it would not have progressed much farther without some kind of a container that could serve for the retention of a variety of molecules that would collectively form a functional unit (simple sugars and amino acids to form polysaccharides and proteins, enzymes to facilitate specific reactions, and naturally occurring phosphates to provide a means of energy transfer for polymerization and other chemical reactions). Fortunately, other molecules that could have formed in the early lakes and oceans, the phospholipids, have the inherent capacity to form laminar and spherical structures comparable to cell membranes. Under conditions of drying and rehydration, phospholipid spheres can surround other molecules and form a cell-like structure (Fig. 1.7). This may bring us close to the level of a regularly self-replicating cell, but much more genetic information would be necessary to reach the level of the most primitive free-living bacteria, which carry a minimum of approximately 450 different genes (Graur and Li, 2000).

No fossils of a bacterial level of evolution are known that represent the earliest stages in the history of life. However, biological activity can be recorded in the sediments in which these early organisms lived. An important aspect of biological processes is that one of the isotopes of carbon, the key to all life, is more likely to be incorporated into these reactions than others. Molecules of calcium carbonate ($CaCO_3$) that

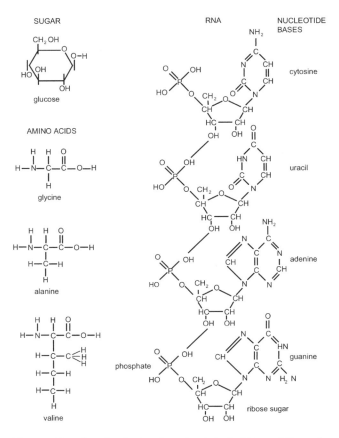

Figure 1.6. Organic molecules that formed early in the history of the earth, and later became incorporated in living cells.

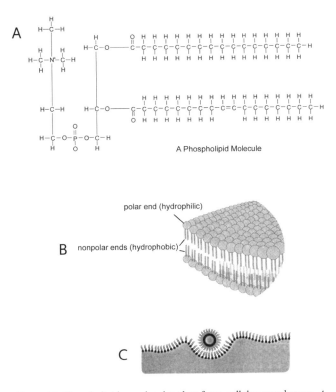

A Phospholipid Molecule

Figure 1.7. Phospholipids: molecules that form cellular membranes. *A,* Phospholipids are characterized by their polarity. The hydrophilic end, to the left, orients toward water, and the hydrophobic tails, on the right, avoid the water. When phospholipid sheets are forced into the water, they form a double sheet (*B*), resembling the cellular or nuclear membranes, with the hydrophobic surfaces facing one another. *C,* Double membranes may also be formed by the incorporation of a water droplet or some cellular element, covered by a single membrane, if it is engulfed by another element covered by a single membrane.

occur in inorganic rocks, such as limestone, have a fixed ratio of the isotopes ^{12}C to ^{13}C of approximately 99 to 1. However, in living organisms, in their fossils, or in the sediments in which they are preserved, the proportion of ^{12}C is increased by 25 to 30 parts per thousand because the lighter isotope is slightly easier to incorporate in biochemical reactions. This small difference, even in rock samples more than 3 billion years old, can be detected by use of a mass spectrometer. This degree of enrichment of ^{12}C in carbon-rich sediments, relative to associated limestones, has been detected in rocks more than 3.7 billion years old indicating the presence of organic molecules (Rosing, 1999). As yet, there is no evidence of cellular structures of this age. What organic molecules there may have been 3.7 billion years ago cannot yet be demonstrated as belonging to self-replicating structures that we would call living organisms. However, they had certainly evolved by 3.5 billion years ago.

BLUE-GREEN ALGAE AND THE ORIGIN OF PHOTOSYNTHESIS

Fossil localities in southern Africa dated at 3.3 to 3.5 billion years old and from the Warrawona Formation in west-

ern Australia dated at 3.458 to 3.471 billion years include a great number of structures that closely resemble those formed by modern-day photosynthetic blue-green algae (cyanobacteria). These structures are termed stromatolites and consist of concentric layers of limestone that are formed by the precipitation of calcium carbonate above colonies of bacteria (Fig. 1.8). The process of photosynthesis removes carbon dioxide from the water, raising the pH, so that calcium carbonate, originally dissolved in the water, precipitates, covering the growing surface. The cyanobacteria pass through the calcium carbonate as it precipitates, so that the stromatolite slowly grows higher and higher, maintaining a position near the surface of the water above the level of accumulating sediments. The close similarity in general appearance and the specific dimensions of the lamellae of modern stromatolites and those known as fossils from the Archean indicate a similar structure and way of life of the organisms that produced them.

It is assumed that, as in more advanced plants, oxygen was produced as a result of removal of carbon from carbon dioxide. Judging from very primitive living bacteria, oxygen was probably toxic to all early organisms but was almost certainly transferred to incompletely oxidized iron, which was common in the early seas. This reaction is evidenced by the extensive deposits of banded iron formation beginning about 3.5 billion years ago, coincident with the first appearance of stromatolites.

Banded ironstone occurs in the form of alternating black and bright red layers of chert (silicon dioxide). The red bands are composed of the mineral hematite (Fe_2O_3), which is soluble in water and so was incorporated in sediments that accumulated on the ocean floor. If there were significant amounts of oxygen in the atmosphere, it would have bound to the iron, forming a completely oxidized molecule that would have immediately precipitated out of solution and not been assimilated in the sediments. The increasing rarity of banded iron formation in the later Proterozoic deposits demonstrates the increasing percentage of oxygen in the oceans and atmosphere, but even in the late Proterozoic it may not have exceeded 1% of its current value.

Blue-green algae belong to the eubacteria, one of the three major divisions of life, the others being archaeobacteria and eukaryotes. The eubacteria and archaeobacteria are thought to have diverged by the early Archean, but even their living relatives retain very primitive structures and ways of life reflecting the environment of their earliest ancestors. They are collectively described as prokaryotes, based on the absence of a nuclear membrane and the small size and circular configuration of their typically single chromosome. Many of the living prokaryotes are either obligatorily anaerobic or restricted to environments with very low oxygen, as was certainly the case of the early earth up to approximately 1.8 billion years ago.

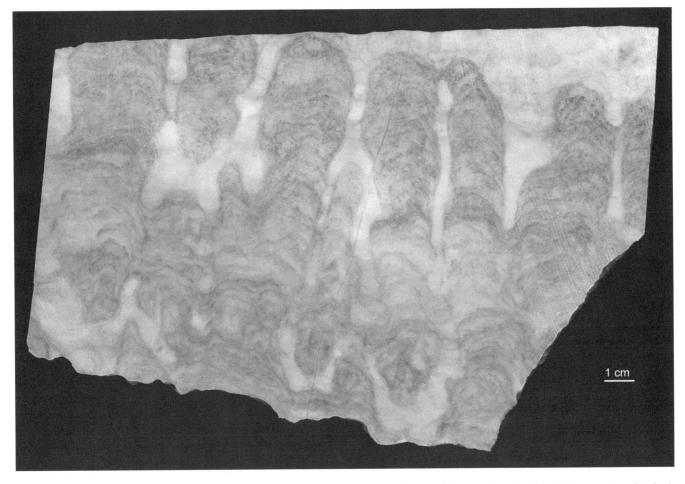

Figure 1.8. The polished surface of a Precambrian stromatolite from northern Canada, showing the succession of layers formed by the deposition of calcium carbonate by blue-green algae. Specimen in the collection of Redpath Museum, McGill University.

Both modern and Archean stromatolites appear as large structures with irregular and sometimes branching columns 2.5 cm or more in diameter. The columns typically occur in great colonies that may extend for long distances in fossil outcrops. Unfortunately, there is little evidence of the microorganisms that were assumed to form the stromatolites seen in Archean rocks. Schopf (1999) described what appear to be fossils of cells from the Warrawona Formation the size of those belonging to modern prokaryotes, but other geologists have argued that they are more likely the results of precipitation of minerals along fine cracks in the sediments (Brasier et al., 2002). More convincing evidence of fossil remains of early Archean organisms is provided by filamentous carbonaceous structures thought to represent microbial mats of photosynthetic prokaryotes from 3.416 billion-year-old sediments from Swaziland, South Africa (Tice and Lowe, 2004).

The mechanism of photosynthesis based on chlorophyll *a*, common to cyanobacteria and photosynthetic eukaryotes, apparently evolved prior to 2.7 billion years ago as indicated by the presence of distinctive lipid molecules, unique to blue-green algae, found in sediments of that age in northwestern Australia (Brocks et al., 1999; Knoll, 2003).

It is not until 1.9 billion years ago that a great diversity of well-preserved individual prokaryotic (lacking a nucleus) cells are known (Fig. 1.9). They are represented by numerous, exquisitely preserved specimens from the Gunflint Chert of northern Ontario (Barghoorn and Tyler, 1965). The gunflint microfossils resemble some living prokaryotes, but most are linked only to rare living counterparts, such as iron-loving bacteria that are only common in isolated iron-rich but oxygen-poor environments of today's oceans. The limited availability of atmospheric oxygen at the time of the Gunflint Formation and most other early Proterozoic localities is demonstrated by the widespread occurrence of banded iron formation, a type of rock that is common in marine deposits dated from 3.5 to 1.85 billion years in age.

What is evident from the well-preserved cells of the Gunflint Formation is that none reached the size or level of complexity common to even the most primitive eukaryotes (identified by the presence of a nuclear membrane). Nor are there any other fossils known prior to the middle of the Proterozoic that can be identified as eukaryotes, although evidence of steranes (chemicals similar to cholesterol), known

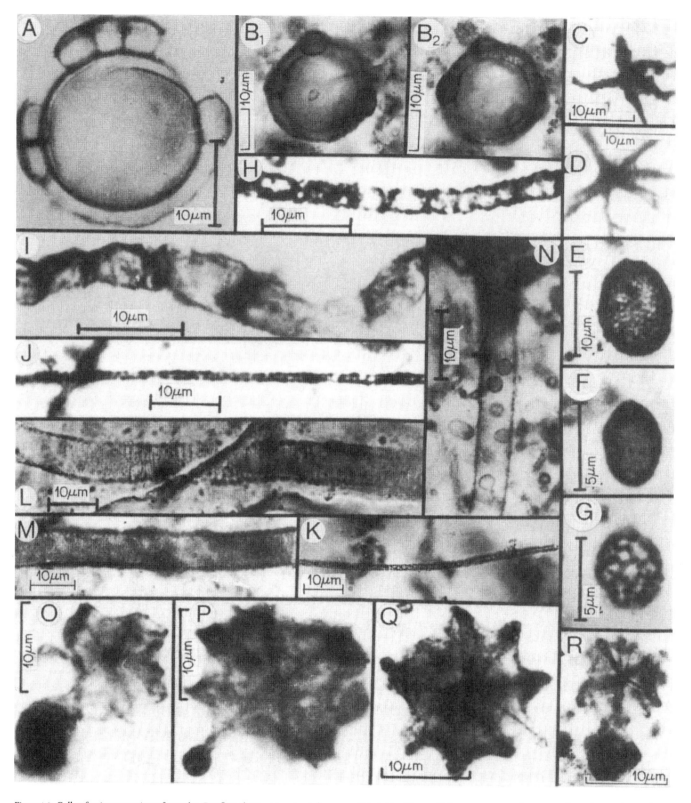

Figure 1.9. Cells of microorganisms from the Gunflint chert, approximately 2 billion years in age. *A* and *B, Eosphaera. C* and *D, Eoastrion. E, F,* and *G, Huroniospora. H–K, Gunflintia. L* and *M, Animikiea. N, Entosphaeroides. O–R, Kakabekia.* None of these genera are known to have living descendants. From Schopf, 1999.

today primarily in eukaryotes, have been determined from 2.7 billion-year-old shales (Brocks et al., 1999). Fossils of cyanobacteria that more closely resemble modern genera are known from beds in the Belcher Islands on the eastern shore of Hudson Bay, which, at 2 billion years, are even older than those of the Gunflint Formation, but presumably represent a different environment that had a locally higher oxygen content.

THE EARLIEST EUKARYOTES

Andrew Knoll (1995) described remains of eukaryotic cells, broadly referred to as acritarchs, from as early as 1.7 to 1.9 billion years ago. These are thought to represent cell walls of vegetative and reproductive cells of protists although they might be reproductive cysts of algae. Other cells attributed to eukaryotes come from 1.5 billion-year-old shales from Australia (Javaux et al., 2001). They are spheroids, comparable in size to some modern eukaryotic cells. They are characterized by having one or more slender tubes extending from their walls, similar to those of some living protists, for passage of reproductive cells.

However, the oldest fossils that can be compared with confidence to any living eukaryote phylum are beautifully preserved red algae, recognized on the basis of diagnostic patterns of cell division, from 1.2 billion-year-old chert deposits from Somerset Island, Canada (Fig. 1.10). These plants already show cell differentiation, with a distinct basal holdfast, a long stem, and a terminal reproductive structure. Differential spore/gamete formation indicates that this plant is the earliest known organism to practice sexual reproduction (Butterfield, 2000). Green algae, which share chlorophyll *a* with red algae, is known by 750 million years ago, and brown algae by 1 billion.

Unicellular eukaryotes that are not photosynthetic (the protozoans) do not appear in the fossil record until near the end of the Proterozoic, although they may have evolved considerably earlier. The earliest known fossils are ciliated forms related to dinoflagellates and vase-shaped structures closely resembling modern testate amoeba that have been collected in large numbers from 750 million-year-old sediments in the Grand Canyon (Porter and Knoll, 2000). Some of the vase-shaped animals display hemispherical perforation that may have been made by other, predatory protozoans, suggesting initial diversification of protozoans sometime in the mid-Neoproterozoic.

While there may have been earlier eukaryotes, it must have required more than 1.5 billion years, since the first appearance of prokaryotes, for eukaryotes to have become sufficiently common and distinctive to be recognizable in the fossil record. By the time modern phyla appeared, they were clearly distinct from all prokaryotes. This implies a long period of evolutionary change, perhaps spent in a restricted environment (such as one unusually rich in oxygen) that was unlikely to be represented in the fossil record.

ORIGIN OF EUKARYOTES

Even the most primitive single-cell eukaryotes, the protists, are clearly distinct from prokaryotes in the presence of a nuclear membrane, much longer, linear chromosomes associated with a mitotic apparatus, mitochondria, and other intracellular organelles. Most striking is the fact that eukaryotes cannot be directly traced to any one of the known groups of eubacteria or archaeobacteria. In fact, eukaryotes did not evolve by the typical Darwinian process of progressive change from an ancestral pattern, but rather by the process of symbiosis, a joining together of cells belonging to very distantly related organisms. A symbiotic origin of the eukaryotic cell was postulated by Konstantin Merezhkovsky in 1905, but was only documented by Lynn Margulis in 1967 through the use of the recently perfected means of electron microscopy (Margulis, 1981). She showed that particular intracellular organelles, the mitochondria of all eukaryotes and the chloroplasts of plants, were surrounded by two cell membranes, one attributable to the organelle and the other to the cell in which it resided. Subsequent studies have shown that the mitochondria share many attributes with proteobacteria among the archaeobacteria, while the chloroplasts in photosynthetic organisms can be closely compared with blue-green algae.

It is more difficult to determine the nature of the host cell that in some way engulfed these initially free-living organisms. In general, the cell membranes of archaeobacteria and eubacteria are too rigid for them to engulf entire cells; they are not predatory, but exchange nutrients and metabolic wastes through the cell membrane. They lack the highly controllable cytoskeleton of eukaryotes, although it has been demonstrated that they possess molecules with similarities to actin, which is a critical component of the eukaryote cytoskeleton (Doolittle, 1995). The ancestral eukaryote may have had an amoeboid form, with cellular machinery to form pseudopodia and a complex of genes associated with cytokinesis and flagellar movement not known in any modern prokaryotes (Richards and Cavalier-Smith, 2005). This suggests that eukaryotes may have evolved from among prokaryotes without any recognized living descendants.

As yet, there is no fossil evidence for the time of occurrence of the initial symbiosis between a relatively large prokaryotic cell with a cytoskeleton and a proteobacterium that was to become a mitochondrion. However, this must have occurred prior to the appearance of the oldest known photosynthetic eukaryotes, perhaps more than 1.9 and possibly as many as 2.7 billion years ago. Yet, the first multicellular plant belonging to a modern phylum did not appear until 1.2 billion years ago, and the oldest eukaryotic animals are not known until 750 million years ago.

There are several factors that may explain the failure of multicellular plants and animals to radiate extensively until near the Precambrian-Cambrian boundary. One was the persistently low levels of atmospheric oxygen, as indicated by the continuing formation of banded ironstone until at least 1.8 billion years ago and the failure of eukaryotic algae to spread into the oceans until about 1.2 billion years ago. The other factor may have been the long time required for the integration of the mitochondria and chloroplasts, and the formation of the mitotic apparatus necessary for controlling the separation and movement of multiple, elongate chro-

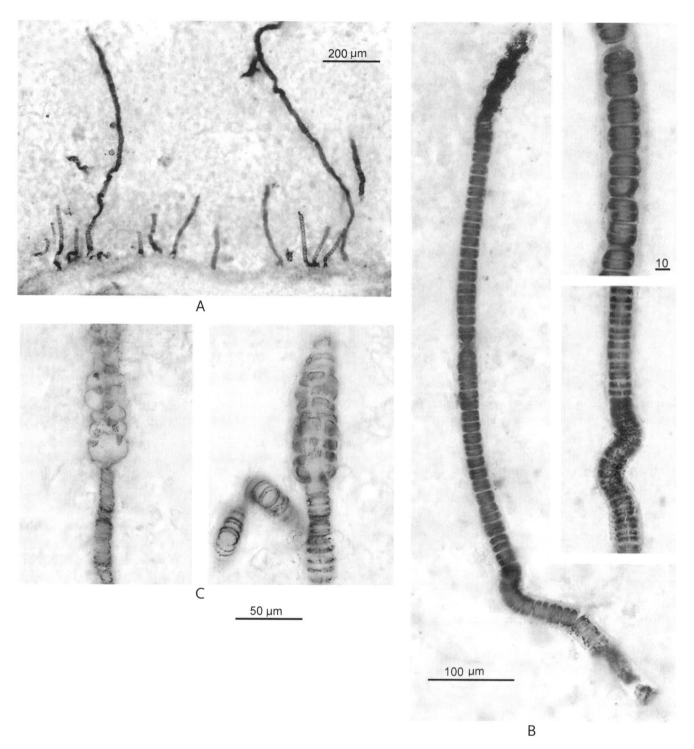

Figure 1.10. Fossils of the red alga *Bangiomorpha pubescens* from 1.2 billion-year-old deposits on Somerset Island in the Canadian Arctic. These are the oldest known multicellular eukaryotes. *A*, A population of plants in the normal position of growth. *B*, Sections of stems of different sized plants, showing the structure of the individual cells. The pairing of the cells results from diffuse transverse intercalary cell division. *C*, Reproductive structures at the apex of the stems. The nature of these structures demonstrates that they reproduced sexually, by separate spore and gamete formation, the oldest fossil evidence for this practice. This genus can be placed within the living family Bangiaceae. It is the oldest known taxonomically resolved eukaryote and marks the onset of a major protistan radiation near the Mesoproterozoic-Neopterozoic boundary. From Butterfield, 2000.

mosomes. Knoll (1995) also documents the protracted evolution of introns (non-coding elements of DNA), necessary for the formation of complex molecules, which can be traced among the living descendants of a sequence of plants represented in the later Proterozoic by acritarchs.

THE EMERGENCE OF MULTICELLULAR ANIMALS

Although the oldest known multicellular plant assignable to a modern phylum is known from 1.2 billion years ago, the

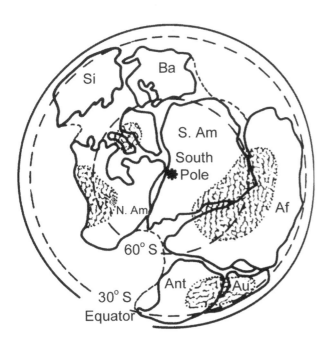

Figure 1.11. Distribution of continents in the late Precambrian, when they were beginning to disperse. All are clustered in the Southern Hemisphere. Their proximity to the South Pole resulted in extensive glaciation, indicated by the wavy shading. Af, Africa. Ant, Antarctica. Au, Australia. Ba, Baltica. N. Am, North America. S. Am, South America; Si, Siberia. From van Andel, 1994.

origin of multicellular animals appears to have been delayed to much later in the Proterozoic, subsequent to major environmental changes that may have come close to wiping out all conspicuous forms of life. During two periods in the Proterozoic, one about 730 million years ago and a second around 600 million years ago, much of the earth's surface was covered by glacial ice, which extended into equatorial latitudes (Fig. 1.11). This is documented by thick layers of tillites (coarse, poorly segregated sediments similar to those formed by modern glaciers), rock surfaces marked by striations like those made by the movement of ice sheets, and large rocks that appear to have been dropped from floating ice. At one time or another, much of the surface of the land and sea may have been covered by ice and snow, greatly reducing the capacity for photosynthesis. This dire hypothesis is supported by the reappearance of extensive deposition of banded ironstone, last seen in significant amounts 1.85 billion years ago, and the apparently very low amount of biological activity reflected in the ratio of carbon isotopes. These periods of time are referred to as the "Snowball Earth," which might have resembled the planet Hoth, seen at the beginning of the second of the initial Star Wars trilogy, *The Empire Strikes Back.* This, of course, was a long time before the emergence of amphibians, but we will see later that their distribution was also influenced by widespread glaciation during two stages of their subsequent evolution.

Fortunately for the history of life, such episodes of gla-

ciation, extending to the equatorial region, have not been repeated. Two features of the early earth may be cited as promoting extensive glaciation toward the end of the Proterozoic. One was the relatively lower luminosity of the sun, whose generation of energy was somewhat less than in the Phanerozoic, and the second was the particular configuration of the continents. Unique to that time, most of the continents were grouped close to the South Pole. As now, the axis of the earth's rotation was tilted, so that the polar regions received much less sunlight than the tropics. Once snow and ice began to accumulate, the sun's light would be reflected, maintaining a low temperature and the continuing spread of glaciers to the edge of land and out to sea.

What eventually brought an end to these periods of extensive glaciation was presumably the gradual accumulation of heat from the core and mantle of the earth, resulting in the occurrence of volcanic eruptions, lava flows, and breaking apart of the previously united continental masses via plate tectonics. After the second period of glaciation, the newly separated continents (with the exception of South America) began a general northward movement that continued throughout the Phanerozoic. At the top of the sediments deposited by the second major period of glaciation is a capping carbonate deposit, below which there is no evidence of metazoan fossils similar to those known at the very end of the Precambrian. This indicates a major gap between typically Proterozoic life forms and a new biota that heralded a great diversity of large, complex, multicellular animals.

Some of the most amazing fossils from the late Proterozoic are found within phosphate beds in China dated from 590 to 600 million years old (Xiao and Knoll, 2000). They consist of spherical structures, ranging from about 0.5 to a little more than 1 mm in diameter. They have a thin surface membrane, within which are one to hundreds of smaller spheres (Fig. 1.12). These fossils have been given different names, depending on their size and the number of contained spheres, but based on comparison with eggs and developing embryos of living sponges, cnidarians (jellyfish, corals, and hydras), and bilaterians (including crustaceans) they can be interpreted as fertilized eggs of metazoans undergoing a succession of cell divisions, up to the point of forming a hollow ball of cells, the morula. In common with these living groups, cell division continues without growth within a membrane of uniform size. Unfortunately, no characters have been recognized that make it possible to identify to which of the living groups these Neoproterozoic fossils might belong. No fossils have yet been found of later-stage embryos that might show the general form of the adult body, although tiny sponges are found in association. The preservation of early stages in development results from a unique mode of deposition, in which a thin film of phosphate covers the fragile embryo before it can decay or be eaten. There is no current evidence as to whether their descendants survived and gave rise to one or other later known phyla, or whether they became extinct without issue.

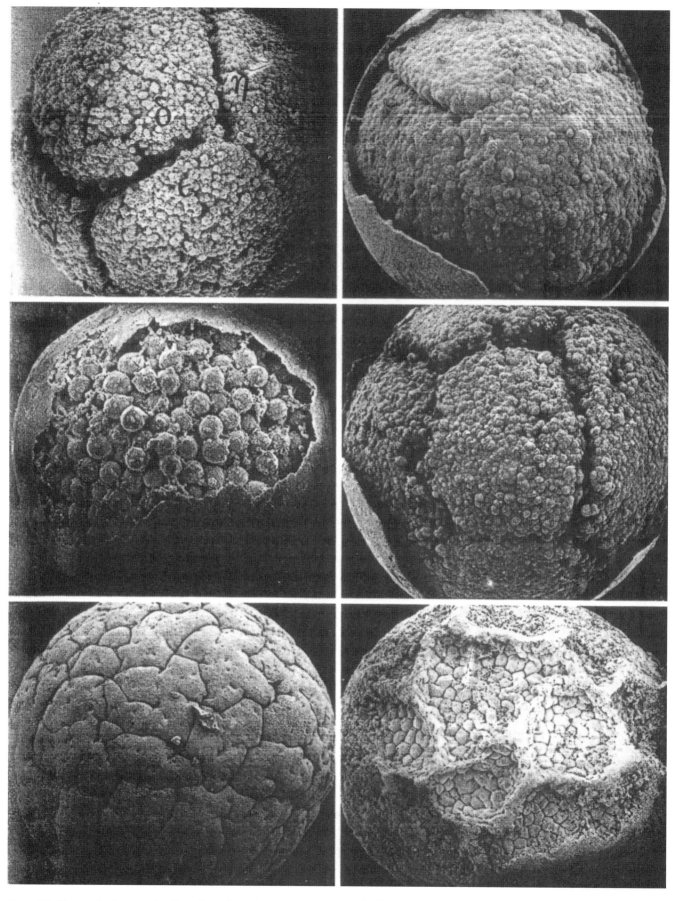

Figure 1.12. Phosphatized eggs and early embryos from the upper Neo-proterozoic of South China. They resemble eggs and embryos of extant sponges, cnidarians, and bilaterians, but none can be specifically allied with any modern groups. All structures are of about the same size, but contain cells of varying size, indicating growth prior to differentiation. The overall size ranges from approximately 400 to 1100 μm in diameter. From Xiao and Knoll, 2000.

A

B

C

D

E

Figure 1.13. Examples of fossils from the Ediacaran assemblage. Impressions of metazoans without preservable elements of an internal or external skeleton, found in many parts of the world in beds deposited near the end of the Neoproterozoic. *A, Tribrachidium,* about 20 mm across. *B, Spriggina,* about 50 mm in length. *C, Parvancorina,* about 30 mm in length. *D, Charnia,* about 150 mm long. The frond-like structure suggests that it might be a plant, but it is more likely a member of the Cnidaria (which includes corals, sea anemones, and jellyfish). *E, Dickinsonia,* about 50 mm in diameter. This biota ranged from about 670 million years ago to 540 million years ago. *A–D* modified from Selden and Nudds, 2004; *E* from Conway Morris, 1998.

The Latest Proterozoic, beginning about 600 million years ago, is distinguished by an assemblage of organisms termed the Ediacaran fauna, named for the Ediacara hills in Australia, from which Glaessner (1983) recognized the uniqueness of these fossils (Fig. 1.13). Nearly all of the Ediacaran species are preserved as impressions, without any evidence of mineralized skeletons, such as are common to later multicellular animals. Neither are the body forms of most of these organisms comparable. Some have been identified, tentatively, as sponges, cniderians, "worms," or arthropods. Others have the form of fronds, which suggest affinities with multicellular algae or some groups of colonial invertebrates, but none show convincing similarities. There are also simple tracks and trails, suggestive of primitive animals with bilateral symmetry, such as arthropods. Perhaps the most advanced of the Ediacaran fossils are *Claudina* and *Namacalathus,* which are represented by calcified structures, not otherwise known until the basal Cambrian (Knoll, 2003). However, with only a few questionable exceptions, Ediacaran fossils provide little evidence of affinities with the modern phyla that began to appear at the base of the Cambrian.

Near the end of the Precambrian, other small animals, vaguely resembling fragments of shells, had evolved the capacity to deposit a mineralized skeleton, but they cannot be attributed to any of the modern metazoan phyla. They are referred to as the "small shelly fauna," and extend into the Early Cambrian, as did some Ediacaran genera, but without conspicuous change.

There is no evidence from the geological record of any significant physical changes in the earth at the very end of the Ediacaran or during the basal Cambrian, with no more than small unconformities in any of the known localities (Rogers and Santosh, 2004). Nor does the fossil record at the very end of the Precambrian indicate a devastating or abrupt extinction event. On the other hand, there is evidence for a very rapid emergence of a great diversity of novel lineages representing the modern metazoan phyla during the first 20 million years of the Cambrian. There is also a major change in the nature of the sediments at the boundary between the Ediacaran and the Cambrian. Rather than just a few simple tracks and trails on the surface, the sediments show extensive bioturbation resulting from many organisms burrowing deeply into the substrate. These are attributed to the emergence of bilaterians (bilaterally symmetrical multicellular animals such as arthropods, annelids, and vertebrates) with advanced body plans, but which still lacked effectively mineralized exoskeletons and so are unlikely to be preserved, except under exceptional conditions of deposition (Seilacher et al., 2005).

2

Advanced Metazoans and the Ancestry of Vertebrates

THE CHENGJIANG AND BURGESS FAUNAS

DURING THE FIRST 20 MILLION years of the Cambrian we observe the first appearance of advanced multicellular animals including arthropods (especially trilobites) with many characteristics of modern genera—an exoskeleton, a head, complex sensory organs, and limbs—as well as coelenterates, mollusks, annelids, brachiopods, and echinoderms. However, there is little evidence of more primitive forms that must have linked them with Precambrian ancestors. Notably rare from this time interval are animals without a mineralized exoskeleton, indicating a major gap in the fossil record. This gap is compensated for by two astounding fossil deposits from later in the Lower Cambrian: the 525 million-year-old Chengjiang deposit in Yunnan Province, China, and the 512 million-year-old Burgess Shale, in British Columbia, Canada (Figs. 2.1 and 2.2). In addition to extensive evidence of animals with exoskeletons, they also contain an extraordinary profusion of soft-bodied forms whose anatomy was preserved in exquisite detail. The fauna from Chengjiang includes the antecedents of more than 12 modern metazoan phyla, embracing at least 90 genera (Hou et al., 2004).

The preservation of animals without mineralized skeletons may be attributed to unusual conditions of burial. In both localities, a wealth of plants and animals had lived in a reef environment in a shallow sea. Either because of an excessive accumulation of sediments or perhaps a minor earthquake, the area in which they lived was suddenly swept down into much deeper waters. The absence of decay and the high degree of articulation indicate that the waters were anoxic, perhaps as a result of the decay of massive amounts of aquatic vegetation. The surrounding sediments later hardened into rock, but it remained soft enough to be readily split and carefully removed to reveal the original anatomy of the animals.

Figure 2.1. Panorama of the Chengjiang biota. Modified from Hou et al., 2004. Note similarity of animals to those from the Burgess Shale (Fig. 2.2).

THE ANCESTRY OF MULTICELLULAR ANIMALS

Who knows what might have been preserved under comparable conditions of deposition earlier in the Cambrian or Late Precambrian? Presumably, the initial radiation of the metazoan phyla must have occurred significantly earlier than the Chengjiang fauna, since most are clearly distinct from one another by that time. However, the absence of bioturbation prior to the base of the Cambrian makes it nearly certain that advanced bilaterians had not evolved before that time. We will see comparable problems of major gaps in the fossil record when we try to establish the pattern of radiation among the amphibian groups in the Early Carboniferous.

Molecular biologists have tried to date the time of divergence of the major groups of multicellular animals based on estimates of the time required for the accumulation of genetic differences between them. However, these estimates commonly suggest dates deep into the Precambrian, some as many as hundreds of millions of years prior to the appearance of any groups in the fossil record (Wray et al., 1996; Valentine, 2004; Welch et al., 2005). This may reflect the divergence of genetic and molecular attributes long prior to observable changes in the external anatomy or to significantly different rates of genetic change over time.

The Genetic Toolbox

Whatever the specific date of divergence of the advanced metazoan phyla, it marks the initiation of major advances in the genetic control of the development and evolution of body form, which will eventually lead to the origin and evolution of amphibians and other vertebrates.

Understanding of the way in which the complex body forms of vertebrates and other multicellular animals developed and evolved was the most important discovery in the field of evolutionary biology made in the late 20th century. Following the recognition of the structure of DNA by Watson and Crick (1953), it soon became possible to understand how DNA coded the synthesis of proteins that form and operate all organisms. A further major achievement, beginning about 1984, was the identification of a particular category of genes that are specifically responsible for establishing the general body form of multicellular animals. This has been chronicled by Walter Gehring (1998) and Sean Carroll (2005) in books that are intended for a broad audience of interested biologists. This discovery revealed an amazingly simple but elegant system, based on a relatively small number of readily recognizable genes common to all organisms from sponges to humans. For the first time, we could recognize a single broad genetic blueprint that determines the basic body form among all metazoans. The entire history of vertebrates, and

A

B

Figure 2.2. Dioramas based on the Burgess Shale Fauna from British Columbia. Modified from Conway Morris, 1998. *A,* Emphasis on swimmers and floaters. 1. Two specimens of the primitive chordate *Pikaia*. 2. The ctenophore *Ctenorhabdotus*. 3. A possible echinoderm, *Eldonia*. 4. The arthropod *Odaraia*. 5. *Nectocaris*, of uncertain affinities. 6. The annelid *Canadia*. 7. *Dinomischus*, of uncertain affinities. 8. The sponge *Pirania*, with symbiotic brachiopods. *B,* Emphasis on bottom-dwelling animals. 9. *Anomalocaris*, a giant arthropod. 10. Hapless trilobite. 11. The arthropod *Marrella*. 13. *Hallucigenia*. 14. The lobopodian *Aysheaia*. 15. The primitive arthropod *Opabinia*. 16. The sponge *Vauxia*.

specifically the changes that have occurred between amphibians and humans, can be understood in terms of modifications in the expression of a small proportion of the 25,000 or so genes common to these groups. Surprisingly, relatively little attention has been given to this discovery in the standard textbooks on evolution.

In unicellular organisms, most of the genes, numbering in the thousands, code for particular proteins that are responsible for the structure, function, and reproduction of a single cell. However, even at the bacterial level, there are a small number of other genes that act as switches that turn on particular enzymes or direct the expression of other genes. This was first recognized in the bacterium *Escherichia coli,* in which the gene for the production of beta-galactosidase could be turned on in the presence of lactose by the release of a specific protein that normally inhibited this process. Attachment of this protein occurred via a particular binding site termed a homeodomain, consisting of a unique sequence of 60 amino acids that corresponded with 180 base pairs of the gene. These occupied a particular area of the gene referred to as a homeobox (Fig. 2.3). Several families of homeobox genes, which have a similar role of regulating the expression of other genes, have been recognized in cells of higher organisms.

A distinct type of homeobox gene is common to fungi (which include both unicellular and multicellular species), vascular land plants, and multicellular animals. In yeast, the best-studied members of the fungi, two homeobox genes, termed Mat alpha-2 and Mat alpha-1, are involved with mating-type recognition and formation of the infectious cell type (Yee and Kronstad, 1998). In vascular plants, the MADS box genes are best known for their influence on the form of the flowers in angiosperms.

In multicellular animals, a particular family of homeobox genes, the *Hox* genes, plays a fundamental role in body formation (S. B. Carroll et al., 2001). The detailed structural and functional similarity of the homeobox genes in these three groups is so great that it suggests that this type of gene evolved prior to their divergence from a common ancestor (Bharathan et al., 1997) (Fig. 2.4).

Hox genes are unique in their position along the chromosome. With only minor exceptions, they are arranged in a linear sequence that is recognizable among all metazoans. Except for vertebrates, all *Hox* genes are restricted to a single chromosome. Even more surprisingly, their linear sequence corresponds with the spatial and temporal sequence of their expression in the embryo. *Hox* genes 1 to 3 begin the sequence and are first expressed early in development and influence the head of the animal, be it an anglerworm, a fruit fly, or an early embryo of a mammal. Other, more downstream genes control the position of expression and nature of the trunk, limbs, and tail.

Hox genes provide the strongest evidence available for the common ancestry of all multicellular animals, since all can be linked by a continuous evolutionary sequence. Sometime in the late Proterozoic there may have been no more than a single homeobox regulatory gene with the properties of later *Hox* genes. There is no way to judge its probable function, although it may have regulated one or more processes within a single cell. By the time of emergence of sponges and cnidarians at the very end of the Precambrian, there were several

1   **20**

Hox-1	Ser	Lys	Arg	Gly	Arg	Thr	Ala	Tyr	Thr	Arg	Pro	Gln	Leu	Val	Glu	Leu	Glu	Lys	Glu	Phe
MM3	Arg	Lys	Arg	Gly	Arg	Gln	Thr	Tyr	Thr	Arg	Tyr	Gln	Thr	Leu	Glu	Leu	Glu	Lys	Glu	Phe
Antp	Arg	Lys	Arg	Gly	Arg	Gln	Thr	Tyr	Thr	Arg	Tyr	Gln	Thr	Leu	Glu	Leu	Glu	Lys	Glu	Phe
ftz	Ser	Lys	Arg	Thr	Arg	Gln	Thr	Tyr	Thr	Arg	Tyr	Gln	Thr	Leu	Glu	Leu	Glu	Lys	Glu	Phe
Ubx	Arg	Arg	Arg	Gly	Arg	Gln	Thr	Tyr	Thr	Arg	Tyr	Gln	Thr	Leu	Glu	Leu	Glu	Lys	Glu	Phe

21   **40**

Hox-1	His	Phe	Asn	Arg	Tyr	Leu	Met	Arg	Pro	Arg	Arg	Val	Glu	Met	Ala	Asn	Leu	Leu	Asn	Leu
MM3	His	Phe	Asn	Arg	Tyr	Leu	Thr	Arg	Arg	Arg	Arg	Ile	Glu	Ile	Ala	His	Val	Leu	Cys	Leu
Antp	His	Phe	Asn	Arg	Tyr	Leu	Thr	Arg	Arg	Arg	Arg	Ile	Glu	Ile	Ala	His	Ala	Leu	Cys	Leu
ftz	His	Phe	Asn	Arg	Tyr	Ile	Thr	Arg	Arg	Arg	Arg	Ile	Asp	Ile	Ala	Asn	Ala	Leu	Ser	Leu
Ubx	His	Thr	Asn	His	Tyr	Leu	Thr	Arg	Arg	Arg	Arg	Ile	Glu	Met	Ala	Tyr	Ala	Leu	Cys	Leu

41   **60**

Hox-1	Thr	Glu	Arg	Gln	Ile	Lys	Ile	Trp	Phe	Gln	Asn	Arg	Arg	Met	Lys	Tyr	Lys	Lys	Asp	Gln
MM3	Thr	Glu	Arg	Gln	Ile	Lys	Ile	Trp	Phe	Gln	Asn	Arg	Arg	Met	Lys	Trp	Lys	Lys	Glu	Asn
Antp	Thr	Glu	Arg	Gln	Ile	Lys	Ile	Trp	Phe	Gln	Asn	Arg	Arg	Met	Lys	Trp	Lys	Lys	Glu	Asn
ftz	Ser	Glu	Arg	Gln	Ile	Lys	Ile	Trp	Phe	Gln	Asn	Arg	Arg	Met	Lys	Ser	Lys	Lys	Asp	Arg
Ubx	Thr	Glu	Arg	Gln	Ile	Lys	Ile	Trp	Phe	Gln	Asn	Arg	Arg	Met	Lys	Leu	Lys	Lys	Glu	Ile

Figure 2.3. Homeodomain sequences of five proteins encoded by homeobox-containing genes. *Antp, ftz,* and *Ubx* are from the fruit fly *Drosophila; Hox-1* from the mouse; and the *MM3* from the frog *Xenopus*. The large boxes surrounded by a light line enclose homologous sequences, using *Antp* (*Antennapedia*) as a standard. The box surrounded by the darker line is the DNA-binding site. The homeodomain serves as a sequence-specific, DNA-binding motif that allows particular proteins to bind to genes that control other aspects of development. From Gehring, 1985.

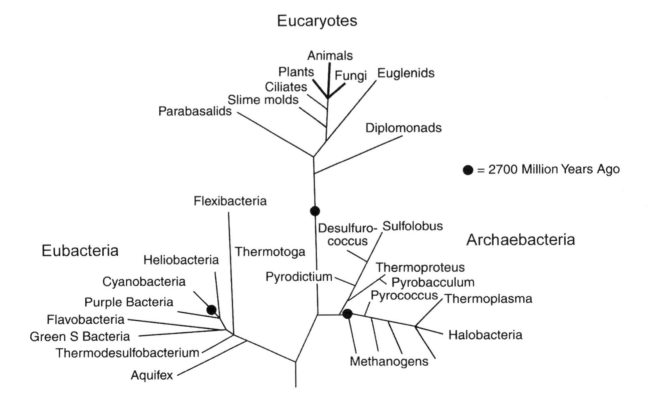

Figure 2.4. Pattern of relationships among major groups of organisms, showing the close common ancestry of plants, animals, and fungi, as determined by molecular analyses of their homeobox genes. The dark circles indicate a particular time (2.7 billion years ago) within the history of the major lineages. The base of the tree can be interpreted as the origin of life. From Knoll, 2003.

homeobox genes, at least some with definite homologies with the *Hox* genes of higher metazoans.

There was apparently a significant advance in the number and functions of *Hox* genes late in the Ediacaran to judge by the presence of tracks and trails that may have been made by bilaterally symmetrical animals such as arthropods and annelids, but which left no skeletal remains until the Lower Cambrian. A further advance can be noted at the very base of the Cambrian, when metazoans gained the capacity to burrow deeply in the sediments. Living bilaterians have three groups of *Hox* genes, anterior, central, and posterior, totaling 8 to 10. The number increased to 14 in the immediate ancestors of vertebrates (Fig. 2.5).

The Cambrian Explosion

Approximately 30 phyla of bilaterian animals are recognized in the modern fauna. Many of the major phyla can be traced to fossils known from the Chengjiang and/or Burgess deposits. Gould (1989) argued that several taxa from the Burgess Shale might belong to separate, long-extinct phyla, but most show some features in common with the long-recognized groups. The broad picture is one of a very rapid ("explosive") appearance of a multitude of distinctive body plans, within 17 to 20 million years after the beginning of the Cambrian, with few, if any, obvious antecedents in the latest Precambrian. This is in strong contrast with the retention of these basic body plans for the next 520 million years. Several question arise. (1) Is this real? (Does the fossil record accurately reflect what actually occurred?) (2) What caused or enabled the initial burst? (3) Why was extensive change then curtailed?

Prior to the Cambrian explosion, there were widespread deposits covering the last 50 million years of the Precambrian that document the clearly more primitive Ediacaran fauna, with little if any evidence of bilaterian fossils and none that are clearly related to modern phyla. The preservation of Ediacaran animals was itself peculiar, since organisms that almost certainly lacked mineralized skeletons are rarely preserved, except in anoxic environments. We know that preservation of soft anatomy in later deposits may be facilitated by certain bacteria, but bacteria living on the surface are typically grazed on by multicellular animals. This suggests the rarity of advanced bilaterians in the late Precambrian, and may explain the scarcity or complete absence of preservation of Ediacaran-type animals in the Lower Cambrian, when there was a wealth of bilaterians of all sizes.

If most of the modern bilaterians, as known from the Chengjiang fauna, did evolve during the previous 20 million years, how can this be explained? As stated by Andrew Knoll (2003, 217): "I believe that oxygen made all the difference." Since nearly all the advanced bilaterian phyla have essentially the same number and general function of *Hox* genes, they must have been present prior to the beginning of their radiation in the Early Cambrian. What we are looking for is some external trigger that enabled their potential to be released. An increase in the amount of oxygen available in seawater would have made it possible for them to increase in size (which depends on the rate and amount of diffusion of O_2 in the tissues) and to form mineralized skeletons. A greater amount of oxygen would also have facilitated the synthesis of collagen, which integrates the cells of all metazoans. The critical level of oxygen is not known exactly, although it is thought to be above 1% of the current amount. However, it need not have required a great increase above some threshold value that was gradually being approached by greater amounts of oxygen being released by plants living in the water and beginning to spread onto land.

Discovery of additional localities that reflect differing levels of oxygen early in the Cambrian or late Precambrian are necessary to establish the actual rate and sequence of events leading to the Cambrian explosion. However, even with the present data, we can be fairly certain that the appearance of a great diversity of relatively large metazoans, many of which did have mineralized skeletons, took place very rapidly, compared with the less fundamental changes that occurred in the body form in later members of the major groups. The relative constancy of general form may be attributed to there being a limit to the number of ways in which animals can move, feed, and form workable body shapes. Poorly designed models would be quickly weeded out by natural selection, while the better models continued to perfect their initial design. There is an astounding diversity of arthropods and frogs, but all can be recognized as having evolved from a single basic body plan.

THE ANCESTRY OF VERTEBRATES

As vertebrates, we may be forgiven for assuming that we are distinct from other multicellular animals, which have been grouped as invertebrates—animals without vertebrae. Nevertheless, since the recognition of the evolutionary relationships among all multicellular animals, we have sought to find our ancestors among a particular, more primitive lineage. The closest living relative of vertebrates, amphioxus (*Branchiostoma*), a tiny animal living in shallow marine waters, has a fishlike body, but lacks vertebrae, other bony elements, jaws, gills, and paired fins (Fig. 2.6A). Like us, amphioxus has a dorsal hollow nerve cord, but the area of the brain is very small and lacks obvious sense organs of smell, sight, and balance. On the other hand, its body is supported by a longitudinal notochord, which is present early in development in vertebrates, and has comparable blocks of muscles, the myotomes, that run along the trunk. The common presence of a notochord unites amphioxus and vertebrates in the phylum Chordata. This phylum is clearly unique among all metazoans in its manner of locomotion, via lateral undulation of the trunk and tail powered by sequential contraction and relaxation of the segmental muscles on either side of the notochord.

In addition, amphioxus also has gill slits in the same position as those in fish, and the circulatory system has the same

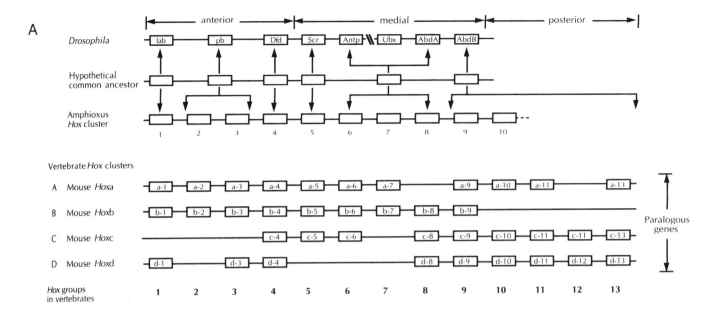

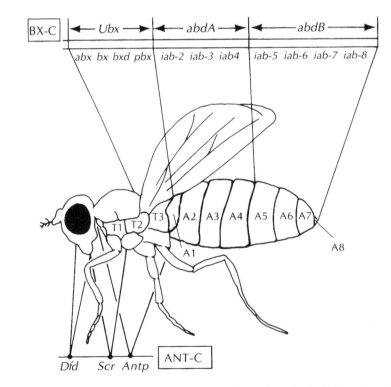

Figure 2.5. Hox gene organization in metazoans. *A*, Beginning at the top: the anterior-to-posterior sequence in the fruit fly *Drosophila;* hypothetical *Hox* genes of the common ancestor of *Drosophila* and other metazoans; amphioxus *Hox* cluster (now known to extend to 14); the four *Hox* clusters of a mouse. Courtesy of Sean Carroll, 1995. The laterally directed arrows indicate the origin of new genes by tandem duplication. Chordates and *Drosophila* have independently duplicated the *Ubx* gene of primitive metazoans. *B*, Regions of expression of the bithorax complex and the antennapedia complex genes in *Drosophila,* showing the colinearity of genes within the *Hox* cluster and their expression along the anterior-posterior axis of the animal. Modified from Gilbert, 1988.

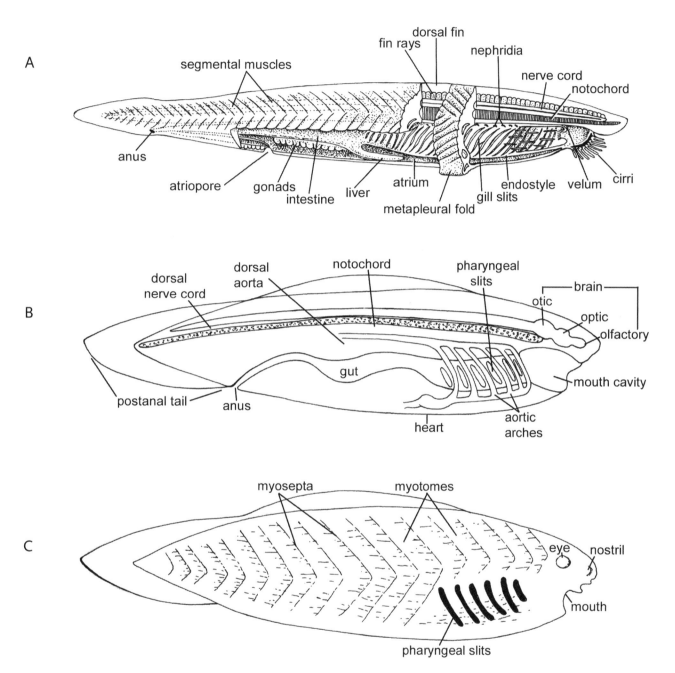

Figure 2.6. A, Diagrammatic view of the primitive chordate *Branchiostoma* (commonly termed Amphioxus), showing the early appearance of many features common to vertebrates. *B, C,* Cutaway and external views of a primitive fish, such as may ultimately have given rise to land vertebrates.

Note addition of newly evolved sensory structures of the head region and reduction in the number of gill slits, compared with those of *Branchiostoma.* From R. L. Carroll, 1988a.

general configuration, although a distinct heart is not present. *Pikaia,* from the Burgess Shale, may be the oldest known fossil relative of amphioxus (Fig. 2.2). It has a similar body form but otherwise shows little structural detail. *Pikaia* and amphioxus are placed together in the Cephalochordata. A further group of chordates are the tunicates, whose larvae resemble amphioxus, but which typically metamorphose to sessile adults that clearly represent a highly divergent way of life from that of vertebrates.

At a higher level of classification, chordates, the worm-like hemichordates, and echinoderms (starfish, sea urchins, crinoids, and sea cucumbers) have been united in the Deuterostomia on the basis of common developmental patterns. This name refers to the formation of the mouth as a second opening for the digestive tract, distinct from the blastopore (a single opening that forms early in development in other bilaterians, the Protostomia). The protostomes include arthropods, nematodes, annelids, and molluscs, as well as many other "minor" phyla.

Vertebrates and other chordates also differ from protostomes in the distribution of major anatomical structures. Among chordates, the neural tube is dorsal in position and the digestive system is ventral, but in arthropods the major neural tracts are ventral and the digestive tract is in a dorsal position (as you can recognize when "de-veining" a shrimp). This suggests the independent evolution of the dorsoventral axis of the body in these two phyla. However, the genetic basis for its orientation is provided by a comparable gene. If the vertebrate gene for the determination of the dorsal surface of the body, *Chordin,* is inserted into an arthropod such as the fruit fly *Drosophila,* that gene will be expressed ventrally, resulting in the appearance of the ventrally situated nerve tracts (Halder et al., 1995). Hence, the functional aspects of the body are homologous in the two groups, even though the position of development is reversed. These observations support a very old hypothesis that vertebrates can best be compared with invertebrates by turning one or the other animal upside down (Geoffroy Saint-Hilaire, 1830).

An even more striking homology between protostomes and deuterostomes, indicating an ultimate common ancestry, is illustrated by the homeobox gene *Pax6,* which initiates the formation of the eye. This gene can be recognized by its specific base sequence among a great number of bilaterian phyla, no matter what particular form their eyes may have, whether the single very complex structure in vertebrates or the multifaceted eyes of insects and other arthropods. What is astounding is that *Pax6* from a mouse can be used to initiate the formation of eyes in the fruit fly *Drosophila,* at whatever place it is inserted. The fruit fly can thus produce typical fruit fly eyes on its wings, limbs, or antennae. These and many other aspects of homeobox genes are described in Walter Gehring's book *Master Control Genes in Development and Evolution: The Homeobox Story* (1998), which also provides a history of their discovery.

Neither amphioxus nor *Pikaia* provides much information regarding the immediate ancestors of chordates. They are linked with echinoderms and the wormlike hemichordates on the basis of comparable genes and common patterns of early development, but no structural intermediates are known. We assume that primitive chordates resembling amphioxus had evolved by the basal Cambrian, and by the time of the Chengjiang fauna had diverged into two lineages, one giving rise to the living *Branchiostoma* and the other to vertebrates.

Several very primitive vertebrates have been described from the 525 million-year-old Chengjiang locality (Fig. 2.7). They were small, inconspicuous members of the Early Cambrian marine fauna, about 30 mm long, nearly transparent, and without bone or scales. They were already clearly distinct from other early multicellular animals in having a fusiform, laterally compressed body, capable of undulatory locomotion.

In addition to having the basic chordate features of a dorsal hollow nerve cord, notochord, gill pouches, and paired myotomes, *Haikouella lanceolata* (also called *Yunnanozoon*) (Fig. 2.7A, B), was advanced above the level of amphioxus in having gill filaments that suggest the need for more oxygen to supply the higher metabolic rate of vertebrates, a brain-like anterior expansion of the nerve cord, and perhaps a heart and eyes that support vertebrate status (Chen et al., 1999). *Haikouichthys ercaicunensis* (Shu et al., 1999), known from more than 500 specimens, shows even more vertebrate characters (Fig. 2.7C). In the head region there are cartilaginous structures in the position of the nasal and otic capsules of a modern jawless fish, the lamprey, and sclerotic cartilage in the area of the eye. Annular and anterior tectal cartilages, common to lampreys, have also been identified. A surprising feature is a series of rod-shaped branchial elements below the six or seven gill pouches that somewhat resemble the gill supports in jawed vertebrates. Small blocks of tissue associated with the notochord have been tentatively identified as vertebral elements and some specimens have dorsal fin rays. The authors recognized a pericardial cavity and a posteroventral fin that might be paired. Paired fins are not known in amphioxus or in better-preserved primitive vertebrates. Such a combination of characters, if correctly identified, suggests a position near the base of the radiation of all later vertebrates, although the cranial structures appear specifically comparable to those of lampreys.

By analogy with the larvae of lampreys, the earliest vertebrates probably fed on microscopic organisms and organic particles drawn into the mouth by currents generated by oral cilia. It is assumed that early vertebrates had a longitudinal glandular structure, the endostyle, at the base of the pharynx that secreted mucus to entrap food particles that were swept into the digestive tract. Excess water was discharged through a series of gill slits on either side of the pharynx. Propulsion was achieved by sinuous movements of the trunk and tail, generated by the alternative contraction and expansion of paired blocks of muscles attached to either side of the notochord. The tail, which extended behind the anus as a median appendage, would have assisted in propulsion and in controlling the direction of movement—up, down, and sideways. Respiration would have occurred through uptake of oxygen from the water through the general body surface and via blood vessels running through gills that were associated with the gill slits.

TRANSITION BETWEEN PRIMITIVE CHORDATES AND VERTEBRATES
Duplication of *Hox* Genes

As yet, we lack any fossils that are intermediate between primitive chordates and the Lower Cambrian vertebrates,

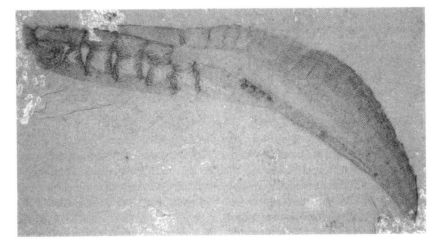

1 mm

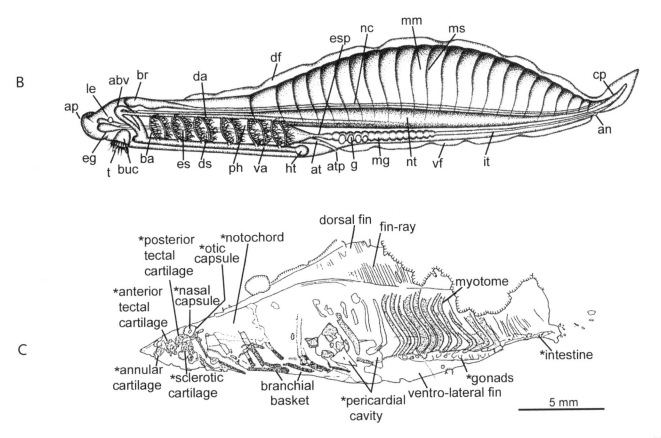

Figure 2.7. The oldest known vertebrates, from the Lower Cambrian of Chengjiang, China. *A*, Photograph of *Yunnanozoon*, also known as *Haik-ouella*. Note small number of pharyngeal gill slits (compared with those of amphioxus) surrounded by gill filaments. Modified from Hou et al., 2004. *B*, Reconstruction of *Haikouella*. From Chen et al., 1999. *C*, The Lower Cam-brian *Haikouichthys*, from China. From Shu et al., 1999. This genus differs significantly from *Haikouella* in the rod-shaped branchial elements, which resemble those of early jawed fish. *Structures of uncertain identity. Abbre-viations used in this and all other figures are listed at the end of the book.

and so have no direct evidence as to the manner of origin of specifically vertebrate characters. However, knowledge of the genetics and mode of development of living vertebrates reveals major changes from the condition seen in amphioxus or other non-vertebrate chordates.

Earlier in this chapter, it was noted that the immediate ancestors of vertebrates, based on knowledge of amphioxus, had 14 *Hox* genes, the highest number of any non-vertebrate (Minguillón et al., 2005). Based on the number and nature of *Hox* genes in all major groups of living vertebrates, it is thought that this number was approximately doubled prior to the divergence of jawed fish from more primitive jawless vertebrates—that is, close to the time of the origin of all vertebrates. This occurred by the duplication of the chromosome bearing the *Hox* cluster and probably all other chromosomes as well (Panopoulou and Poustka, 2005). A second episode of *Hox* gene duplication apparently occurred later, near the time of origin of each of the vertebrate lineages with living descendants. However, the number and nature of the *Hox* genes recognized in each of the two groups of living jawless fish, the lamprey and hagfish, and those of the jawed vertebrates, differ, strongly suggesting that duplication occurred separately in each. It also occurred independently among advanced ray-finned fish, which have up to seven *Hox* clusters.

The duplication of *Hox* genes in primitive vertebrates presumably occurred via polyploidy (duplication of reproductive cells without segregation of daughter chromosomes), which is a common phenomenon among both plants and animals. However, very few polyploid organisms retain duplicate sets of *Hox* genes, probably because the presence of multiple copies of regulatory genes would be detrimental to normal development. Hence, selection would normally result in the loss or silencing of such duplicates. Perhaps extensive duplication of *Hox* genes could have been more readily tolerated in the anatomically simpler early members of the vertebrate clades in the latest Precambrian or Early Cambrian. Loss of individual *Hox* genes also occurred in many groups, none of which retain all of the genes expected to result from chromosome duplication.

The duplication of most of the original set of *Hox* genes in early vertebrates resulted in a greater degree of control of development of all the areas of expression of the initial *Hox* genes. This would have enabled each anatomical element of the body to achieve a higher level of complexity. The original *Hox* gene could continue to direct the position and general morphology of particular structures, while new copies of the gene, as they accumulated minor genetic changes, could result in modification or addition of new attributes. The importance of this greater degree of control over development

will become obvious as we consider the origin of the limbs of land vertebrates.

Neural Crest and Placodes

In addition to the duplication of *Hox* genes, an entirely new type of tissue, the neural crest, evolved between primitive chordates and the first vertebrates (Hall, 1999, 2005). The evolution of multicellular animals has proceeded by the progressive addition of major tissue types. Their ancestors had but a single cell, but at the level of sponges several distinct cell types are apparent, of which the most obvious, the ectoderm, formed an envelope around the entire body. The cnidarians, such as jellyfish and corals, added a second layer of tissue, the endoderm, which constituted the lining of the digestive tract. More advanced metazoans, the bilaterians, evolved a third layer, the mesoderm, which forms the muscles and most internal organs, and in vertebrates, the bony skeleton. These three layers are also present in amphioxus, but vertebrates evolved a fourth type of developmental tissue, derived from neural crest cells. These are cells, unique to vertebrates, which proliferate along the line of contact between the ectoderm of the trunk and the neural tube (Fig. 2.8). They are motile, and spread anteriorly into the head region and back into the trunk and tail. They have the capacity to induce or otherwise contribute to the formation of a variety of tissues and structures throughout the body. In the head, they form the upper and lower jaws, sensory neurons, glial cells, osteoblasts, fibroblasts, odontoblasts, and the ganglia of the cranial nerves and facial muscles. In the trunk, they form the spinal ganglia, the sympathetic and parasympathetic nervous systems, and in fish, scales. They also proliferate melanocytes that result in the pigmentation of the skin.

A comparable type of embryonic tissue that is involved in the development of the head region alone is represented by the placodes, specialized tissue condensations associated with the sensory structures of balance, sight, smell, and taste, and the formation of nerve ganglia. In common with neural crest cells, placodal ectoderm arises from the lateral neural folds, but affects only the head region rather than migrating throughout the body. Without the evolution of neural crest cells and sensory placodes, most of the advanced features of vertebrates would never have evolved and chordates would probably never have emerged from the water.

THE EARLY RADIATION OF FISH

No vertebrates have been recognized from the Burgess Shale. Fragments of tissue resembling bone have been described from the Middle and Upper Cambrian (Repetski, 1978; G. C. Young et al., 1996), but their identity as vertebrates

Figure 2.8. (opposite) Neural crest cells and placodes. *A,* Position of the neural crest cells, black, between the open neural tube, and the closing neural fold (below), between the neural ectoderm (stippled) and the epidermal ectoderm. *B–F,* Successive stages in the migration of cranial neural crest cells in the snapping turtle. Head shown in cutaway lateral views. *G,* Placodes. Dorsal view of head region showing early development of the brain, which still retains a dorsal fontanelle (Fon). Lateral to the brain are the paired placodes, shown in black. *A* and *G,* from Hall, 1999; *B–F,* from Meier and Packard, 1984.

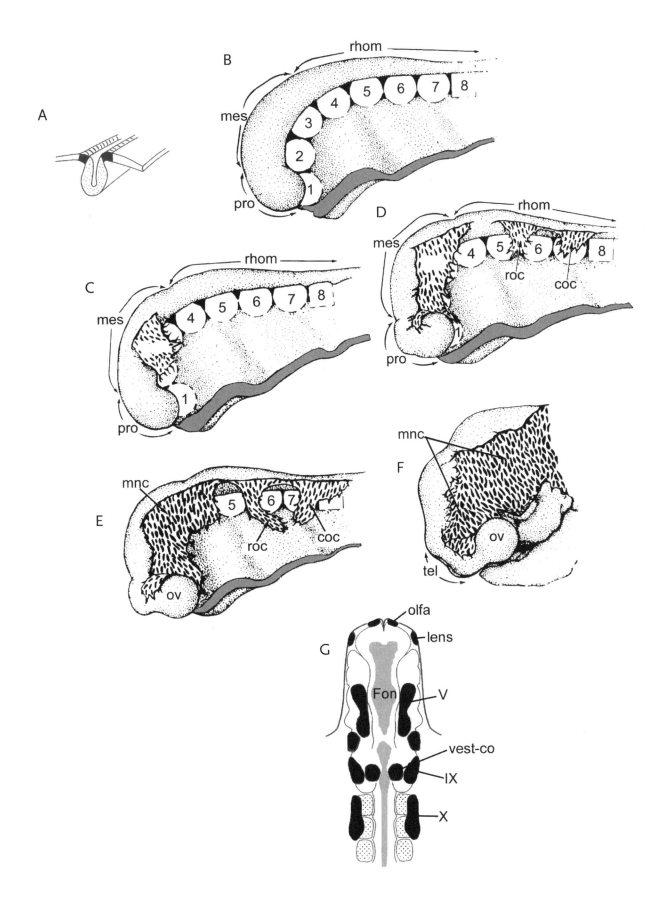

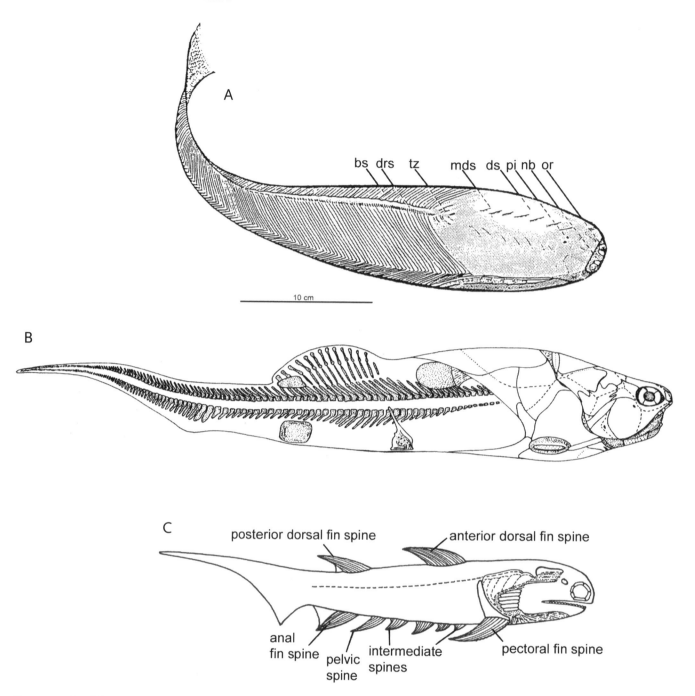

Figure 2.9. Archaic fish. *A, Sacabambaspis janvieri,* from the Ordovician of Bolivia, the oldest known vertebrate for which the entire external skeleton can be described. Only the exoskeleton is ossified. It is assumed that this animal lacked jaws, and fed by filtering tiny prey through the gill filaments. From Gagnier, 1991. *B,* The Middle Devonian placoderm *Coccosteus.* The region of the head, gill cover, and shoulder girdle is covered with dermal bone, but the more posterior portion of the body is without armor and shows the vertebral column. Modified from Miles and Westoll, 1968. *C,* The primitive jawed fish, *Climatius,* an acanthodian. Modified from Moy-Thomas and Miles, 1971. There is a series of paired fins between the pectoral and pelvic fins, but the jaw apparatus resembles that of primitive Chondrichthyes and Osteichthyes.

has not been confirmed. This leaves a gap of approximately 50 million years between the Chengjiang fauna and the next oldest vertebrates from the Middle Ordovician of Bolivia, found and described by Pierre-Yves Gagnier (1991).

The Bolivian genus *Sacabambaspis* is known from numerous specimens, from which almost the entire external anatomy can be studied (Fig. 2.9A). It is strikingly different from *Haikouella* and *Haikouichthys* in being entirely covered

with bony plates and scales, except for small openings for the mouth, nose, eyes, paired pineal structures, and a single pair of posterior openings for the gill chambers. Two rows of lateral line canals extend along the large dermal plate of the skull. The internal skeleton was not ossified, but the flexible plates beneath the oral chamber suggest that it was a filter feeder, without jaws. *Sacabambaspis* is the oldest known member of a highly diverse assemblage of fish termed ostracoderms for their massive bony covering that were common until the end of the Devonian. None are known to have had jaws, and one group (the anaspids) is thought to be ancestors of the modern lampreys, which also lack jaws. The ancestry of the other living group of jawless fish, the hagfish, has not been established.

The position and mechanics of the gill supports differ significantly between jawless fish (agnathans) and jawed vertebrates (gnathostomes). The gill supports in the jawless lamprey are in the form of a flexible framework that is external to the filamentous gills and is compressed by the circular muscles that surround the gill region. In jawed vertebrates, the gill supports are rod-shaped structures that are medial to the gills and hinged posteriorly. They are opened and closed by specialized mandibular and pharyngeal muscles. It is possible that gill supports in agnathans and gnathostomes evolved independently, after both lineages had diverged and then reached a large enough size that filamentous gills and specialized gill supports and musculature became necessary because of increased metabolic requirements.

Early Jawed Fish

Unfortunately, no fossils are yet known that document the origin of jaws, although the rod-shaped structures in the hyoid region of *Haikouichthys* could be precursors. Four groups of jawed vertebrates are known in the Paleozoic (R. L. Carroll, 1988a). The placoderms, known primarily from the Devonian, were heavily armored with primitive genera looking somewhat like ostracoderms (Fig. 2.9B). Their jaws are not like those of other jawed fish, and may have evolved independently. The acanthodians, known from the Late Ordovician into the Permian, are the most primitive fish with jaws that resembled those of advanced jawed fish, the sharks and their relatives, and the bony fish. In all these groups, the upper and lower jaws appear as the most anterior elements of a series of bones that include the gill supports (Fig. 2.10A). These include the palatoquadrate and Meckel's cartilage, which form the upper and lower jaws, and the hyomandibular, which links the braincase to the area of the jaw articulation. The hyomandibular presumably evolved from a previously existing gill support, but the palatoquadrate and Meckel's cartilage seem to have evolved in relationship with active feeding, rather than having been derived from pre-existing gill supports, as was once hypothesized.

The early history of advanced jawed fishes is still very poorly known. Small bony structures resembling the scales of later sharks (Chondrichthyes) are known as early as the Llandovery (the lower part of the Silurian) in Siberia and isolated shark teeth by the end of the Silurian, but the first articulated shark skeleton with associated teeth is not known until the Lower Devonian (R. F. Miller et al., 2003). This very early shark, *Doliodus*, is surprisingly primitive in retaining pectoral fin spines resembling those of acanthodians (Turner and Miller, 2005). The other major group of fish, the Osteichthyes, or bony fish, can first be recognized on the basis of distinctive scales from the Late Silurian and body fossils from the Early Devonian (Cloutier and Arratia, 2004).

Osteichthyes, plus the sharks and their allies, have a basically similar structure of the jaws and their relationship with the braincase (Fig.2.10A) that suggests a close common ancestry. In contrast, they were highly divergent in their external appearance and ways of life from the first occurrence of fairly complete specimens in the fossil record of the Early Devonian.

Osteichthyes and Chondrichthyes may be thought of as representing different adaptive strategies among actively swimming predaceous vertebrates. From the first appearance of well-preserved bony fish, we see characters that will eventually enable their descendants to conquer the land. Bone, as a specialized body tissue, evolved early in the history of vertebrates, as seen in the mid-Ordovician ostracoderms. It served several functions at that time. As in ourselves, it provided support and protection of the body and its internal organs. It could also have served as a reserve for calcium and phosphate, both vital for metabolic activities. In addition, there are regularly shaped spaces within the skull bones of various primitive vertebrates that resemble structures in modern fish that contain electrosensory organs. They may have been especially important for fish living on the bottom or in turbid water for both capture of prey and avoidance of predators, which would have emitted electrical charges in association with their muscular activity and discharge of nerve impulses.

The role of bone for protection seems paramount in the massively ossified ostracoderms and placoderms, although it is lost in the living jawless fish. But, while bone has been very important throughout the history of most vertebrate groups, it has a significant drawback: it is very heavy, about twice the weight (or specific gravity) of other body tissues, which are only slightly heavier than water. This explains why most ostracoderms and many placoderms lived on or close to the bottom, and must have been slow and awkward swimmers.

The capacity to form bone was a heritage of sharks, as can be seen from the histology of the teeth, denticles, and fin spines, but it is never known to occur as a general body covering or in the formation of the braincase, jaws, vertebrae, or internal fin supports. These structures are instead strengthened by calcification of the cartilage and covered by prismatic calcite ($CaCO_3$). In the absence of extensive bony scales or ossification of the internal skeleton, cartilaginous fish have a specific gravity only slightly greater than that of water, but must swim to keep their body suspended in the water column. Constant swimming is also necessary in sharks to maintain the flow of water through their gills for respiration.

Cartilaginous fish are divided into two categories: the fusiform, actively swimming sharks; and the dorsoventrally flattened skates and rays, which live on or close to the bottom and must use muscular expansion and contraction of the pharyngeal region to pump water in via the dorsally positioned spiracle (the most anterior gill slit) and out through the more posterior gill slits that are ventrally located. The

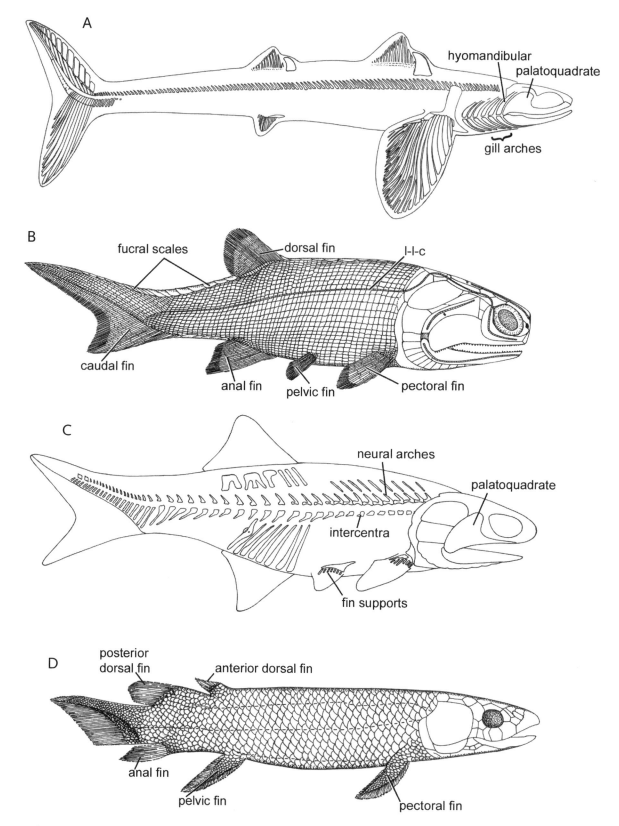

Figure 2.10. Paleozoic antecedents of modern fish groups. *A,* The Upper Devonian shark, *Cladoselache.* Modified from Zangerl, 1981. *B* and *C,* The Upper Devonian palaeoniscoid *Moy-thomasia,* a primitive actinopterygian fish, showing the external surface of the dermal bones and scales, and the underlying endochondral bones of the skull, axial skeleton, and internal fin supports. *D,* The Devonian lungfish *Dipterus,* a sarcopterygian fish; note the presence of two dorsal fins. *B* and *D* modified from Moy-Thomas and Miles, 1971; *C,* from Gardiner, 1984.

divergence between these two distinctive body shapes can be observed throughout the history of cartilaginous fish.

In the other and much more diverse assemblage of modern fish, the Osteichthyes, the internal skeleton is formed largely of bone, which requires that they must either put more effort into swimming to maintain their position in the water, or evolve other ways to reduce their total body weight (Fig. 2.10C). Bone is approximately twice the weight of water. This was an especially serious problem in Paleozoic bony fish, which had a very heavy covering of bony scales over the entire body. They served not only for protection against other early fish, such as the large and highly divergent placoderms and the early sharks, but also for stiffening of the trunk in the absence of ossified vertebrae centra.

Early bony fish, as exemplified by palaeoniscoids, were generally small, only a few centimeters in length, but with relatively large jaws. Based on the anatomy and behavior of primitive living bony fish, the ancestral Osteichthyes almost certainly had paired sacs, the swim bladders, extending ventrally from the esophagus. You can see how these are used by watching goldfish in an aquarium. They frequently rise to the surface and swallow bubbles of air. The air passes back into the swim bladder and increases the fish's buoyancy. Since bony fish were initially quite small, they did not require much air to achieve neutral buoyancy. In this manner they could control, behaviorally, their optimal depth in the water for feeding and avoiding predation, without having to swim constantly. Feeding near the surface and taking in air bubbles were combined in a unified behavioral pattern.

Although the ancestral bony fish were relatively small and retained a conservative body form, they eventually achieved a much greater diversity than any other group of vertebrates, with more than 30,000 living species. This may be attributed to the structural characteristics of bone. While cartilage is light and enabled the sharks to be highly successful predators in the waters of all oceans for the past 400 million years, they never achieved a high degree of structural or behavioral diversity, and only about 820 species are known today (Turner and Miller, 2005). In contrast, bone (with a compressive strength of 1330–2100 kg/cm², four times that of concrete) is much stronger and even very small elements can be incorporated into complex mechanical structures. The differences between cartilage and bone as elements of vertebrate structural complexes may be compared with the difference between the human use of wood or metal for making machinery. Very large wooden gears were used in medieval clocks, but metal was necessary for wristwatches.

The use of bone in early Osteichthyes enabled them to evolve complex pumping systems for feeding and respiration that made it possible for them to extend their feeding habits over a wide range of environments and types of prey. These functional complexes also enabled their descendants to feed and breathe on land. The functions of feeding and respiration are separated in cartilaginous fish. Feeding involves primarily the jaws, which lie beneath the braincase. The most anterior

portion of the branchial complex, the hyoid arch, is linked to the back of the skull, but the more posterior gill arches are not associated with the capture of prey. In contrast, the early bony fish evolved close functional connections between movements of the jaws and the hyoid apparatus that coupled feeding and respiration. Because of their small size, early bony fish must have fed on even smaller prey and/or particles of organic matter suspended in the water. They fed, not only by grasping prey in their jaws, but by sucking in large amounts of water, within which food was suspended. Contraction of muscles that ran from the ventral portion of the trunk and attached to the lower surface of the hyoid apparatus and the front of the lower jaws expanded the area of the pharynx and back of the jaw both ventrally and laterally so as to create a vacuum. As the mouth was opened by these and other muscles, the water rushed in, carrying food. As the jaws were closed, the water was forced into the pharynx and, as in the most primitive vertebrates, passed out the gill slits, and so carried oxygen to the gills. Larval salamanders retain this functional complex for both feeding and respiration. Specialized modes of aquatic locomotion, making use of closely integrated internal fin supports, also formed the basis for the limbs of terrestrial vertebrates.

SARCOPTERYGIAN FISH

The earliest known fossils of bony fish, consisting primarily of isolated scales, appeared in the Upper Silurian, but by very early in the Devonian two major groups can be distinguished (Zhu and Yu, 2004). One, whose early members retained the general body form that has just been described, are termed actinopterygians, or ray-finned fish, because the internal supports of their fins are arranged as a series of parallel rays extending from the shoulder girdle and pelvis (Fig. 2.10C). These formed a fairly wide base of the fins for effective propulsion, but may have restricted their maneuverability. In primitive living actinopterygians, the muscles to move the fins are largely in the body wall, and the fins themselves consist primarily of internal bony supports and a covering of specialized scales, termed lepidotrichia. The other group of early bony fish, the sarcopterygians, or lobe-finned fish, had a more massive, but restricted bony axis and muscles within the fin to move the elements relative to one another. The base of the fin could rotate, to a degree, in all directions (Fig. 2.10D).

Most groups of lobe-finned fish soon reached larger size than the early actinopterygians, but some appear to have adapted to an environment close to the bottom rather than swimming in open water. The more massive, lobate fins of the sarcopterygians may originally have evolved from the flatter fins of ancestral actinopterygians to enable them to push against the substrate at the bottom of the water bodies in which they lived. Life near the bottom may also be reflected in the relatively larger size and greater degree of ossification of the olfactory capsule compared with the eye, and the elaboration of structures thought to function as electrosensory organs, as in modern fish living in deep and/or sediment-filled waters. These characteristics may have evolved initially in fairly deep oceanic waters, but would also have served in shallower waters close to shore. While most Paleozoic ray-finned fish retained a relatively conservative body form and are classified in a single large assemblage, the Palaeoniscoidea, the lobe-finned fish diverged very rapidly with several major types being recognized by the mid-Devonian. Their rapid radiation makes it difficult to establish their specific relationships with one another.

A striking characteristic of early sarcopterygians was their multipartite braincase (Fig. 2.11). This seems an unusual feature in view of the solid, unitary braincase in mammals, but was an important feature of feeding and respiration in several groups of sarcopterygians, including the lineage that gave rise to land vertebrates. The evolution of the vertebrate braincase can be traced back to their Cambrian ancestors, in which the head region was dominated by paired sensory structures. The first elements to evolve were the nasal, optic, and otic capsules (Fig. 2.7C). Behind the otic capsule developed an occipital plate that formed the surface for insertion of the notochord and later for articulation with the vertebral column. Between these elements, the braincase remained unossified in the early bony fish. In later fossils, these areas appear as gaps or fissures. A lateral cranial fissure was retained between the occipital plate and the otic capsule, and a ventral cranial fissure between the otic capsule and the more anterior portion of the braincase, termed the ethmoid, which supported the eyes and the nasal capsules.

Early in sarcopterygian evolution, the bone behind the ventral cranial fissure lost the area of ossification that had surrounded the notochord, leaving an elongate anteroposterior gap. In contrast, the dorsal portion of the otic-occipital formed an area of articulation with the anterior ethmoid element. The presence of this line of mobility within the braincase long puzzled paleontologists. Mechanics of this system were worked out by Keith Thomson (1967) based on comparison between fossils closely related to the group that gave rise to tetrapods and the living coelacanth *Latimeria*. Paleontologists had long hypothesized the presence of a pair of subcephalic muscles, running on either side of the notochord between the posterior, otic-occipital portion of the braincase and the ethmoid. Their contraction would lower the anterior portion of the braincase and the attached portion of the palate, which bears very large teeth. Although the two parts of the braincase could have moved relative to one another no more than about 15 degrees, this would have been enough to force the palatal fangs through the thick scales of their probable prey among other bony fish. The recently described subcephalic muscles in the living coelacanth are in the exact position that was hypothesized for early sarcopterygians. Surprisingly, this pattern of intracranial mobility was retained in species close to the immediate ancestors of land vertebrates.

The high degree of similarity of the feeding mechanism of living coelacanths (the genus *Latimeria*) and the ancestors of tetrapods was once thought to indicate their close relationship, to the exclusion of other groups of lobe-finned fish. However, the fossil record now shows that this mechanism appeared very early in the evolution of lobe-finned fish, although it was quickly lost in the early members of the best-known living sarcopterygians, the lungfish.

Before the discovery of a living coelacanth, lungfish, now living in South America, Africa, and Australia, were long thought to represent the group that was ancestral to land vertebrates. This was primarily because of the retention of lungs that function in the same manner as those of tetrapods. In fact, the South American and African lungfish will drown if they cannot make use of atmospheric oxygen. These lungfish are also capable of aestivating. If the water in which they normally live dries up, they can survive for months at a time living in burrows they dig with their tails in the bottom mud. They emerge as soon as the water returns. Fossils of lungfish burrows are known as early as the Carboniferous. Modern lungfish have a pattern of behavior that is well suited for animals living at the water-land interface, but neither the fossil nor the living lungfish have limbs suitable for locomotion out of the water.

While adaptation to shallow freshwater, reliance on lung respiration, and the capacity to aestivate might logically be associated with transition between aquatic and terrestrial ways of life, close relationship of lungfish with early land vertebrates is clearly excluded by their highly specialized dentition and the nature of the braincase. Instead of a multitude of individual teeth along the margins of the upper and lower jaws, lungfish, early in their evolution, lost the marginal teeth and fused the dentition of the palate and medial surface of the lower jaws into large tooth plates—two pairs attached to the bones of the palate, and one to the inside surface of the lower jaws that served as crushing and shearing structures. While the internal bone of the palate (the palatoquadrate) articulated loosely with the braincase in early bony fish, it became fused to the base of the braincase in all lungfish after the Early Devonian (Chang, 2004). These specializations of the feeding apparatus preclude lungfish from close affinities with terrestrial vertebrates.

CHOANATE FISH

Although the fin structure and anatomical similarities of the skull link ancestral sarcopterygians to amphibians, most lacked a key characteristic of tetrapods and their immediate ancestors—an internal naris. Sharks and their relatives and actinopterygian bony fish have two external openings into the nasal capsule: an anterior incurrent nostril and a posterior, excurrent nostril, so that water flows into the nasal capsule at the front and passes out at the back as the fish swims through the water. This provides optimal flow across the olfactory epithelium. In contrast, all terrestrial vertebrates have a single external opening through which air is inhaled.

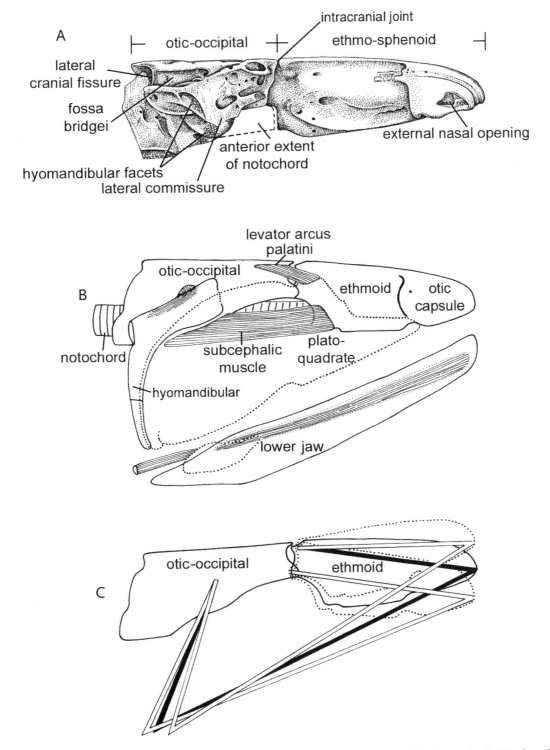

A

intracranial joint

otic-occipital

ethmo-sphenoid

lateral
cranial fissure

fossa
bridgei

external nasal opening

anterior extent
of notochord

hyomandibular facets

lateral commissure

levator arcus
palatini

B

otic-occipital

ethmoid

otic
capsule

notochord

subcephalic
muscle

plato-
quadrate

hyomandibular

lower jaw

C

otic-occipital

ethmoid

Figure 2.11. Braincase of the choanate fish *Eusthenopteron* from the Upper Devonian, showing the multipartite braincase that is characteristic of primitive sarcopterygian fish. This structure retains the separation of the otic-occipital region and the anterior ethmoid portion of the braincase that evolved to support the discrete elements of the sensory system of the brain in early vertebrates. *A,* Lateral view of the endochondral braincase, showing the position of the intracranial joint and the gap below the anterior portion of the otic-occipital that was occupied by the notochord. *B,* Outline of the braincase, hyomandibular, and lower jaw showing the position of muscles observed in the living coelacanth *Latimeria* that serve to raise and lower the ethmoid region relative to the otic-occipital. *C,* Diagram demonstrating the linkage between the palatoquadrate, whose outline is indicated by the triangular lines, showing how the snout is raised as the mouth is opened and lowered as it is closed. From Thomson, 1967.

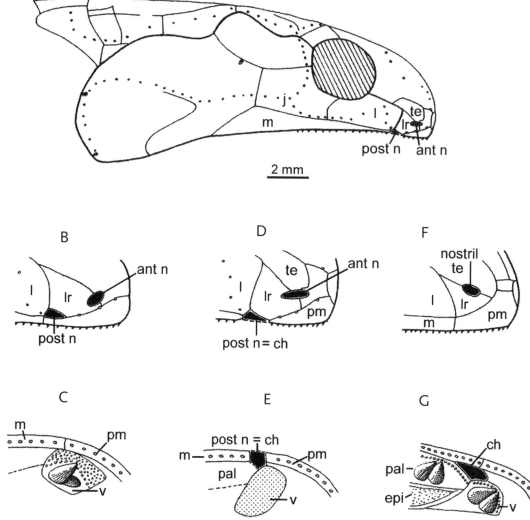

Figure 2.12. Diagram of the transition between the posterior, excurrent narial opening, and the internally located choana. *A,* Reconstruction of the head and cheek of *Kenichthys cambelli* in lateral view. *B, C,* Lateral and palatal views of the snout region of *Youngolepis;* the posterior narial opening cannot be seen in view *C. D, E,* Lateral and palatal views of *Kenichthys,* in which the posterior narial opening is between the premaxilla and maxilla and faces ventrally. *F, G,* Lateral and palatal views of the choanate fish *Eusthenopteron,* in which the posterior narial opening (now recognizable as a choana) lies medial to the sutural connection between the premaxilla and the maxilla. From Zhu and Ahlberg, 2004.

It then passes through the nasal sac and down into the mouth via the internal naris, or choana. This was the first specifically tetrapod trait to evolve in their fish ancestors. The oldest fossils illustrating the origin of the internal naris were recently studied by Zhu and Ahlberg (2004). They described fish from the Lower to Upper Devonian that show a succession of positions of the posterior naris from the side of the snout to the roof of the mouth, via an intermediate position along the tooth row between the premaxilla and the maxilla (Fig. 2.12).

The initial adaptive value of this change is difficult to imagine. It may have been the result of an accident in the course of development of the nasal capsule and the bones of the skull roof and palate resulting from minor genetic changes. Similar developmental accidents produce the human birth defect cleft palate. Once an internal naris was established, it provided a major step in the evolution of one group of sarcopterygians that enabled them to inhale atmospheric oxygen without opening the mouth.

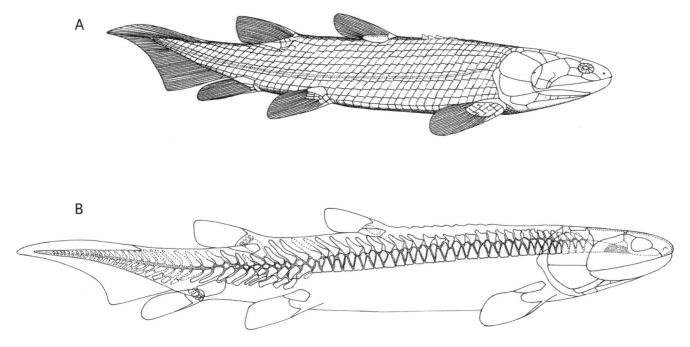

Figure 2.13. The Middle Devonian *Osteolepis* showing the primitive body form of choanate fish. *A,* External surface. Note presence of two dorsal fins, an asymmetrical (heterocercal) caudal fin, and thick, lobate paired fins, behind which is the anal fin. *B,* Internal skeleton, showing high degree of ossification of vertebral elements, but little is known of the internal structure of the pelvic and pectoral fins. Modified from Moy-Thomas and Miles, 1971.

The presence of an internal nostril is the primary means for recognizing the specific group of lobe-finned fish that gave rise to tetrapods, for which they are referred to as choanate fish. Choanate fish were common and diverse during the Upper Devonian and Carboniferous and continued into the Permian. The best-known families that have an internal naris are the Osteolepidae, Tristichopteridae, and Panderichthyidae, which document the transition between fish and tetrapods. *Osteolepis* is a well-known choanate fish that illustrates the primitive morphology of this group (Fig. 2.13).

The most informative genus is *Eusthenopteron,* known from thousands of specimens collected from rocks exposed along the Baie des Chaleurs at the base of the Gaspé Peninsula in Canada. This genus has been intensively studied by several generations of paleontologists who have described and illustrated every bone in the body. In addition, the sequence of development has been analyzed on the basis of growth stages from 3 cm hatchlings to adults approaching 2 m in length (Schultze and Cloutier, 1996; Cote et al., 2002) (Fig. 2.14; see also Fig. 3.1).

Superficially, *Eusthenopteron* appears entirely fishlike, with close resemblance in size and shape to living pike such as the muskellunge. However, it is distinguished in having two dorsal fins, rather than a single one, and in the trifid nature of the tail fin. The scaly external surface of the paired fins resembles that of other early bony fish, but the internal skeleton is entirely different in the possession of a strong central axis. The external bones of the skull differ in their specific pattern, as do the bones that form the vertebral column. In modern bony fish they are cylindrical, but in *Eusthenopteron*

they are composed of several triangular elements. As a basis for analysis of the transition between choanate fish and early tetrapods, it is necessary to describe the skeleton of *Eusthenopteron* in considerable detail.

Skeletal Anatomy of *Eusthenopteron*

There may seem to be a very large number of bone names to keep track of, but the progressive loss and fusion of skull bones throughout the history of amphibians provides an effective means of classifying nearly all the intermediate forms leading up to the modern frogs, salamanders, and caecilians. In fact, many of these bones are expressed in the human skull. In addition to loss and fusion of bones, changes in the shape and function of all skeletal elements are associated with adaptation to various environments and ways of life.

As in other bony fish, the skull is composed of two sets of bones: the external dermal bones (initially formed by the coalescence of small scales in their ancestors) that make up the surface, and the internal, or endochondral, bones that constitute the braincase, the underlying bones of the upper and lower jaws, and the hyoid apparatus.

The dermal bones that cover the skull can be recognized as belonging to several functional units that move relative to one another to provide the flexibility necessary for feeding and respiration in the water. Mobility of the parts of the braincase discussed in relationship to the cranial mechanics of feeding is dependent on comparable movement between groups of dermal bones. Both the dermal bones and the underlying braincase become progressively more strongly integrated during the transition toward a terrestrial way of life.

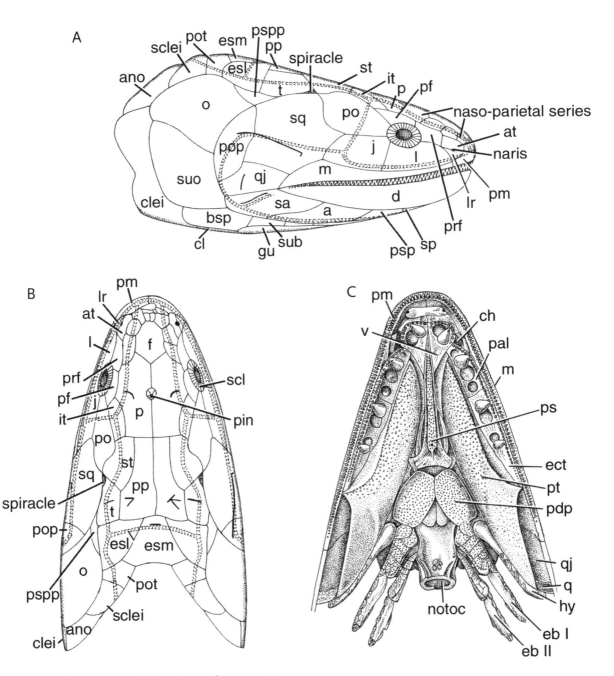

Figure 2.14. The skull of *Eusthenopteron,* the best-known of sarcopterygian fish, close to the ancestry of land vertebrates, in lateral, dorsal, and palatal views. Modified from Moy-Thomas and Miles, 1971.

If we look first at the dorsal surface of the snout, we see large paired bones, the parietals, which surround a small pineal opening for the light-sensitive pineal eye. This organ, which is retained in modern lizards and *Sphenodon,* is involved with sensing seasonal changes in the amount of sunlight that affect reproduction and other activities. Anterior to the parietal is the medial frontal bone and several smaller elements that are too variable in number and configuration to justify individual names. They gradually coalesce in more derived choanates. The anterior and lateral margins of the ethmoid region are made up of the premaxillae, anteriorly, and the maxillae. The external naris is bordered by the anterior tectal and lateral rostral. The margins of the opening for the eye, termed the orbit, are formed by the lacrimal, jugal, pre- and postfrontal, and postorbital. There is a narrow gap between the margins of the prefrontal and lateral rostral dorsally, and the lacrimal ventrally, which allows the plate below the eye to move relative to the snout. The posterior portion of the cheek is formed by the squamosal, preopercular, and quadratojugal.

Dorsally, there is a hinge line behind the parietal that separates it from the back of the skull table consisting of the postparietal, supratemporal, and tabular. These elements are separated from the posterior portion of the cheek by a narrow cleft that widens to accommodate the spiracular opening for the most anterior gill slit. The most posterior bones of the skull roof are the medial and lateral extrascapulars that link the skull and the dorsal trunk. Behind the cheek is the operculum, composed of the opercular and subopercular bones, whose lateral movement expands the gill chamber and allows the posterior discharge of water from the pharynx.

The central portion of the palate is made up of the median parasphenoid and the paired palatoquadrates and laterally by the vomer, palatine, and ectopterygoid, all of which bear large fangs. The parasphenoid goes back no farther than the level of the ventral cranial fissure, behind which are smaller denticle-covered plates.

The lower jaw consists of the tooth-bearing dentary and a series of infradentaries—laterally two splenials, an angular and a surangular. The bone at the back of the lower jaw articulates with the palatoquadrate of the upper jaw. Beneath the lower jaws, the intermandibular area is covered by a series of gular plates.

Beneath and behind the skull is the hyobranchial apparatus, which acts in feeding and respiration. This functional complex follows almost exactly the pattern already seen in primitive actinopterygian fish, and remains little changed even in the larvae of primitive living salamanders. Behind the skull proper are the dermal bones of the shoulder girdle including, from top to bottom, anocleithrum, cleithrum, clavicle, and a median ventral interclavicle.

The notochord, present in the most primitive vertebrates, is sheathed with several paired vertebral elements; dorsally the neural arches, laterally the pleurocentra, and ventrally the intercentra. There are also small homologues of tetrapod ribs that link the arches and centra but do not extend over the flanks.

Superficially, the paired fins of *Eusthenopteron* resemble those of large ray-finned fish but the internal skeleton is highly distinctive. Beneath the base of the dermal shoulder girdle is the small scapulocoracoid with a posteriorly facing glenoid articulation for the rounded head of the humerus. The internal support of the limbs is small compared with the size of the body, but they share with terrestrial vertebrates the presence of a single proximal element, comparable with the humerus in the forelimb and the femur in the hind. Two elements occur more distally, the ulna and radius in the forelimb and the tibia and fibula in the hind, plus bones in the position of the wrist and ankle. All of these bones are sheathed with dermal scales, and there is nothing that presages hands or feet.

All in all, *Eusthenopteron* looks and must have behaved in a fully fishlike manner, with no evidence whatsoever of the terrestrial future of its descendants. Yet, within 15 million years they had evolved all the basic anatomical features necessary for life on land.

3

The Origin of Amphibians

FOR NEARLY 170 MILLION YEARS after the origin of vertebrates in the Early Cambrian, they remained confined to the water. Their entire anatomy was adapted to feeding, respiration, locomotion, and reproduction in an aquatic environment, just as are modern fish. Then, within a geologically short period of about 15 million years, animals arose that were capable of walking and breathing on land. This occurred between approximately 380 and 365 million years ago. The age of the oldest known amphibians is easy to remember because it happens to coincide with the number of days in the year.

The emergence of terrestrial vertebrates is not only a very important event in the history of life, but also one for which many unanswered questions remain. The most difficult to answer is *why* animals that were highly successful within a strictly aquatic environment would leave the water and achieve a terrestrial way of life. What would have been the adaptive advantage of undergoing a multitude of major changes in their anatomy and behavior in order to survive on land?

More specific questions involve *what* advances occurred in various parts of the skeleton that enabled a more or less progressive sequence of changes in behavior from swimming to support and walking on land, and from exchange of respiratory gases with the water to exchanging oxygen and carbon dioxide with the air.

Other longstanding problems concern *where* the transition between fish and amphibians occurred, both in terms of geography, whether in the northern or southern continents, and in terms of the environment, whether from saltwater or freshwater and if into an arid or humid environment.

The general limits as to *when* the transition occurred are broadly bracketed between the last occurrence of immediate ancestors that were still obligatorily aquatic, and the first appearance of animals with girdles and limbs capable of support and locomotion on land. However, there is still much controversy over when and for what adaptive reasons hands with fingers and feet with toes like our own evolved. Although the hands and feet of later land vertebrates are well adapted for terrestrial locomotion, the earliest known amphibians had appendages that can be interpreted as being more suitable for locomotion in the water.

Another question involving the origin of land vertebrates that may also be asked in regard to any major anatomical change is *how* it occurred, in terms of modifications in the underlying genes. This question can now be approached through study of the nature of the specific genes that govern the patterns of development of bones and associated soft tissue.

The other questions of why, what, where, and when can be answered primarily through study of the geological and fossil record. During the past 20 years, a great number of fossils have been discovered that illustrate critical stages in the transformation between obligatorily aquatic fish and early land vertebrates. Of special importance has been the research of Per Ahlberg, Jennifer Clack, Michael Coates, Eric Jarvik, Hans-Peter Schultze, and Emilia Vorobyeva. Jennifer Clack has played an especially large role in investigating the origin of amphibians through her own fieldwork in Greenland; the extremely detailed preparation and description of the most primitive known amphibians *Acanthostega, Ichthyostega,* and *Tulerpeton;* the training of students and colleagues who are further extending this work; and her introduction of this very exciting evolutionary event to a more general audience (Clack, 2002a, 2005b).

During the last decade, we have also gained much information regarding the configuration of the continents and the probable environment of the world, before, during, and after these events. In addition, we are continuing to learn more and more about the anatomy, physiology, and behavior of living animals that can serve as models for comparison with their fish and amphibian ancestors.

ANCESTORS AND DESCENDANTS

In Chapter 2, we saw that the choanate or osteolepiform fish possessed many skeletal features similar to those of land vertebrates. Of particular significance were the presence of internal nostrils and paired fins supported by an internal skeleton comparable to that of our own arms and legs. All of the major skull bones of primitive amphibians can also be recognized in these Middle and Late Devonian fish. On the other hand, the general body form of *Eusthenopteron* and *Panderichthys,* with paired and caudal fins, indicates not only an effective means of aquatic locomotion, but the total absence of any features suggestive of movement on land (Fig. 3.1 A, B).

Although these fish would have had the capacity to breathe from the air, the hyoid apparatus and the presence of a large, bony operculum indicate that they depended primarily on aquatic respiration. The hyoid apparatus is also well adapted for aquatic gape-and-suck feeding, common to most fish.

Eusthenopteron is the best-known fish close to the ancestry of amphibians, but the somewhat earlier *Panderichthys* is more informative in illustrating an intermediate stage, showing the loss of some typical fish structures and the evolution of other characteristics common to land vertebrates. *Panderichthys,* known from the Frasnian and even earlier Givetian of the Baltic region (Fig. 3.2), has completely lost the two dorsal fins present in other lobe-finned fish, as well as the dorsal and ventral portions of the caudal fin. The median ventral anal fin, at the base of the tail, is also lost. Together, the loss of both dorsal and ventral fins strongly suggests that *Panderichthys* was adapted to swimming near the bottom in very shallow water (Plate 1). In contrast with *Eusthenopteron,* the skull of *Panderichthys* is dorsoventrally flattened, with the eye sockets on the top but facing laterally. This indicates that it would have had a field of vision including both the water and the air above it. The skull of *Panderichthys* is also better consolidated than that of *Eusthenopteron,* with the bones anterior to the eyes incorporated into paired frontals, as in terrestrial vertebrates, and the parietals and postparietals sutured together, closing the joint that existed in most osteolepiform fish (Fig. 3.3).

Integration of these skull bones would have restricted movement between the units of the skull that were involved in cranial kinesis in *Eusthenopteron* (discussed in Chapter 2). Solid connections between the skull bones would have facilitated movement of the head in the air, without support of the water.

We may picture *Panderichthys* swimming in very shallow water, keeping an eye on possible prey and predators from the interface between the water and the air. Prey, such as fish, could have been caught and manipulated in either medium. Although fish such as *Panderichthys* may occasionally have been trapped on land as the tide went out, they show no structures of the fins that would have enabled more than erratic flopping around on the beach as a means of getting back in the water.

In fact, the paired fins and the pelvic girdle of *Panderichthys,* as described by Boisvert in 2005, appear less similar to those of terrestrial vertebrates than those of *Eusthenopteron* (Fig. 3.4) in the absence of small bones extending distally from the ulnare and fibulare. In contrast, a subsequently discovered specimen (Boisvert et al., 2008) has four small radials distal to the ulnare that are arranged in the position of the metacarpals at the base of the digits in the earliest tetrapods. This implies their essentially common timing of development. This is in strong contrast with the longstanding hypothesis of the progressive appearance of digits in a posterior to anterior sequence across the "digital arch."

A second fish, *Elpistostega,* described from the same locality as the best-known specimens of *Eusthenopteron* in eastern Canada, has been placed in the same family as *Panderichthys.* The skull is very similar, but the scales covering the body are thick and rhomboidal rather than thin and rounded (Schultze, 1996). Knowledge of *Panderichthys* and *Elpistostega* demonstrates the presence of additional advanced lobe-finned fish approximately 10 million years prior to the first appearance of unquestioned tetrapods. However, they remain in-

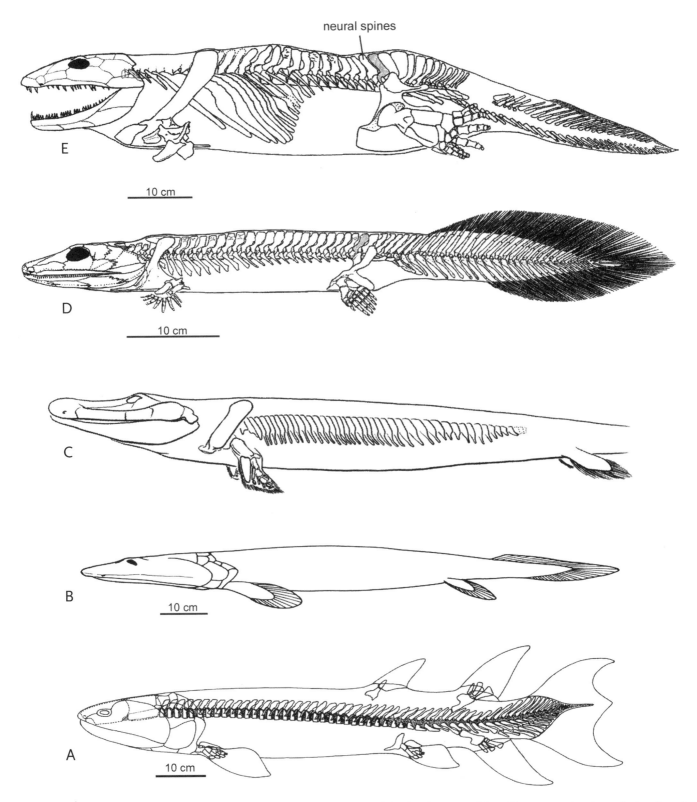

neural spines

E

10 cm

D

10 cm

C

B

10 cm

A

10 cm

Figure 3.1. Sequence of changes in the skeleton leading from the obligatorily aquatic fish *Eusthenopteron* (*A*) to the facultatively terrestrial amphibians *Acanthostega* (*D*) and *Ichthyostega* (*E*). Note that the tails of *Acanthostega* and *Ichthyostega* would be effective in aquatic locomotion. *B*, *Panderichthys*, a choanate fish, in which the loss of the medial dorsal and anal fins would permit locomotion in very shallow water. The very low profile of the skull and the dorsal position of the orbits suggest swimming at the interface of water and air. *C*, Preliminary sketch of *Tiktaalik*, a Late Devonian choanate fish recently discovered in the Canadian Arctic. *A*, modified from Andrews and Westoll, 1970; *B*, from Vorobyeva and Schultze, 1991; *C*, modified from Daeschler et al., 2006; *D* and *E* from Ahlberg et al., 2005.

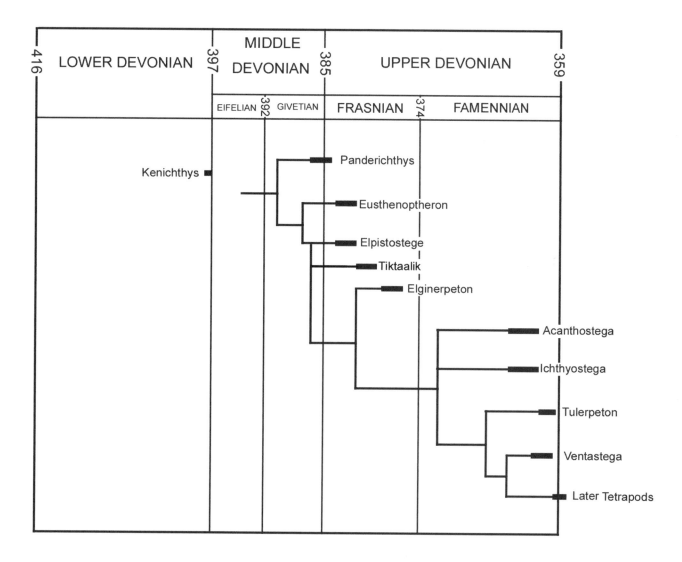

Figure 3.2. Geological time scale and putative phylogeny of selected choanate fish and early tetrapods. Modified from Clack, 2005a.

completely known, and indicate an early stage in the transition toward terrestrial vertebrates.

TIKTAALIK

Suddenly, in the summer of 2005, understanding of the fish-amphibian transition was greatly enhanced by the discovery of a new genus, *Tiktaalik,* with many more tetrapod characters than *Panderichthys* or *Elpistostega.* Numerous well-preserved specimens were found on Ellesmere Island in the Canadian Arctic by a field party led by Edward Daeschler, Neil Shubin, and Farish Jenkins. Not only does *Tiktaalik* exhibit more tetrapod features of the skull and forelimb, but its presence in sediments deposited in a river channel indicates a major shift from marine to fresh water. It is also 2 to 3 million years younger than *Panderichthys.*

Preliminary descriptions of the skeleton and nature of preservation of *Tiktaalik* were published by Daeschler et al. (2006) and Shubin et al. (2006), together with a discussion by

Ahlberg and Clack. In common with *Panderichthys* and *Elpistostega,* the skull (Fig. 3.3C) is dorsoventrally compressed but greatly widened laterally compared with *Eusthenopteron.* The eyes are near the midline, with a protective ridge on the postfrontal. In contrast with *Panderichthys,* all the bony supports for the operculum are lost, as well as the extrascapular bones at the back of the skull table and the suprascapular series that had linked the skull and the pectoral girdle. Thus, the skull was able to move in all directions, independent of the trunk, as in terrestrial vertebrates.

Another significant advance was the further expansion of the spiracular cleft. In primitive bony fish this opening served for the exit of water flowing from the mouth and through the gill cavity associated with aquatic respiration. Its expansion in *Panderichthys* and *Tiktaalik* and its retention as the squamosal notch in primitive tetrapods such as *Acanthostega* may be attributed to a specialized role associated with the intermediate environment inhabited during the fish-amphibian transition. *Panderichthys, Elpistostega,* and *Tiktaalik* possessed greatly flat-

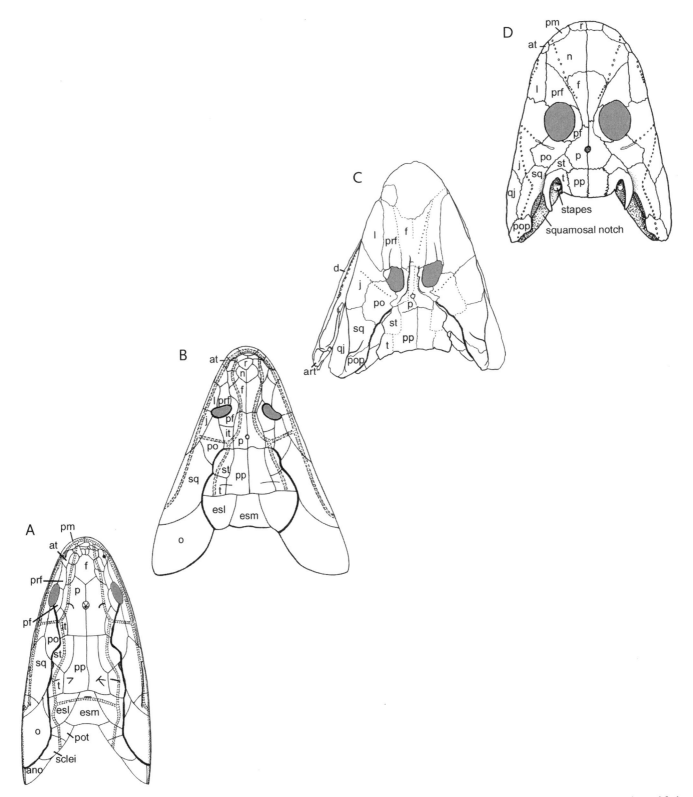

Figure 3.3. Changes in the structure of the skull roof seen in the transition from fish to amphibians. *A, Eusthenopteron,* a typical lobe-finned fish from the early Upper Devonian. Modified from Andrews and Westoll, 1970. *B, Panderichthys,* an intermediate form from the late Middle Devonian and early Upper Devonian. From Vorobyeva and Schultze, 1991. *C, Tiktaalik,* a newly discovered link between panderichthyids and the earliest tetrapods. Modified from Daeschler et al., 2006. *D, Acanthostega,* an amphibian from the latest Devonian. From Clack, 2002c. Thicker lines in *Eusthenopteron, Panderichthys,* and *Tiktaalik* indicate mobility between areas of the skull roof that became progressively consolidated from the Middle through the Upper Devonian.

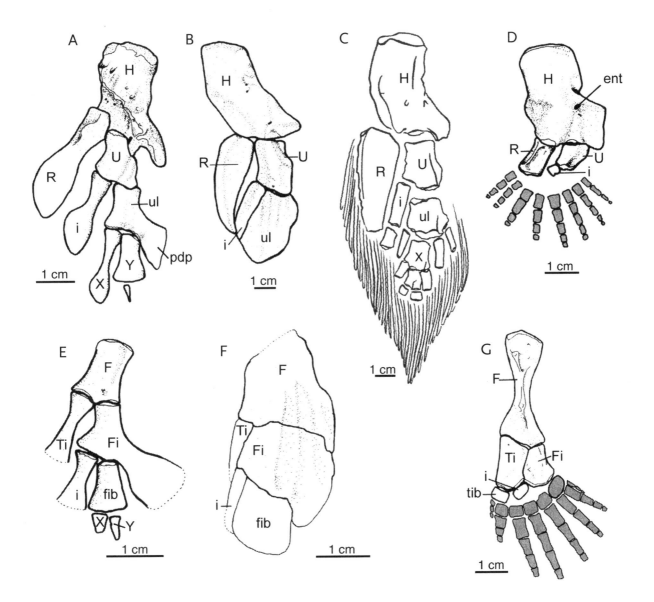

Figure 3.4. Bones of the front and hind limbs in choanate fish and a primitive amphibian. *A–D,* anterior limbs; *E–G,* hind limbs. *A* and *E, Eusthenopteron; B* and *F, Panderichthys; C, Tiktaalik; D* and *G, Acanthostega.* The limb bones of *Panderichthys* are specialized in a manner that is divergent from those of *Eusthenopteron* and early tetrapods in the flattening or loss of distal elements. *A* and *E,* modified from Andrews and Westoll, 1970; *B* and *F,* modified from Boisvert, 2005; *C,* from Shubin et al., 2006 (original illustration by Kolliopi Monoyios); *D* and *G,* from Coates, 1996.

tened skulls and apparently lacked both dorsal and anal fins. This suggests life in very shallow water, in which the top of the skull may have been exposed to the air as in their structural analogues, the crocodiles. When these fish were near the bottom, it would have been difficult for them to lower their jaws, and dislodged bottom sediments may have interfered with aquatic respiration. As argued by Brazeau and Ahlberg (2006), this could have led to selection for a change in the function of the spiracular opening. Rather than allowing for discharge of water from the oropharyngeal cavity as it was compressed, it could have served for inhalation of oxygen from the air as the cavity was expanded.

Both *Panderichthys* and *Tiktaalik* also show reduction in

the height of the pterygoid lateral to the spiracular passage and shortening of the hyomandibular, which linked the operculum and the area of jaw articulation in *Eusthenopteron.* Together, the configuration of these structures documents transition toward early tetrapods, and initiates the changes that will eventually lead to the origin of the middle ear cavity in amniotes and frogs. On the other hand, perhaps as a result of life near the bottom, *Tiktaalik* retained the large gular plates that lie between the lower jaws in most bony fish but are completely absent in the earliest tetrapods.

While the skull of *Tiktaalik* suggests close affinities with early amphibians, there are no ossified remains of the vertebrae. Perhaps they remained cartilaginous to increase the fish's buoyancy in the water or to decrease its weight on land. But without bone, they would not have served effectively to support the trunk against the force of gravity. On the other hand, the ribs were much larger than those of other fish associated with the transition toward life on land, with wide, overlapping posterior flanges. The ribs outline the area of the unossified vertebrae, but suggest a very large number of segments, up to 45, compared with about 30 in *Eusthenopteron* and *Acanthostega*, or about 26 in *Ichthyostega* (Fig. 3.1C).

Of the appendicular skeleton, only the pectoral girdle and fin have so far been described. The scapulocoracoid is expanded relative to the surrounding dermal elements, approaching in extent that of early tetrapods, and the surface for articulation with the humerus is intermediate in orientation and shape. The endochondral bones of the fin are more numerous and distally extended than those of *Panderichthys* or *Eusthenopteron*, but remain sheathed in lepidotrichia, precluding the expression of any digit-like structures. Their individual configuration and relationships to one another are also very unlike those of primitive tetrapods. The radius extends far beyond the end of the ulna, precluding a hinge-like bending in the area of the wrist. The bones distal to the ulna form a successively bifurcating median axis, unlike that of any tetrapods.

While *Tiktaalik* illustrates a plausible, intermediate way of life between marine fish and terrestrial amphibians, the absence of ossification of the vertebrae, their great number, and the divergent specialization of the forelimb suggest that this genus was not an immediate sister-taxon of any known tetrapods.

THE OLDEST KNOWN TETRAPODS

The best-known of Late Devonian amphibians are *Acanthostega* and *Ichthyostega* from East Greenland, most recently described by Per Ahlberg et al. (2005), Jennifer Clack (2002c), and Michael Coates (1996) (Plate 2). These animals are clearly amphibious in having means of locomotion appropriate for both water and land (Fig. 3.1 D, E). The tail is unquestionably like that of fish, with both dorsal and ventral fin rays. The skulls have lateral line canals and pit lines that would have functioned to detect movement in the water (Fig. 3.5). The hands and feet are more primitive in many respects than those of later, Carboniferous amphibians, but the structure of the pectoral and pelvic girdles indicates that the trunk was held above the ground, against the force of gravity. Remains of several other amphibians have been described from the Upper Devonian, but none are as completely known as *Acanthostega* or *Ichthyostega*.

When in articulation, the skulls of these genera are immediately distinguishable from those of *Eusthenopteron* and other choanate fish in their separation from the dermal shoulder girdle. This results from the loss of all elements of the opercular series except for the preopercular, which is actually a part of the cheek, and the extrascapular series at the back of the skull. The skull is now free to move in any plane relative to the trunk. Attachment is maintained by the extension of the notochord through the occiput, as indicated by the absence of bones in the area of the basisphenoid and basioccipital in later tetrapods. There is no defined area of articulation between the lower portion of the occiput and the cervical centra. The many lineages of later tetrapods can be distinguished from one another by the evolution of divergent means of articulating the occipital surface with the anterior elements of the vertebral column.

Acanthostega and *Ichthyostega* retain some small bones in the anterior area of the snout between the premaxilla and the nasal that are subsequently lost, but lose the intertemporal bone that lay between the parietal and postorbital in *Eusthenopteron*, while this bone was retained in *Tulerpeton* and several lineages of later tetrapods. All the Upper Devonian tetrapods retain a notch at the back of the squamosal, beneath the tabular, which is more or less in the position of the spiracular cleft of their fish ancestors. In later tetrapods, this has been referred to as an otic notch, on the assumption that it held a tympanum, as in modern frogs. *Acanthostega* is unique in having a second notch, facing dorsally, between a longer lateral extension of the tabular (the tabular "horn") and the medial surface that connects to the postparietal.

The palate of *Acanthostega* and *Ichthyostega* retains the general pattern of their fish ancestors, with a median parasphenoid, large pterygoids immediately lateral to it, and three more lateral palatal bones, the ectopterygoid posteriorly, the palatine, and the vomer at the front. As in *Eusthenopteron* and some primitive tetrapods, they retain large fangs on the palatine and vomers, but in contrast with those groups, there are no fangs on the ectopterygoid. All of these forms have smaller denticles on all three of these bones, but can be distinguished from choanate fish in the absence of a row of tiny denticles lateral to the main tooth row of the premaxilla and maxilla. The parasphenoid, which forms the base of the braincase, has a movable articulation with the pterygoids via paired basipterygoid processes. The quadrate is situated well behind the occipital surface.

The braincase of *Acanthostega* has been thoroughly described by Jennifer Clack (1998b). As described by Ahlberg et al. (1996) the braincase in *Panderichthys*, like that of other osteolepiform (choanate) fish, retains the mobility between the anterior sphenethmoid portion of the braincase and the otic occipital portion just as was described in *Eusthenopteron*, but the two portions became united in *Acanthostega* and other early tetrapods, thus forming a solid strut down the middle of the skull (Fig. 3.6).

The lateral surface of the otic capsule is also significantly modified in *Acanthostega* and later tetrapods, relative to their fish ancestors. Instead of being formed by a bony lateral commissure into which the head of the hyomandibular attached (Fig. 2.11A), this area has a large opening, the fenestra ovalis, into which the expanded footplate of the stapes is inserted. These bones are homologous, in the sense of having evolved

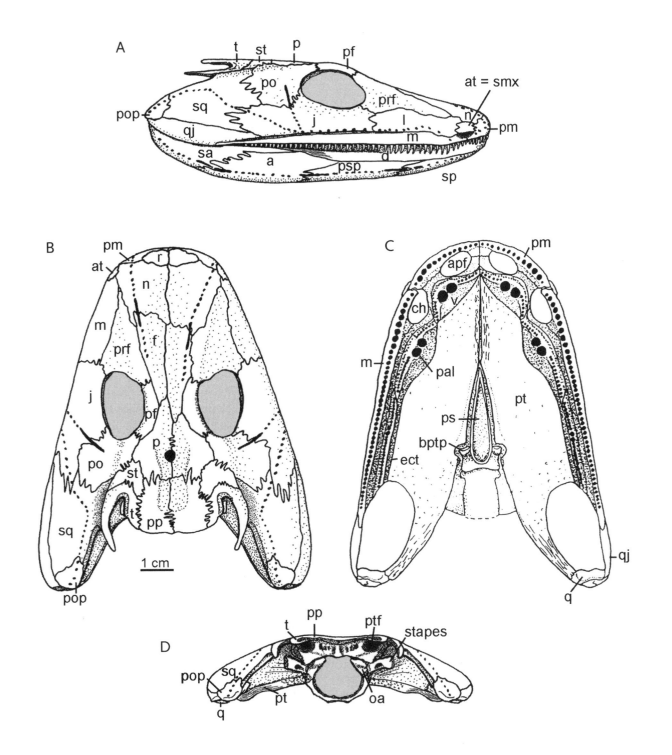

Figure 3.5. Skull of *Acanthostega*, a Late Devonian amphibian in *A*, Lateral; *B*, Dorsal; *C*, Palatal; and *D*, Occipital views. *A*, *B*, and *D* from Clack, 2002c; *C* from Clack 2002a. This pattern is close to that from which all later tetrapods evolved. Note that the lateral line canal system, common to later aquatic tetrapods, is expressed primarily as pit lines.

from one another, but are given different names because of their different functions. The hyomandibular of fish serves as a mechanical link between the braincase and the palatoquadrate in relation to the mechanics of feeding and respiration (as discussed in Chapter 2), but the main role of the stapes,

at least in more derived terrestrial vertebrates, is for the conduction of airborne vibrations to the inner ear. The stapes in *Acanthostega* may represent an early stage in the origin of the impedance matching middle ear common to frogs, reptiles, mammals, and birds. However, its relatively large size prob-

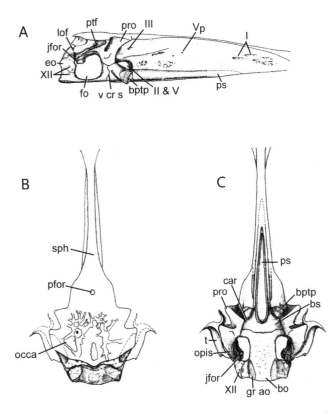

Figure 3.6. Braincase of *Acanthostega*, in A, Lateral; B, Dorsal; and C, Ventral views. From Clack, 1998b.

ably precludes its response to airborne vibrations. Clack and Ahlberg (1998) suggest that *Ichthyostega* (in which the stapes does not penetrate the braincase) may have used its uniquely derived otic capsule for detecting underwater vibrations.

The large, blade-shaped stem of the stapes in *Acanthostega* and other early tetrapods has been suggested as being associated with support of the braincase on the back of the palate (R. L. Carroll, 1980), but this is now questioned (Clack, 2003). The close proximity of the stapes and the area of the spiracular cleft in fish and the "otic" notch in early tetrapods has also been suggested as being associated with passage of either water or air through this opening in forms transitional between fish and tetrapods. Skates and rays use their spiracles to draw in water dorsally, when their gill slits on the ventral surface of the head are appressed to the bottom. Early tetrapods, in contrast, may have used dorsally located "otic notches" for drawing in air for respiration when their snouts were underwater. This was supported by Brazeau and Ahlberg (2006) on the basis of the configuration of the palatoquadrate. Use of the spiracle or otic notch for inhaling air is also suggested by the distribution of the lateral line canals and orbits in *Panderichthys* and some early tetrapods, in which they are conspicuously absent from the skull table (the highest part of the skull) but present on the snout and cheek, which could remain underwater while the eyes and otic notch were exposed to the air while basking.

Acanthostega may represent the most primitive configuration of the vertebrae and ribs of any tetrapod, but it already

shows some advances over its aquatic ancestors in supporting the head on the trunk and the body above the ground. As in *Eusthenopteron*, the vertebrae consist of three elements, all paired ancestrally and during early developments: the neural arches, anterior intercentra, and posterior pleurocentra.

From the time of its original discovery and description by Gunnar Säve-Söderbergh (1932) and Eric Jarvik (1952, 1980, 1996), the Upper Devonian genus *Ichthyostega* was hailed as the oldest known amphibian. Despite the fishlike tail, the presence of limbs with fingers and toes spoke only of terrestrial locomotion. However, Jennifer Clack's discovery of even more completely preserved skeletons of a second amphibian genus, *Acanthostega,* from the same locality in East Greenland revealed unexpected features of the limbs that raised questions as to whether they had evolved for locomotion on land or in the water. Coates and Clack (1990) found that both *Acanthostega* and *Ichthyostega* had more than the customary number of five fingers and five toes of most primitive vertebrates—eight fingers and toes in *Acanthostega* and seven toes in *Ichthyostega*. In addition, grooves on the major bones of the hyoid apparatus suggest the presence of blood vessels serving the gills, as in fish. Although neither of these amphibians had a large bony operculum lateral to the gills, the major external bone of the shoulder girdle, the cleithrum, was embayed in the same way as in related fish, to form the posterior wall of the gill chamber. Only animals that live habitually in the water can make use of the gills for respiration because they must be continuously wet for exchange of oxygen to occur. For these reasons, Clack and Coates (1995) argued that *Acanthostega* and *Ichthyostega* were probably aquatic, and that their hands and feet had evolved for locomotion in the water, rather than on land.

While the presence of a fishlike caudal fin and the likely retention of gills, at least in juvenile animals, strongly support the aquatic behavior of *Acanthostega* and *Ichthyostega,* the structure of the shoulder and pelvic girdles is so similar to that of unquestioned terrestrial amphibians from later deposits that it is very difficult to understand how they might have evolved for life in an aquatic environment (Figs. 3.7 and 3.8).

If one compares the skeletons of *Ichthyostega* and *Acanthostega* with *Eusthenopteron,* the most conspicuous differences are in the size and structure of the bones supporting the front and rear limbs. In the fish, the pelvic girdle is a small, paired structure well below the vertebral column. The pelvis served for articulation with the head of the femur, as well as an area for the origin of muscles to move the limb. The concave, posteriorly facing area of articulation permitted the femur to rotate freely, for steering and paddling.

In contrast, the pelvises of *Acanthostega* and *Ichthyostega* were much larger, with an extensive ventral puboischiadic plate and a dorsally oriented iliac process that reached to the level of the vertebral column. The iliac process was linked to the vertebral column via a sacral rib. The surface for articulation with the femur, termed the acetabulum, is near the mid-length of the puboischiadic plate. This articulation al-

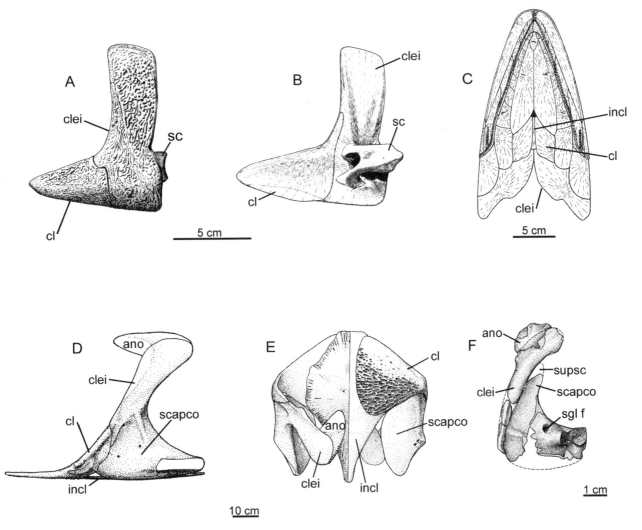

Figure 3.7. Pectoral girdle in *Eusthenopteron* and Upper Devonian tetrapods. A–C, Lateral, medial, and ventral views of the shoulder girdle of *Eusthenopteron*. Modified from Jarvik, 1980. D–E, Lateral and composite dorsal and ventral views of shoulder girdle in *Acanthostega*. From Coates, 1996. F, Lateral view of the shoulder girdle of *Tulerpeton*, which establishes the basic pattern for basal Carboniferous tetrapods. From Lebedev and Coates, 1995.

lowed the leg to move in a limited anterior-posterior arc, as well as ventrally, and to a limited extent dorsally. The largest area of muscle attachment was on the ventral surface of the puboischiadic plate. Studies of primitive living amphibians and reptiles show that muscles originating from this plate and attaching to the ventral surface of the femur would lower the distal end of the femur, and thus lift the trunk above the ground.

Neither a sacral rib connecting the pelvis to the vertebral column nor extensive ventral musculature to raise the body on the limb is necessary in a fish, since the body of aquatic animals is supported by the buoyancy of the water, and only limited movements of the fins are necessary to raise it in the water column. It is hence difficult to imagine why the structure of the rear limb and girdle of *Ichthyostega* and *Acanthostega* would have evolved if these animals were obligatorily aquatic.

The pectoral girdle in early amphibians is more compli-

cated than the pelvic girdle in having an external covering of dermal bones, which in their fish ancestors had been linked to the skull. This link is broken in *Ichthyostega* and *Acanthostega* by the loss of the opercular bones. This was an extremely important change since it allowed the head to move much more freely relative to the trunk. In fish, the head and trunk move as a unit. In order to move the head from side to side or up and down in feeding, the orientation of the entire body is altered by the movement of the fins and tail. In terrestrial animals, the head bends relative to the trunk. Lateral and dorsoventral movements of the head may have been limited in the earliest tetrapods, but the potential for such movement was initiated by the loss of the operculum and other bones that once joined the shoulder girdle to the skull.

Acanthostega and *Ichthyostega* retained three external bones of the shoulder girdle present in fish: the paired dorsal cleithra, the lateral clavicles, and the median ventral interclavicle. The interclavicle is much enlarged from its size in

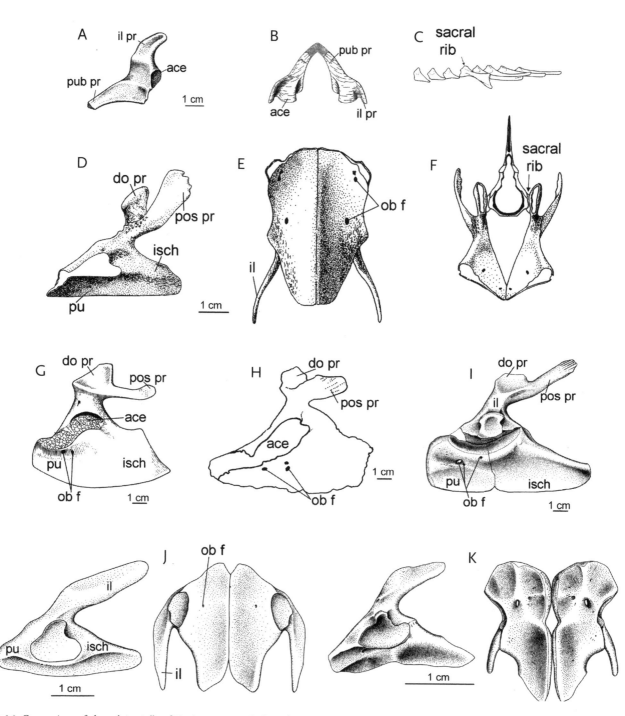

Figure 3.8. Comparison of the pelvic girdle of *Eusthenopteron* with that of primitive tetrapods. *A* and *B,* Lateral and ventral views of the pelvis of *Eusthenopteron.* Modified from Andrews and Westoll, 1970. See also Figure 3.1, which shows the position of the pelvis relative to the vertebral column in *Eusthenopteron. C,* Posterior trunk ribs, the sacral rib, and caudal ribs in *Acanthostega.* From Coates, 1996. *D–F,* Pelvic girdle of *Acanthostega* in lateral, ventral, and anterior views. From Coates, 1996. *G,* Lateral view of the pelvic girdle of *Ichthyostega.* Modified from Jarvik, 1980. *H,* The Lower Carboniferous whatcheerid *Whatcheeria.* After Lombard and Bolt, 1995. *I,* The anthracosaur *Proterogyrinus.* From Holmes, 1984. *J,* Lateral and ventral views of the pelvis of the Upper Carboniferous temnospondyl *Dendrerpeton,* from Holmes et al., 1998. *K,* Lateral and ventral views of the pelvis of the Lower Permian amniote *Captorhinus.* From Holmes, 2003. Note the consistency of structure from *Acanthostega* to *Captorhinus.*

Eusthenopteron. It has been suggested that this bone served to protect the area of the heart and lungs when the primitive amphibians dragged themselves along the ground. Medially to the cleithrum and clavicle are bones more comparable to those of the pelvic girdle, the scapulocoracoids, which provide articulation and areas for muscle attachment to move the limbs. In common with the bones of the pelvis, the area of articulation for the forelimb faced more laterally in the amphibians, and the ventral area of the scapulocoracoid was much expanded for the attachment of muscles to lower the distal end of the humerus and so raise the trunk and head above the ground.

Further evidence for support of the body without the buoyancy of the water is provided by the vertebral column and ribs. The vertebrae in *Ichthyostega* and *Acanthostega* retain the many separate elements present in osteolepiform fish, but have evolved three specialized areas of attachment that are critical for life on land. In their fish ancestors, in which the head functioned as an extension of the trunk, there was not a bony articulation between the anterior vertebrae and the skull. Instead, the internal stiffening rod of the vertebral column, the notochord, extended into the back of the skull to provide attachment and allow limited movement in all three planes. This condition is retained in *Ichthyostega* and *Acanthostega,* but anterior articulating surfaces on the paired neural arches of the first cervical vertebra of *Acanthostega* may have served for the attachment of a paired proatlas, which in other early tetrapods has areas of attachment that would have linked the head and the neck. The proatlas both strengthens the connection and limits the degree of rotation of the head.

In the rest of the vertebral column, the neural arches have evolved specialized surfaces of articulation, the zygapophyses, to provide controlled movement and additional support throughout the column. The vertebra medial to the iliac processes of the pelvis is enlarged, and bears specialized sacral ribs that attach to the iliac blade.

Ahlberg et al. (2005) recently re-described the vertebral column of *Ichthyostega,* which shows a specialized means for supporting the trunk against the forces of gravity. While the neural spines of *Acanthostega* are nearly uniform throughout the trunk, those of *Ichthyostega* show regional specialization, similar to that of terrestrial mammals. The neural spines of the anterior trunk are angled posteriorly, but those in the lumbar region and those just above the sacrum are angled anteriorly. Because muscles produce their maximum force when acting at right angles to the structures to which they are attached, the varied geometry of the neural spines indicates that the dorsal axial muscles could have raised the central portion of the trunk, lifting it above the ground. This configuration is analogous to that of a suspension bridge, with the trunk lifted by muscles (acting like cables) running from elevated supports (the pelvic and pectoral girdles and limbs). In contrast, most later amphibians strengthened the backbone by increasing the degree of ossification of the vertebral centra.

Eusthenopteron had small bones that linked the neural arches and the centra, perhaps to maintain their relative position. These bones had the structure and position of the heads of our ribs, but without the long shafts that assist us in breathing and help to support our trunk. *Ichthyostega* and *Acanthostega* had more orthodox ribs. Except for the more anterior ribs, those of *Acanthostega* were quite short. The anterior ribs of *Ichthyostega* were extremely long and flattened, and probably restricted the lateral bending of the trunk. The elaboration of the shafts of the ribs would have served to support and protect the lungs and viscera of the early amphibians when they were on land and not supported by the buoyancy of the water. It would have been difficult, if not impossible, for fish such as *Eusthenopteron* or *Panderichthys*, without elongate ribs, to fill their lungs if they were out of the water.

In contrast with the clear demonstration of support for the body on land provided by the pectoral and pelvic girdles, the bones of the wrists, ankles, hands, and feet of *Acanthostega* and *Ichthyostega* are clearly different from those of later amphibians and almost certainly less effective for terrestrial locomotion (Figs. 3.9 and 3.10). Jennifer Clack (2002a) illustrated footprints that are thought to have been made by Late Devonian amphibians, and at least some appear to indicate that the toes were oriented laterally rather than anteriorly, suggesting that the limbs may have been better adapted for paddling in the water than walking on land. In contrast, footprints from the Lower Carboniferous, about 10 million years later than the Upper Devonian amphibians, are like those of most later tetrapods in having the toes pointing forward (Fig. 3.11).

Not only do Late Devonian amphibians have a greater number of fingers and toes than do later tetrapods, but the areas of the wrist and ankle are commonly less well ossified, have fewer bones, and do not appear to have formed an effective hinge joint between the feet and the lower part of the limbs. In these features, the evolution of the feet appears to have lagged behind that of the girdles. This suggests that the primary force of selection in the fish-amphibian transition acted upon the capacity to support the body on land, rather than on fully effective terrestrial locomotion. This further complicates the question of *why* the immediate ancestors of amphibians would have come out on land.

THE ADVANTAGES OF TERRESTRIALITY

Many hypotheses have been proposed to explain why the aquatic ancestors of amphibians might have moved onto land, including the following (based on Clack, 2002b):

1. To get back into the water under arid conditions
2. "Limbs" initially for burrowing in the mud
3. Competition and/or predation with other aquatic forms
4. To escape oxygen-depleted water
5. Feeding on terrestrial or semi-terrestrial food sources
6. Increase in body temperature
   a. Increase in rate of digestion
   b. Speed development
7. Spawning on land
8. Limbs evolved originally for amplexus in the water

One of the more fully elaborated ideas was that of Alfred Romer (1958). He argued that the highly oxidized nature

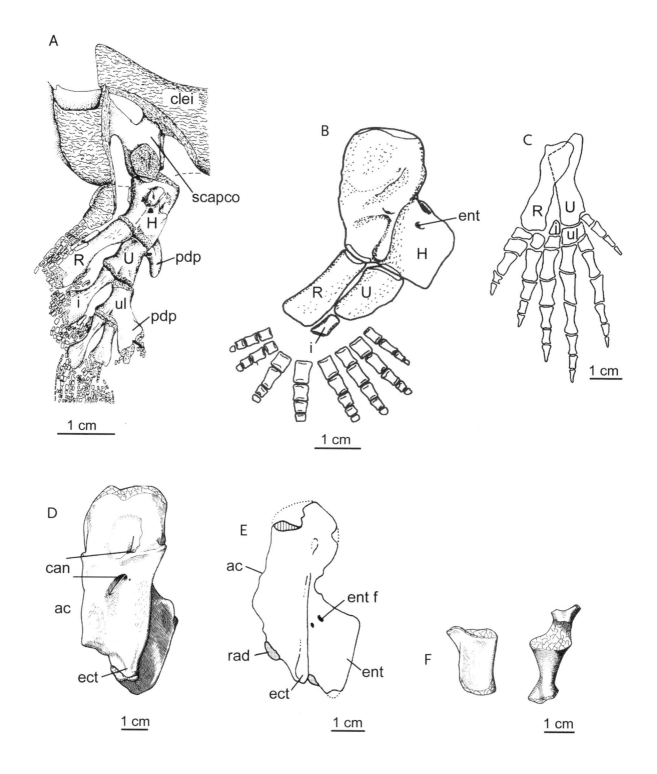

Figure 3.9. Forelimbs of *Eusthenopteron* and Upper Devonian tetrapods. *A,* Forelimb of *Eusthenopteron.* Modified from Andrews and Westoll, 1970. *B,* Forelimb of *Acanthostega.* From Coates, 1996. *C,* Forelimb of *Tulerpeton.* From Lebedev and Coates, 1995. *D, E,* Humeri of *Ichthyostega* (modified from Jarvik, 1980) and *Tulerpeton* (Lebedev and Coates, 1995). *F,* Radius and ulna of *Ichthyostega,* carpus and manus not known in this genus. Modified from Jarvik, 1980.

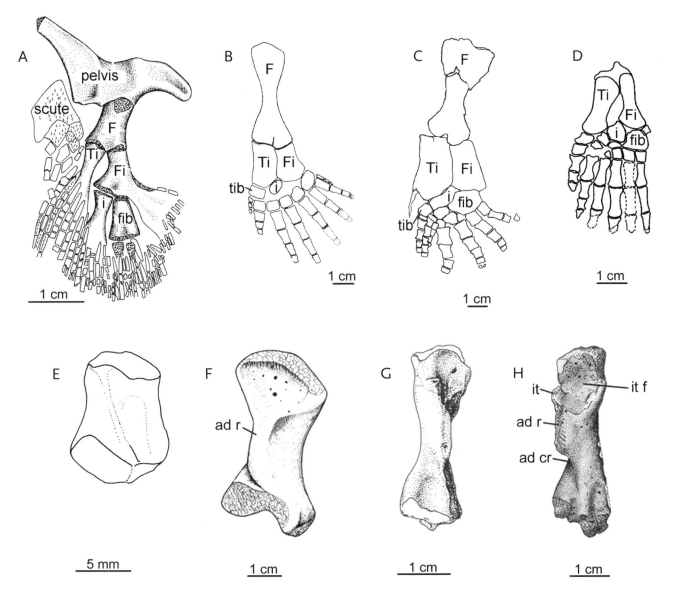

Figure 3.10. A, Hind limb of *Eusthenopteron.* Modified from Andrews and Westoll, 1970. Note presence of external dermal scales (lepidotrichia). *B–D,* Hind limbs of *Acanthostega, Ichthyostega,* and *Tulerpeton.* From R. L. Carroll and Holmes, 2007. *E–H,* Ventral views of femora. *E, Eusthenopteron.* From Coates, 1996. *F, Ichthyostega.* Modified from Jarvik, 1980. *G, Acanthostega.* From Coates, 1996. *H, Tulerpeton.* From Lebedev and Coates, 1995.

of the red sediments in which were preserved the fossils of many Carboniferous amphibians indicated that the ancestors of amphibians may have lived in near-shore environments that were subject to periodic desiccation. Lobe-finned fish such as *Eusthenopteron* may have fed on other fish trapped in shallow pools, separated from larger water bodies. They would have been at an advantage over ray-finned fish because of their more muscular fins, which would have enabled them to escape back to deeper water. Selection would have favored any changes in limb structure leading to more effective terrestrial locomotion. While plausible, the specific basis for this theory has been discounted since it is now recognized that most of the Late Devonian through Early Permian amphibians lived in warm, humid environments.

Other theories have focused on more direct advantages for moving onto land, such as feeding, respiration, or breeding. In the modern world, there are many food sources on land that would be available for new invaders. Even in the Late Devonian, there were a host of primitive plants as well as a diversity of invertebrates. However, few, if any, of these organisms would have been appropriate sources of food for the early amphibians. It is almost certain that they could not have digested any plant material, since none of the living amphibians are herbivorous as adults. This is because of the absence of appropriate digestive enzymes that are provided by symbiotic bacteria in modern herbivorous reptiles and mammals. Frog tadpoles feed primarily on microscopic plant material, but this food source is acquired in the water, prior to

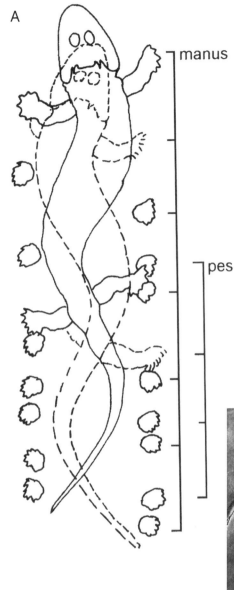

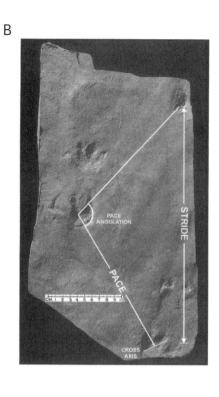

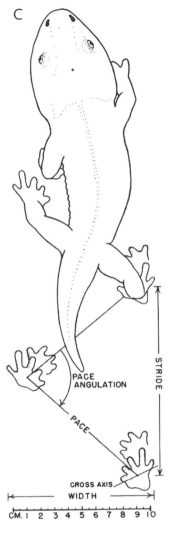

Figure 3.11. Footprints. *A,* Diagram from Clack, 2002a, showing how an *Acanthostega*-like tetrapod might produce tracks resembling those known from the Upper Devonian Genoa Locality in Australia. *B,* Amphibian trackway (Redpath Museum no. 206777) from the Lower Carboniferous (Tournaisian) of the Horton Bluff Formation, Hantsport, Nova Scotia. Tail drag indicates that the trackway was made on land. Note similarity with diagram of a trackway from the Upper Carboniferous (*C.* Modified from Baird, 1952). *D,* "Swimsporen," a trackway from the Horton Bluff Formation made by an animal dog-paddling in shallow water (Redpath Museum number 206836).

their emergence as terrestrial adults, and requires a unique, highly complicated feeding apparatus that will be discussed in Chapter 10.

As pointed out by William DiMichele and Robert Hook (1992) and Jane Gray and William Shear (1992), the fossil record of terrestrial nonvertebrate animals is poorly known from the critical period of the emergence of terrestrial vertebrates. What is known consists primarily of small arthropods, most of which feed on detritus in the soil. These included fungus-eating microarthropods, mites, spiders, millipedes, and centipedes. Collembolans (springtails) appeared in the Early Devonian and wingless insects appeared in the mid- to Late Devonian, but all were of small size. The arthropleurids, relatives of scorpions, reached gigantic size by the Upper Carboniferous, but are known from only very small relatives in the Devonian. There were extremely large eurypterids by the end of the Lower Carboniferous (Plate 4), but these were primarily, although not entirely, aquatic. Few of these animals appear as probable food for the early tetrapods, all of which were of large size, with a dentition suited for large prey. Stomach contents have not been described for any of the Late Devonian amphibians, but *Eusthenopteron* is known to have eaten fish as an adult.

In summary, no obvious sources of food for large tetrapods are known from the Upper Devonian that would have enticed the ancestors of *Acanthostega* and *Ichthyostega* to invade the land.

A further problem that would have been faced by early tetrapods attempting to feed on land is the fact that feeding in most bony fish, and presumably the immediate ancestors of amphibians, depends on support of the prey by the water. Adult individuals of large choanate fish such as *Eusthenopteron* probably captured medium and large prey by open-mouthed lunging, but to judge by the similarity of the hyoid apparatus to that of living bony fish, small prey was drawn into the mouth by suction, created by rapid expansion of the oropharyngeal cavity. Aquatic feeding occurs by the same mechanism in the larvae of living salamanders and was certainly a heritage of all land vertebrates. It is very unlikely that early amphibians had a tongue that could be used in capture or manipulations of prey, since such a structure evolved independently in frogs and salamanders after their divergence in the Carboniferous (see Chapters 10 and 11). Use of the jaws for capture of prey may have evolved fairly early in land vertebrates, but the behavioral patterns would have had to evolve after they left the water.

It has also been suggested that early tetrapods came onto land to lay their eggs and so protect them from the many aquatic predators. This is conceivable, but the primitive members of all three modern amphibian orders lay their eggs in the water and/or have aquatic larvae, as did many Paleozoic amphibians. It seems highly improbable that a capacity for hatchling amphibians to develop on land would have been lost, once it had been achieved.

Recently, Robert Martin (2002) suggested that hands and feet might have evolved in obligatorily aquatic ancestors of amphibians as a means for the males to hold on to the females at the time of fertilization (a behavior termed amplexus), as is the practice in most frogs. However, it seems improbable that a behavioral pattern that evolved for grasping the female in the water would function for terrestrial locomotion. The importance of an anterior grasp by the forelimbs during mating in modern amphibians also appears at odds with the fact that the hind limb evolved at a faster pace than the forelimb during the origin of tetrapods, as demonstrated by Michael Coates (1996).

None of these hypotheses involving the behavior and ecology of ancestral amphibians are subject to actual testing. There were, however, other advantages for the descendants of large choanate fish to come out of the water. Jennifer Clack (2002b) noted that a rise in body temperature would increase both the rate of digestion and the speed of development. All localities from which Devonian and Early Carboniferous tetrapods have been found were then in tropical or subtropical regions, with oceanic water temperatures of 15 to 20 degrees centigrade. On land, however, the radiant heat of the sun could raise the body temperature of amphibians to 30 to 35 degrees centigrade, depending on their size and environmental factors such as wind speed, the nature of the substrate, and the like (Fig. 3.12).

Small animals, such as the young of early tetrapods, could raise their body temperature quickly, but the added heat would have been rapidly lost when the sun went behind a cloud or the animal went back into the water. The larger adults, however, would have been able to raise their body temperature higher and retain it for a longer period of time (R. L. Carroll et al., 2005).

In terms of heat gain and loss, the closest analogues to ancestral tetrapods are provided by living reptiles, especially crocodiles and lizards, which are ectothermic and highly dependent on the radiant energy of the sun for maintaining a high metabolic rate. In contrast, modern amphibians have little capacity to make use of radiant heat to maintain a high

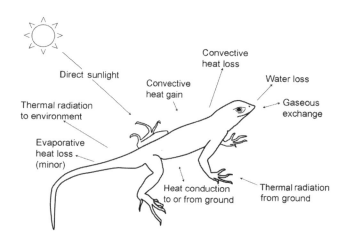

Figure 3.12. Modes of heat gain and loss for a basking lizard. From Pianka and Vitt, 2003.

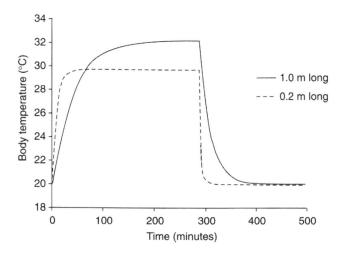

Figure 3.13. Rate and amount of heat gain and loss in animals the size of the adults of *Eusthenopteron, Panderichthys, Acanthostega,* and *Ichthyostega* (approximately 1 m in length), and smaller tetrapods. Ambient water temperature is 20 degrees centigrade, and ambient air temperature is 30 degrees centigrade. On the left, time of emergence from the water; on the right, return to the water. From R. L. Carroll et al., 2005.

body temperature because of their small size and moist body surface, both of which result in rapid heat loss.

But what were the selective advantages of warming up? Studies of modern amphibians and reptiles by Robert Gatten and his colleagues (1992) show that they have a Q_{10} of approximately 2; that is, all metabolic processes roughly double with a 10 degree increase in body temperature—twice the rate of digestion, elaboration of reproductive tissue, respiration, and increased capacity for locomotion. An amphibian that had warmed its body to 30 degrees centigrade by basking would have twice the metabolic rate of one habitually swimming in the water at 20 degrees centigrade, until it cooled to water temperature. For several minutes, however, it would have a great advantage in the capture of aquatic prey or the avoidance of predators (Fig. 3.13).

Whatever other selective advantages there may have been for stem tetrapods to have come onto land, increased control over their body temperature must have been a major factor. The structure of the earliest tetrapods indicates that they remained primarily aquatic animals, most likely preying upon fish and other aquatic organisms, as do modern crocodiles. Their attainment of primitive limbs does not necessarily imply that they were highly adapted to life on land. Rather, the ancestors of tetrapods may have emerged from the water in order to take advantage of an abundant but previously unexploited resource—the radiant heat of the sun—which enabled them to function better as shallow water, predatory fish. The thermal advantage of basking would have given them faster reaction times and higher metabolic and assimilation rates, leading to faster growth and ultimately enhanced reproductive rates. This set the stage for the attainment of a truly terrestrial way of life.

Panderichthys and *Tiktaalik* appear as plausible intermediates in the potential for utilizing solar energy. Although they

show no evidence of the capacity to support their bodies on land, they were presumably able to raise their temperature behaviorally by swimming in shallow water that was itself warmed by the radiant heat of the sun. The position of the eye sockets suggests that the dorsal surface of the body was exposed to the air, so that they could have basked while partially submerged. *Panderichthys* was tantalizingly close to an amphibious mode of behavior, but it was another 15 million years before effective limbs evolved.

CHANGES IN MODE OF LOCOMOTION

While the early ancestors of terrestrial vertebrates may have needed relatively little change in their skeletons to bask in shallow water and on the margins of the shore, effective terrestrial locomotion required many more advances in the structure of their limbs and the behavioral aspects of their locomotion.

In fish the size and shape of *Eusthenopteron,* most of the power for swimming is produced by lateral undulation of the muscles of the trunk and tail. The paired fins are small and used for turning and controlling the depth of the fish in the water, as well as pitch and yaw. Their configuration and directions of movement are related to their passage through the water, and as such, are unsuitable as a basis for terrestrial locomotion. Nonetheless, the bony supports of these fins were the primary building blocks of the tetrapod limbs.

The earliest and most conspicuous changes in the evolution from fins to limbs definitely occurred prior to the emergence of *Ichthyostega* and *Acanthostega.* These included the loss of the heavy scales that formed the surface of the fins and contributed greatly to their effectiveness in aquatic locomotion and the increase in size of the posterior as opposed to the anterior paired fin.

Among early bony fish, the pectoral fins had a very important role in keeping the fish on an even keel. In contrast with modern bony fish, such as salmon, goldfish, and tuna, which have a symmetrical caudal fin that propels the body straight forward, the tail fin of primitive ray-finned and lobe-finned fish was asymmetrical, with the force of propulsion acting to raise the tail. This force was counteracted by the pectoral fin, which acted as a hydrofoil to raise the head and anterior trunk. Although *Eusthenopteron* and *Panderichthys* had evolved a more symmetrical tail, the weight of their thick scales still required a large pectoral fin to assist in raising the body in the water column.

Ichthyostega and *Acanthostega* may have still spent much of their lives in the water, but their rear limbs were already longer than their forelimbs, and the internal bony supports were far larger than those of their fish ancestors. However, the early land vertebrates and all their terrestrial descendants retained some of the structural and functional differences between the front and hind limbs that were evident in their antecedents (R. L. Carroll and Holmes, 2007). The articulating surface between the pectoral girdle and the humerus in early tetrapods (and to a degree in *Panderichthys*) is distinctive in being elongate and may have enabled both hinging and rotational movement, while the articulating surface between the pelvis and femur forms a simpler ball-and-socket joint.

In the forelimb of primitive tetrapods, the wrist acts primarily as a hinge joint, but the elbow both bends and rotates. In the rear limb, the knee acts primarily as a hinge, but both rotation and bending occur in the ankle. This may be traced to differences in the structure and function of the fore- and hind fins of choanate fish. In *Eusthenopteron*, the bones of the fore- and hind limbs are clearly comparable with those of one another in the presence of a single bone proximally, the humerus and the femur, followed by two bones, the radius and ulna in the forelimb and the tibia and fibula in the hind. However, the humerus of *Eusthenopteron* resembles the fibula, not the femur, in having a long posterior process, indicating differences in the function of the elbow and knee joints in osteolepiform fish and early tetrapods. The ulnare also has a long posterior process, but there is no equivalent on the fibulare, foreshadowing differences in the wrist and ankle joints.

The great increase in the relative size of the limbs in early tetrapods, compared with the fins of *Eusthenopteron* and *Panderichthys*, would have significantly reduced the effectiveness of aquatic locomotion because of increased drag. The presence of limbs rather than fins would have been especially awkward for the obligatorily aquatic hatchlings, for whom terrestrial limbs would have been of no conceivable use. Judging from the larvae of both Paleozoic and modern amphibians, the early larval stages of the immediate ancestors of tetrapods probably lacked fins or limbs through delay in early development. Even if the internal skeleton of the limbs had begun to form, the most effective portion of juvenile fins, the external fin rays, had probably been completely lost prior to the appearance of *Acanthostega* and *Ichthyostega*.

It seems difficult to understand how a complex functional unit like the tetrapod limb could have evolved through a series of intermediate stages, but there is certainly strong evidence of slow, sequential change between the appendages of Upper Devonian fish and Lower Carboniferous tetrapods. The strong central axis of the fins of *Eusthenopteron* and *Panderichthys* would have already enabled them to push against the substrate much more effectively than contemporary ray-finned fish. The bones that were to form the base of the most posterior digit(s) were already present, although one or two fingers or toes would not have served the function of a hand or foot very effectively. Multiplication of toes had occurred by the end of the Late Devonian, but neither their particular number nor their specific identity was fully determined until the Lower Carboniferous.

Scales covered the entire body in *Eusthenopteron*—the trunk and base of the fins were covered by overlapping circular scales and the ends of the paired fins by elongate, jointed scales termed lepidotrichia. In contrast, *Acanthostega* and *Ichthyostega* lost most of the scales over the dorsal surface of the body, but the scales covering the ventral surface between

the limbs, termed gastralia, were thickened and elongated, as in Carboniferous amphibians, presumably to protect the internal organs as the trunk was dragged over the ground. More important, the lepidotrichia that had formed the distal surface of the paired fins were completely lost, thus exposing the internal skeleton of the toes. A few rectangular scutes are present in *Acanthostega*, but they do not at all resemble lepidotrichia. Coates (1996) suggests that they may have strengthened the palm of the hand and the sole of the foot.

The development of scales and other superficial bony elements is regulated by a particular type of tissue, the neural crest cells, which extend throughout the body in early embryos (see Chapter 2). It is possible that mutations affecting some stage in the distribution or expression of neural crest cells acted to suppress most, if not all, scale formation. This made possible the expression of digits in early tetrapods. Later mutation may have enabled the reappearance of dorsal scales, present in most early tetrapods, but without the redevelopment of the lepidotrichia.

GENETIC BASIS OF CHANGE

The historical sequence of amphibian evolution has been established primarily on the basis of the fossil record. We are fortunate in having highly informative fossils of plausible ancestors from the early part of the Late Devonian and primitive tetrapods from the very end of the Devonian. However, only a few, tantalizingly incomplete remains of potential intermediates have been collected from the 15 million-year gap between these well-known horizons. Eventually, a more complete sequence will be established as other fossil localities of Late Devonian age are discovered. On the other hand, even a step-by-step sequence of fossils will not be able to answer questions regarding *how* these changes occurred in terms of the sequence of underlying genetic modifications.

From the standpoint of adaptive change, the main difference from their fish ancestors was the capacity of tetrapods to use their limbs for support and locomotion on land. The fish required fins that were capable of a degree of bending, but without the necessity for clearly defined joints. The entire structure acted as a unit to move the body forward against the resistance of water and to control its orientation. In order to move effectively on land, the end of the limb had to bend at a narrowly defined joint so that it could remain in contact with the substrate while the proximal portions of the limb supported the body and moved it forward. Hinging was achieved by reducing the length of bony elements near the end of the internal skeleton of the fin. The external fin rays were also eliminated during this transition.

One might hypothesize that there was some degree of genetically based variability in the length of the more distal fin elements in the immediate antecedents of tetrapods that was acted upon by natural selection to give rise to an effective wrist and ankle. However, it is more difficult to account for the entirely novel appearance of digits. These differ in being aligned in a parallel array, very different from the succession of bifurcating elements that characterized the more proximal bones of the fin. Can the appearance of these entirely novel structures be accounted for by the standard explanation of the differential selection of alternative alleles governing the shape and size of individual bones that is basic to selection theory, or might more specific explanations be found through knowledge of the genetic basis for development?

As was discussed in Chapter 2, the past fifteen years have brought about a revolution in our understanding of the genetic basis for the control of development in multicellular plants and animals. Knowledge of the genome has increased to the point where we can understand the way in which particular genes control the development of the general body form and specific anatomical structures. From knowledge of genes that regulate development in living animals, it should be possible to determine what specific genetic changes may have resulted in particular types of anatomical modification documented by the fossil record.

Not surprisingly, the integration of information from the study of molecular genetics and developmental biology has brought with it many concepts and technical terms that were not known by the past generation of evolutionary biologists and paleontologists. However, the great explanatory value of developmental data for understanding major anatomical changes in all taxonomic groups justifies acquisition of this new vocabulary.

Although the specific number of *Hox* genes has not been determined in any of the modern amphibian groups, it is assumed that it is the same as in reptiles and mammals—39 total, arrayed on four chromosomes. No chromosome has all 13. Homologous genes on the four chromosomes are distinguished by the letters *a, b, c, d*. As discussed in Chapter 2, the *Hox* genes are expressed in linear fashion from the head to the end of the tail in the developing embryo. *Hox9–13* are also expressed in the fins or limbs. As shown by Nelson and his colleagues (1996) and Blanco and co-authors (1998), the genes that contribute to the formation of the limbs are similar in the chick, the mouse, and the frog. This suggests that very similar genes with comparable functions were probably present in the common ancestors of these groups, going back to the Carboniferous, some 350 million years ago, and probably in the early tetrapods of the Upper Devonian.

Much less work has been done on *Hox* or other key genes in fish, but Sordino and his colleagues (1995) have demonstrated the nature and function of some *Hox* genes that are involved in development of the fin of the zebrafish *Danio*. As shown by Blanco and his colleagues and by Wagner and Chiu (2001), differences in the actions of these genes in fish and land vertebrates can be associated with differences in limb function and development.

As we have seen from the fossils, lobe-finned fish such as *Eusthenopteron,* in common with the ray-finned fish *Danio,* differ from land vertebrates in having fins that function as a single flexible structure, without conspicuous joints. In contrast, the limbs of even the earliest known tetrapods have three clearly distinct elements: the upper limb, the lower limb, and the hand or foot. The bones of the upper and lower limb of tetrapods are comparable to those of *Eusthenopteron* (although not those of *Danio*), but there is nothing comparable with the joint between the lower limb and the hand and foot, nor with the digits of tetrapods. These are clearly evolutionary novelties that cannot be attributed to simple modification of fish structures, but must have evolved de novo. Can this be associated with recognizable changes in the nature and function of the genes controlling general limb structure among vertebrates?

The *Hox* genes that have been most extensively studied in relationship to limb development in tetrapods are those in clusters *a* and *d*. Early work by Davis and Capecchi (1996) showed a close correlation between the sequence of *Hox* genes associated with each of the units of the fore- and hind limbs. In the mouse and chick, the major limb segments can be distinguished by the successive expression of different *Hox* genes: *Hox9* in the shoulder and pelvic girdles, *Hox10* in the upper arm and leg, *Hox11* in the lower limb, *Hox12* in the wrist and ankle, and *Hox13* in the hand or foot. If both *Hoxa* and *Hoxd* genes are knocked out (that is, manipulated genetically so as to render them inoperative), those segments of the limb in which they are normally expressed either do not develop or develop abnormally.

In relationship to the common expression of the same *Hox* genes in both the fore- and hind limbs, they have a broad developmental and structural similarity. This is clearly evident, even in modern vertebrates such as ourselves, in which there is a common sequence of basic elements. Both the fore- and hind limbs have a single bone at the base, the humerus anteriorly and the femur posteriorly, succeeded by two major bones, the ulna and radius of the arm and the tibia and fibula of the leg. These are followed by a mosaic of smaller bones that make up the wrist and ankle, and finally by the fingers and toes. Because of the basic similarity between the major units of the front and hind limbs, they have been designated by terms applicable to both. The most proximal unit, formed by the humerus or femur, is termed the stylopodium. The next unit, consisting of the ulna and radius or the tibia and fibula, is referred to as the zeugopodium. The most distal portion of the limb is called the autopodium. The autopodium is unique to terrestrial vertebrates. It includes a succession of structurally and functionally distinct bones, those of the wrist (carpals) and ankle (tarsals) collectively designated the mesopodials, the elongate metapodials (the metacarpals of the forelimb and the metatarsals of the hind limb), and the digits (fingers and toes), themselves consisting of a series of phalanges (Fig. 3.14A).

Work by Nelson and his colleagues (1996) demonstrated phases of expression of *Hoxa* and *Hoxd* in the development of the limbs of chicks and mice that parallel the evolutionary distinction between the long bones of the stylopodium and zeugopodium on one hand, and the shorter elements of the

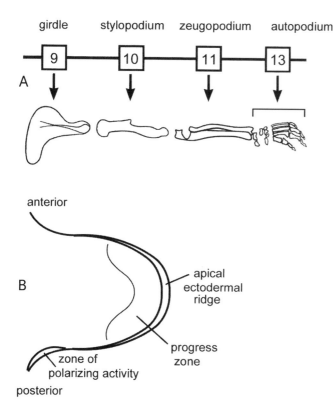

girdle   stylopodium   zeugopodium   autopodium

A

anterior

B

apical
ectodermal
ridge

progress
zone

zone of
polarizing activity

posterior

Figure 3.14. A, Terminology for limb segments applicable to both fore- and hind limbs, and the areas of *Hoxa* gene expression in all terrestrial vertebrates. The development of the autopodium is the result of the separation of the areas of expression of *Hoxa11* and *Hoxa13,* which are both expressed at the end of the fin in fish. *B,* Embryological structures associated with limb development. From R. L. Carroll, 1997.

wrist, ankles, fingers, and toes (autopodium) on the other. Whereas fish show overlap in the expression of *Hoxa11* and *Hoxa13,* these areas of expression are mutually exclusive in tetrapods. Each appears to be mutually suppressive of the other. If *Hoxa13* is artificially expressed in the area where *Hoxa11* is normally active, small nodular bones resembling those that normally make up the carpus or tarsus develop, in contrast to the elongate bones of the zeugopodium.

The distinction between the influence of *Hoxa11* and *Hoxa13* is further supported by observations on the wrists and ankles of frogs. Uniquely among amphibians, two proximal bones of the ankle are elongated, resembling the tibia and fibula in their general configuration rather than the more distal ankle bones. As shown by Blanco and his colleagues (1998), frogs are unusual in having the area of expression of *Hoxa11* encompassing these proximal tarsals. This is not the case in the forelimb, in which the proximal carpals are short, nodular bones that develop within the normal area of expression of *Hoxa13.* What happened in the origin of frogs provides an informative model of what may have occurred in the origin of tetrapods. Instead of changing the configuration of bones in the position of the tarsus to the shape of bones in the position of the tibia and fibula, the ancestors of tetrapods must have changed the shape of bones distal to the zeugopodium that were initially elongate, to the short, rounded form

of carpals and tarsals. This presumably required a previous change from the condition in primitive fish, exemplified by *Danio,* in which the areas of expression and activity of *Hoxa11* and *Hoxa13* overlapped, to the modern tetrapod condition in which these areas of expression are separated.

We can assume that *Eusthenopteron* retained the condition seen in *Danio,* in which the influence of *Hoxa11* continued to the end of the internal skeleton of the fin. By the appearance of *Acanthostega* and *Ichthyostega,* the area of expression of *Hoxa11* presumably terminated at the end of the zeugopodium, so that the next series of bones were much shorter and so functioned as a line of flexion for the hand or foot. All of the proximate bones of the hands and feet are relatively short in *Acanthostega* and *Ichthyostega,* making it difficult to distinguish carpals and tarsals from metacarpals and metatarsals. A clear distinction between roughly three rows of rounded carpals or tarsals, followed by elongate metapodials and proximal phalanges, is not well established until the Carboniferous (Figs. 3.9 and 3.10).

The achievement of a distinct developmental field in the autopodium may explain the origin of the functional separation of the base of the autopodium from the zeugopodium, but it does not explain the origin of the digits. No lobe-finned fish has anything comparable to digits. Shubin and Alberg (1986) suggested that digits might have evolved from a series of posterior fin rays, such as occur in lungfish. They hypothesized that the main axis of the limb, which initially extended laterally from the trunk, became bent so that the distal end extended more or less anteriorly, with the posterior fin rays extending from it at right angles. Since the limb as a whole develops from proximal to distal, the bending of the distal portion would explain why the digits in most tetrapods develop in a posterior to anterior sequence. The relatively large number of fin rays in such sarcopterygians as lungfish would explain why primitive tetrapods had a large number of digits, which was later reduced. *Eusthenopteron* and *Panderichthys,* which are otherwise similar to early tetrapods, do not have posterior fin rays, but *Tiktaalik* does. Wagner and Larsson (2007) argue that the origin of digits may be traced back to *Eusthenopteron,* in the sense that the antecedents of one or two posterior digits (labeled X and Y in Fig. 3.4*A, E*) may already be present. The early appearance of at least one digit is also supported by experiments in which knockouts of *Hoxa13* and *Hoxd13* genes leave one posterior digit, while the rest are lost.

How, then, might we account for the addition of the more anterior digits? Wagner and Larsson (2007) suggest that they represent a reiteration (or duplication) of the previously existing posterior digits, or perhaps the formation of separate digital fields. The latter idea is supported by the two groups of digits of different sizes in the hind limb of *Ichthyostega,* and the ease with which duplication of digits can be achieved experimentally (Fig. 3.1*E*).

A multitude of digits were present in *Acanthostega* and *Ichthyostega,* but there seems little control over their number or specific identity. In the absence of any fossil evidence of the mode of elaboration of the metapodials and digits, we must look to developmental data from modern species regarding the sequence of ossification and the control of digit number and digit identity. A study of foot development in the chick by Drossopoulu and his colleagues (2000) provides a plausible model for the origin of the tetrapod pattern within the Late Devonian and Early Carboniferous.

An important developmental feature involved with the ontogeny (growth and development) of both the paired fins of bony fish and the limbs of most land vertebrates is the zone of polarizing activity located at the posterior proximal edge of the developing appendage (Fig. 3.14B). Its importance in development is demonstrated by its capacity to generate extra digits if it is transplanted to other parts of the limb. A major role of the zone of polarizing activity is to secrete the signaling protein Sonic hedgehog (Shh). In the chick, Shh has two phases of expression. The first, which lasts about 24 hours, serves to prime the tissue to form digits and thus controls digit number. Later, Shh acts over a short range to induce expression of bone morphogenetic protein 2 (bmp2), which Drossopoulu et al. argue specifies some aspects of digit identity. In relationship to the posterior position of the zone of polarizing activity, the digits typically chondrify (form in cartilage) in a posterior to anterior sequence. This we know is an ancient attribute of tetrapod limbs, for it occurs in both living amniotes (reptiles, birds, and mammals) and frogs whose time of divergence goes back to at least the Lower Carboniferous. Growth series of several reptilian groups described by Michael Caldwell (1994) from the Upper Permian show a comparable sequence of ossification.

One of the most conspicuous attributes of individual digits is the number of phalanges, the bones between each of the joints in the fingers and toes. The hand of *Acanthostega* has several adjacent digits with the same phalangeal count, which suggests there was little capacity to distinguish the identity of each digit. However, most Paleozoic amphibians have a set number of digits, each with different numbers of phalanges or differences in their size. Sanz-Ezquerro and Tickle (2003) described how the numbers may be controlled in the chick. This involves an additional signaling molecule, *Fgf8* (fibroblast growth factor 8), and another structure common to vertebrate limb development, the apical ectodermal ridge, which forms at the leading edge of the developing limb bud.

As in the case of regulation of the presence and number of digits, Sonic hedgehog is the key signaling molecule. If it is applied between developing digits, additional phalanges will be induced. This results from the activation of *Fgf8* within the ectodermal ridge overlying each digit. New phalanges are produced by elongation and subsequent segmentation of the penultimate phalanx. As long as *Fgf8* is expressed in the apical ectodermal ridge, the chondrogenic precursor of the digit will continue to elongate. The number of phalanges is regulated by the duration of *Fgf8* signaling. The terminal phalanx or digit tip (whose development is controlled by a number of other genes) forms when *Fgf8* signaling stops.

Much more information regarding the genetic control of development in the limbs of vertebrates is discussed in the recently published book *Fins into Limbs* (Hall, 2007).

We now have a general idea of the various genes and signaling molecules that influence the development of the autopodium in modern terrestrial vertebrates. Their similarity in the frog, the mouse, and the chick strongly suggests that incorporation of these genes in the development of the distal end of the limb occurred in a stepwise fashion during the origin of tetrapods in the Late Devonian. The large number of genes and other regulatory elements that were involved and their complex interrelationships with one another and the developing tissues of the limbs help to explain the long time span over which the fins of choanate fish evolved into the fully effective tetrapod limbs of the Lower Carboniferous. The same genes and their close homologues also play many other roles in vertebrate development (S. B. Carroll, 2005).

ADDITIONAL FISH AND TETRAPODS FROM THE LATE DEVONIAN

We have concentrated on *Panderichthys*, *Tiktaalik*, *Ichthyostega*, and *Acanthostega* in establishing the changes that occurred in the fish-amphibian transition because these are the best known of advanced choanate fish and early tetrapods. However, there are several other fish and putative tetrapods, known from less complete skeletal remains, which suggest a diversity of lineages that are in some way related to this transition. Unfortunately, few are sufficiently well known to establish their specific relationships with the better-known species.

Close to the fish side, *Elginerpeton*, from the Frasnian (the lower stage in the Upper Devonian) is known from a dissociated jaw and scattered postcranial remains from Scotland (Ahlberg, 1998). The lower jaw shows a combination of primitive choanate features and others shared with early tetrapods, but is divergent from all other genera associated with the fish-amphibian transition in its great size (44 cm in length) and its very narrow width. It is distinct from *Acanthostega* and *Ichthyostega* in retaining a row of tiny teeth lateral to the primary dental row common to *Eusthenopteron* (Ahlberg and Clack, 1998). All more certain Devonian tetrapods are restricted to the Fammenian, the last stage of the Upper Devonian.

Of particular interest as the only evidence of a probable tetrapod from the Late Devonian of Australia is an isolated jaw designated *Metaxygnathus* (Campbell and Bell, 1977). As in tetrapods, the surangular precludes contact between the dentary and the articular and there is not a row of denticles lateral to the primary marginal dentition (A. Warren and Turner, 2004). Another possible tetrapod jaw, designated *Sinostega*, has been described from Upper Devonian beds in central China (Zhu et al., 2002).

Ventastega is an unquestioned tetrapod from the latest Devonian of the Baltic region. Much of the skull and lower jaw is known, as well as dermal elements of the shoulder girdle and an ilium differing from those of any other Devonian or Lower Carboniferous tetrapod in having a single, posterodorsally directed, paddle-shaped iliac process, rather than distinct dorsal and posterior processes (Ahlberg et al., 1994). More recently, Ahlberg et al. (2008) have shown that *Ventastega* lacks some of the specialized features of *Acanthostega* and *Ichthyostega,* and so may be closer to the ancestry of more advanced amphibians from the Lower Carboniferous that also retain the primitive state of these characters.

As many as three tetrapod taxa have been described from the Upper Devonian of Red Hill, Pennsylvania: a shoulder girdle designated *Hynerpeton,* with a possibly associated jaw; a second type of jaw designated *Densignathus;* and an unnamed humerus, highly distinct from any others that have been discovered (Shubin et al., 2004).

The most informative tetrapod that has been described from the Late Devonian, aside from *Ichthyostega* and *Acanthostega,* is *Tulerpeton* from the Tula region, central Russia. *Tulerpeton* is the only Devonian amphibian that is a plausible sister-taxon of any of the later tetrapods. Surprisingly, it comes from an estuarine or occasionally marine horizon (Lebedev and Clack, 1993; Lebedev and Coates, 1995). The most complete specimen consists of scattered skull bones, the left side of the pectoral girdle, the right fore- and hind limbs, and ventral scales (Figs. 3.7, 3.9, and 3.10). These bones indicate an animal that was more like Carboniferous amphibians than *Acanthostega* and *Ichthyostega,* and some suggest specific affinities with a group discussed in the following chapter, the anthracosaurian labyrinthodonts. As in *Acanthostega,* the lateral line canal of the jugal shows two modes of expression. One section, as in later tetrapods, is an open groove, while the other opens to the surface as a row of ovoid pores, a typical feature of fish. The presence of an intertemporal bone, lost in *Acanthostega* and *Ichythostega,* is important in showing affinities with major groups of Carboniferous amphibians.

The vertebral intercentra are fully crescentic, in contrast with most of those in *Acanthostega,* in which all but those of the axis and the sacral vertebra are paired. The pleurocentra (largely unossified in *Acanthostega*) are robust and show a clear area for articulation with the neural arch dorsally. These bones are comparable to those of the major amphibian groups from the Carboniferous. The shoulder girdle (Fig. 3.7F) is well preserved. As in *Acanthostega,* an anocleithrum that links the shoulder girdle with the head in *Eusthenopteron* is retained. The cleithrum forms the anterior margin of the shoulder girdle, but in contrast with *Acanthostega* and *Ichthyostega,* it is not fused to the scapulocoracoid, which extends far dorsally as in Carboniferous amphibians. The scapulocoracoid lacks the many openings for nerves and blood vessels seen in other Devonian amphibians.

The humerus broadly resembles that of other Devonian and Lower Carboniferous tetrapods. It is expanded both proximally and distally, forming articulating surfaces with the scapulocoracoid and ulna and radius. These surfaces are in nearly the same plane and there is no obvious shaft. Other structural details are identified in Figure 3.9E. The ulna and radius are immediately recognizable as such, with most of the features of later Paleozoic amphibians and reptiles. The ulna has a distinct olecranon to articulate with the humerus, and both have elongate shafts.

The wrist is well articulated, with only a short gap between it and the long bones. Eight, short, squarish bones can be recognized as carpals, proximal to the elongate metacarpals. This number is smaller than in many Carboniferous amphibians; nevertheless, this carpus is much more like that of later amphibians than is that of *Acanthostega.* Six digits are in nearly their normal position, continuing the process of reduction from the numbers in *Acanthostega* (eight) and (for the foot) *Ichthyostega* (seven). The phalangeal count (the number of bones in each digit, beginning with the medial finger) is 2, 3, 4, 5, 4, 2. The sixth digit is unlike the rest in that its small distal carpal articulates directly with the ulna, rather than with a more proximal carpal as in Carboniferous amphibians. Based on its position, and the phalangeal count of later amphibians, this is almost certainly the digit that was subsequently lost.

The tarsus resembles that of later amphibians in having three large proximal bones, the tibiale, intermedium, and fibulare, a number of centralia, and apparently similar sized, small, squarish distal tarsals. Only five digits are clearly evident, but a sixth may be represented by an incomplete metatarsal. Only the first two digits are complete, with a phalangeal count of 2 and 3. By analogy with the hand, the estimated count for the remainder would be 4, 5, ?, 4.

Many very thin, sub-elliptical scales were found in place along the ventral surface of the trunk of *Tulerpeton.* Similar scales were also found with *Ichthyostega.* Smaller scales also covered the limbs and trunk in *Ichthyostega,* but in contrast with *Acanthostega,* which seems to lack dorsal scales but had thicker, more rod-shaped scales restricted to the ventral surface of the trunk.

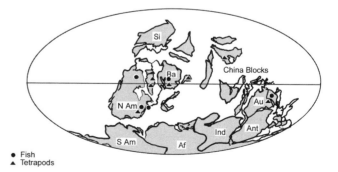

Figure 3.15. Map showing the distribution of continents in the Middle to Late Devonian and distribution of advanced choanate fish (0) and Upper Devonian tetrapods (Δ). Af, Africa. Ant, Antarctica. Au, Australia. Ba, Baltica. Ind, India. N. Am, North America. S. Am, South America. Si, Siberia.

THE GEOGRAPHY OF THE EARLIEST TETRAPODS

While the first discovered and still best-known Late Devonian amphibians were found in East Greenland, other species are now known from the Baltic, central Russia, Scotland, Belgium, eastern North America, Australia, and China. Close relatives among choanate fish also had a worldwide distribution, as determined by John Long (1995) (Fig. 3.15). The transition between these two groups may have occurred on any of these land masses and various stages probably appeared more than once. It was initially assumed that the immediate ancestors of amphibians were likely freshwater fishes since modern amphibians are almost entirely limited to that environment, but more and more Devonian localities, including those containing *Eusthenopteron, Panderichthys,* and *Tulerpeton,* are now thought to be at least marginally marine or estuarine. These fish and the early amphibians were probably euryhaline (having tolerance for a wide range of salinity). Because of the close proximity of the main northern and southern continents, Laurasia and Gondwana, in the Late Devonian, both the advanced choanate fish and their amphibian descendants may have found it easy to migrate from one to the other.

The great geographic range of these animals suggests a considerable variety of local environments that may have imposed an assortment of adaptive challenges to animals close to the fish-amphibian transition. For example, areas of upwelling such as those around the Galapagos Islands may have had especially cold but biologically rich waters that put a special value on basking, as is the case for the modern marine iguana.

Fossils from the very end of the Devonian appear on the verge of a new way of life, lacking only a few changes in the limb structure necessary for them to become the lords of the land. However, the Lower Carboniferous beds, where we would expect to find the first fully terrestrial vertebrates, provide frustratingly little evidence of the next step in vertebrate evolution, but a variety of peculiar amphibians that have left no descendants and tantalizing fragments of others that may point the way to advanced amphibians and also reptiles.

4

The Radiation of Carboniferous Amphibians

ROMER'S GAP

IN CONTRAST WITH OUR detailed knowledge of *Acanthostega* and *Ichthyostega*, the fossil record for the first 30 million years of the Carboniferous is scant and confusing (Fig. 4.1). The dearth of amphibians from this time led Michael Coates and Jennifer Clack (1995) to refer to it as "Romer's Gap," in reference to the long search for fossils from the Lower Carboniferous by Dr. Alfred Romer, the 20th century's most notable vertebrate paleontologist. All the major lineages from the later Carboniferous and Permian presumably evolved within this period of time, but few early representatives of the major lineages are known until much later.

Many specimens are known only from incomplete skeletons (sometimes by only one or two distinctive bones) but those that are known exhibit a very wide range of body forms—from animals closely resembling the large amphibians of the Upper Devonian to animals only a few centimeters in length but with well-ossified skeletons indicative of adult individuals. Some have totally lost their limbs or have evolved very different skull proportions. Because of our very incomplete knowledge of the fossil record for much of the Early Carboniferous, the specific origin and interrelationships among the many amphibians known from the later Paleozoic remain highly contentious.

Two major types of Paleozoic amphibians have been recognized since the late 19th century (Miall, 1875; Zittel, 1890). The labyrinthodonts, named for the labyrinthine infolding of the dentine of their teeth (Fig. 4.2), were generally large animals, up to a meter or more in length, which retained many of the skeletal characteristics of their Late Devonian fish ancestors. The lepospondyls were much smaller and their anatomy was highly modified relative to that of the earliest tetrapods. None retained the large palatal fangs common to the choanate fish and labyrinthodonts, and their vertebral centra were cylindrical rather than composed of a number of

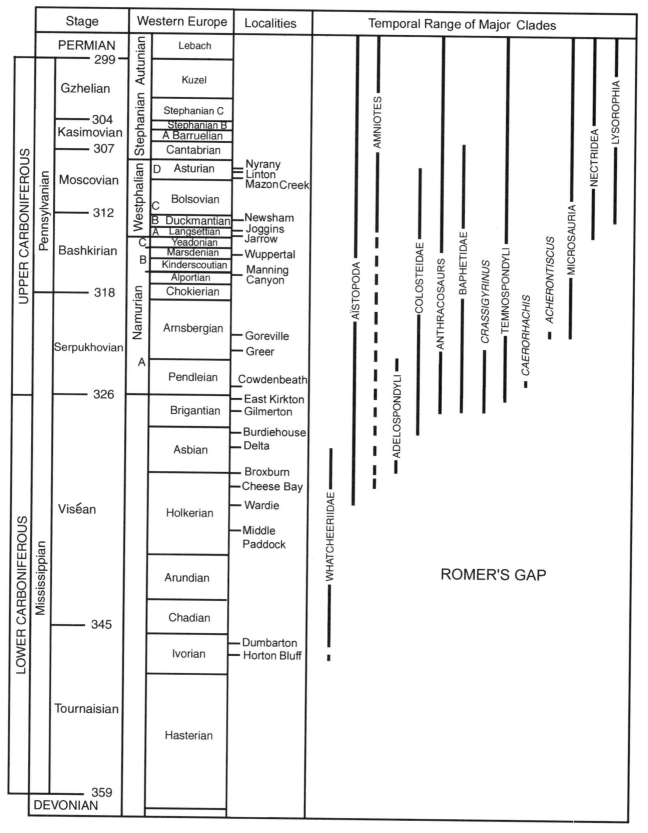

Figure 4.1. Temporal ranges of Carboniferous tetrapods. Geological time scale based on Gradstein et al., 2004. Relative sequence of Scottish amphibian localities from Smithson, 1985b, updated by more recent descriptive pa- pers. Romer's Gap refers to the paucity of amphibian fossils from the first half of the Carboniferous. Ages given in millions of years.

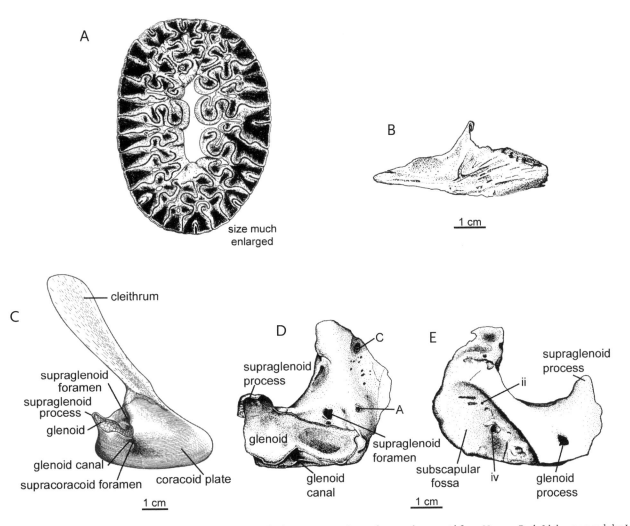

A

size much
enlarged

B

1 cm

C

cleithrum

supraglenoid
foramen
supraglenoid
process
glenoid

glenoid canal

supracoracoid foramen

coracoid plate

1 cm

D

C

supraglenoid
process

A

glenoid

supraglenoid
foramen

glenoid
canal

1 cm

E

supraglenoid
process

ii

subscapular
fossa

iv

glenoid
process

1 cm

Figure 4.2. A, Cross section of a palatine fang of a primitive labyrinthodont, *Eogyrinus,* illustrating labyrinthine infolding of the dentine. From Panchen, 1970. Labyrinthine teeth are found in only the largest of the lepospondyls. *B–E,* Isolated bones from the Lower Tourniasian of Horton Bluff, Nova Scotia. *B,* Jugal, broadly resembling that of *Proterogyrinus,* an embolomere from the Namurian B of Greer, West Virginia. *C,* Lateral view of the scapulocoracoid of *Ichthyostega.* Modified from Jarvik, 1996. *D* and *E,* Lateral and medial surfaces of a scapulocoracoid from Horton. Both *Ichthyostega* and the Horton specimen have a supraglenoid process, which is not present in other early tetrapods. The Horton scapulocoracoid is more primitive in the presence of numerous foramina, indicated by *A, C,* ii and iv, in addition to those common to later tetrapods. These are also present in *Acanthostega* (Coates, 1996), but were not illustrated in *Ichthyostega* by Jarvik (1980, 1996).

small crescents. These categories are very easy to distinguish from one another, but almost no fossil evidence is available to establish how the lineages within either the labyrinthodonts or the lepospondyls were related to one another, or how any of the lepospondyl lineages might be related to any particular labyrinthodonts.

The usual way to treat these groups is to discuss all the labyrinthodonts together, followed by all the lepospondyls. However, this approach is based on an untested assumption that both labyrinthodonts and lepospondyls were monophyletic groups; that is, that each had a single common ancestor and that the group includes all the descendants of that ancestor (Hennig, 1966). This implies that there was a simple dichotomy early in the history of tetrapods: one ramus included all the labyrinthodonts and any potential descendants, and the other included all the lepospondyls and their descendants.

On the other hand, what we know from the fossil record strongly supports the hypothesis that labyrinthodonts evolved directly from particular sarcopterygian fish and are the only plausible ancestors of any of the lepospondyl groups. The question then arises as to the nature of the ancestry of lepospondyls. Are they monophyletic, in the sense of having a single common ancestor, or might they be polyphyletic, in having evolved from more than one lineage of labyrinthodonts?

Another way to investigate the nature of relationships among labyrinthodonts and lepospondyls is by discussing each of the individual lineages of Carboniferous tetrapods in a temporal sequence based on the time of first occurrence of each taxon in the fossil record, as recognized by the appearance of new anatomical and/or adaptive characteristics. This will be done on the basis of a succession of fossil localities that include one or more taxa that are not known from earlier in the Carboniferous. This approach assumes that the time of first appearance in the fossil record broadly reflects the time of divergence of different lineages that can be distinguished from one another by a succession of evolutionary changes associated with increased adaptation for life on land or by divergent specializations reflecting adaptation to varied environments and modes of feeding, locomotion, and reproduction.

THE LOWER CARBONIFEROUS OF HORTON BLUFF

Our first view of Lower Carboniferous land vertebrates is from the Horton Bluff locality in the Minas basin of the Bay of Fundy in Nova Scotia, approximately 10 million years after the end of the Devonian (Clack and Carroll, 2000). This locality is tantalizing in showing a glimpse of a diversity of amphibians represented by isolated bones and footprints, few of which can be linked closely to either Devonian or later Paleozoic species (Plate 3).

The most extensive evidence for truly terrestrial amphibians is provided by hundreds of footprints and long trackways that could only have been made by animals walking on land. The trackways show that the actual movements of the limbs were comparable to those made by tetrapods throughout the rest of the Paleozoic (Fig. 3.11). Unfortunately, few individual footprints are sufficiently well preserved to show the number of digits with assurance. None show definite evidence of more than five digits on either the front or the hind limbs or the assemblage of small and large digits of *Ichthyostega*. Some of the best-preserved footprints appear to show the skin surface ventrally, with a patchwork of small scale impressions and rather stubby digits. Well-preserved trackways clearly illustrate the pace and stride and impressions of a tail drag, providing clear evidence that they were made on land rather than in the water, where the tail would have floated above the sediments.

Other trackways were as assuredly made by animals walking or "dog paddling" in shallow water. What one sees are the marks made by the ends of toes as they moved across the sediments. They represent the actual direction of movements, like a short film clip of the limbs moving from anterior to posterior in a shallow arc. This is clearly a modification of a basically terrestrial gait, as opposed to the more or less laterally directed paddles of an aquatic animal. The most striking footprints are within trackways 20 m or more in length, with individual footprints up to more than 20 cm across (Mossman and Sarjeant, 1980; Sarjeant, 1988). Unfortunately, the individual footprints had long been exposed to the tides and show

little detail. No skeletal remains of amphibians this large are known until the very end of the Lower Carboniferous (Godfrey, 1988).

Unfortunately, none of these footprints can be specifically associated with the skeletal remains at Horton Bluff, none of which include bones of the hands or feet. In fact, very few of the bones are found in association with one another. The sediments at Horton Bluff were deposited in streams or rivers that flowed into a larger body of water. The amphibians may have died on land or in shallow water. The soft anatomy decayed or was consumed by predators or scavengers, and the bones fell apart and were washed away. Most of the bones are well preserved, showing much anatomical detail, but they cannot be confidently assembled into skeletons.

One can readily identify what parts of the skeleton are represented, but several different shapes of humeri, femora, and elements of the shoulder girdle can be recognized, indicating several different kinds of amphibians. There is little evidence to indicate which humeri belong with which femora, or which of either belong to the four types of interclavicles. Most of the remains are of fairly large animals, broadly resembling better-known labyrinthodonts from later horizons, but there is not enough evidence to associate them with specific lineages.

The most important part of the skeleton for the identification of vertebrates is the skull, but the conditions of preservation and deposition at Horton are such that we have as yet found no more than isolated bones—a single jugal and parts of the jaw and dentition. Some appear comparable to later known groups of large amphibians, the anthracosaurs and temnospondyls, but do not provide enough evidence for positive identification. Vertebral elements are equally equivocal.

One well-preserved bone from the shoulder girdle, the scapulocoracoid (Fig. 4.2D, E), has several features resembling that of the Upper Devonian *Ichthyostega*. In particular, a unique structure at the back of the glenoid, termed by Jarvik (1996) the supraglenoid process, has not previously been reported in any other tetrapods. Another feature, common to *Acanthostega* and *Ichthyostega* but not *Tulerpeton*, is the presence of several more foramina in the scapulocoracoid than are retained in later tetrapods. Other scapulocoracoids resemble those of later tetrapods in having a smaller number of openings, differing from one specimen to another, but generally comparable. A similar process of reduction in the number of foramina is also evident in the pelvic girdles of Carboniferous tetrapods, but no examples have yet been described from Horton.

The process of reduction in the number of foramina penetrating the endochondral bones of the pectoral and pelvic girdles between Upper Devonian and later Paleozoic amphibians can be explained in terms of accommodating the many blood vessels and nerves running between the body and the evolving limbs. In their fish ancestors, the scapulocoracoid and pelvic girdle were very small and were not a barrier to their passage. With the rapid expansion of the girdles for the elaboration of articulating surfaces for the limb bones and for the origin of the muscle masses that were necessary to lift the body against the force of gravity, the scapulocoracoid and the puboischiadic plate came to occupy the areas through which the nerves and blood vessels had previously passed. Presumably the nerves and blood vessels developed more or less in their original positions and were accommodated by the later ossifying bone in a more or less random manner. Frequently, passages from one side of the bone to the other may be forked, with one hole on one side and two holes on the other, or one bone may have two openings close to one another, but a similar bone has only one in the same area. This initially resulted in the formation of a large number of foramina, essentially one each for every nerve or blood vessel serving the limbs. This was later rationalized by reduction to a smaller number of openings, each occupied by a more or less fixed number of blood vessels or nerves.

Among later Carboniferous and Permian tetrapods, each group has a more or less distinct pattern of openings, but it is not sufficiently constrained to serve as a basis for classification. The several patterns that are observed at Horton cannot be used by themselves for the identification of different taxa. On the other hand, the scapulocoracoids with a reduced number are clearly distinct from those suggestive of *Ichthyostega*.

The humeri and femora are sufficiently massive to be preserved intact. One humerus (Fig. 4.3) clearly fits the pattern of later anthracosaurs, but also retains features in common with the Upper Devonian *Tulerpeton*. A second humerus and also a pelvic girdle resemble those from an articulated skeleton of *Pederpes,* which occurs in the next younger Carboniferous deposit, discussed in the following section of this chapter. Another, of much smaller size and lower degree of ossification, is similar to those of a later Carboniferous lineage, the colosteids (Fig. 4.14). The femora, while reasonably well preserved, do not have sufficiently distinctive features for specific comparison with later taxa.

Clearly, tetrapods were both common and somewhat diverse at the Horton locality, suggesting an early stage in their radiation that was to dominate the later Paleozoic, but it may take many years of diligent collection to reveal their specific relationships. Nearly all the specimens from the Horton locality have come from a few hundred feet of sea cliff and adjacent beach below the lighthouse. Collection is limited to a few hours each day, between tides, which, at 15 m, are among the highest in the world. With every tide, the rocks that have fallen from the cliff are rearranged and the beach may be swept clean or covered with sediments. The beach is exposed for hundreds of meters into the bay at low tide, but the most distant rocks are exposed only briefly, and then covered by the next tide.

Several other exposures of the Horton Bluff Formation are mapped along the seashore in Nova Scotia, but all that have been investigated are somewhat metamorphosed by the folding and heating of the sediments, and all traces of fossils have been lost. Fortunately, there is another locality, at Albert Mines in New Brunswick, which has beds of the same age with comparable plants and fish, but so far no remains of amphibians have been found. Nowhere else in the world have tetrapods been found this far back in the Carboniferous.

The fauna at Horton, like that of the Upper Devonian proto-tetrapods and tetrapods, is at the water-land interface. What we will see as we proceed up the Carboniferous is an increasingly greater distinction between clearly aquatic and clearly terrestrial tetrapods, as distinguished by general body form, nature of limbs, sensory structures, and reproductive practices. What types of animals are more likely to be preserved at a given locality depends on the particular manner of deposition.

In addition to the early tetrapods at Horton, there were also a variety of fish with whom they must have shared the water. One group, the rhizodontids, supports the hypothesis that the ancestors of amphibians might have come on land to avoid predators (Plate 3). The largest specimens are estimated to be about 2.3 m long, about twice the estimated length of the largest known amphibians. The skull alone was at least 40 cm long, with giant stabbing teeth. A second group of fish were the acanthodians, discussed in Chapter 2, known primarily from fin spines. Those of one group, the gyracanths, reached lengths of up to 20 cm, indicating fairly large fish. Groups of spines are found together, but little else of the body is known. The spines of other, smaller acanthodians are very common at Horton. There are also several spines, about 10 cm long, from primitive sharks. The most common remains are those of primitive ray-finned fish, the palaeoniscoids, relatives of the modern gar and sturgeons. The largest may have been about 30 cm long. Young of these fish may have been the prey of adult amphibians, while the adult fish could have preyed upon juvenile amphibians. In some layers there are thousands of tiny palaeoniscoid scales and occasional jawbones, recognizable by their peculiar patterns of sculpturing. The presence of these small fossils demonstrates that equally small tetrapod bones could have been preserved and identified from the Horton sediments, but none have yet been recognized.

To complete the biota, a diversity of plants are also known from Horton, including giant lycopods, relatives of the tiny modern "ground pine," large sphenopsids, related to modern horsetails, and plants with fernlike foliage. Small trails and burrows in fine-grained shale were presumably made by a variety of worms and arthropods.

THE WHATCHEERIDAE

The next locality is Dumbarton in western Scotland, about 5 million years later than Horton Bluff but still within the upper Tournaisian. These localities are now separated by thousands of miles of the Atlantic Ocean, but in the Lower Carboniferous, before the tectonic separation of Europe and

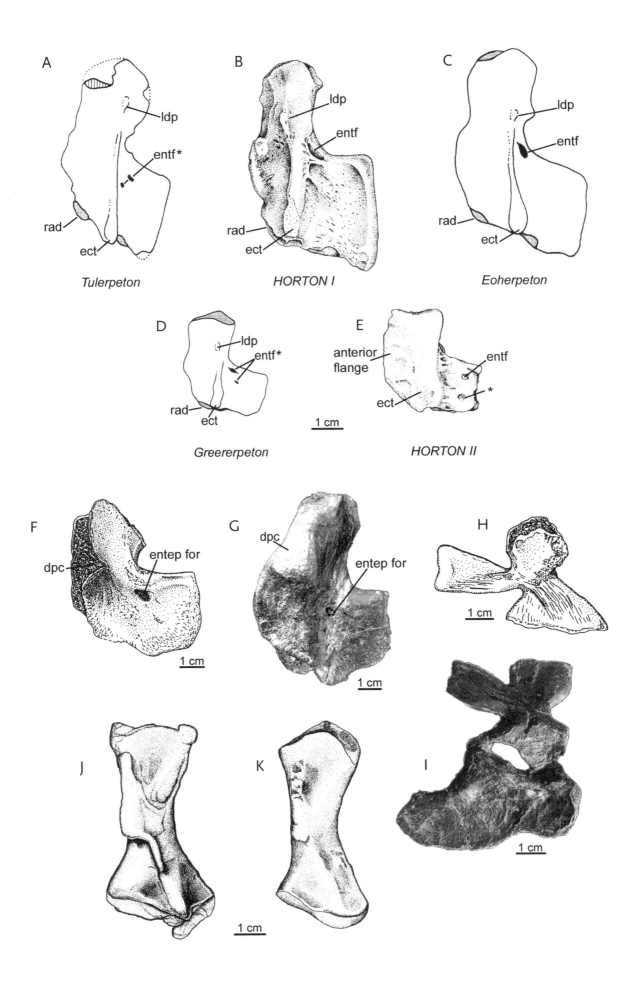

A

ldp

entf*

rad

ect

Tulerpeton

B

ldp

entf

rad

ect

HORTON I

C

ldp

entf

rad

ect

Eoherpeton

D

ldp

entf*

rad

ect

Greererpeton

E

anterior
flange

entf

ect

*

1 cm

HORTON II

F

dpc

entep for

1 cm

G

dpc

entep for

1 cm

H

1 cm

J

K

1 cm

I

1 cm

North America, they may have been only a few hundred kilometers apart. In contrast with the diversity at Horton, only a single specimen has been discovered at Dumbarton; however, it consists of an almost complete skeleton, recently described by Clack and Finney (2005). This animal, named *Pederpes,* would have been about a meter long, including the probable length of its missing tail (Fig. 4.4).

The general appearance of *Pederpes* resembles that of *Acanthostega* and *Ichthyostega,* with a large skull compared with a moderate-length trunk. The vertebrae are multipartite, with crescentic intercentra, paired pleurocentra, and large neural arches. The limbs and girdles are massive, like those of *Ichthyostega,* in contrast with the relatively smaller size in *Acanthostega* (Fig. 3.1). On the other hand, more detailed study shows that *Pederpes* exhibits a puzzling mixture of anatomical characters. Some, such as the extension of lappets of the tabular and postparietal over the dorsal surface of the occiput, separation of the scapulocoracoid from the cleithrum, absence of the postbranchial lamina of the cleithrum, loss of the anocleithrum, and the modernization of the foot structure, are clearly advanced relative to these Upper Devonian genera. Others, however, including the retention of a large fang on the ectopterygoid bone of the palate and an intertemporal bone common to *Eusthenopteron,* indicate that the lineage leading to *Pederpes* had diverged prior to those including *Ichthyostega* and *Acanthostega,* which had lost these structures. *Pederpes* also resembles later tetrapods in the rectangular configuration of the tabular, compared with the complex structure of that bone in *Acanthostega,* and the presence of paired postparietals, in contrast with their medial fusion in *Ichthyostega.* Yet other structures, including the retention of a preoperculum, lost in nearly all later tetrapods, and a predominance of pit lines rather than open lateral line canals, indicate a position close to the base of all later land vertebrates, but are of no use in establishing relationships to any specific lineage.

Other, derived characters, that are first recognized in *Pederpes,* do imply affinities with one or two other genera that occur later in the Lower Carboniferous, suggesting their inclusion in the same family, the Whatcheeridae, which had a very wide geographical distribution. Almost certainly related to *Pederpes* is *Whatcheeria,* known only from the Delta locality in Iowa, dated as equivalent in age to the European Asbian in the later Viséan (Lombard and Bolt, 1995; Bolt and Lombard, 2000) (Figs. 4.5 and 4.6). This genus is known from hundreds of specimens, including articulated skeletons, more than any other Lower Carboniferous tetrapod, but the great wealth of material still requires considerably more preparation and analysis.

Another genus, *Ossinodus,* comes from the Middle Paddock locality in Queensland, Australia, equivalent in age to the Holkerian of Europe (A. Warren and Turner, 2004) (Fig. 4.4C). *Ossinodus* is known from only very incomplete remains and its taxonomic position is still uncertain, but *Pederpes* can be logically included in the Whatcheeridae, indicating the duration of a particular body form for at least 18 million years.

The habitat and way of life of these genera may have been fairly similar. The remains of all three genera are found in shallow water deposits ranging from marine to freshwater, described as lacustrine, fluvial, lagoonal, or swamp environments. The associated fauna closely resemble that at Horton, including gyracanths and other acanthodians, sharks, palaeoniscoids, rhizodontid sarcopterygians, and some evidence of lungfish. The disarticulated bones of *Ossinodus,* belonging to at least four different individuals, were described as being deposited by a flood event on a tidal channel floor. In contrast, Clack and Finney (2005) suggest that *Pederpes* may have initially been preserved as a mummified carcass. Hundreds of specimens of *Whatcheeria* were preserved in sink holes in a limestone deposit, along with two more advanced labyrinthodonts.

We will concentrate on *Pederpes,* which provides the best currently available information on the skeletal anatomy of a stem tetrapod near the base of the Carboniferous. Of the Upper Devonian tetrapods, *Acanthostega* provides the most complete basis for comparison, as a result of recent detailed descriptions of all aspects of the skeleton by Jennifer Clack and her colleagues. But, whether comparison is made with *Acanthostega* or *Ichthyostega,* there are many significant changes between the latest known Devonian and earliest Carboniferous taxa. As best seen in occipital view (Fig. 3.5D) the skull of *Acanthostega* is low and the cheek and table form a continuous arc. In marked contrast, the skulls of *Pederpes* and *Whatcheeria* are higher than they are wide, and the cheek forms a sharp angle with the skull table. In contrast with several other groups of later Carboniferous tetrapods, the supratemporal forms a strong, interdigitating suture with the squamosal, integrating the skull table and cheek.

The configuration of the bones at the back of the skull table in whatcheerids, the postparietals and tabulars, resembles that of the early members of several later amphibian lineages in that the postparietals are wider and longer than the tabulars, precluding the tabular's articulation with the parietals. This distinction seems trivial from the standpoint of geometry, but is important in terms of the mechanics of the skull and serves as a means of classifying major groups. The tabular lacks the dorsal embayment of *Acanthostega,* but overhangs the squamosal notch as in *Ichthyostega.* There is

Figure 4.3. (opposite) Elements of appendicular skeleton of amphibians from Horton, compared with tetrapods from other horizons. *A,* Left humerus of the Upper Devonian *Tulerpeton.* From Lebedev and Coates, 1995. *B,* Left humerus from Horton Bluff. From Clack and Carroll, 2000. *C,* Left humerus of *Eoherpeton,* an embolomere from the lower Namurian A. From Smithson, 1985a. *D,* Left humerus of the colosteus *Greererpeton,* from the Namurian A. From Godfrey, 1989. *E,* Horton humerus II. *F,* Humerus of *Pederpes,* from the Lower Tournaisian. From Clack and Finney, 2005. *G,* Humerus III from Horton. *H,* Ilium of *Pederpes.* From Clack and Finney, 2005. *I,* Pelvic girdle from Horton. *J* and *K,* Two femora in ventral view from Horton.

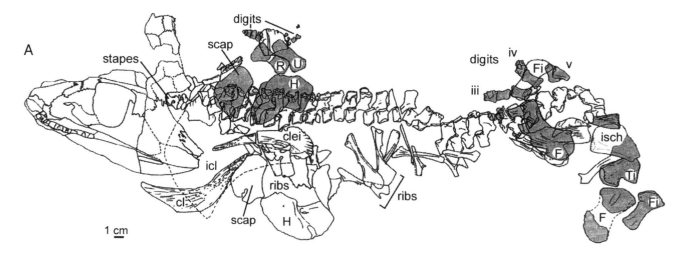

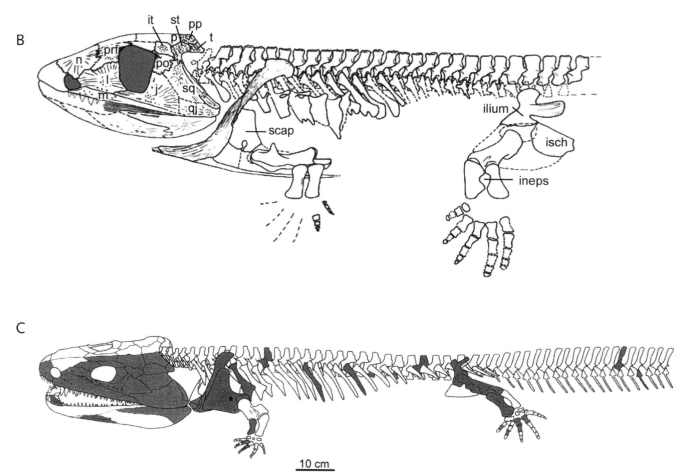

Figure 4.4. Skeletons of whatcheerids. *A, B, Pederpes finneyae*, from the Lower Tournaisian, north of Dumbarton, Scotland. From Clack and Finney, 2005. *A*, Skeletal as prepared. The shaded areas are of the bones visible on the opposite side of the block. *B*, Reconstruction. *C*, Reconstruction of *Ossinodus pueri* from the Lower Carboniferous of Australia. From A. Warren, 2007, assembled from disarticulated bones (black) of several individuals.

no evidence that it played any role in supporting the poorly known braincase. Neither the otic capsule nor the basioccipital is preserved in *Pederpes*, and the sphenethmoid was apparently unossified in *Whatcheeria.*

A striking resemblance between *Pederpes* and *Whatcheeria* is the peculiar shape of the orbital margin, which extends far medially on the dorsal skull roof and very far ventrally on the cheek. No other amphibians have such a configuration. Another unusual feature is the low degree of expression of superficial ornamentation of the dermal bones of the skull and the dermal shoulder girdle. In most primitive amphibians the bones are almost uniformly marked with a pattern of small polygonal pits or a combination of pits and long grooves. *Whatcheeria* shows almost no sculpturing of the skull or dermal shoulder girdle, while *Pederpes* has only a muted pattern of pits and grooves. This suggests a progressive reduction of ornamentation in the Northern Hemisphere genera, but the Australian genus *Ossinodus* shows an almost uniform pattern of small polygonal pits, common to many other groups of Paleozoic amphibians.

Very few of the sensory canals on the skull in *Pederpes* were exposed in open grooves. None are apparent on the skull table, and those on the lacrimal, jugal, and postorbital are expressed as pit lines. The only groove is on the surangular. There is also a groove on this bone in *Whatcheeria*, but that genus also has conspicuous open grooves on the ventral circumorbital bones, although not on the skull table. Neither pit lines nor continuous grooves are visible in *Ossinodus*, but few bones that carried sensory lines in the other genera are known for that genus.

Only the dorsal margin of the occipital surface is exposed in *Pederpes*, formed by unsculptured extensions of the tabulars and postparietal. As in *Whatcheeria*, it forms a roughly 90 degree angle with the skull table. This is accentuated by the presence of a dorsal ridge in *Pederpes*, but not in *Whatcheeria*. Neither the otic capsule, the basioccipital, nor the sphenethmoid is known in *Pederpes*, but the otic capsules fused at the midline in *Whatcheeria*, roofing the neural canal. The stapes in *Pederpes* resembles that of *Acanthostega*, but is proportionately smaller, with the blade hardly wider than the footplate. In contrast with *Acanthostega*, there is no evidence of the hyoid apparatus in any of the whatcheerids.

All the whatcheerids have enlarged maxillary teeth toward the anterior end of the bone, in the same general position as the canines in early amniotes, as well as enlarged teeth toward the posterior extremity of the premaxilla. However, the taxonomic significance of this similarity is uncertain since enlarged teeth are illustrated in approximately this same position in *Acanthostega* (Clack, 2002c).

The lower jaws, much better exposed in *Whatcheeria*, are primitive in the retention of an ossified Meckelian bone and a toothed adsymphysial adjacent to the symphysis, as in other early labyrinthodonts (Ahlberg and Clack, 1998).

The vertebral column in *Pederpes* is more advanced than *Acanthostega* in the greater degree of ossification and integration of the vertebral elements, but there were apparently nearly the same number of presacral vertebrae, approximately 28. The intercentra are median crescentic elements throughout the column and the pleurocentra are paired, more dorsal elements, a pattern termed rhachitomous. The neural arches are not fully consolidated at the midline. A supraneural canal is retained above the passage for the spinal cord. *Whatcheeria* differs strikingly from all other primitive tetrapods in that the pleurocentra are fused dorsally, and some have two pairs of articulating surfaces that were attached to the pedicles of two successive neural arches. It had

about 30 presacral vertebrae. The anterior ribs of all three genera have much expanded distal ends, but of a different pattern than those of *Ichthyostega*.

In contrast with *Ichthyostega* and *Acanthostega*, but in common with *Tulerpeton*, the scapula of whatcheerids is not fused with the cleithrum and extends farther dorsally. An anocleithrum has not been recognized, but the cleithrum, clavicle, and kite-shaped interclavicle are massive. On the other hand, the scapulocoracoid of *Pederpes* and *Whatcheeria* differs from those of the Devonian genera in the absence or delay of ossification of the ventral, coracoid area. In contrast, the area of ossification in *Ossinodus* continues below the glenoid, as in most other groups of primitive tetrapods.

The humeri in all the genera assigned to the Whatcheeriidae are less well ossified than those of either *Tulerpeton* or the bones from the Horton locality suggested as belonging to an anthracosaur, but they do resemble a second large humerus from Horton in the shortness of the shaft and the general outline (Fig. 4.3G). There is little evidence of ossified carpals. One stout digit of the manus with three phalanges is recognized in *Pederpes*. A single much smaller digit has also been found, which Clack and Finney (2005) hypothesize might represent an accessory digit, comparable to those in excess of five in *Ichthyostega*.

The ilium in all three genera resembles that of the Devonian tetrapods in having distinct dorsal and posterodorsal processes, with the dorsal blade offset medially relative to the posterior process, but the posterior process is shorter. The posterior process in *Whatcheeria* extends laterally beyond the base of the dorsal process, but not in *Pederpes*. The acetabular area is not known in *Pederpes*, but in *Whatcheeria*, as in *Ichthyostega* and *Acanthostega*, it extends anteriorly to the anterior margin of the pubis. Three openings of the obturator canal pierce the puboischiadic plate. This feature cannot be determined in the other two genera. The pubis was apparently not ossified in *Pederpes*, and is incompletely known in *Ossinodus*.

The femur of *Pederpes* is short compared with the Devonian genera, without a clearly defined shaft between the poorly ossified extremities. It is unique in the head being in continuity with the proximal end of the adductor blade. On the other hand, the tibia and fibula are more derived in their elongation and narrowing of the shaft, resulting in the interepipodial space common to later tetrapods. Five toes are preserved, although more might have been present. Only a single possible tarsal has been identified. Most important, the oblique termination of the metatarsals resembles that of later amphibians that definitely had the toes facing forward in the typical tetrapod manner. The phalangeal count is restored as 2, 3, 4, 4, (> 2). The pes of *Whatcheeria* is relatively broader and more paddle-like than in *Pederpes*, suggesting specialization toward more effective aquatic locomotion.

The ventral scales are highly developed in *Pederpes*, but have not been reported in *Whatcheeria*. They would not be

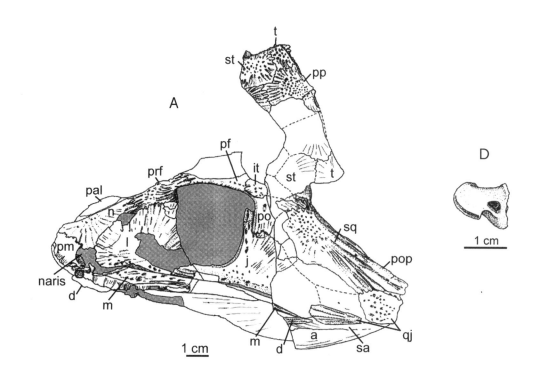

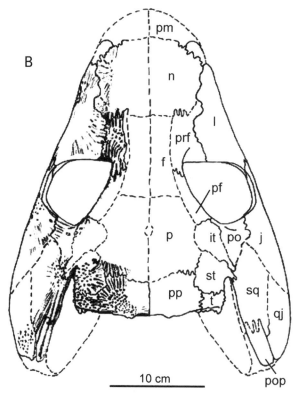
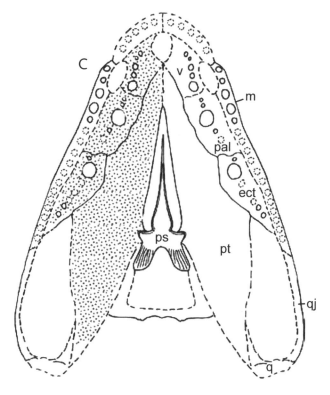

Figure 4.5. Skull of *Pederpes.* From Clack and Finney, 2005. *A,* Lateral view of the skull as preserved. *B,* Reconstruction of skull in dorsal view. *C,* Reconstruction of palate. *D,* Reconstruction of left stapes.

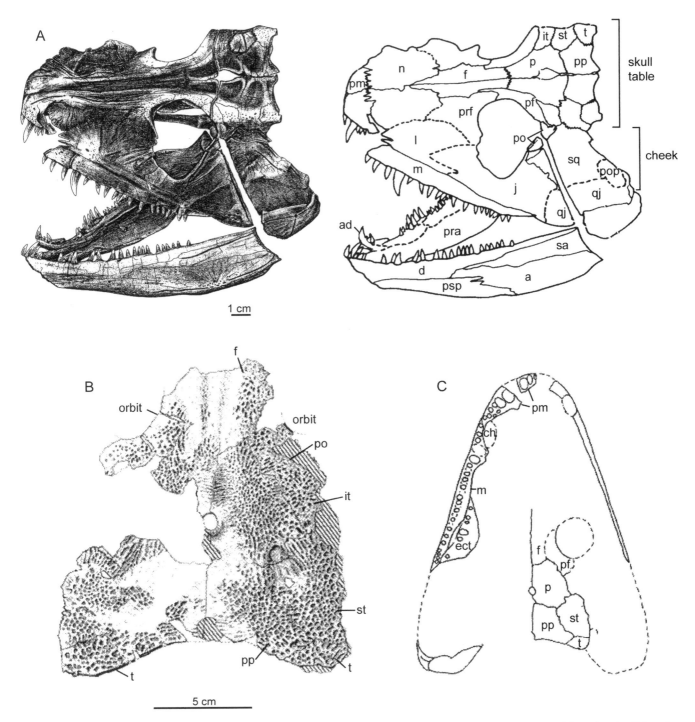

1 cm

5 cm

Figure 4.6. Skulls of *Whatcheeria* and *Ossinodus*. A, Skull of *Whatcheeria deltae*, from the Lower Carboniferous of Iowa. From Lombard and Bolt, 1995. B, Articulated skull table of *Ossinodus pueri* from the mid-Viséan of Queensland, Australia. C, *Ossinodus pueri*, reconstruction of the right side of the palate and the skull table. B, C, from Warren and Turner, 2004.

expected to be preserved with the disarticulated elements of *Ossinodus*.

From all the material that is known of *Pederpes* and *Whatcheeria*, a strong case can be made for their belonging to the same lineage. The low degree of ossification of the endochondral shoulder girdle, the pelvic girdle, and the carpals and tarsals in a specimen the size of *Pederpes* suggests that it was habitually, if not obligatorily, aquatic. This is supported by the retention of sensory pit lines on the skull. On

the other hand, the unique anatomical specializations of these genera clearly distinguish them from other aquatic or terrestrial groups that appeared in the fossil record later in the Carboniferous. They appear to have left no descendants, and show no evidence of having been the sister-taxa of any later groups.

The taxonomic position of *Ossinodus* is much harder to establish because of our very incomplete knowledge of the skeleton, including the nature of the cheek, vertebral cen-

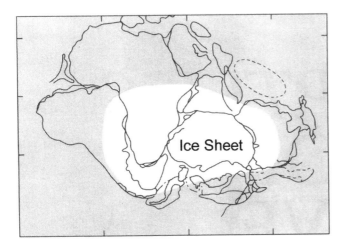

Figure 4.7. Extent of the ice sheet in Gondwana at about the time of the Carboniferous-Permian boundary, 286 million years ago. From White, 1990.

tra, and distal limb elements. Even among the incomplete remains of *Ossinodus,* there are several characters that suggest that close relationship is not likely. Most obvious is the presence of conspicuous sculpturing of the dermal skull in *Ossinodus* and its near absence in the other genera, but also important is the continuity of ossification of the scapulocoracoid and perhaps of the pelvic girdle, which would imply evolutionary reversal. The retention of many primitive features of the skeleton confirms the position of *Ossinodus* near the base of tetrapods, and its divergence prior to the loss of the intertemporal bone and other derived features of *Ichthyostega* and *Acanthostega.* Only the discovery of more complete skeletons will make it possible to determine whether *Ossinodus* should be retained as a member of the Whatcheeridae, or be recognized as a separate basal lineage. There is no evidence to support its affinities with any other groups known at the present. However, *Ossinodus* remains informative as the only Carboniferous tetrapod known from the Southern Hemisphere. The absence of amphibian fossils from Gondwanaland for the remainder of the Carboniferous may be attributed to climatic deterioration that led to very extensive glaciation by the end of the Paleozoic (Fig. 4.7).

Whatever their specific affinities, *Pederpes, Ossinodus,* and *Whatcheeria* all broadly resemble the Upper Devonian amphibians and their immediate choanate antecedents in their large size (a meter or more in length), large head to trunk ratio, jaw articulation behind the occiput, and large, fanglike teeth borne on some, if not all, of the lateral dermal bones of the palate.

LETHISCUS

After the first occurrence of whatcheerids, the next tetrapod lineage to appear in the fossil record is very different from the crocodile-like animals we have just seen. The genus *Lethiscus* is known only from a single specimen coming from the mid-Viséan Wardie locality near Edinburgh, Scot-

land (Fig. 4.8). It was first described by Carl Wellstead (1982) and later studied via X-ray tomography by Jason Anderson and his colleagues (2003). *Lethiscus* consists of an entire skeleton with a very long vertebral column but no trace of limbs. Seventy-eight vertebrae occur in sequence, with little evidence of reduction in size posteriorly. All bear ribs and so can be recognized as from the trunk region. A single, much smaller, disarticulated vertebra is assumed to have come from the tail. The head is relatively small, with the eyes in an anterior position. There is no squamosal notch, but a very large temporal opening in the cheek.

These highly distinctive features are characteristic of a diverse order of amphibians that extends into the Lower Permian, the Aïstopoda. The only trace of the appendicular skeleton they exhibit is a small angled bone behind the skull that has been identified as a cleithrum and (in one other specimen) a possible interclavicle. In contrast with any earlier tetrapods, the vertebrae consist of a single cylindrical centrum fused to the neural arch. This is termed the holospondylus configuration.

Lethiscus also shows unique modifications of the dermal integument from that of the fish ancestors of tetrapods. *Eusthenopteron* shows a complete covering of more or less circular scales over the surface of the trunk and tail. Early tetrapods have, to a variable degree, differentiated the dorsal and ventral scalation. The ventral scales become elliptical or more clearly elongated and are arranged in a chevron pattern with the apex pointed anteriorly. They initially overlap one another extensively, but in various groups, including both some amphibians and early amniotes, they narrow to a rod-shaped configuration and are referred to as gastralia. The occurrence of gastralia is a highly distinctive character of later aïstopods that was already initiated in *Lethiscus,* as revealed by X-ray tomography. *Lethiscus* and many later Carboniferous aïstopods are also unique in the possession of specialized dorsal ossifications in the form of numerous tiny polygonal osteoderms that distinguish them from all other Paleozoic tetrapods.

A more detailed description of aïstopods will be given later in this chapter on the basis of much better-known later Carboniferous and Early Permian species.

Lethiscus is the only amphibian known from a locality teeming with fish.

CASINERIA

The most tantalizing tetrapod from the Lower Carboniferous is the sole specimen of *Casineria kiddi* from the shrimp beds of Cheese Bay, on the coast of Scotland, dating from about the middle of the Viséan (Paton et al., 1999). It consists of a nearly complete skeleton, but lacking the head and most of the tail (Fig. 4.9). It is the only known tetrapod from a deposit with a fauna otherwise consisting of excellently preserved shrimp, scorpions, and, less commonly, fish, similar to those of other Viséan localities. It is the smallest known

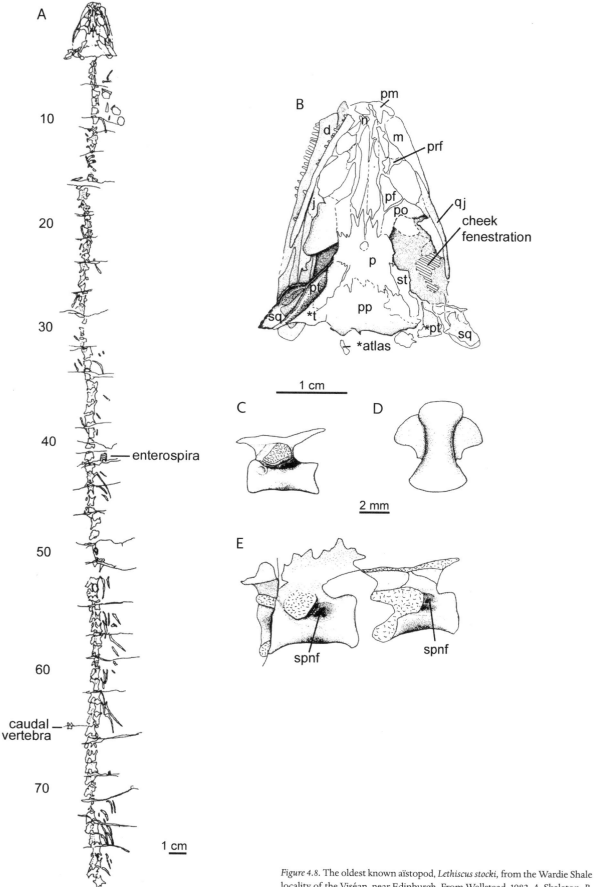

Figure 4.8. The oldest known aïstopod, *Lethiscus stocki*, from the Wardie Shale locality of the Viséan, near Edinburgh. From Wellstead, 1982. *A,* Skeleton. *B,* Skull. *C* and *D,* Anterior vertebra in lateral and ventral views. *E,* Left lateral view of vertebrae 57 and 58 showing openings for spinal nerves. *Bones of uncertain identity.

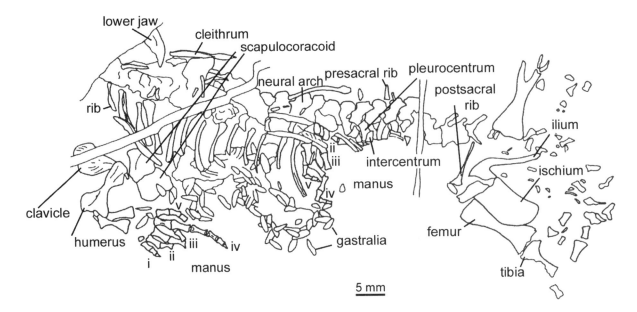

Figure 4.9. Casineria kiddi, from the Cheese Bay Shrimp Bed, upper Viséan, Scotland, i–v, digits in manus. From Paton et al., 1999.

terrestrial vertebrate from the Lower Carboniferous, with an approximate length, from the neck to the base of the tail, of only about 80 mm. Yet, the high degree of ossification of the limbs and girdles suggests that it was a fairly mature animal, although the carpals in the otherwise very well-articulated forelimb remain unossified. By its size alone, this animal represents a biologically highly significant departure from all the tetrapods we know from earlier deposits.

Of the head region, only the back of the lower jaw is preserved. The cervical region is poorly known, but the more posterior trunk vertebrae are preserved in a fairly long series, suggesting a presacral count of about 25. The pleurocentra are nearly cylindrical and longer than the crescentic intercentra. Both articulate with the neural arch. The scapular and coracoid portions of the scapulocoracoid show evidence of having ossified separately, as in *Pederpes* and *Whatcheeria,* in contrast with other early tetrapods, in which they are ossified as a single element. The clavicle retains a very large blade. The humerus has a distinct shaft between the expanded proximal and distal areas of articulation. They are twisted relative to one another at about a 90 degree angle as in many more advanced Paleozoic amphibians. The hand, known from both sides, can be reconstructed as having a phalangeal count of 2, 3, 4, 5, 4, which is roughly comparable with that of *Tulerpeton.* Uniquely, *Casineria* has curved terminal phalanges (unguals) forming claws, as in many early amniotes.

The ilium appears to have a relatively longer posterior process than in earlier tetrapods, but there is no evidence of a dorsal process. The femur is about 30% longer than the humerus. Narrow, elliptical gastralia are scattered along the ventral side of the trunk. The feet and end of the tail are missing. The significance of *Casineria* will be discussed further in Chapter 7.

ADELOSPONDYLI

The next amphibian morphotype to appear in the fossil record are the adelospondylids, whose oldest known fossils occur in the locality of Broxburn, near Edinburgh (Andrews and Carroll, 1991). In fact, all fossils of this group have been discovered within 30 km of that city and nowhere else in the world. All are associated with a diverse fish fauna and some were accompanied by other aquatic amphibians. Most were found in sediments deposited in the large, freshwater Lake Cadell, which was located close to the equator, not far from the sea. They have been collected from nine different localities, spanning about 10 million years. Four genera have been named, but they are all very similar to one another. Like the aïstopods, adelspondyls have no trace of limbs. However, the entire dermal shoulder girdle is retained, but nothing of the scapulocoracoid or the pelvis, even in well-articulated skeletons (Figs. 4.10 and 4.11).

Adelospondyls were clearly aquatic as seen from retention of lateral line canals running along the tabular-squamosal, postparietal, postorbital, and lower margin of the lower jaw of most specimens, as well as the presence of massive hyoid elements.

Despite the common absence of limbs and great elongation of the vertebral column, other characters of adelospondyls differ strongly from aïstopods and from other early tetrapods. In contrast with aïstopods, the dorsal surface of the skull is solidly ossified, with no trace of fenestration. However, it differs from other early tetrapods in the loss and/or fusion of many of the bones present in whatcheerids, who have the most primitive complement. Only a single bone is found in the position initially occupied by the supratemporal, intertemporal, tabular, and squamosal. It is not possible to determine which have been totally lost and which may have fused to others. The single composite unit is termed the tabular-squamosal, for two of the larger bones that occupied this area in other early amphibians. Ventrally, it is in contact with the quadratojugal and quadrate, as would be the case of the original squamosal. The posterior dorsal portion occupies the position of the tabular and has a tabular-like "horn," or process, which overhangs the area of the squamosal or otic notch in whatcheerids and other primitive tetrapods. There is, however, no line of modility between the skull table and the cheek, as least in the adults.

Whatever the original function of the squamosal notch in early tetrapods, the area beneath the posterior process of the tabular-squamosal in adelospondyls appears unique in its geometry, and has been hypothesized as supporting a depressor mandibulae muscle attached to a retroarticular process at the back of the lower jaw (otherwise very uncommon among primitive tetrapods) for opening the mouth.

The postparietals extend lappets ventrally down the occipital surface, where they may have supported the back of the braincase. Lateral to the lappets are openings below the dorsal extremity of the tabular-squamosal in the position of the post-temporal fenestrae of some other early tetrapods, including microsaurs and temnospondyls. They are thought to have been occupied by the muscles extending from the neck, but there is no evidence as to whether or not these openings are homologous.

The orbits are far forward in adelogyrinids, which may be associated with exclusion of the postorbital from the margin of the orbit, and its reduction in size in most specimens or complete absence in *Adelospondylus*. The small size of the skull and the relatively larger size of the eye may explain the apparent entrance of the maxilla into the orbital margin. Another surprising feature of the adelogyrinids is the configuration of the dentition. The marginal teeth are columnar, rather than pointed, and end in a laterally compressed tip, hooked posteriorly. Nearly all are in place at one time rather than being rapidly replaced with about one half missing at any one time, and show little if any wear. In contrast with choanate fish and Devonian tetrapods, the teeth show no evidence of labyrinthine infolding. The shape of the jaw articulation suggests that the mouth could have been opened very widely.

Another feature of adelogyrinids that distinguishes them very strongly from other tetrapods and suggests a unique manner of feeding is the presence of a particularly massive hyoid apparatus (Fig. 4.12F). The elements are so large that they were originally identified as limb bones! None show the grooves that are present in *Acanthostega* that are assumed to have carried blood vessels associated with the external gills.

The parasphenoid has a long, narrow cultriform process and a narrowly triangular basal plate, without obvious areas of articulation with the base of the braincase. The ectopterygoid, palatine, and vomer are fairly slender bones that bear small denticles but no trace of the large fangs common to the Devonian tetrapods and whatcheerids.

Little is known of the braincase. The sphenethmoid is represented by narrow strips of bone on either side of the cultriform process of the parasphenoid. There is no evidence of the dorsal portion of the occipital plate, but the basioccipital, not ossified in either *Ichthyostega* or *Acanthostega*, appears to be represented by a circular bone, recessed to receive the odontoid process of the atlas.

In common with aïstopods, the vertebral centra consist of a single, elongate cylinder, recessed at each end. They differ, however, in that they are only loosely attached to the neural arches. In none of the specimens is the entire column preserved, but the number of centra in articulation runs from 50 to 70 with no sign of a pelvic girdle. The anterior ribs resemble those of *Ichthyostega* and whatcheerids in having conspicuous flanges.

Of the shoulder girdle, the cleithrum, clavicle, and interclavicle are present in most specimens, but no trace of the scapulocoracoid or pelvis has been seen. The dermal bones broadly resemble those of whatcheerids and other large early tetrapods.

A single larval specimen of an adelospondylid, apparently from the same horizon as *Adelospondylus watsoni*, includes a partial skull and lower jaw, along with 70 tiny, cylindrical centra and a clavicle. There is no trace of external gills.

In contrast with aïstopods, no species comparable to adelospondyls are known from later in the fossil record.

COLOSTEIDS

After looking at three highly divergent taxa, with no obvious affinities with one another or with any particular group of stem tetrapods, we now return, in the chronological order of the beds in which they are preserved, to another labyrinthodont taxon. Only slightly later than the Delta and Broxburn localities is the Burdiehouse Limestone (Asbian), a freshwater lake deposit in Scotland, from which has come a single skull representing the first occurrence of another, highly distinct group of amphibians, the colosteids. The colosteids are also known from a series of later horizons in Great Britain and eastern North America, going up into the Westphalian D. They are clearly derived from among the basal labyrinthodonts, lacking any of the distinctive features of the *Lethiscus*, *Casineria*, or adelospondylids.

Colosteids were large animals, with adult skull lengths in the successive genera as follows: *Pholidogaster*—17 cm, *Greererpeton*—18 cm, *Colosteus*—9.5 cm. The total length of the largest specimen of *Greererpeton* reached 1.4 m. Again, their cranial anatomy and body proportions, with approximately 40 presacral vertebrae and relatively small limbs, separate them clearly from any of the Devonian genera, the

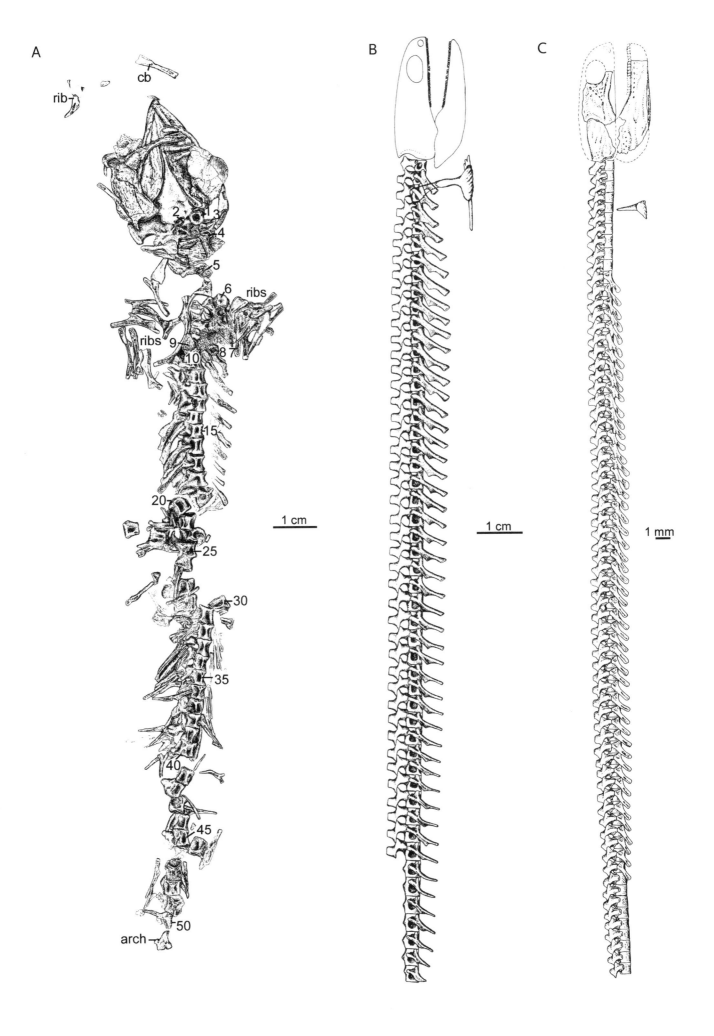

A

rib

cb

2 1 3

4

5

6 ribs

ribs 9 8 7

10

15

20

25

30

35

40

45

50

arch

1 cm

B

1 cm

C

1 mm

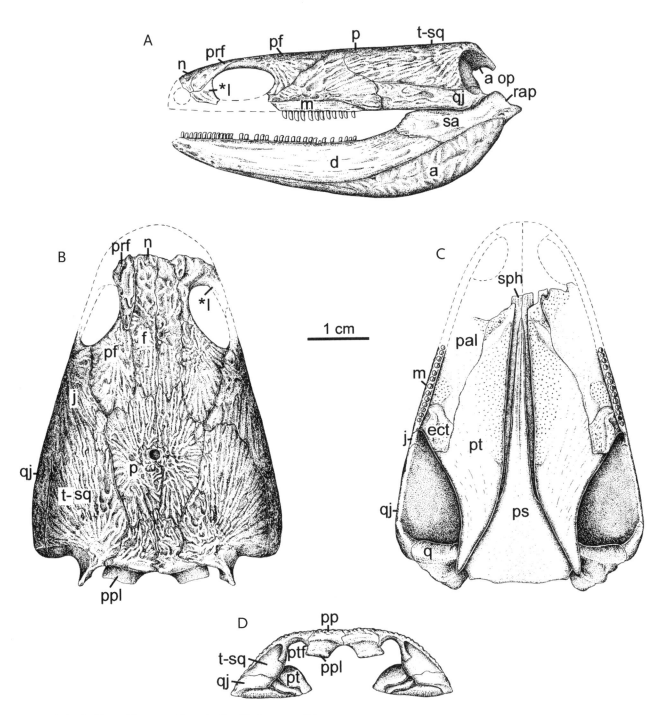

Figure 4.11. Skull of *Adelospondylus watsoni* from the Rumbles Ironstone, Edinburgh, in lateral, dorsal, palatal, and occipital views. From Andrews and Carroll, 1991. *Bone possibly identified as lacrimal.

Figure 4.10. (*opposite*) Skeletons of adelospondylids. *A,* Skeleton of the oldest known adelospondylid, *Palaeomolgophis,* from the Viséan Broxburn locality. *B,* Reconstruction of *Palaeomolgophis. C,* Reconstruction of a larval adelospondyl from the lower Namurian of the Loanhead locality. Both localities near Edinburgh, Scotland. All drawings reproduced from Andrews and Carroll, 1991.

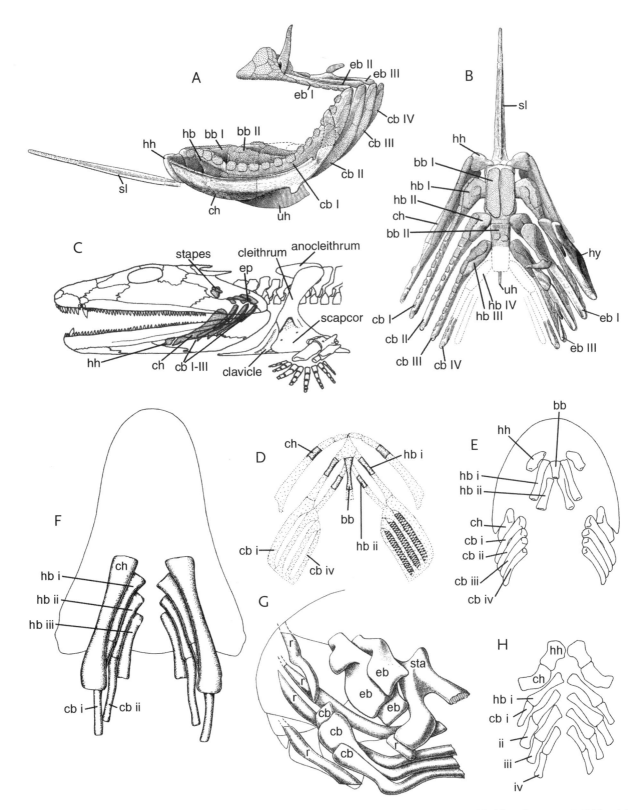

Figure 4.12. Hyoid apparatus of sarcopterygian fish and Paleozoic tetrapods. *A* and *B,* The osteolepiform *Eusthenopteron* in lateral and dorsal views. Modified from Jarvik, 1954. *C,* Lateral view of *Acanthostega.* From Clack, 2000. *D,* A branchiosaurid. From Boy and Sues, 2000. *E,* The neotenic Upper Permian labyrinthodont *Dvinosaurus.* Modified from Bystrow, 1938. *F,* The adelospondyl *Adelogyrinus.* From Andrews and Carroll, 1991. *G,* The Lower Permian microsaur *Pantylus.* Modified from Romer, 1969. *H,* The lysorophid *Brachydectes elongatus.* From Wellstead, 1991.

whatcheerids, and most later Paleozoic lineages. Retention of conspicuous lateral line canals and limited ossification of the ends of the limb bones, carpals, and tarsals all suggest habitually aquatic locomotion (Figs. 4.13–4.15).

The oldest known colosteid (initially named *Otocratia*) is known only from the ventral surface of a poorly preserved skull roof. However, the margins of the skull are well defined as are the sutures between most of the bones. Their patterns accord very closely to those of a second skull, apparently from the somewhat higher Gilmerton ironstone (Brigantian), which Alec Panchen (1970) placed in the same genus as the earlier named *Pholidogaster* (Fig. 4.13). Much more complete material is known from the Greer locality (equivalent to the European Arnsbergian), in West Virginia, from which 60 specimens are known, including extremely well-preserved complete skeletons described by Smithson (1982) and Godfrey (1989). The latest known genus, *Colosteus,* most recently described by Robert Hook (1983), comes from the Westphalian D of Linton, Ohio.

As opposed to the uncertain interrelationships of the three genera attributed to the Whatcheeridae, the high degree of consistency of the colosteid skull over approximately 21 million years very strongly supports membership within a single family. The differences over that period of time are primarily proportional, with the eyes becoming progressively smaller relative to overall skull length and more anterior in position.

In marked contrast with members of the Whatcheeridae, the skull is low and broad and the squamosal notches are closed, except in the most immature specimens. As seen in occipital view, the skull table and cheek form a continuous arch, but some degree of mobility between the skull table and the cheek may be retained. A very small intertemporal bone is present in *Greererpeton,* but is not retained in the other genera. As in all more primitive labyrinthodonts, the supratemporal remains in broad contact with the postparietal. Anteriorly, the skull is unique among early labyrinthodonts in the prefrontal extending between the premaxilla and maxilla to reach the margin of the naris. Lateral line canal grooves form a conspicuous pattern on the snout and cheek, but not on the skull table.

In contrast with any amphibians so far discussed, the pterygoids do not extend close to the midline, but are separated from one another to form what is called the interpterygoid vacuity. All have very long fangs on the ectopterygoid and palatine, but the vomerine fangs are reduced during the course of their evolution to the size of the smaller teeth on the palatine and ectopterygoid. The most striking autapomorphy (unique shared derived character) is the expansion of the most posterior premaxillary tooth into a large fang that fits into a groove in the lateral surface of the dentary bone. These teeth would have been effective in the capture of fairly large, aquatic prey. Ceratobranchials with attached gill rakers are present in smaller specimens of *Colosteus.*

The individual vertebrae (Fig. 4.13) retain the primitive, rhachitomous pattern of *Ichthyostega* and *Pederpes,* with clearly distinct neural arches, crescentic intercentra, and paired, more dorsal pleurocentra. The configuration of the atlas is similar to that of all later Paleozoic labyrinthodonts in having a paired proatlas (apparently already present in *Acanthostega*), linking the occiput with the atlas neural arch (also paired). The median axis arch is much higher, providing an extensive surface for the attachment of muscles and/or ligaments that inserted on the occiput. Beneath the atlas and axis arches are paired pleurocentra and an intercentrum. That of the atlas is paired in juveniles but median in adults.

The ventral surfaces of the clavicles are greatly expanded, but the stem is short. The interclavicle is clearly distinct from that of Devonian tetrapods and the whatcheerids in lacking a posterior stem, but has a wide anterior process. As in Devonian tetrapods, the scapulocoracoid is ossified as a single unit, even in fairly small individuals, in contrast with the separation of the scapula and coracoid in whatcheerids. However, the unfinished surface at the top of the scapula and its wide separation from the top of the cleithrum suggest the presence of a large, cartilaginous suprascapula.

The humerus is proportionately much smaller than in other Paleozoic labyrinthodonts and the edges are less well defined, suggesting incomplete ossification. Its size and general configuration resemble a single, disarticulated humerus from the Horton Bluff Formation (Fig. 4.3E). It also retains one of the accessory foramina present in the oldest known tetrapods, including *Acanthostega* (Coates, 1996, 385), indicating a very early divergence from basal labyrinthodonts. The ulna and radius, however, are long and gracile. Some of the carpus and manus is preserved in one specimen of *Greererpeton,* but they are not fully articulated and a phalangeal formula cannot be established. In *Colosteus,* the formula is 2, 2, 3, 3.

The pelvis is strikingly different from that of all previously described labyrinthodonts in the presence of only a single, posterodorsally directed iliac blade, without a dorsally directed process. Sutures are clearly defined between the ilium and the pubis and ischium. There is only a single obturator foramen. The femur is long and slender, with a much shorter adductor blade than other early labyrinthodonts. The tibia and fibular appear fairly gracile. The proximal portion of the tarsus can be reconstructed (Fig. 4.14), but the four posterior distal tarsal are not preserved. The phalangeal count is 2, 2, 3, 4, 3+.

Colosteids cannot be linked directly to any of the more primitive labyrinthodonts, nor did they leave any recognizable descendants.

GILMERTON AND COWDENBEATH

Some of the most striking amphibian remains for their size and taxonomic diversity come from localities near the boundary of the Lower and Upper Carboniferous. The Gilmerton Ironstone of late Viséan age is exposed near

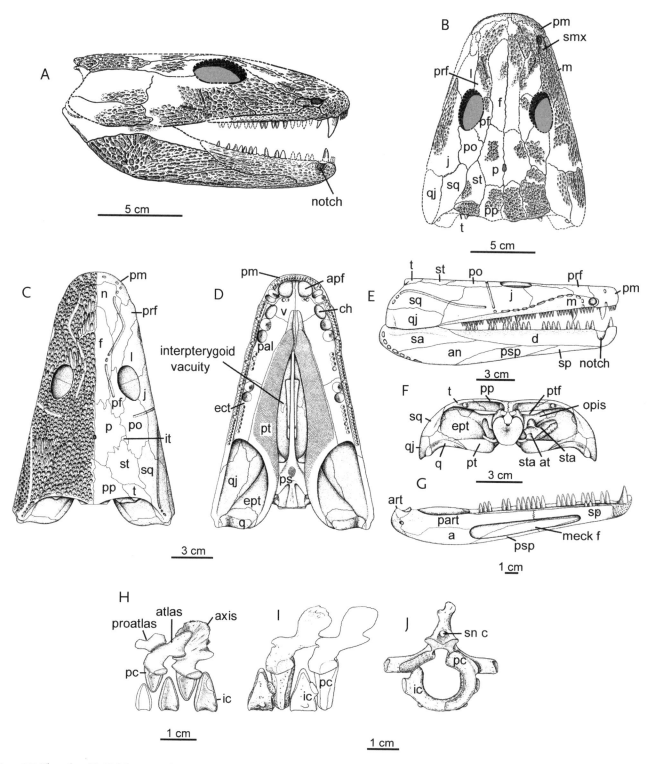

Figure 4.13. The colosteids *Pholidogaster* and *Greererpeton*. *A* and *B*, The skull of *Pholidogaster pisciformis* in lateral and dorsal views, from the Upper Viséan Gilmerton Ironstone, near Edinburgh. *C–J*, *Greererpeton* from the Namurian A of Greer, West Virginia. *C–F*, Skull in dorsal, palatal, lateral, and occipital views. *G*, Jaw in medial view. *H–J*, Anterior cervicals and trunk vertebrae in lateral and posterior views. *A* and *B* from Panchen 1975; *C–G* from Smithson, 1982; *H–J* from Godfrey, 1989.

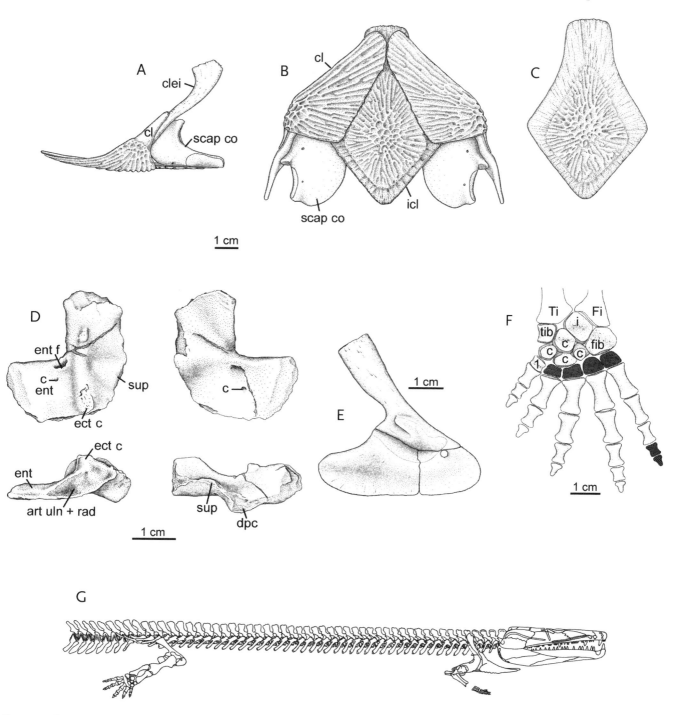

Figure 4.14. *Greererpeton* postcranial elements and restoration of skeleton. *A–C*, Shoulder girdle in lateral and ventral views, interclavicle in ventral view. *D*, Humerus in dorsal, ventral, distal, and anterior views. *E*, Lateral view of ilium. *F*, Tarsus and pes. *G*, Reconstruction of skeleton. From Godfrey, 1989.

Edinburgh. In addition to the colosteid *Pholidogaster*, it also contains the oldest known remains of three other highly distinctive groups, the anthracosaurs, the baphetids, and the crassigyrinids. All of these taxa have also been recovered from the Dora opencast site, near Cowdenbeath, Fife just 2 or 3 million years later, in the basal Namurian. Both of these localities must represent deposition in fairly deep water, judging from the size and nature of the animals collected (Plate 4). All of these labyrinthodonts are represented by specimens well over a meter in total length. The skulls alone range from 15 cm to 32 cm in length. *Pholidogaster* and the baphetid may have been habitually aquatic and *Crassigyrinus* probably never left the water.

Anthracosaurs

Most of the amphibians so far discussed belonged to groups that showed little diversity and became extinct within the Lower Carboniferous. At last, after nearly 35 million

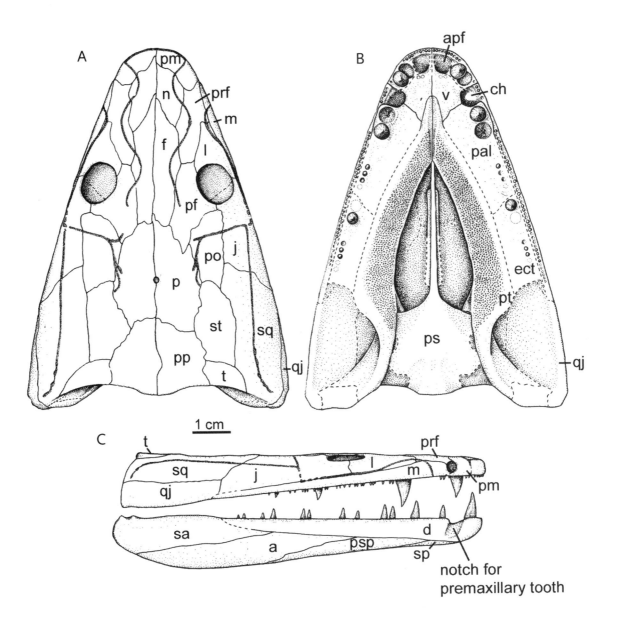

Figure 4.15. Skull of *Colosteus scutellatus* from the Middle Pennsylvanian of Linton, Ohio. Modified from Hook, 1983.

years of the Carboniferous, we finally encounter a number of lineages that radiated extensively and extended into the later Paleozoic. One lineage, the order Anthracosauria, may be related at least distantly to tetrapods living today (Smithson, 2000). The oldest known fossil is the Gilmerton genus *Eoherpeton,* recognized from a single, nearly complete skull about 15 cm in length, described by Alec Panchen (1975) and Tim Smithson (1985a) (Fig. 4.16).

The pattern of bones at the back of the skull table immediately distinguishes *Eoherpeton* from any other amphibian so far observed in the Carboniferous. In contrast with

Upper Devonian amphibians and whatcheerids, the posterolateral corner of the parietal has extended posteriorly beyond the shorter postparietals to reach the tabular, thus eliminating the primitive sutural connection between the postparietal and supratemporal. The occipital lappet of the tabular has also increased in depth relative to that of the postparietal so that it forms a large area in articulation with the massive, medially fused otic capsules. This would have provided firm support for the back of the braincase not seen in earlier labyrinthodonts. The intertemporal, lost in many other groups, is retained in all anthracosaurs. The

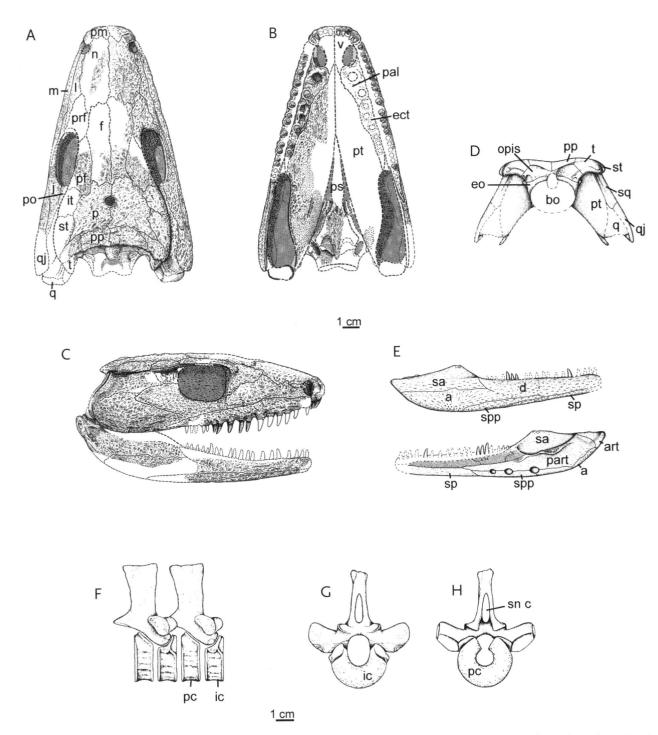

1 cm

1 cm

Figure 4.16. The earliest known anthracosaur, *Eoherpeton* from the upper Viséan Gilmerton Ironstone, near Edinburgh. *A–C*, Skull in dorsal, palatal, and
lateral views. From Panchen 1975. *D, E*, Occipital view of skull, and lateral
and medial views of lower jaw. *F–H*, Lateral, anterior, and posterior views
of trunk vertebrae. *D–F* from Smithson, 1985a.

prefrontals are not extended far anteriorly, as they are in
colosteids.

As in whatcheerids, the cheeks are at a high angle relative to the skull table, but the two surfaces are not strongly
sutured to one another. Rather, the rectangular skull table,
whose lateral margin somewhat overlaps the cheek, appears

as a clearly distinct area of the skull. As in other anthracosaurs, it is frequently found disarticulated from the cheek.
There is a shallow notch in the posterodorsal corner of the
squamosal, just beneath the posterior extension (horn) of the
tabular. Smithson (1985a) argued from the small degree of
indentation of the cheek and the massive nature of the sta-

pes that anthracosaurs did not have an impedance matching middle ear.

Although the end of the snout is not well preserved, the narrowness of the space for the internal nares suggests that the vomers were reduced and did not support vomerine fangs, an absence characteristic of all well-known embolomerous anthracosaurs. The palatine and elongate ectopterygoid both carry a row of teeth, as large or larger than those of the marginal dentition. No teeth in the premaxilla or maxilla are enlarged as fangs, in contrast with whatcheerids. The base of the parasphenoid has extensive basipterygoid processes that articulate loosely with the pterygoid. The entire pterygoid has an extensive covering of denticles.

The lower jaws retain the same complement of bones as more primitive labyrinthodonts. One primitive feature of *Eoherpeton* is the retention of a series of small openings between the prearticular and the more ventral jawbones. The three coronoid bones bear a uniform covering of denticles, with three small teeth on the second coronoid bone, but no prominent fangs. A parasymphyseal tusk is known in some later anthracosaurs, but this portion of the lower jaw is not known in *Eoherpeton*.

Panchen (1975) did not recognize any lateral line canal grooves on the skull, suggesting that *Eoherpeton* was essentially terrestrial, at least as an adult. Another factor indicative of the way of life of these animals is their overall body proportions. Smithson (1985a) noted that the appendicular skeleton of the Cowdenbeath specimen is very large relative to those of later anthracosaurs of roughly the same overall size. This can be quantified by comparison of the length of the major limb elements with the length of the jaw, which is known in both specimens. Enough is known of the lower jaw in the Cowdenbeath specimen to estimate that its length (which is essentially comparable to that of the skull in the type) was about 15 cm. The femur is approximately 14 cm in length. The femur of the only slightly later anthracosaur, *Proterogyrinus scheelei*, is only about one-half the length of the skull.

Of the postcranial skeleton, only fragments of the dermal shoulder girdle of *Eoherpeton* are known from the Gilmerton specimen. However, much more has been discovered from the slightly later Namurian A locality of Cowdenbeath, near Fife, Scotland. The most distinct aspect of the postcranial skeleton of anthracosaurs is the configuration of the vertebral centra (Fig. 4.16F–H). In contrast with ancestral labyrinthodonts, in which the intercentra appear as the dominant central elements, anthracosaurs have fused the primitively paired pleurocentra ventrally to form horseshoe-shaped structures, extending above the intercentra to help support the neural arches. Above the level of the zygopophyses is an opening within the neural spine, the supraneural canal, which presumably contained a ligament helping to support the vertebral column. In contrast with whatcheeriids, most of the ribs are long and cylindrical, except for the cervical region, where they are flattened and slightly expanded distally.

Unfortunately, the specimens from Cowdenbeath are not sufficiently well known for the length of the vertebral column to be even estimated.

The scapulocoracoid has a long scapular blade, a broad helical glenoid, and a fully integrated coracoid area; this is a common feature of anthracosaurs that continues into the Permian genera. The dermal shoulder girdle is not well known in *Eoherpeton*, but shows a consistent pattern in later genera with extensive areas of ventral expansion of the clavicular blades, an interclavicle with a moderately narrow stem, and a narrow flattened cleithrum. The humerus is similar to that of *Tulerpeton* and the large isolated humerus from Horton (Fig. 4.3B). The remainder of the forelimb is not known in *Eoherpeton*. The pelvis has a greatly elongate puboischiadic plate, fused to the ilium. The ilium is similar to that of the Upper Devonian tetrapod in having both dorsal and posterodorsal processes. The posterior process appears to be about as long relative to other dimensions as in the whatcheerids, but it is broken distally, and probably more closely resembled the longer process seen in better-preserved later anthracosaurs. All the elements in the pelvis are co-ossified, without evidence of sutures. Only a single obturator foramen is visible, but the smaller openings described in the contemporary North American species *Proterogyrinus scheelei* would probably not be visible, judging from the fractured surface of the single, poorly preserved specimen. The puboischiadic plate also appears relatively longer in *Eoherpeton*. The proximal head of the femur is not known, but the shaft is relatively long, with the adductor crest extending nearly to the level of the distal surfaces of articulation. The tibia and fibula are closely comparable to those of *Tulerpeton*. The foot is not known.

Smithson (1986) described a second anthracosaur from Cowdenbeath as *Proterogyrinus pancheni*. On the basis of the skull it is closely related to a species previously described by Robert Holmes (1984) as *Proterogyrinus scheelei* from the North American locality of Greer, West Virginia, but, despite its essentially equivalent age, it differs significantly from *Eoherpeton*. *Proterogyrinus scheelei* is known from at least 11 specimens, several of which consist of well-preserved, nearly complete skeletons, showing nearly every bone in the body (Figs. 4.17 and 4.18). The general anatomy of the skull is essentially the same in the two genera, but the proportions, number of teeth, and jaw structure show they are distinct as least at the generic level. The lower jaw is particularly distinctive in having two very large foramina on the inside surface, between the prearticular bone and the splenials. This feature has long been accepted as a diagnostic character of a particular group of anthracosaurs, the embolomeres, which were dominant labyrinthodonts of the later Carboniferous. The anterior coronoid bears at least one large tooth, in contrast with teeth on the middle coronoid in *Eoherpeton*.

Although only the posterior portion of the braincase is known in *Eoherpeton*, it is better ossified than comparable elements in *Proterogyrinus*. In *Eoherpeton*, the otic capsule is ossified as a massive, medial structure closely adhering to the back of the dermal skull roof. In *Proterogyrinus*, the medial and dorsal portions of this bone are not ossified, but leave a gap between the paired opisthotic bones and the ventral lappet of the postparietals, above the foramen magnum. The nature of the posterior portion of the parasphenoid and basisphenoid is similar in the two genera. The sphenethmoid is not known in whatcheerids or *Eoherpeton*, but in *Proterogyrinus* it is much less well ossified than in *Acanthostega* or the Lower Permian embolomere *Archeria* (Clack and Holmes, 1988). It is difficult to postulate either an adaptive or a phylogenetic significance for the variable degree of ossification of the braincase in early tetrapods, which appears to differ as much within as between groups.

Much more detail of the postcranial skeleton is known in *Proterogyrinus* than can be learned from either of the specimens of *Eoherpeton*, which enables us to gain a very complete knowledge of early anthracosaurs.

Proterogyrinus has 32 presacral vertebrae, compared with 30 in the equally well-known *Acanthostega* and an estimate of approximately 28 in whatcheerids. The atlas-axis complex provides a model for the pattern in all later Paleozoic labyrinthodonts (Fig. 4.18). Although a proatlas has not been identified, it is obvious from the facets for its articulation on the exoccipitals and on the atlas arch that it consisted of paired elements, resembling abbreviated neural arches, which bridge the gap between the occiput and the first cervical. The axis arch is paired and articulated loosely with the paired pleurocentra of the first cervical. The axis arch is much expanded anteroposteriorly, and fused at the midline. It articulates with a large, crescentic pleurocentrum. The intercentra of the atlas and axis are smaller, crescentic elements.

The trunk vertebrae consist of neural arches with tall, rectangular spines that articulate with large, nearly cylindrical pleurocentra. The intercentra are smaller and crescentic. The specific configuration and relative size of the vertebral centra have long been used as a major character in the classification of labyrinthodonts. Using the osteolepiform fish *Eusthenopteron* and the early tetrapod *Acanthostega* as outgroups and considering the ontogeny of living taxa, it has been generally accepted that the primitive condition for tetrapods, ontogenetically and/or phylogenetically, was for all the elements of the vertebrae to be paired. In *Acanthostega* only the atlas and the sacral intercentra are fused at the midline. In most early labyrinthodonts, all the intercentra are crescentic, but the pleurocentra remain paired. The degree of fusion of the neural arches changes ontogenetically, and some groups retain the paired condition. The pattern seen in *Pederpes*, in which the intercentra are crescentic, the pleurocentra are paired, and the arches are not fused to either centra, is termed rhachitomous and has been considered primitive for labyrinthodonts above the stage of *Acanthostega*.

Anthracosaurs, beginning with *Eoherpeton*, were derived in the fusion of the two halves of the pleurocentra. In this genus, the pleurocentra and intercentra are both crescentic

and of nearly equal size, and may have served equally for support of the neural arch. *Eoherpeton* has been placed in a separate family or even infraorder from the common European Upper Carboniferous anthracosaurs, termed embolomeres, because the later group had fully cylindrical pleurocentra. However, the intermediate condition shown by *Proterogyrinus*, which otherwise closely resembles later embolomeres, suggests a continuity between these Lower and Upper Carboniferous clades that does not support this level of taxonomic distinction.

The ribs beneath the shoulder girdle are short, but broadened distally, while those more posterior are longer and circular. The immediately postsacral ribs are longer and extend posteriorly.

Although the cleithrum is not known in *Eoherpeton*, the general proportions of the scapulocoracoid in *Proterogyrinus* suggest that it was much less extensively ossified dorsally, leaving a large gap for a cartilaginous suprascapula. The configuration of the dermal elements forms the model for later embolomeres. The interclavicle has a long stem and the clavicles retain wide ventral blades. The humerus retains the general proportions of *Tulerpeton* and the best-known element from Horton (Fig. 4.3B).

The ulna and radius are considerably more gracile than those of *Pederpes*. Only two carpals, 1 and 2, are known in *Proterogyrinus*. Four have been described in the later genus, *Archeria*, but they are not in articulation (Romer, 1957). No complete manus has been discovered in *Proterogyrinus*, but the phalangeal count of the later embolomere *Archeria* has been reconstructed as 2, 3, 4, 5, 4, on the basis of nearly complete material. The pelvis, as in all later embolomeres, has a long posterior process, as well as the dorsal process common to all primitive tetrapods. The unfinished bone surface of the acetabulum does not extend over the pubis as in whatcheerids, but accessory obturator foramina are still expressed. Sutures separating the three elements of the pelvis are retained in adults.

The shape of the femur is very similar to that of *Eoherpeton*, with the adductor ridge extending far distally, as is also the case for *Archeria*. The tibia and fibula are very much like those of *Eoherpeton* and much more gracile than those of *Pederpes*. All the proximal bones of the tarsus are ossified and include a fibulare, an intermedium, tibiale partially fused to a proximal centrale, and three distal centralia. Only the first distal tarsal is preserved. The phalangeal count is reconstructed as 2, 3, 4, 5, 5, as in *Archeria*, with uncertainties for only digits 3 and 4.

Upper Carboniferous Embolomeres

Fossils of *Proterogyrinus* are known from the very end of the Lower Carboniferous. The next sites from which have come most of the main anthracosaur lineages are at the base of the Upper Carboniferous in Great Britain. Nearly all are from deposits associated with coal seams, indicating a common environment of coal swamps and locally large areas of

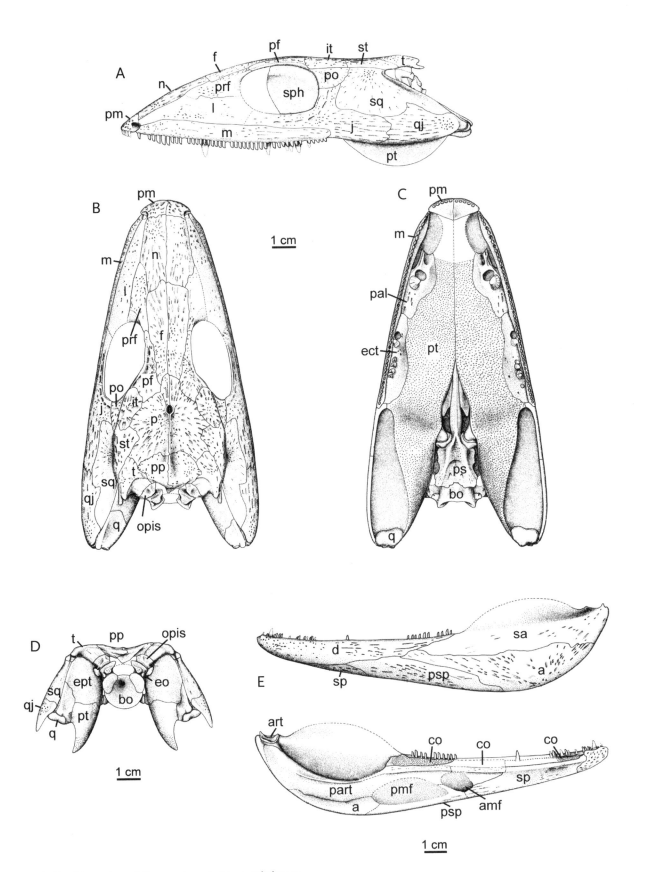

Figure 4.17. Skull and lower jaws of *Proterogyrinus scheelei*, an embolomere
from the Namurian A of Greer, West Virginia. *A–D,* Skull in lateral, dorsal,
palatal, and occipital views. *E,* Lower jaws in lateral and medial views. From
Holmes, 1984.

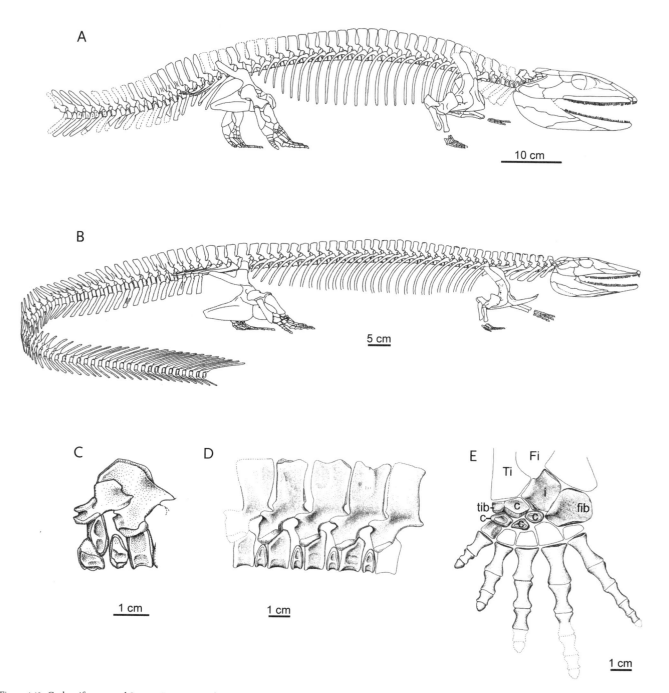

Figure 4.18. Carboniferous and Lower Permian anthracosaurs. *A*, Skeleton of *Proterogyrinus* from the Namurian A of Greer, West Virginia. *B*, Skeleton of *Archeria* from the Lower Permian of Texas. *C–E*, Cervical and trunk ver-tebrae and foot of *Proterogyrinus*. *A, C–E*, from Holmes, 1984; *B*, modified from Romer, 1957.

water, lakes, or oxbow lakes of large rivers, typically with an-aerobic bottom layers.

By this time, both the intercentra and pleurocentra have become fully cylindrical, which has long been considered the hallmark of embolomeres. However, both this and other characteristics form a continuum between the Upper and Lower Carboniferous members of this lineage.

Most of the embolomeres from the Upper Carbonifer-ous of Great Britain are known from disarticulated mate-rial, with the most distinctive element being the skull, with a considerable degree of variation in size, proportions, and dentition, indicating a variety of diets, but retaining the general configuration and number of skull bones character-istic of earlier embolomeres (Fig. 4.19). The largest known skulls were those of *Anthracosaurus*, at 38 cm in length, also distinguished by the posterior position of the orbits and the huge palatal fangs (Panchen, 1970, 1977). Our extensive knowledge of British embolomeres results from the wide

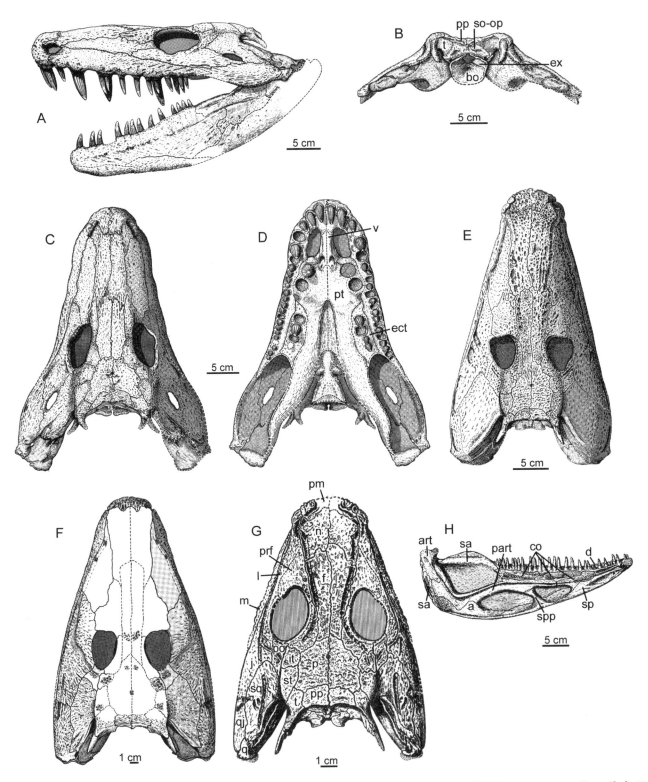

Figure 4.19. Upper Carboniferous embolomere skulls. *A–D,* Lateral, occipital, dorsal, and palatal views of *Anthracosaurus russelli.* From Panchen, 1977. Note the very large teeth on the pterygoid, in the general position of the vomerine fangs of temnospondyls. *E,* Dorsal view of *Eogyrinus attheyi.* From Panchen, 1972. *F,* Dorsal view of *Pholiderpeton scutigerum.* From Clack, 1987. *G,* Dorsal view of *Palaeoherpeton decorum.* From Panchen, 1970. *H,* Medial view of lower jaw of *Eogyrinus attheyi.* From Panchen, 1970. Note the very large openings that are characteristic of embolomeres.

extent of commercial coal deposits and the nature of mining in the 19th century, during which time much work was done by hand. Specimens could be collected as they were exposed and in heaps of discarded rocks removed while mining the coal.

Smaller, less complete remains of embolomeres were also found in what is now the Czech Republic. Coal mined near Pilzen was used in heating the National Museum in Prague but was occasionally delivered to the director's office, where it would be searched for fossils. These coals were also an important source of other Paleozoic amphibians and reptiles. Large North America embolomeres from this time include *Neopteroplax*, with a skull length of 34 cm (Romer, 1963).

Several embolomere specimens were also found in a very different preservational environment at Joggins, Nova Scotia, equivalent to the European Westphalian A. Nearly all the fossils from this deposit are preserved within the stumps of upright lycopods. This deposit will be discussed in a later section.

Lower Permian Embolomeres

The latest known embolomeres from the Lower Permian were preserved under very different conditions from those in Europe. The Texas and Oklahoma red beds where they lived included bog deposits with a rich flora within an extensive deltaic environment. The genus *Archeria* is sufficiently different from all the Carboniferous genera to be placed in a separate family, although most of the basic cranial features indicate descent from the same lineage (Fig. 4.18*B*). In one specimen, the lower jaws retained the primitive features of a parasymphyseal plate and parasymphyseal tusks on the dentary, although the coronoid fangs of *Eoherpeton* were not observed. The most striking difference from the best-known Carboniferous genus *Proterogyrinus* is in the body proportions, with approximately 40 presacral vertebrae and relatively much smaller limbs. Elongation of the trunk continues the process of aquatic adaptation begun subsequent to *Eoherpeton*.

On the other hand, the marginal teeth of *Archeria* show a unique configuration for the group. They were small and close set with chisel-shaped tips. The individual teeth closely resemble those of the adelogyrinids and one of the baphetids (a group to be discussed shortly), but the anatomy of the skull was very different in these three groups, making a common function for such teeth seem unlikely. Like other anthracosaurs, they lived only a few degrees from the Paleozoic equator.

Because of the general continuity of cranial anatomy and large size in all the genera from *Eoherpeton* to *Archeria*, all may be included in a single assemblage, the embolomeres, despite progressive increase in trunk length, relative shortening of the limbs, and variation of the dentition. This reflects the lack of evidence for any fundamental changes in the anatomy or divergence of clearly separate lineages.

Baphetidae

In common with the earliest known anthracosaur, the oldest member of a second labyrinthodont group, the family Baphetidae, also occurs at Gilmerton, where it is represented by a single skull of *Loxomma allmanni*. Baphetids are immediately recognizable by the shape of their orbits. Rather than a simple circular outline, the anterior margin extends far anteriorly to form a keyhole-shaped opening, termed the antorbital fenestra. Several hypotheses have been suggested to explain this area of expansion, including space for an electric organ or a glandular structure. Eileen Beaumont, who reviewed this group in 1977, suggested it might provide an enlarged space for the expansion of the pterygoideus muscle that would have been a major element in closing the jaw, but none of these hypotheses are subject to testing or verification.

Baphetids are also unusual in being known almost entirely by the skull. What little has been discovered of the postcranial skeleton indicates that the limbs were small, indicative of aquatic amphibians, as supported by the conspicuous lateral line canal grooves at the front of the head. Oddly, there are no grooves on the skull table, which has led to the suggestion that they swam with the back of the head out of the water and used the dorsally located squamosal notches as spiracles to inhale oxygen from the air. Preferential preservation of the skull may be attributed to its very massive construction, associated with its very large marginal and palatal teeth.

Aside from their keyhole-shaped orbits, most of the features of the skull remain primitive. *Loxomma allmanni* and *L. acuterhinus* both retain an intertemporal bone, but it is lost in more advanced genera, as illustrated by *Megalocephalus* (Fig. 4.20). Other features of the skull remain primitive throughout the history of the group. These include the presence of internarial or rostral bones, presumably a relict of stem tetrapods such as *Ichthyostega* and *Acanthostega*. The configuration of the back of the skull table is clearly more primitive than that of anthracosaurs in the small size of the tabular and its lack of contact with the parietal. On the other hand, the supratemporal appears to form a solid sutural joint with the squamosal, with little likelihood of any mobility of the cheek. Both the tabular and the posterior portion of the supratemporal overhang the temporal notch. The occipital surface, as restored in *Megalocephalus*, shows a solid sutural connection between the paroccipital process of the opisthotic and the ventral process of the tabular, in contrast with whatcheerids, in which this area remains unossified. The opisthotic, however, seems to be fused medially in both groups, as in *Acanthostega*.

The braincase is well ossified in *Loxomma*, as in *Acanthostega* and most large anthracosaurs, but unlike that of some later and smaller taxa. The palate resembles that of whatcheerids, rather than Devonian tetrapods or anthracosaurs, in the retention of large fangs on all the marginal palatal bones—the vomers, palatine, and ectopterygoid. Beau-

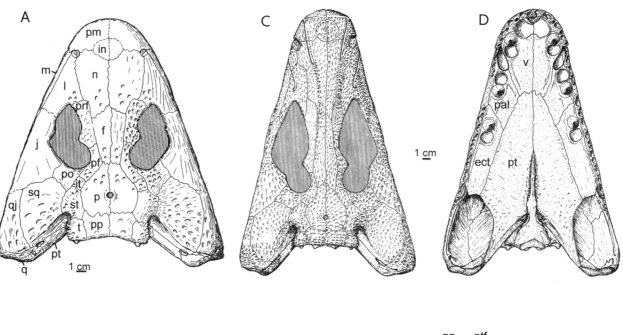

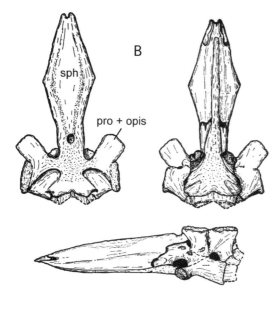

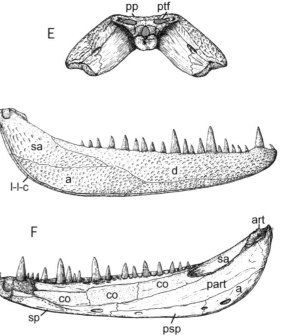

Figure 4.20. Skulls of baphetids. *A,* The primitive baphetid, *Loxomma acutirhinus,* from the Westphalian B of Scotland. *B,* Braincase of *Loxomma acutirhinus,* in dorsal, ventral, and lateral views. *C–F,* Dorsal, palatal, and occipital views of skull and lower jaws of *Megalocephalus pachycephalus* from the Westphalian A-C of Great Britain and Irish Republic. All from Beaumont, 1977.

mont (1977) suggested that the pterygoids might be movable relative to the basicranial articulation with the parasphenoid and the lateral and anterior bones of the palate in *Megalocephalus,* but in the genus *Kyrinion,* recently described by Clack (2003), the parasphenoid, pterygoid, and marginal palatal bones are united by interdigitating sutures, locking the basicranial articulation and rendering the whole skull akinetic.

The lower jaws are clearly distinct from anthracosaurs,

but resemble basal tetrapods in lacking conspicuous fenestrae on the medial surface (Ahlberg and Clack, 1998). There are very large parasymphysial fangs that fit into recesses in the anterior end of the palate. Paired parasymphysial plates are incorporated into the mandibular symphysis, as in some anthracosaurs, including specimens of *Archeria.*

Very little postcranial material has been described in association with the numerous large skulls. The most important exception is that of *Baphetes kirbyi* from the Westphalian B in the area of Manchester, England, described by Angela Milner and William Lindsay (1998). The elements are disarticulated, but closely associated with an incomplete, but well-preserved skull. Associated jaws show paired fangs and replacing pits mesial to the marginal tooth row as in *Megalocephalus*.

A single bone is identified as a ceratobranchial, suggestive of a persistently aquatic form, although it is not grooved as are those of *Acanthostega*. Of the vertebral column, several scattered and poorly ossified crescentic intercentra are recognized. As in other primitive tetrapods, they are wedge-shaped in lateral view. One haemal arch has been identified, but there is no trace of pleurocentra.

The description makes no reference to the scapulocoracoid, suggesting its low degree of ossification relative to the dermal elements of the shoulder girdle that lie close to the back of the jaw. The clavicle is said to have a narrow area of ventral expansion, which is deeply sculptured, as is the skull. The cleithrum is long and narrow. The stem has a thin flange about halfway along its length that is compared with the postbranchial lamina described in *Acanthostega* and that was also identified in *Greererpeton* by Lebedev and Coates (1995). This is a very primitive feature that would place baphetids near the base of the tetrapod stem. The interclavicle is known for both *Megalocephalus* and *Spathicephalus*, in which it is rhomboidal as in temnospondyls, rather than having a long stem as in anthracosaurs. The humerus is the most informative element of the limb bones. Its length of 8 cm is only about 25% of the roughly 32 cm associated skull. As in stem tetrapods, the proximal and distal areas of expansion show only about 30 degrees of torsion compared with 90 degrees in advanced labyrinthodonts. The anterior crest is extensive, but with little evidence for a distinct suppinator process. *Baphetes* is advanced over the most primitive tetrapods in having only a single entepicondylar foramen, in contrast with several foramina in this area in *Acanthostega*. Milner and Lindsay (1998) compare the nearly complete radius with those of *Greererpeton* and embolomeres, but it provides no evidence of specific affinities.

The pelvic girdle retains the clearly distinct dorsal and posterodorsal processes of the ilium, which closely resemble those of Devonian tetrapods and anthracosaurs. A disarticulated ischium was reported but not the pubis. No femur was found; the tibia and fibula were both intact but not taxonomically informative.

Elements from the pes include a more or less articulated digit that may have had a phalangeal count of 5, suggesting affinities with stem tetrapods and anthracosaurs, but not with temnospondyls. Elements that were identified as possible metacarpals or metatarsals were mentioned, but not carpals or tarsals, suggesting the low degree of ossification associated with aquatic labyrinthodonts.

Beaumont (1977) recognized four genera of baphetids, not counting *Spathicephalus* (described below), to which Clack (2001, 2003) has added two more. *Loxomma acutirhinus*

provides the most informative basis for comparison of the cranium with plausible antecedents, while *Baphetes kirkbyi* provides the best available information on the postcranial skeleton.

Other baphetid species, also discussed by Beaumont, are known from the Namurian to the end of the Westphalian in North America as well as Great Britain and Czechoslovakia. Most retain the basic features of the earliest known genus. However, early in this radiation there also evolved a more specialized lineage known only from a single genus, *Spathicephalus*, characterized by a very broad and flattened skull, up to 22 cm in length (Fig. 4.21). The least distorted skull is only 2.1 cm high. The orbits are very close to the midline, widely separating the pre- and postfrontal bones. The posttemporal fossa, present in more primitive baphetids, is apparently closed. The dentition differs greatly from that of any other baphetid, consisting of very close-set chisel-shaped marginal teeth, superficially resembling those of adelospondyls, but without large palatal fangs. Beaumont and Smithson (1998) suggest that they were sluggish bottom-dwelling filter-feeders of small, soft invertebrates.

Spathicephalus mirus is known from 10 specimens from three areas in Scotland, the Ramsay and Burghlee collieries (both tapping the Burghlee or Rumbles Ironstone), Loanhead District near Edinburgh, and the Dora Bone Bed, Cowdenbeath, Fife, both at the base of the Namurian. A second species, *S. pereger*, is represented by a partial skull showing the distinctive orbital region, from beds of similar age in Nova Scotia, Canada. This specimen, described by Donald Baird (1962), is distinguished in having the reticulate dermal ornamentation of most baphetids, rather than the pustular ornament of *S. mirus*.

Beaumont (1977) argued that *Spathicephalus mirus* should be placed in a separate family, Spathicephalidae, within the superfamily Loxommatoidea. This may appear justifiable on the basis of the degree of morphological distinction, but would render the "parental" family Loxommatidae (now Baphetidae) paraphyletic (that is, an ancestral group artificially separated from its immediate descendants).

Eucritta

Another genus, *Eucritta melanolimnetes* (meaning "the creature from the black lagoon"), from the late Viséan locality of East Kirkton, was assigned to the Baphetidae by Jennifer Clack (2001, 2003). It is known from five specimens, including a nearly complete skeleton, in which the anterior margin of the orbit is extended in a similar direction to the antorbital fenestra of *Loxomma, Baphetes,* and *Megalocephalus,* but to a much lesser extent (Fig. 4.21). In contrast with previously described baphetids, it is known only from small specimens. The type consists of a nearly complete skeleton, 14 cm in length, but lacking the tail. This is only about one-half the length of the skull alone in most other genera. The skull table of *Eucritta* resembles that of primitive baphetids in retaining an intertemporal and a small tabular that does

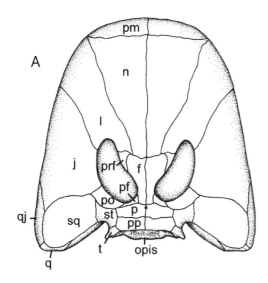

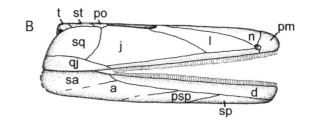

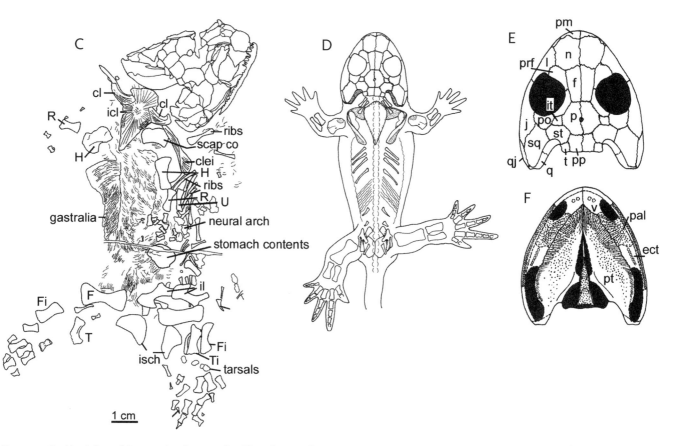

Figure 4.21. Spathicephalus and *Eucritta*. A and B, Dorsal and lateral views of the skull of the aberrant baphetid *Spathicephalus mirus,* from the Namurian A, near Edinburgh. From Beaumont and Smithson, 1998. C–E, the putative baphetid *Eucritta,* from the late Viséan of East Kirkton, near Edinburgh. Skeleton as exposed, reconstruction, reconstruction of the skull in dorsal and palatal views. From Clack, 2001.

not reach the parietal. The bones of the appendicular skeleton also have comparable dimensions, relative to the skull, as those observed in *Baphetes kirkbyi,* the only other baphetid to have significant postcranial material. Very little is known of the vertebrae. In the type, a longitudinal series of diaphanous shapes probably represent the neural arches. There may have been approximately 24 trunk vertebrae, judging by the

number of ribs and the distance between the head and the pelvic girdle.

Two anatomical features of the postcranial skeleton appear to distinguish *Eucritta* from other genera assigned to the Baphetidae. The interclavicle of the type has a relatively long posterior stem, giving the bone as a whole the shape of a kite, similar to those of whatcheerids and anthracosaurs, but the interclavicles of *Megalocephalus* and *Spathicephalus* are rhomboidal. The pes of *Eucritta* is illustrated as having a phalangeal count of 2, 2, 3, ?4, ?, resembling that of temnospondyls (to be described shortly along with the other elements of the East Kirkton fauna) but *Baphetes kirkbyi* appears to have one digit of the pes with five phalanges, a characteristic of *Tulerpeton* and anthracosaurs.

The most critical question regarding the affinities of *Eucritta* is the possible presence of an antorbital fenestra homologous with that of typical baphetids. *Eucritta* is certainly smaller than most other baphetids, with a skull length of approximately 10 cm in the largest known specimens but lengths of more than 30 cm in *Baphetes*. From this, one might suggest that the configuration of the antorbital fenestra changed during ontogeny and that a fully grown specimen of *Eucritta* would resemble *Loxomma* or *Baphetes*. This, however, is apparently refuted by a specimen of *Baphetes lintonensis* from the Westphalian D of Ohio, described by Eileen Beaumont (1977), in which a skull less than 8 cm in length has a fully formed fenestra. The homology of the slight extension of the anterior margin of the orbital opening with the typical baphetid antorbital fenestra may also be questioned in association with the palatal dentition of *Eucritta,* which differs from all baphetids except the highly aberrant *Spathicephalus* in lacking large palatal fangs. If Beaumont's hypothesis is correct—that extension of the orbital margin was necessary to accommodate enlargement of the pterygoideus muscle for driving the huge palatal fangs into the prey—it makes no sense to elaborate such an opening in the short-skulled *Eucritta,* which apparently has no fangs at all.

There seems adequate evidence to question the close relationship of *Eucritta* with baphetids. However, it might be viewed as a smaller-sized sister-taxon of baphetids that had persisted from a more primitive morphotype that diverged prior to known baphetids. The other possibility is that *Eucritta* represents another isolated taxon, like *Caerorhachis* and *Acherontiscus* (to be described shortly), known from only a single horizon, whose specific affinities cannot be established on the basis of our current knowledge of the fossil record. *Eucritta* otherwise expresses a mosaic of characters, some of which resemble stem tetrapods, others that are suggestive of anthracosaurs, and some that are common to temnospondyls.

The origin and relationships of baphetids remain uncertain, whether or not *Eucritta* is included. Broadly, they are characterized by features that are either uniquely derived characters (autapomorphies), such as the anterior extension of the orbit and the extremely massive skull that clearly distinguish them from all other early tetrapods, or a host of primitive characters (plesiomorphies) that suggest an origin near the base of tetrapod radiation.

Crassigyrinus

In addition to the oldest known anthracosaur, *Eoherpeton,* and the earliest baphetid, *Loxomma,* a third genus from Gilmerton, *Crassigyrinus,* represents yet another highly distinctive lineage. In contrast with the many anthracosaurs and numerous baphetids, *Crassigyrinus* is the only recognized representative of a family of its own, Crassigyrinidae. In addition to Gilmerton, near the top of the Viséan, it also occurs in the lower Namurian within a bone bed in the Dora opencast site near Cowenbeath, Fife. Both localities are near Edinburgh. A number of ribs of similar size to those of *Crassigyrinus* and bearing similar striae were also reported by Stephen Godfrey (1988) from the Greer locality, in West Virginia, nearly contemporary with the two Scottish localities and also accompanied by a primitive embolomere and a colosteid. *Crassigyrinus* provides a further, especially striking example of an isolated lineage with no obvious shared derived characters that link it with any other group of Paleozoic amphibians (Figs. 4.22 and 4.23).

Our knowledge of *Crassigyrinus* results primarily from the research of Alec Panchen, who not only prepared and described previously recognized specimens, but also spearheaded a major excavation and collecting effort at Cowenbeath that resulted in the discovery of new material that provided knowledge of almost the entire skeleton (Panchen, 1973, 1985; Panchen and Smithson, 1990; Clack, 1998a).

Crassigyrinus is immediately impressive in terms of its great size. The skull alone is 30 cm in length and the skeleton to the base of the tail approaches 90 cm. The tooth row extends for almost 20 cm and the palatal dentition includes huge fangs. *Crassigyrinus* was certainly the most ferocious freshwater predator of its time. It was undoubtedly tied to an aquatic way of life, for the forelimbs are ludicrously short—so short that the humerus (Fig. 4.23) was initially identified as a stapes! The rear limbs, however, were long enough to serve for paddling and steering. Many elements of the postcranial skeleton, including the pleurocentra, scapulocoracoid, pubis, carpals, and tarsals, are presumed by their absence to have remained cartilaginous, which would have increased buoyancy in the water. The great relative size of the orbits in such a large skull may have been associated with huge eyes, adapted, like those of ichthyosaurs, to gaining as much light as possible in deep, dark waters.

The classification of *Crassigyrinus* rests primarily with the skull, which is very well known. What has long impressed paleontologists is the clear distinction of the skull table from the cheeks that results in a configuration that superficially resembles that of anthracosaurs. However, comparison can also be made with *Eusthenopteron* and *Panderichthys* (Fig. 3.3), in which there is a persistent line of mobility that includes the spiracular cleft, and like *Crassigyrinus,* retains the primitive pattern of dermal bones at the back of the skull table. The tabular remains small and is separated from the small parietal by an extensive supratemporal bone. In contrast, the orbits of *Crassigyrinus* are very close to one another and the postfrontals apparently meet at the midline, separating the frontals from the parietal. *Crassigyrinus* does resemble anthracosaurs in the elaborated ventral surface of

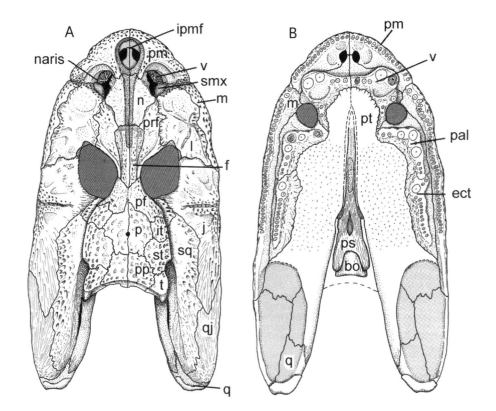

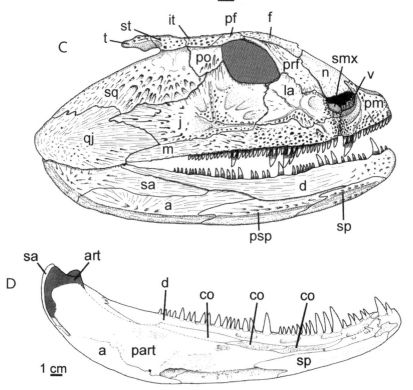

Figure 4.22. A–C, Skull of *Crassigyrinus* from the upper Viséan and lower Namurian of Scotland, in dorsal, palatal, and lateral views. From Clack, 1998a.
D, Medial view of lower jaw. From Ahlberg and Clack, 1998.

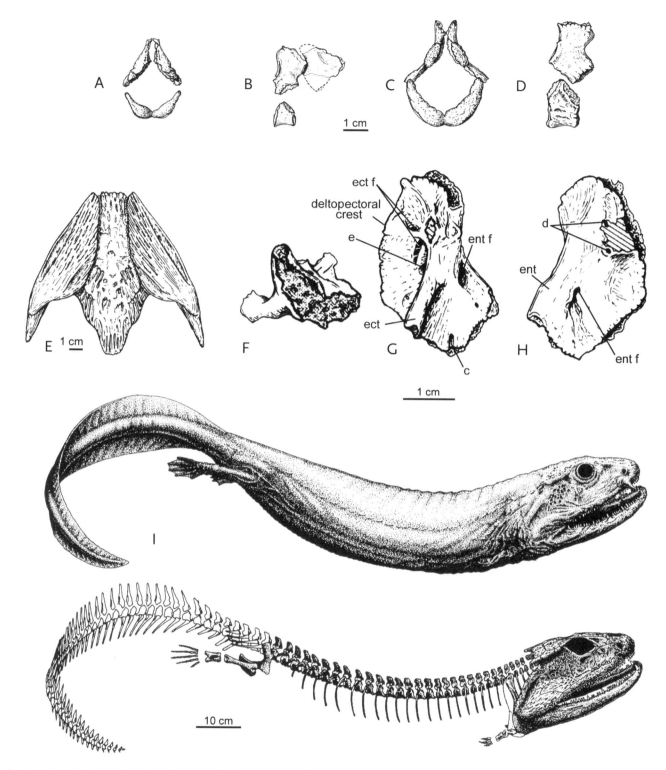

Figure 4.23. Crassigyrinus scoticus. From Panchen, 1985; Panchen and Smithson, 1990. *A*, Paired atlas arch and weakly fused left and right halves of intercentrum in anterior view. *B*, Lateral view of atlas and axis arches. *C*, *D*, 14th presacral vertebra in anterior and lateral views. Note absence of posterior zygapophyses and paired nature of arch. *E*, Ventral view of dermal shoulder girdle. *F–H*, Proximal, dorsal, and ventral views of left humerus. *I*, Flesh and skeletal reconstructions of *Crassigyrinus*. The tail is conjectural. c, d, and e formamina not found in other Carboniferous tetrapods but recognized by Jarvic (1980) in *Ichthyostega*.

the tabular, suggesting at least a cartilaginous attachment with the otic capsule. Clack (1998a) was unable to find a separate preopercular ossification, as had been postulated by Panchen (1985).

In contrast with the skull table, the anterior portion of the skull is sui generis. The nares face dorsally as well as laterally, and there is a large median opening, the interpremaxillary fenestra, in the position of the paired internasal bones of baphetids. Its lateral margins extend far posteriorly, to the level of the anterior extremities of the very large orbits. A similar median groove or gap in the antorbital portion of the dermal skull was also reported in *Acanthostega* by Clack (2003).

The jaw articulation extended far behind the occiput to give the mouth an extremely wide gape. In contrast with Panchen's 1985 reconstruction of the palate, which indicated an embolomere like narrowing of the vomers, Clack's 1998a reconstruction, following additional preparation, shows very large vomerine fangs and transverse rows of smaller teeth both anterior and posterior to the internal nares. The configuration of this region somewhat resembles that of *Eusthenopteron* (Fig. 2.14C).

The basicranial articulation, formed by the basisphenoid and extensions from the parasphenoid, resembles that of anthracosaurs, suggesting a similar pattern of cranial kinesis. The epipterygoid, as in baphetids and primitive anthracosaurs, remains a large element situated above the pterygoid. It is associated with the basicranial articulation and extends dorsally toward the skull roof and posteriorly toward the quadrate. The cultiform process of the parasphenoid remains free from the pterygoid, in contrast with advanced baphetids. The occipital surface is poorly known, but the basioccipital is ossified only ventrally, suggesting that there remained a fishlike notochordal connection between the skull and the trunk. The exoccipitals have not been identified, but areas for their articulation are evident on the dorsolateral surfaces of the basioccipital. They may, as in baphetids, have been only loosely attached to the otic capsule.

As shown by Ahlberg and Clack (1998), the lower jaws are primitive in the presence of a long ventral slit between the prearticular and the postsplenial and an elongate prearticular, placing them just above the phylogenetic level of *Acanthostega*. The medial margin of the dentary has a series of recesses that presumably accommodated the tips of the large palatal fangs. While the absence of pleurocentra may be attributed to their having been unossified, this is also a primitive characteristic of much of the vertebral column in *Acanthostega*. In addition, the intercentra are constricted medially, as if they were still in the process of fusion from a primitively paired condition. The neural arches are also paired and the prezygapophyses remain ill defined. Postzygapophyses appear to be absent. There are approximately 31 presacral vertebrae.

Numerous ribs are associated with the Cowdenbeath skeleton. Most are circular or elliptical in cross section and long

enough to curve around the trunk, as are the ribs of anthracosaurs, but the heads of the ribs and the surface of the intercentra provide little surface for attachment. The thoracic ribs are shorter and somewhat expanded distally.

The dermal bones of the shoulder girdle resemble those of other labyrinthodonts in their conspicuous sculpturing and the wide expansion of the clavicular blades. The extent of expansion of the blade of the cleithrum is not known and an anocleithrum has not been recognized. The interclavicle is kite-shaped, with a widely triangular posterior process. The humerus (Fig. 4.23) is highly distinctive in its small size, very incomplete ossification, and presence of more than one foramina, as is also the case in *Ichthyostega* and *Acanthostega*. It differs greatly from those of other Paleozoic tetrapods in general appearance, but has most of the basic structures common to other early labyrinthodonts: ectepicondylar ridge, entepicondyle, anterior ridge, and limited torsion of the proximal and distal areas of articulation. The radius and ulna are comparable to those of other primitive tetrapods.

The pelvic girdle resembles that of *Acanthostega*, whatcheerids, and anthracosaurs in there being both dorsal and posterior processes of the ilium. The area of the acetabulum is not well defined. The pubis is not ossified, and the ischium is not firmly sutured to the ilium. The femur, tibia, and fibula are not well ossified, but follow the general pattern of other primitive tetrapods. Bones that may be metatarsals and/or phalanges are preserved with the skeleton, but neither a digital nor a phalangeal count can be restored.

Panchen (1985, 550) described *Crassigyrinus* as "a uniquely primitive and aberrant amphibian." On the other hand (562) he argued that crassigyrinids were the sister-group of anthracosaurs. It is informative that *Crassigyrinus* and *Eoherpeton*, the oldest known anthracosaurs, occur first in the same horizon and locality (Gilmerton), approximately 30 million years after the end of the Devonian. The recovery of their remains, along with those of the oldest known baphetid, may be indicative of an unusual environment of deposition and preservation, not yet encountered earlier in the Carboniferous.

Most, if not all, of the characteristics of *Crassigyrinus* appear to be either primitive for tetrapods or unique to this genus. The loss of the preopercular bone might be thought of as an advanced character shared with most Carboniferous tetrapods, suggesting divergence above the level of whatcheerids, but considering the diversity of other Carboniferous lineages that lack this bone, its loss can be readily attributed to convergence.

Crassigyrinus may share more similarities with primitive anthracosaurs (for example, *Eoherpeton*) than any other labyrinthodont groups (whatcheeriids, colosteids, baphetids, temnospondyls, and *Caerorhachis*), but few if any are *unique* shared derived characters. Another approach to establishing the ancestry of crassigyrinids is to note the number of traits of *Crassigyrinus* that are more primitive than all these groups: lack of postzygapophyses, incomplete medial fusion of intercentra, absence of ossified pleurocentra, retention of notochordal articulation between the column and the skull, nature of the lower jaw, and extremely primitive humerus.

All these features suggest that the *Crassigyrinus* lineage diverged prior to any of the other known Carboniferous families. More specific affinities cannot be determined without evidence from new discoveries of Paleozoic tetrapods. Until that time, the relationship of *Crassigyrinus* can be considered in terms of the Scottish judicial system (applicable in the case of a fossil known only from Scotland) as "unproven," in contrast to being either guilty or innocent of particular relationships.

EAST KIRKTON FAUNA

We now jump from some of the deepest water deposits with amphibian remains to an essentially terrestrial assemblage, but without leaving Scotland. For most of the Lower Carboniferous, the fossil record consists of temporally and geographically isolated localities, each of which has yielded only a small number of taxa with little evidence of their phylogenetic or ecological relationships with one another. Finally, at the very end of the Viséan, near the top of the Brigantian, we find a real community of vertebrates, aquatic and terrestrial (Plate 5). This locality occupies the East Kirkton limestone quarry, last worked commercially prior to 1844, near the town of Bathgate, 27 km west of Edinburgh. A few fossils were collected during the intervening years, but the locality was effectively rediscovered by a very knowledgeable commercial collector, Stan Woods, in 1984. The locality was subsequently studied in great detail by groups working with the Royal Scottish Museum in Edinburgh, leading to an extensive report published in the Transactions of the Royal Society of Edinburgh (Rolfe et al., 1994).

The East Kirkton locality is unique for the Carboniferous in being associated with volcanic activity. This area of Scotland was then located just south of the equator in a zone of continental rifting, accompanied by the formation of a number of cinder cones that contributed ash falls to the area of a small lake in which limestone and shale beds were gradually accumulating. Hot water springs may have affected the temperature of the lake, although the preservation of the fossils does not seem to have been directly related to geothermal activity.

The best-preserved fossils, including nearly complete, articulated specimens, have come from a single layer of black shale in unit 82, near the bottom of the sedimentary sequence. Other, less well-preserved skeletons and isolated bones are known from the base of the quarry up to unit 52, while many fish, but no tetrapods, are known in beds 24 to 37. This seems to have been a small lake, not closely connected to other, larger bodies of water, since no fossils have been found of large aquatic tetrapods, such as whatcheeriids, embolomeres, or *Crassigyrinus.* What is known of the tetra-

pods suggests that most were more terrestrial in habits than those found in earlier beds. Many had well-developed limbs, and none are known to have possessed lateral line canals. A variety of terrestrial invertebrates are also known, as well as a rich assortment of land plants.

Six distinct tetrapod taxa have been found at East Kirkton. One is an aïstopod, *Ophiderpeton kirktonense,* clearly related to the much earlier *Lethiscus* on the basis of a long vertebral column consisting of holospondylus vertebrae, but no evidence of limbs. A total of five specimens have been recognized. The skull is poorly known, but the orbits are far forward and there is a large opening in the cheek region. Angela Milner (1994) argued that the absence of girdles and limbs points to a snakelike mode of locomotion, most effective on land. Better-known aïstopods will be discussed in more detail later in this chapter, together with the fauna of a number of coal swamp localities. A further genus at East Kirkton, *Eucritta,* was discussed with the baphetids.

East Kirkton Anthracosaurs

Two additional East Kirkton genera have been recognized as anthracosaurs on the basis of the large size of the tabular and its contact with the parietal, but both differ significantly from members of the embolomere lineage described earlier in this chapter (Clack, 1994; Smithson, 1994). Both are considerably smaller, and, while not fully ossified, may be close to adults. In neither is the skull well preserved, but *Eldeceeon* (Fig. 4.24) provides the better overall comparison with previously described anthracosaurs, of which *Eoherpeton* is the most similar.

The two fairly well-articulated specimens were preserved on limestone blocks but do not show good surface detail. Nevertheless, the outlines of most bones can be determined. The most striking difference from the anthracosaurs previously described is the body size of approximately 35 cm total length, about one-quarter the length of *Proterogyrinus,* but the limbs are much more robust. In addition, the number of trunk vertebrae in *Eldeceeon* is approximately 25, while that of *Proterogyrinus* is 32. These proportions suggest that the East Kirkton genus was more terrestrial in its habits than any of the embolomeres, but might have diverged from earlier members of that lineage. The vertebrae most closely match those of *Eoherpeton,* with horseshoe-shaped intercentra and pleurocenta, but the intercentra are relatively smaller. Ribs appear to be missing in the lumbar region in both specimens.

The stem of the interclavicle is somewhat more narrow than that of embolomeres. The scapulocoracoid and pelvis generally resemble those of *Proterogyrinus,* although the pubis may have been unossified. Little detail of the long bones is visible. Most of the tarsals are ossified but not articulated. The carpals were not ossified. The phalangeal count for both the manus and the pes is 2, 3, 4, 5, 4. Of the animals we have seen so far, only anthracosaurs share a similar count.

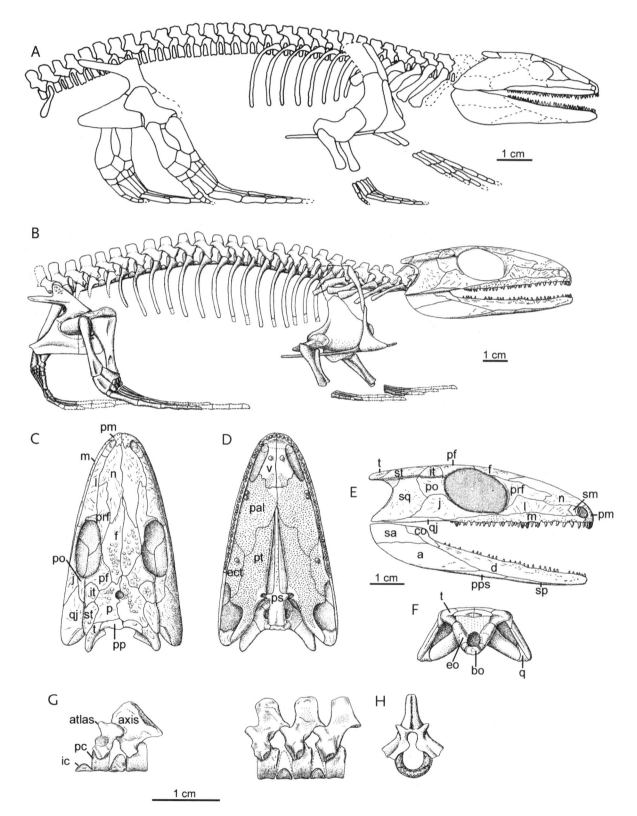

Figure 4.24. Small, terrestrial anthracosaurs from the upper Viséan of East Kirkton and the Westphalian D of the Czech Republic. *A, Eldeceeon rolfei* from East Kirkton. From Smithson, 1994. *B–H, Gephyrostegus bohemicus.*

From R. L. Carroll, 1970a. *C–F,* Skull of *Gephyrostegus* in dorsal, palatal, lateral, and occipital views. *G, H,* Cervical and trunk vertebrae in lateral and anterior views.

Clearly, *Eldeceeon* had a way of life very different from that of the larger embolomeres, although details of skeletal anatomy are similar.

Silvanerpeton, also known from two specimens, is even less well preserved than *Eldeceeon*. One specimen is from the black shale of unit 82, but the surface of the bone is split between the two counterparts. The other is preserved in a limestone block and shows very little surface detail. Neither carpals nor tarsals are ossified. The individual vertebrae resemble those of *Eldeceeon* but there are possibly more than 30. The pleurocentra are in the form of almost complete hoops and the intercentra as shallow crescents. One anatomical detail that is clearly visible is the phalangeal count of the pes: 2, 3, 4, 5, 5. This is one more phalanx on the fifth digit than that of *Eldeceeon,* but it matches that of the few embolomeres in which the pes is adequately known. There are a maximum of 42 teeth in the maxilla, compared with 35 to 40 in *Eldeceeon.* Denticles are visible on the coronoids, but not the larger teeth seen in other early anthracosaurs. In contrast with all earlier anthracosaurs except *Eoherpeton, Silvanerpeton* lacks a large opening on the inside surface of the lower jaw. In contrast with *Eldeceeon,* the stem of the interclavicle is wide, as in other anthracosaurs. As with *Eldeceeon,* the most important characters are the small size of these specimens and their relatively large limbs compared with the more aquatic embolomeres. Clack (1994) described *Silvanerpeton* as being apparently less terrestrial in its adaptation than *Eldeceeon,* based on the apparently cartilaginous nature of the coracoid and pubis and the lack of ossification of the mesopodials, but also notes that no lateral line canals are visible on the surface of any dermal bones of the skull.

While *Eldeceeon* and *Silvanerpeton* can certainly be distinguished at the generic level, they are sufficiently similar to one another and clearly distinct from all the previously described embolomeres to suggest that they belong to a separate anthracosaurian lineage. The greatest similarity of these two genera lies with the most primitive known anthracosaur, *Eoherpeton,* in the retention of primitive features of the skull and vertebrae, but proportionately large limbs. This suggests a common ancestry of these two lineages sometime below the Brigantian. The primary distinctions are the subsequent adaptation of the embolomeres to a more aquatic way of life, accompanied by increase in the length of the trunk and reduction in the relative length of the limbs, compared with overall size reduction in the lineage leading to *Eldeceeon* and *Silvanerpeton.*

Later Gephyrostegids

Taxa from a series of subsequent horizons appear as a logical continuation of the group including *Eldeceeon* and *Silvanerpeton.* These include specimens from the Namurian B of Wupperfal, Germany; Westphalian A of Joggins, Nova Scotia; and Westphalian D, of Linton, Ohio, and Nýřany, in the Czech Republic (Plate 8).

The best known of these genera is *Gephyrostegus,* from Nýřany (Fig. 4.24B–H), which is the highest in the stratigraphic sequence (R. L. Carroll, 1970a). Several specimens are known, representing nearly all of the anatomy. The configuration of the bones at the back of the skull table resembles that of *Eoherpeton,* indicating close affinities within the Anthracosauria, although the skull is relatively smaller and somewhat lower and more elongate. As a result of the small size of the skull, the frontals enter the margin of the orbits. The squamosal notch is somewhat deeper. Although a ventral extension of the supratemporal reaches the posterior margin of the squamosal, there does not seem to be a strong union between the skull table and the cheek. A more important distinction from previously discussed anthracosaurs is the retention of two vomerine fangs, as large as those extending from the palatine and ectopterygoid. This is a key feature in which *Gephyrostegus* and presumably *Eldeceeon* and *Silvanerpeton* are more primitive than the earliest known embolomeres, and are hence thought to have diverged at an earlier time. Other primitive features of *Gephyrostegus* can be seen in the lower jaw (Ahlberg and Clack, 1998). It has a parasymphyseal tusk, as in some other anthracosaurs, but lacks the parasymphyseal bone. A large tooth as well as small denticles are retained on one or another of the coronoid bones. Most important, *Gephyrostegus* lacks the two large Meckelian fossae present in most embolomeres, but not *Eoherpeton.*

The intercentra of the vertebrae are apparently more primitive, or at least less extensively ossified, than in *Eoherpeton,* forming a very low crescent. Both have horseshoe-shaped pleurocentra. There are 24 presacral vertebrae, lower than the number in any amphibian yet described. The absence of lateral line canals and the high degree of ossification of the girdles and limbs (except for the carpals) suggest a primarily terrestrial mode of locomotion. This is further accentuated by the partial fusion of the intermedium, tibiale, and proximal centrale into an analogue of the amniote astragalus.

The interclavicle is kite-shaped, as in other anthracosaurs. The humerus has a smaller entepicondyle and less robust anterior crest than either *Eoherpeton* or *Proterogyrinus,* suggesting a more agile limb. The pelvis differs from that of embolomeres in the relative shortness of the posterior process. The femur is robust, but the adductor ridge is shorter than that of early embolomeres. The phalangeal formula of the manus is 2, 3, 4, 5, 3+ and that of the pes, 2, ?, ?, ?, 5. The five phalanges in the fifth digit match those of the few embolomeres in which it is known, as well as *Silvanerpeton* but not *Eldeceeon.*

A second genus that is probably closely related to *Gephyrostegus* is *Eusauropleura* from Linton Ohio, like the former genus, preserved in a coal shale that was probably deposited in an oxbow lake. It lacks a skull, but the proportions of the trunk appear very similar. The manus has a phalangeal count of 2, 3, 4, 5, 4. *Eusauropleura* is less well ossified in the absence of the pubis.

The next older specimen that might be related to *Gephy-*

rostegus is a single lower jaw from Joggins, Nova Scotia, a locality that will be discussed in more detail in a later section.

The oldest of the small-sized anthracosaurs subsequent to those from East Kirkton is *Bruktererpeton,* from the Namurian B. It is known from an almost complete skeleton, found in a marine, plant-bearing shale (Boy and Bandel, 1973). The proportions are very similar to those of *Eldeceeon,* with 24 presacral vertebrae and well-developed limbs. The skull is somewhat shorter compared with the length of the vertebral column. The ulna, radius, tibia, and fibula are very long relative to the expansion of the proximal and distal area of articulation, suggesting adaptation to rapid terrestrial locomotion, but the proximal tarsals are not integrated in the manner of *Gephyrostegus.*

All of these small terrestrial amphibians may be grouped in a single family, the Gephyrostegidae, which is thought to have diverged from the same ancestral lineage as the embolomeres. Both may be included in the order Anthracosauria.

Westlothiana

The anthracosaurs have often been classified within a larger but ill-defined group, the reptiliamorphs, which have been thought to include the ancestors of reptiles, or more inclusively, the amniotes. Gephyrostegids were previously suggested as being especially close to the ancestry of reptiles, but they lack many definitive features of that group and the best-known genus only occurs long after the appearance of unquestioned amniotes. A further taxon in the East Kirkton fauna, *Westlothiana,* has many more amniote features, but is an improbable ancestor for other reasons.

Westlothiana is known from several specimens, from which much of the skeleton can be reconstructed (Smithson et al., 1994) (Fig.4.25). The skull broadly resembles that of gephyrostegids, although it is much smaller, with a length of approximately 2 cm. The arrangement of bones at the back of the skull can be readily derived from that of anthracosaurs, but not that of any other early tetrapods. The much widened parietal articulates with the enlarged tabular, but there is no trace of the intertemporal bone. The tabular overhangs the back of the squamosal for a short distance, but there is no obvious squamosal notch. The supratemporal, as in early amniotes, extends as a narrow splint lateral to the tabular. It is probable that the skull table was not closely integrated with the cheek. Canine teeth, which are a diagnostic feature of the oldest known amniotes, are not developed in *Westlothiana.* The sclerotic plates that surround the eye are very conspicuous, but they were probably present in most early tetrapods, although rarely preserved.

The palate is less well known than the skull roof, but fangs are not evident on any of the lateral bones. This is an advance over all labyrinthodonts. There is also no sign of labyrinthodont enfolding of the dentine, which is indicated by a superficial wrinkling of the enamel at the base of the tooth in large labyrinthodonts. However, it is not expressed in small or larval labyrinthodonts, but only appears in large adult forms. Hence, its apparent absence in *Westlothiana* does not preclude close affinities with anthracosaurs. The pterygoid retains the configuration of typical anthracosaurs in the absence of a transverse flange, which is a diagnostic feature of early amniotes (Fig. 4.41).

Unfortunately, the occipital surface of the skull is not exposed in any specimen, precluding the identification of several features that might be diagnostic of early amniotes. The lower jaw is only exposed in lateral view, which shows no features by which amniotes can be distinguished from gephyrostegids.

Amniotes show a diagnostic configuration of the anterior cervical vertebrae, but their structure is not known in *Westlothiana.* The trunk vertebrae, in contrast, are very similar to those of undeniable amniotes. Unlike those of any of the previously described labyrinthodonts, the pleurocentra have fused to the neural arches and the intercentra are small, crescentic structures. Individually, they would not be distinguishable from those of specific early amniotes. On the other hand, the number of presacral vertebrae, revealed by articulated skeletons, is greater than that of any Paleozoic reptiles—36. The highest number among early amniotes is 32 in *Paleothyris* (R. L. Carroll, 1969b). The elongation of the truck is further accentuated by the relative shortness of the limbs, especially the forelimbs.

Little detail is known of the scapulocoracoid, although the internal surface shows that the two elements were ossified without a suture. The blade of the clavicle is much wider than in early amniotes. The humerus, in contrast, is nearly identical to that of the oldest known amniote, despite its small size (Fig. 4.40E). Of special significance are the presence of a long shaft between the proximal and distal areas of expansion, the loss of most of the anterior ridge, and the presence of an incipient supinator process at its distal extremity. The ulna, bearing a conspicuous olecranon, and the radius have slender, elongate shafts. Several carpals are ossified, but disarticulated. The manus cannot be restored.

The pelvic girdle can be immediately differentiated from that of anthracosaurs in the reduction of the dorsal process of the ilium to a small protuberance on the narrow, posteriodorsally directed iliac blade. All three elements are well ossified, with only a trace of sutures along their lines of contact, but the ischium is unusually long, as was that of *Eoherpeton.* In isolation, the femur appears long and robust, but it is only one-half the length of the skull, compared with the earliest known amniote, *Hylonomus,* in which the femur is equal in length to that of the skull. The tibia and fibula have approximately the proportions of those of *Gephyrostegus.* The tarsus is disarticulated but can be restored in accordance with the pattern of gephyrostegids, not early amniotes, with a persistent separation of the intermedium, tibiale, and proximal centrale (Fig. 4.40). Elements of the pes are present on

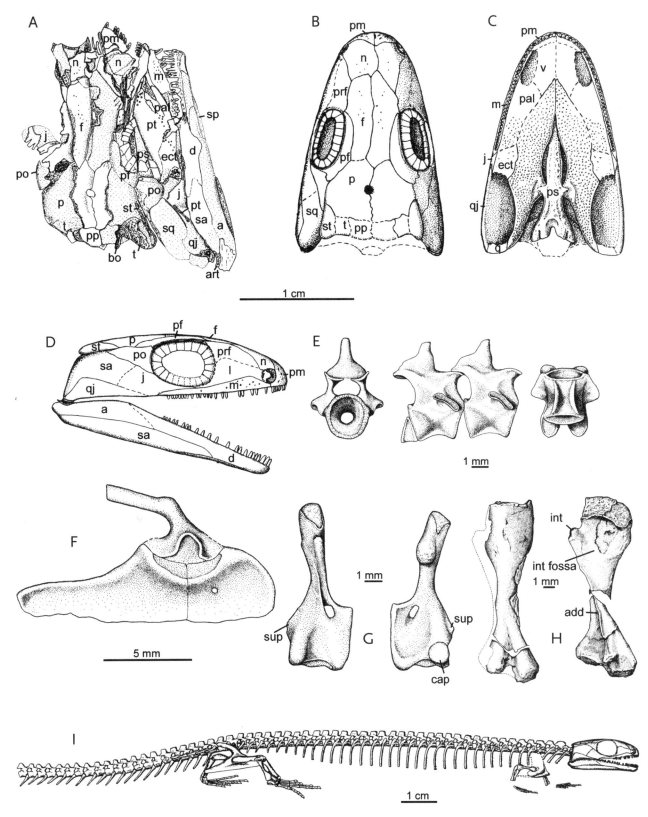

Figure 4.25. Westlothiana, a possible reptiliomorph from the upper Viséan of East Kirkton. From Smithson et al., 1994. *A*, Skull as preserved; uniform stippling indicates impression of missing bones. *B–D*, Reconstruction of skull in dorsal, palatal, and lateral views. *E*, Reconstruction of trunk vertebrae in anterior, lateral, and ventral views. *F*, Pelvic girdle in lateral view; note remnant of anterior dorsal process of ilium. *G*, Left humerus in dorsal and ventral views. *H*, Right femur in dorsal and ventral views. *I*, Reconstruction of skeleton.

both sides in the type and reveal a phalangeal count of 2, 3, 4, 5, 4, as in early amniotes and *Eldeceeon,* but not other anthracosaurs.

Westlothiana has many features of the skeleton that are closer to those of early amniotes than any other Lower Carboniferous tetrapods, but the more primitive palate and tarsus and the great elongation of the trunk relative to the limbs suggest that it belongs to an earlier and divergent lineage.

Temnospondyls

Of all the tetrapod taxa at East Kirkton, the most common is *Balanerpeton,* represented by approximately thirty specimens, several of which are preserved in the highly informative manner of skeletons from unit 82 (Milner and Sequeira, 1994). It is the oldest known member of another major group, the temnospondyls (Figs. 4.26 and 4.27). *Balenerpeton* is recognized as a temnospondyl principally by the large palatal (or interpterygoid) vacuities—extensive spaces between the pterygoids and the cultiform process of the parasphenoid. In *Balanerpeton,* they reach anteriorly to the vomers, and the pterygoid bones do not meet at the midline. The presence of large interpterygoid vacuities may be associated with a distinct pattern of articulation between the parasphenoid and the pterygoid. Unlike anthracosaurs, the denticle-covered anterior margin of the medial extension of the pterygoid in *Balanerpeton* and other temnospondyls extends along the transversely oriented anterior margin of the basicranial process of the parasphenoid. In later and larger temnospondyls these elements become fused so that movement between the palate and the base of the braincase is precluded.

In contrast with embolomeres, there are fangs on the vomer as well as the palatine and ectopterygoid, as is also the case in primitive whatcheerids. The largest skull is approximately 5 cm in length, about 30% the length of the skull plus trunk. The largest known femur, belonging to a specimen without a skull, suggests a skull length of 6.2 cm. There is no trace of lateral line canals or pit lines on any of the skull bones. Sclerotic plates are fully displayed in the type. The pattern of the bones at the back of the skull table is primitive in retaining the intertemporal and a small tabular that is not in contact with the parietal and lacks a posterior horn. It may be considered advanced, however, in the sutural attachment of the skull table and cheek and the large size of the squamosal notch. In association with this notch is a stapes that differs from those of any earlier described tetrapods, including *Acanthostega,* colosteids, and anthracosaurs, in its small size and cylindrical shape of the stem (Fig. 4.26D). Assuming the presence of a tympanic membrane in the otic notch, these elements have the relative size and configuration of the structures associated with the impedance matching middle ear of modern anurans. No Paleozoic group other than the temnospondyls shows evidence of the capacity to respond to high-frequency airborne vibrations, with the possible exception of the seymouriamorphs.

The extent of the epipterygoid ossification is much reduced relative to those of anthracosaurs, baphetids, and colosteids (Fig. 4.26E). There remains only a small triangular base associated with the basicranial articulation and a narrow dorsal process. Because of the extensive interpterygoid vacuity, we can see the sphenethmoid portion of the braincase above the cultiform process of the parasphenoid. Only a portion is ossified, compared with the very large area of ossification in large baphetids, most anthracosaurs, and *Acanthostega.* The basioccipital is well ossified, in contrast with that of the earliest tetrapods, and, together with the base of the exoccipitals forms a recessed occipital cotyle.

The lower jaws are shallow and without any large teeth on the denticulate coronoids. There are a number of small foramina in the postsplenial, which extends far dorsally in marked contrast with whatcheerids, anthracosaurs, baphetids, or *Caerorhachis* as a result of the relatively short prearticular. There is no trace of the hyobranchial skeleton.

The vertebrae are primitive in the presence of paired pleurocentra and small crescentic intercentra. The atlas-axis complex of temnospondyls otherwise resembles that of anthracosaurs in the presence of paired proatlantes, a paired atlas arch, and loosely articulated central elements, unfortunately not in place in the otherwise well-preserved type specimen. There are approximately 24 presacral vertebrae. The ribs are all short. Those in the area of the shoulder are flattened and somewhat widened distally, but lack the uncinate processes of some later and larger temnospondyls such as *Eryops.* The neural arches in the trunk region are not fully fused medially.

The dermal shoulder girdle of *Balenerpeton,* like that of temnospondyls in general, can be distinguished from that of Devonian labyrinthodonts and anthracosaurs by the relative shortness of the posterior portion of the rhomboidal interclavicle. The clavicle retains a large ventral blade and the cleithrum has a broadened dorsal portion. The scapulocoracoid is poorly preserved, but the illustrations suggest that the scapula and coracoid were co-ossified. The humerus is also not well known, but it appears to have a slender, if short shaft between the proximal and distal surfaces of articulation, and bears an entepicondylar foramen. The ulna and radius are small, but with elongate shafts. The bones of the carpus are individually well ossified, but in no specimen are they fully articulated. In contrast to most early labyrinthodonts, there are only four digits in the manus, with a phalangeal count of 2, 2, 3, 3, a pattern that remains highly consistent throughout temnospondyls.

In the adults, the three elements of the pelvis are well ossified and closely sutured to one another, but smaller specimens show that the pubis is slow to ossify. The most important feature is that the iliac blade is reduced to a single posterodorsally directed process, coincidently resembling that of *West-*

lothiana. The femur is about one-half the length of the skull, and the fibula about 70% the length of the femur. The tarsals are fully ossified and sufficiently well articulated that nearly all can be identified (Fig. 4.27). The phalangeal count is 2, 2, 3, 4, 3, close to that of *Eucritta* as far as it is known, and (not counting the number of digits) closer to that of *Acanthostega* than to *Tulerpeton* (Table 5.1).

Taking all the osteological characteristics of *Balanerpeton* into consideration, it cannot be associated with any of the other clades of Carboniferous tetrapods so far discussed on the basis of obvious synapomorphies. It appears as one of the pioneering small tetrapods, and at the same time, a close stem taxa of the hoard of later temnospondyls. There is, however, some evidence from isolated bones of the presence of other temnospondyl lineages from the Late Carboniferous. From East Kirkton, Milner and Sequeira (1994) have also described three large isolated anterior ribs possessing uncinate processes that resemble those of temnospondyls 1 to 2 m in length such as the Lower Permian genera *Edops* and *Eryops.* In addition, Stephen Godfrey (1988) has described trunk vertebrae from the Greer locality in West Virginia, of about the same age as East Kirkton, which also resemble in their general form those of large temnospondyls, although they retain a supraneural canal comparable to that of stem tetrapods.

One particularly important fossil of *Balanerpeton* is a well-preserved advanced larva (Fig. 4.27F). This is the oldest known amphibian fossil that illustrates specifically larval features of the anatomy that clearly distinguish it from adults, in contrast with simply small size. This specimen is unique among Lower Carboniferous labyrinthodonts in documenting the sequence of ossification of different bones in the skeleton. All the dermal bones of the skull roof are already fully formed and close to adult proportions, but endochondral bones, specifically the vertebral centra, the scapula, pubis, and bones of the wrist and ankle are not ossified. On the other hand, the neural arches are formed throughout much of the column, but the right and left halves are not co-ossified and appear as separate pairs.

This sequence of ossification of the bones is clearly distinct from that seen in the juvenile specimen of an adelospondyl illustrated in Figure 4.10. The skull in that specimen is only about 8 mm long, compared with approximately 30 mm in the larva of *Balanerpeton,* and yet the vertebral centra are fully cylindrical to the very end of the column as preserved (R. L. Carroll, 1989). Comparably rapid ossification of the vertebral column is also seen in all other groups referred to as lepospondyls.

We do not know when a distinct larval stage first evolved among labyrinthodonts. This was almost certainly not a heritage of amphibians, for we know that one of their closely related antecedents, *Eusthenopteron,* lacked a larval stage (Cote et al., 2002). *Balanerpeton* currently provides the earliest known evidence of an initial stage in the evolution of the pattern of development seen in modern frogs and salamanders.

CAERORHACHIS

Leaving the diverse East Kirkton fauna, we move up to the Namurian A, just above the level of the Dora opencast site, and evaluate another enigmatic genus, *Caerorhachis,* described initially by Holmes and Carroll (1977) and later by Ruta et al. (2002). It is represented by a single specimen from the Ramsay colliery, Loanhead near Edinburgh (Fig. 4.28). *Caerorhachis* is clearly a labyrinthodont, with conspicuous labyrinthine infolding of the fangs on the vomers, palatine, and ectopterygoid, and the parasymphyseal tusks of the lower jaw. Four characters suggest affinities with temnospondyls: a relatively large interpterygoid vacuity (although more primitive than *Balanerpeton* in the anterior union of the pterygoids), the phalangeal count of the pes, the absence of larger teeth on the coronoids, and an ilium in which the dorsal process is reduced to a small protuberance.

On the other hand, both the structure and the number of the vertebrae are divergent from those of early temnospondyls. The presence of high, crescentic pleurocentra as well as intercentra resembles primitive embolomeres and gephyrostegids. *Caerorhachis* has 30 to 32 presacral vertebrae, compared with 24 in *Balanerpeton.* Body proportions are also distinct. The ratio of the length of the skull to that of the presacral length is approximately 22% in *Caerorhachis,* versus 30% in *Balanerpeton* and 32% in the next later temnospondyl, *Dendrerpeton,* and the limbs are relatively shorter. All these factors suggest differences in locomotion from those seen in early temnospondyls of comparable size.

The medial surface of the lower jaw is also distinct from that in *Balanerpeton,* but somewhat resembles that of *Eoherpeton* in the presence of a row of openings between the relatively longer prearticular and the very narrow postsplenial. If, as argued by Ruta et al. (2002), a parasymphyseal bone is retained, it would be another common feature of primitive anthracosaurs. However, further study of the silicone casts from which the specimen has been described fails to confirm its presence. On the other hand, the pattern of the bones at the back of the skull table, with the tabular small and not in contact with the parietal, is more primitive than that of any anthracosaur. An even more primitive feature, not encountered in either anthracosaurs or temnospondyls, is the structure of the pelvis, in which the unfinished surface of the acetabulum extends anteriorly to the front margin of the pubis. This is otherwise observed only in Devonian amphibians and whatcheerids.

The combination of characters exhibited by *Caerorhachis* may be explained by its being a relict of a very early Carboniferous lineage that precociously adapted to a terrestrial way of life. The latter is supported by the absence of lateral line canals and a high degree of ossification of the pleurocentra, tarsus,

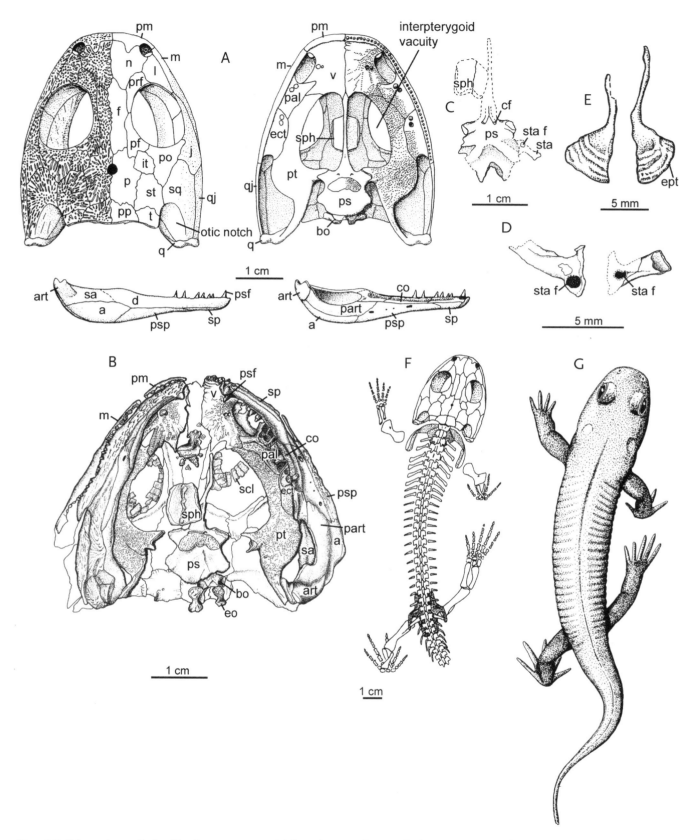

Figure 4.26. Balanerpeton woodi, the oldest known temnospondyl, from the late Viséan of East Kirkton. *A*, Reconstruction of the skull in dorsal and palatal views, and lateral and medial views of the lower jaw. *B*, Specimen drawing of palate. *C*, Association of stapes with parasphenoid. *D*, Stapes. *E*, Epipterygoid. *F*, Reconstruction of skeleton. *G*, Flesh reconstruction. From A. R. Milner and Sequeira, 1994.

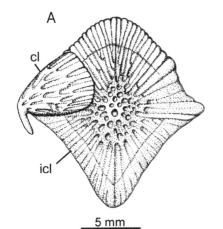

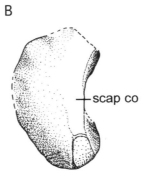

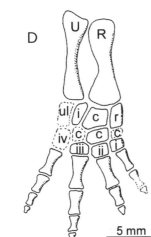

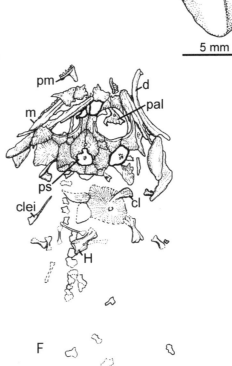

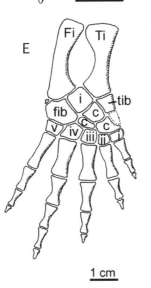

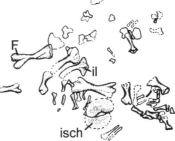

Figure 4.27. Postcranial skeleton of *Balanerpeton woodi*. From A. R. Milner
and Sequeira, 1994. *A*, Dermal shoulder girdle. *B*, Scapula. *C*, Ilium. *D*, Lower
forelimb. *E*, Lower hind limb. *F*, Immature specimen.

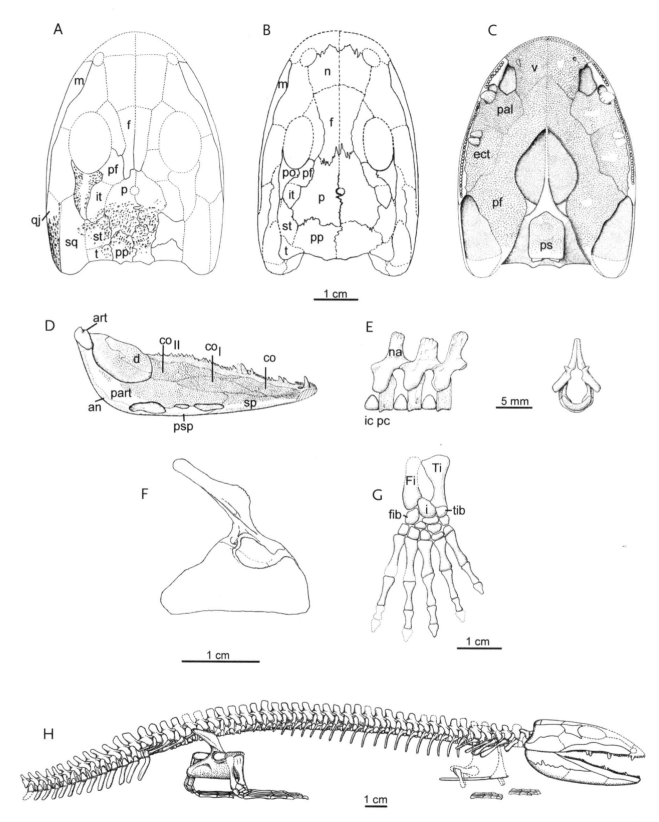

Figure 4.28. Caerorhachis bairdi, from the Namurian A of Scotland. *A,* Reconstruction of skull in dorsal view. From Holmes and Carroll, 1977. *B,* Reconstruction of skull. Modified from Ruta et al., 2002. *C* and *D,* Reconstructions of palate and medial surface of the lower jaw. From Holmes and Carroll, 1977. *E,* Reconstruction of trunk vertebrae in lateral and anterior views. From Holmes and Carroll, 1977. *F,* Lateral view of pelvis. Modified from Ruta et al., 2002. *G,* Reconstructions of lower rear limb. *H,* Reconstruction of skeleton. From Holmes and Carroll, 1977.

and pelvis. However, primitive attributes of the lower jaw and the extension of the acetabular surface to the anterior margin of the pubis suggest a very early divergence from primitive tetrapods. Its relatively late appearance in the fossil record may be attributed to the overall rarity of amphibians indicative of a terrestrial way of life in the Lower Carboniferous.

ACHERONTISCUS

Another genus, *Acherontiscus,* which I described in 1969 from about the same geological horizon as *Caerorhachis,* is represented by a single specimen from the collections of the Royal Scottish Museum (Fig. 4.29). It is even more enigmatic in being without any data as to time or place of collection, as well as representing a unique amphibian morphology. The surrounding matrix contains spores indicating a horizon from the upper part of the Viséan to the middle of the Namurian, suggesting an age sometime in the Arnsbergian. The specimen consists of a nearly entire skeleton of a very elongate animal. It has a small skull and a vertebral column with approximately 64 segments.

After a short gap behind the skull, the first preserved element is a narrow crescentic pleurocentrum. The remaining segments consist of both intercentra and pleurocenta. Beginning with the fifth, both central elements appear to be cylindrical. In the more posterior segments, the pleurocentra are clearly longer. At least 36 of the preserved vertebrae are from the trunk region. The most posterior are definitely caudals, distinguished by their haemal arches, missing in aïstopods.

While the vertebrae are vaguely similar to those of embolomeres, the skull is very small, and resembles that of some microsaurs (discussed in the following section) with the number of bones reduced from that of any labyrinthodont. There are relatively few teeth. Those in the middle of the jaw are bulbous and (as revealed in cross section) devoid of labyrinthine infolding. The skull shows conspicuous lateral-line canal grooves, but no evidence of an otic notch. The eyes are far forward. Just behind the lower jaw are a number of thin rectangular plates that may be identified as elements of the hyoid apparatus, although they cannot be compared closely with those of any other known amphibians.

There is no trace of limbs. All that is known of the shoulder girdle are the dermal clavicle and interclavicle. There is no evidence of a pelvic girdle. The absence of limbs and the endochondral elements of the girdles suggests affinities with adelogyrinids, but the nature of the teeth and vertebrae makes close affinities unlikely. Just another ghost in the mists of time.

MICROSAURS
The Goreville Microsaur

Roughly equivalent in time (somewhere in the middle Namurian) appears another, highly distinctive assemblage of Paleozoic amphibians, the order Microsauria. This group appears to be primarily terrestrial, but with secondarily aquatic adaptation in several families. Microsaurs are the most diverse of all the groups termed lepospondyls, in both numbers of species and adaptive and anatomical specializations (Plate 6). They are also the only clade that retains several skeletal attributes of early labyrinthodonts that suggest possible affinities (R. L. Carroll and Gaskill, 1978).

The oldest known microsaur is represented by eight specimens, preserved together in a mudstone nodule collected from an inactive limestone quarry near the town of Goreville in Illinois. From the same quarry have come specimens of the lungfish *Tranodis* and a colosteid similar to *Greererpeton.* These genera indicate the presence of permanent water. In North American terminology this horizon is placed within the Elvirian, which is broadly equivalent to the Namurian E2 or Arnsbergian of European usage. These specimens are currently under study by Eric Lombard and John Bolt, who published a preliminary report in 1999. More detailed description awaits the very difficult preparation necessary to disentangle the small and closely integrated skeletons (Fig. 4.30). Because it is still incompletely known, the Goreville microsaur remains unnamed.

In common with the aïstopods, adelogyrinids, and *Acherontiscus,* the earliest known microsaur is relatively long bodied, with approximately 34 presacral vertebrae. This, however, is not a general character of the group, as indicated by other skeletons from only slightly higher in the Carboniferous of Utah, and many later taxa that have much shorter vertebral columns and proportionately larger girdles and limbs. Other families of later microsaurs have reduced their limbs, but none have lost them entirely.

In contrast with most early labyrinthodonts, the jaw articulation in this and most other microsaurs does not extend behind the occiput. The skull surface lacks the conspicuous sculpturing of most labyrinthodonts and there is no trace of lateral line canals. The eyes are about midway in the length of the skull and the surrounding bones are ridged along their circumference, suggesting a relatively thick layer of soft tissue above the bone. The frontal appears to enter the margin of the orbit. Only a few of the bones of the skull roof are sufficiently well exposed to show their relationships with one another. No overall pattern can be discerned.

The palate is not clearly visible in any of the skulls, but the surface shows an almost continuous covering of small denticle. One larger tooth extends from the palatine. The occiput is not well exposed, but the configuration of the surface for articulation with the atlas can be reconstructed on the basis of the latter bone. For a small skull, the endochondral bones are well ossified, with the otic capsule showing the impression of the dorsal semicircular canals. The sphenethmoid is also ossified and fused to the cultriform process of the parasphenoid. The articulation between the basisphenoid and the epipterygoid and pterygoid of the palate appears to resemble

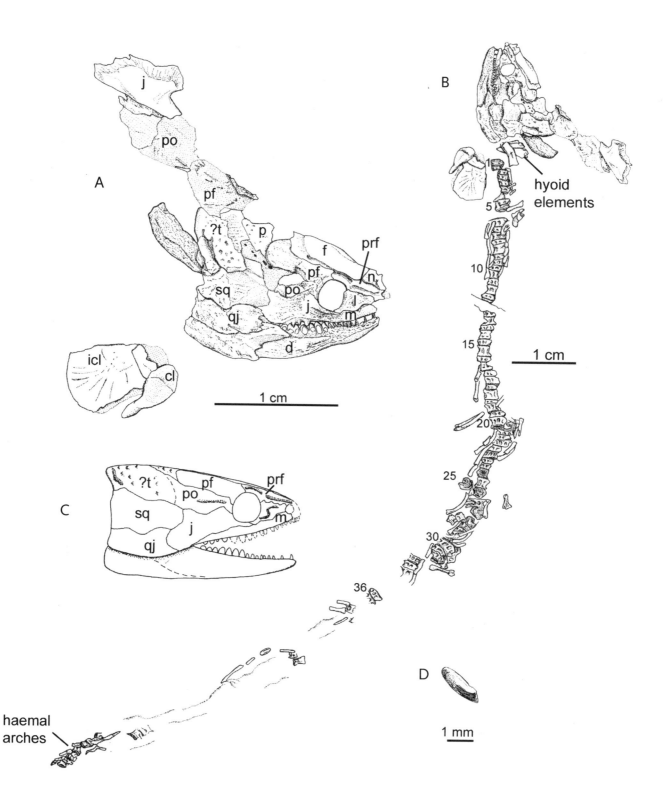

Figure 4.29. *Acherontiscus caledoniae.* Only known specimen of an isolated clade from the Namurian A of Scotland. *A,* Skull and dermal shoulder girdle. *B,* Skeleton (postcranium drawn from a rubber mold of a natural cast). *C,* Reconstruction of skull. *D,* Single ventral scale in medial view. From R. L. Carroll, 1969d.

that of later microsaurs. No stapes can be seen with any of the skulls.

The lower jaw is not sculptured. The coronoid bones bear two rows of teeth. A comparable pattern of coronoid teeth is seen in Carboniferous and Permian gymnarthrid microsaurs. As in nearly all other specimens referred to as lepospondyls, none of the teeth show labyrinthine infolding.

The structure of the atlas (Fig. 4.31) provides the stron-

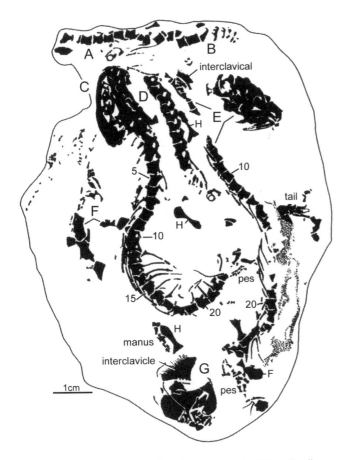

Figure 4.30. Unnamed microsaur from the Namurian A of Goreville, Illinois. From Lombard and Bolt, 1999. Large letters *A–G* indicate associated bones of different specimens; small letters *F* and *H* identify femur and humerus. From Lombard and Bolt, 1999.

gest evidence for the recognition of this genus as a microsaur. Uniquely among Paleozoic amphibians, this bone has a surface for articulation with the skull that is more than twice the width of the more posterior centra. It consists of a conspicuous medial odontoid process and lateral extensions that articulate respectively with the concave basioccipital and the laterally placed exoccipitals. The configuration of the occipital condyles and the anterior surface of the atlas in microsaurs would favor dorsoventral flexion of the head on the trunk and greatly limit lateral bending or rotation. In terms of its functional anatomy, this structure is highly divergent from that of labyrinthodonts and early amniotes, which have a multipartite atlas that would have allowed some degree of bending or rotation in all directions. No microsaur has an atlas intercentrum.

The atlas arch bears prezygapophyses for the paired proatlantes. These structures have not been seen in other microsaurs, but provide an important link to primitive labyrinthodonts such as *Acanthostega*, anthracosaurs, and early temnospondyls. The neural arches of the atlas are fused to the centrum without trace of sutures, but appear to be separated dorsally. There are also two very clearly developed and separate areas for articulation of the two heads of the atlas ribs, in common with numerous later microsaurs.

In contrast with labyrinthodonts, the second cervical, or axis, vertebra of microsaurs and other lepospondyls typically resembles the more posterior cervical vertebrae rather than having a distinct structural and functional anatomy. As is the case for the remaining trunk vertebrae, the axis has a low, crescentic intercentrum as well as a large cylindrical and amphicoelous (recessed at both ends) pleurocentrum, and is suturally attached to the arch rather than being fused as are the atlas and the caudal vertebrae. Among later microsaurs, long-bodied genera tend to retain the intercentra, but they are missing in animals with a shorter presacral column. This distinction can be observed within a single family, implying functional rather than phylogenetic significance (R. L. Carroll, 1988b). The intercentra are succeeded by haemal arches behind the sacrum.

Ribs are visible throughout the trunk. Behind the atlas, the tuberculum attaches to the transverse process of the neural arch and the capitulum attaches to the top of the intercentra. The shafts are long and cylindrical.

Of the shoulder girdle, only the interclavicle is preserved. The anterior margin is fimbriate, that is, appearing like the teeth of a comb. This characteristic occurs sporadically among both labyrinthodonts and lepospondyls. The humerus is not fully exposed in any specimen, but appears to be about the length of three trunk centra and is pierced by the entepicondylar foramen. In contrast with most of the amphibians so far described, the proximal and distal ends are twisted at about 90 degrees and there is a well-developed deltopectoral crest above a distinct shaft. The ulna and radius have not been recognized. Some carpals can be seen, along with evidence of three digits, but there may have been more.

The pelvis cannot be reconstructed. The femur has a long, slender shaft, distinguishing it from most of the large labyrinthodonts of the Lower Carboniferous, a clearly developed intertrochanteric fossa, and an elongate adductor crest. The more distal portion of the limb is not known.

Except for the common retention of some primitive features, no characters specifically support the relationships of microsaurs within any of the other groups of Carboniferous tetrapods. The retention of the proatlas and crescentic intercentra suggests that microsaurs are more likely to have evolved from primitive labyrinthodonts than from any of the other lepospondyl groups. Evidence from later microsaurs provides additional information for evaluating relationships.

Utaherpeton

The next younger microsaur is *Utaherpeton*, from the Manning Canyon Formation of Utah, equivalent to the Namurian B of Europe. *Utaherpeton* in known from one juvenile and one more mature specimen, together showing much of the skeleton (Figs. 4.32 and 4.33).

Like the Goreville microsaur, the Utah specimens lack some key features necessary for establishing the nature of their ancestry and relationships, but provide information re-

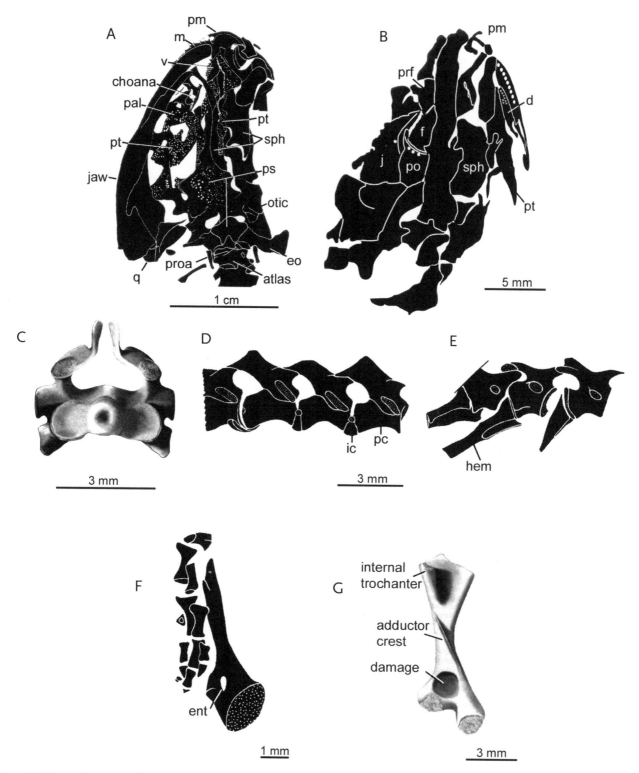

Figure 4.31. Unnamed microsaurs from Goreville, Illinois. *A* and *B*, Views of two skulls. *C*, Atlas vertebra in anterior view. *D*, Trunk vertebrae in lateral view. *E*, Caudal vertebrae in lateral view. *F*, Humerus and bones of manus. *G*, Femur. From Lombard and Bolt, 1999.

garding the sequence of ossification of their vertebrae that contributes to analysis of possible affinities with other lepospondyl groups. Both specimens of *Utaherpeton* are preserved as counterparts, exposing both bones and impressions of their surface that can be cast so as to replicate their original configuration.

In contrast with the fossils of whatcheerids, colosteids, and anthracosaurs, for which specimens from several horizons and localities retain a generally similar body form, the Goreville microsaur and *Utaherpeton* differ strikingly in their proportions. The Goreville microsaur had an elongate trunk and limbs that are proportionately small, compared with both the length of individual vertebrae and the length of the skull. The presacral vertebrae of *Utaherpeton* number 26 (in both specimens) compared with approximately 34 in the earlier genus. The femur is the length of three trunk centra in the Goreville genus and almost four in *Utaherpeton*. Clearly, these animals had very different modes of locomotion and probably different ways of life. The environment of deposition at Goreville and the presence of large colosteids and a lungfish in the same deposits suggest a fairly large and deep body of water. In contrast, *Utaherpeton* was preserved in what has been interpreted as a moist lowland with small ponds and riparian habitats (pertaining to river banks), as indicated by accompanying fossils including ferns, fernlike foliage, lycopods, calamites, and primitive relatives of conifers, all of which lived close to the water.

Much of the skull roof and scattered elements of the palate are preserved, but not critical areas that are necessary for establishing the position of *Utaherpeton* among later microsaurs. Its identity as a microsaur is based on the configuration of the basioccipital, which has lateral areas for articulation with the exoccipitals that document the wide surface of articulation characteristic of all later microsaurs. The atlas is present in the type specimen, but cannot be exposed without damage to the underlying interclavicle. The other key to identification as a microsaur is the configuration of the pelvis, which is clearly shown in the juvenile specimen as having bifurcate dorsal iliac processes. This structure is common to many primitive labyrinthodonts, but is distinct from that of other limbed lepospondyls in which the ilium has only a single, posterodorsal process.

The skull, as reconstructed (Fig. 4.33), broadly resembles that of most later microsaurs (R. L. Carroll and Gaskill, 1978) in the location of the orbit midway in the length of the skull and its relatively large size (expected in small animals). In contrast with the Goreville microsaur, the frontal does not enter the margin of the orbit but, as in a few later microsaurs, the maxilla does. The posterior margin of the squamosal is close to vertical, with no evidence of a squamosal notch. In contrast with primitive members of all labyrinthodont groups, the jaw articulation does not extend behind the posterior margin of the occiput. As in other primitive microsaurs, the marginal teeth are small and numerous (approaching 30 in the maxilla) and nearly uniform in length. The snout region

differs from most early tetrapods (with the exception of colosteids, nectrideans, aïstopods, and lysorophids, with very short distances between the orbits and nares) in the extension of the prefrontal into the margin of the external nostril.

Unfortunately, the posterior bones of the skull table are not well preserved, precluding knowledge of their number and configuration, which is critical for understanding the specific affinities of this genus with later microsaurs and with their possible sister-taxa among other Paleozoic orders. Later microsaurs have been grouped into two suborders on the basis of distinct patterns of the bones adjacent to the parietal (Fig. 4.34). The tuditanomorphs include seven of the microsaur families and are considered the more primitive because of their generally conservative cranial anatomy. The remaining four families, with a more divergent morphology, are grouped as the microbrachomorphs. The early members of these suborders have the same number of skull bones, but their arrangement differs in a consistent manner. In contrast with all of the adequately known labyrinthodont amphibians we have seen, a single large bone fills the entire space initially occupied by the intertemporal, supratemporal, and tabular. This bone is termed the tabular because it occupies the posterolateral corner of the skull table, as does the bone of this name in all more primitive groups. The microsaur tabular may have evolved by spreading its area of ossification through this entire space, or via fusion with one or both of the other bones. The difference between tuditanomorphs and microbrachomorphs may be attributed to different patterns of incorporation of the area initially occupied by the intertemporal and supratemporal. This assumes that both of these bones were present in their immediate ancestors as suggested by their presence in the early members of most labyrinthodont groups. The pattern in the tuditanomorphs can be explained by the incorporation of both of these areas of ossification into the tabular, which extended anteriorly to reach the postorbital and postfrontal. In contrast, the configuration of these bones in the microbrachomorph *Microbrachus* can be attributed to an anterior expansion of the tabular, as in anthracosaurs, so that it reached the parietal, and the incorporation of the intertemporal and supratemporal into the parietal, which then extended laterally to the postorbital.

If the microsaurs are monophyletic, it suggests an early dichotomy between the two suborders, whose timing might be established by additional knowledge of the skulls of the two early genera. An early anterior expansion of the tabular to reach the parietal in all primitive microsaurs suggests affinities with anthracosaurs, but other features, such as the configuration of the occiput, question the homology of this similarity. While the occiput cannot be fully reconstructed in either of the two early microsaurs, this is possible in members of many of the later families (Fig. 4.35). They show that the postparietals meet the exoccipitals lateral to the foramen magnum, in clear contrast with the anthracosaurs, in which the otic capsule is interposed between these bones. Most

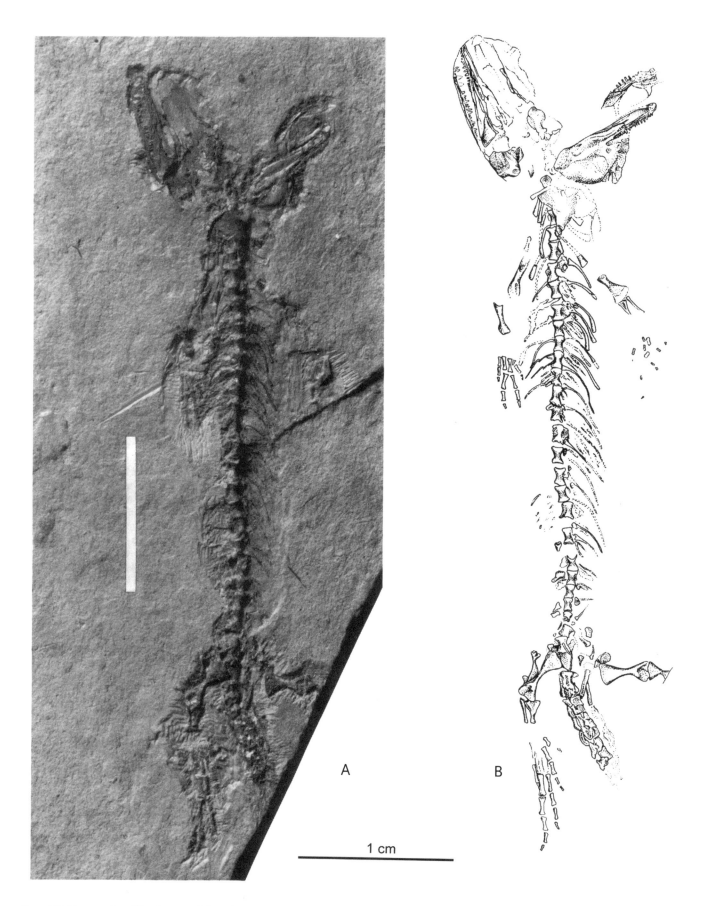

A   B

1 cm

Figure 4.32. Photograph and line drawing of the early microsaur *Utaherpeton franklini* from the Mississippian-Pennsylvanian boundary of Utah, equivalent to the Namurian B of Europe. From R. L. Carroll et al., 1991.

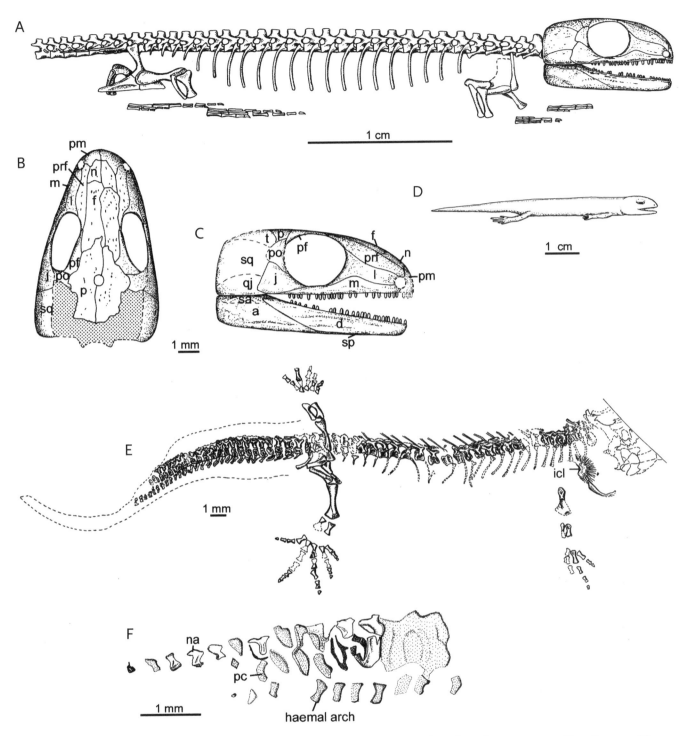

Figure 4.33. *Utaherpeton franklini. A,* Reconstruction of skeleton based on the type. *B* and *C,* Reconstruction of the skull in dorsal and lateral views. *D,* Flesh reconstruction, approximately life size. *E,* Juvenile specimen. *F,* Detail of caudal vertebrae, drawn from counterpart of juvenile, reversed for comparison with *E.* From R. L. Carroll et al., 1991.

later microsaurs show the evolution of a new bone in a median position above the foramen magnum, the supraoccipital. This combination of features separates microsaurs from all other early tetrapods. There is no evidence of the nature of the anterior portion of the braincase or the configuration of the stapes in the Namurian microsaurs.

The squamosal in *Utaherpeton* extends a narrow surface beneath the skull table and another posteromedially to form the lateral rim of the occiput. There is not a strong sutural connection between the squamosal and the lateral bones of the skull table, as is the case in early temnospondyls, but neither is there evidence of movement between these two portions of the skull, which appears to be the case in anthracosaurs.

The palate of all microsaurs is distinct from labyrinthodonts in the absence of the fang and pit pairs seen most

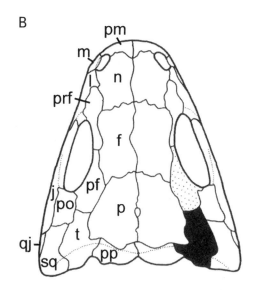

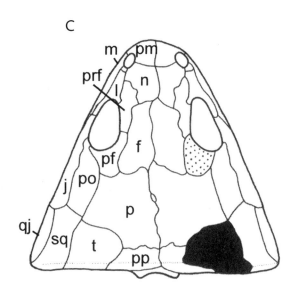

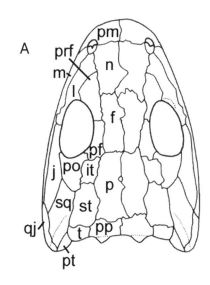

Figure 4.34. Putative origin of the pattern of bones in the skull table of microsaurs. *A,* The temnospondyl *Dendrerpeton* as a model of the pattern in primitive labyrinthodonts, most of which retain the intertemporal bone (it). *B,* Skull of the oldest adequately known tuditanomorph microsaur, *Asaphestera,* from the Westphalian A of Joggins, Nova Scotia. *C, Microbrachis,* one of the oldest known microbrachomorph microsaurs, from the Westphalian D of the Czech Republic. The bone termed tabular in microsaur is shaded black; the postfrontal is evenly stippled. From R. L. Carroll and Gaskill, 1978.

clearly in the large whatcheerids, anthracosaurs, baphetids, and *Crassigyrinus,* and to a lesser degree in temnospondyls, although many have an extensive covering of denticles and/or large teeth. On the other hand, *Crinodon* (Fig. 4.35), from later in the Carboniferous, shows a pattern of large palatal teeth on the vomer, palatine, ectopterygoid, and even the pterygoid, which could have evolved from those of early labyrinthodonts, although without the conspicuous adjacent pit for a replacement tooth.

The vertebral column is clearly visible in both of the Namurian microsaurs. In contrast with the genus from Goreville, there are certainly no intercentra in *Utaherpeton,* in

which the entire column can be seen in ventral view in the type and in lateral view in the juvenile. There are, however, haemal arches, their serial homologues, in the tail. The presence or absence of intercentra might be thought to be evidence for a wide taxonomic gap between the two Namurian microsaurs, but this can also be seen among different members of a single family, the Gymnarthridae, known from the Pennsylvanian (Upper Carboniferous) and Permian. In the Carboniferous species *Sparodus,* with a short trunk of about 25 vertebrae, there are no intercentra, but in the Lower Permian *Cardiocephalus,* with 38 presacral pleurocentra, they are accompanied by intercentra throughout the column.

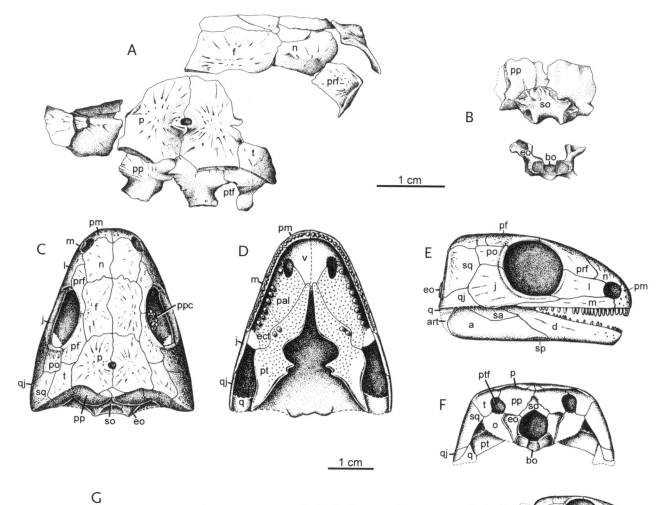

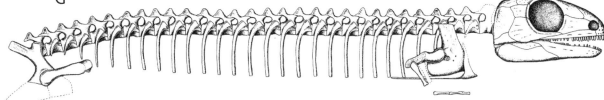

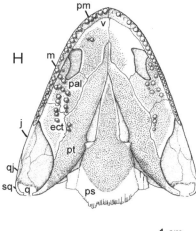

Figure 4.35. Microsaurs. *A* and *B*, Specimen drawings of the tuditanid microsaur *Asaphestera intermedium* from the Westphalian A of the tree stump fauna of Joggins, Nova Scotia; skull roof in dorsal view, postparietals and supraoccipital in ventral view, and surface for articulation with the atlas in posterior view. *C–F*, Reconstruction of the skull of *Asaphestera* in dorsal, palatal, lateral, and occipital views. *G*, Reconstruction of the known elements of the skeleton. The number of vertebrae is estimated. *H*, Palate of the tuditanid microsaur *Crinodon* from the Westphalian D coal swamp fauna of the Czech Republic. Note that the large teeth on the lateral palatal bones are in the position of the large fangs in primitive labyrinthodonts, but lack the associated replacement pits. From R. L. Carroll, 1966.

The skulls of these genera are well known and very similar in most features. That of *Sparodus* appears much larger, being 40% of the total length of the presacral vertebral column, compared with 22% for *Cardiocephalus,* but relative to the length of the individual vertebrae, they are more comparable. The skull is equal to the length of about 10 presacral centra in *Sparodus,* but between eight and nine in *Cardiocephalus* (R. L. Carroll, 1988b). In the Goreville microsaur, the skull is equal in length to six anterior vertebrae, while that of the mature specimen of *Utaherpeton* is equal to eight.

The orientation of the trunk vertebrae in the mature specimen of *Utaherpeton* precludes determining whether the arch and centra were suturally attached or fully fused with one another, but those of the anterior caudals appear to be fused. The central are fully cylindrical. The transverse processes are clearly defined, but there is no evidence of an area for articulation with the capitulum, which attached to the intercentrum in the Goreville genus. The first five ribs have widely expanded distal ends. The more posterior are simple cylinders. There is no trace of uncinate processes, as seen in some other lepospondyl groups.

The scapulocoracoid in the originally described specimen (termed the type) is represented by only a poorly defined impression, but in the juvenile the scapular portion is massive but the coracoid has not yet ossified. The interclavicle is diamond-shaped, without the clearly distinct posterior stem of anthracosaurs or the rhomboidal configuration of temnospondyls. The clavicle has very narrow blades. Neither of these bones is sculptured. The cleithrum was not identified. The humerus is of a tetrahedral shape, with the expanded ends at right angles to one another, in contrast with any of the large, primitive labyrinthodonts. The entepicondylar foramen is retained. The radius and ulna are lightly built. The olecranon is well defined, but the carpals are not ossified. The manus has four digits and the phalangeal count is 2, 3, 3, 2, both similar to those of temnospondyls and most later microsaurs.

In the type, the pelvic girdle is well exposed ventrally with all three elements present but distinguished by clearly defined sutures. The ilium of the juvenile is bifurcate, as in other primitive microsaurs and anthracosaurs. The femur of the type is equal in length to three and a half anterior centra. The tibia and fibula are less than one-half this length, but in none of the bones are the distal extremities fully ossified. The tarsals lack ossification, but the phalangeal count, especially clearly shown in the juvenile, is 2, 3, 4, 5, 3. Surprisingly, this is close to that of anthracosaurs and early amniotes, in contrast with that of the manus, which corresponds more closely with that of temnospondyls.

The well-preserved dermal scales in the juvenile clearly show the chevron pattern of the ventral elements common to all early tetrapods. This specimen, whose linear dimensions are approximately 60% those of the adult, also demonstrates changes in the osteology that occur during ontogeny. In contrast with the type, the anterior margin of the interclavicle remains fimbriated as in the Goreville microsaur and

several later members of this order. The vertebral column shows several features of immaturity. The arches and centra are not fused, but have become separated from the centra throughout the caudal region. The neural spines remain paired throughout the column. The posterior caudals are still in the process of ossifying. The crescentic shape of the pleurocentra is evident only to the level of the 25th segment, beyond which they were apparently still paired, as in the most primitive tetrapods and in most temnospondyls. The last trace of both the pleurocentra and haemal arches is evident at the level of the 30th neural arch. The last five neural arches have already ossified. *Utaherpeton* is the only lepospondyl to show ossification of the pleurocentra from paired elements.

A period of approximately 8 million years separates *Utaherpeton* from the next known microsaurs, from the Westphalian A of Joggins, Nova Scotia. In that locality, which is discussed in the following section, five or six microsaurs are recognized. They are the oldest genera from which the cranial anatomy is known in detail, but they come from a very different environment.

THE JOGGINS LOCALITY

We now turn from the aquatic habitat of early microsaurs to a locality that is famous for fully terrestrial tetrapods, the Joggins Formation on the shore of the Bay of Fundy in Nova Scotia (Plate 7). Joggins has been famous since the early 19th century for the presence of a succession of nearly 40 fossil forests exposed in high cliffs running for miles along the coastline. In the summer of 1852, William Dawson, principal of McGill University, accompanied by Charles Lyell, the father of modern geology, investigated the fossilized forests. In the course of this study, they came upon a stump that had fallen onto the beach and broken open. It contained numerous bones that were described as belonging to the earliest known reptiles. Subsequently, Dawson explained the occurrence of vertebrates within the stumps in the following words (1891): "A forest or grove of large ribbed trees known as Sigillariae, was either submerged by subsidence, or, growing on low ground, was invaded with the muddy waters of an inundation, or successive inundations, so that the trunks were buried to a depth of several feet. The projecting tops having been removed by sub-aerial decay, the buried stumps became hollow, while their hard outer bark remained intact. They thus became hollow cylinders in a vertical position and open at the top. The surface having become dry land, covered with vegetation, was haunted by small quadrupeds and other land animals, which from time to time fell into the open holes, in some cases nine feet deep, and could not extricate themselves" (Fig. 7.10). This is still accepted as the most probable explanation for the presence of terrestrial vertebrates in the Joggins stumps. Scores of *Sigillaria* stumps were collected by Dawson and others over the years, and this practice is still continuing, resulting in the discovery of hundreds of skeletons in various stages of disarticulation.

Joggins Microsaurs

The most diverse assemblage were the microsaurs, with at least six genera, differing greatly in their size and dentition. The critical posterior portion of the skull is preserved in the gymnarthid *Leiocephalikon* and the tuditanid *Asaphestera*, confirming the link between *Utaherepeton* and later microsaurs (Figs. 4.35 and 4.36). The number and arrangement of the bones at the back of the skull table follow the pattern of later members of the Tuditanomorpha, with the tabular extending forward to the postorbital and postfrontal (R. L. Carroll, 1963, 1966). The exoccipitals and basioccipital form the broadly concave articulating surface for the atlas vertebra, as in all other microsaurs. The more dorsal portion of the occipital surface is distinctive in the great ventral extension of the posterior lappets of the postparietal bones that must have overlapped the dorsal portion of the exoccipitals, as well as forming the medial margin of large post-temporal fenestrae. This configuration is most closely matched by the limbless adelospondyls (Fig. 4.11), but *Asaphestera* and other microsaurs differ from adelospondyls and most other early amphibians in the presence of a conspicuous supraoccipital bone.

The skeletons of all the microsaurs at Joggins are disarticulated, as are those of other taxa, presumably as a result of scavenging by other animals caught in the tree stumps, so that the skeletons cannot be properly reconstructed. We have no idea of the number of presacral vertebrae, although the individual vertebrae are well preserved. The arches and centra of *Asaphestera* are fused, but there is no evidence of intercentra. The scapulocoracoid is ossified as a single unit, with a tall scapular blade. The clavicle blades are narrow, and the interclavicle has a posterior stem.

The ilium, like that of *Utaherpeton* and most other Carboniferous microsaurs, has both a dorsally and a posterodorsally directed process. The humerus and femur are stout bones, with expanded extremities separated by a slender shaft. The humerus is pierced by an entepicondylar foramen as in other early tetrapods, but resembles those of temnospondyls rather than anthracosaurs in the 90 degree torsion between the proximal and distal articulating surfaces. The marginal teeth of *Asaphestera* are slender and numerous, as are those of several other microsaurs represented only by fragmentary remains.

Three other genera are represented by jaws with a smaller number of large, conical teeth, increasing in size with that of the jaws and skulls (Fig. 4.36). The smallest is *Leiocephalikon*, with a skull about 13 mm in length, with several medial rows of teeth on the coronoid bones. The jaw of *Hylerpeton* is about 35 mm in length, with a few denticles on the narrow coronoids. Only the front of the skull is known in *Trachystegos*, but the teeth in the upper and lower jaws are much larger than those of *Hylerpeton*, and there are also numerous large,

flattened teeth on one or more of the palatal bones. These jaws are not well preserved, but subsequent microsaurs are known to have retained all three coronoids and both splenials (R. L. Carroll and Gaskill, 1978).

The scapulocoracoid, humerus, and femur associated with *Trachystegos* resemble those of *Asaphestera*, with the extremities of the humerus at right angles to one another. Other, smaller bones may represent additional microsaurs, based on their similarity to those of genera known from later horizons.

Microsaurs appear to have adapted to a wider range of environments than most of the Lower Carboniferous groups. The Goreville microsaur and *Utaherpeton* come from definitely aquatic deposits, while all those from Joggins appear primarily terrestrial. As we move up the geological column we will encounter this order in the coal swamp deposits of the late Westphalian, in the Lower Permian red beds deposited in a deltaic environment, and in shallow lakes.

Joggins Temnospondyls

The most common genus found in the Joggins tree stumps was the temnospondyl *Dendrerpeton*, known from approximately a hundred specimens, from which almost the entire skeleton can be reconstructed (Fig. 4.37). It resembles the oldest known temnospondyl *Balanerpeton* in general body form and proportions, but was considerably longer, with the skull up to 11 cm in length. Mature specimens were fully ossified, including the carpals and tarsals. The absence of lateral line canal grooves and presence of stout limbs indicate a basically terrestrial way of life.

Further evidence of terrestriality is provided by the area of the middle ear, recently described by Robinson and his colleagues (2005). X-ray tomography revealed the structure of the stapes and its relationship with the otic capsule and the area of the squamosal, or otic notch (Fig. 4.38). This corresponds sufficiently closely with the middle ear of modern frogs, as discussed by Bolt and Lombard (1985), to indicate a comparable capacity for the reception of airborne vibrations. *Dendrerpeton* provides a good model for the ancestry of most later temnospondyls, which were the dominant labyrinthodonts in the later Paleozoic.

A smaller number of specimens have been recovered of the anthracosaur *Calligenethlon*, which is represented by a long growth series. The largest specimens are the size of small embolomeres, with cylindrical intercentra and nearly cylindrical pleurocentra, but their occurrence in the tree stumps suggests more terrestrial habits than the large European genera that seem to have been limited to large bodies of water.

Amniotes

Despite the numerical dominance of microsaurs and labyrinthodonts, the most distinctive fossils found at Joggins were not of amphibians, but those of the oldest known reptiles, or more inclusively, amniotes, the common ancestors

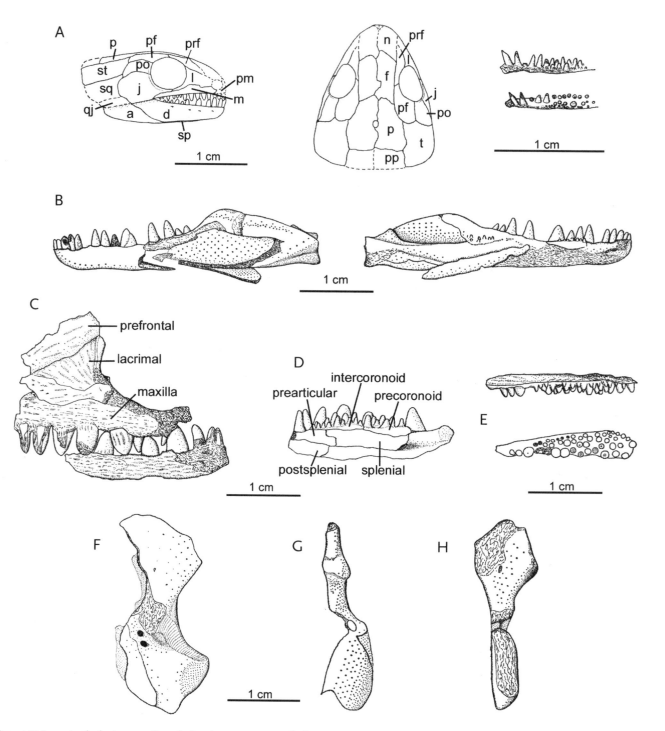

Figure 4.36. Large-toothed microsaurs from the Joggins tree stumps. *A,* Skull of the gymnarthid *Leiocephalikon problematicum* in lateral and dorsal views, plus medial and dorsal views of the dentition of the lower jaw. *B,* Lateral and medial views of the lower jaw of *Hylerpeton dawsoni. C–E, Trachystegos* *megalodon. C,* Antorbital region. *D,* Medial surface of the lower jaw. *E,* Lateral and ventral views of palatal bone; some teeth are flattened by wear. *F–H, Trachystegos,* left scapulocoracoid and two views of left humerus. From R. L. Carroll, 1966.

of modern reptiles, birds, and mammals. The best-known of the Joggins reptiles is *Hylonomus,* which represents a major morphological and adaptive advance from any adequately known tetrapods from earlier in the Carboniferous (Figs. 4.39 and 4.40). The general body form would have closely

resembled that of small, agile iguanid lizards. From the specimens at Joggins and a second, somewhat younger tree stump locality near Sydney, Nova Scotia, almost every bone in the body of the early amniotes is known (R. L. Carroll, 1964c, 1969b). These early amniotes were small, with a skull

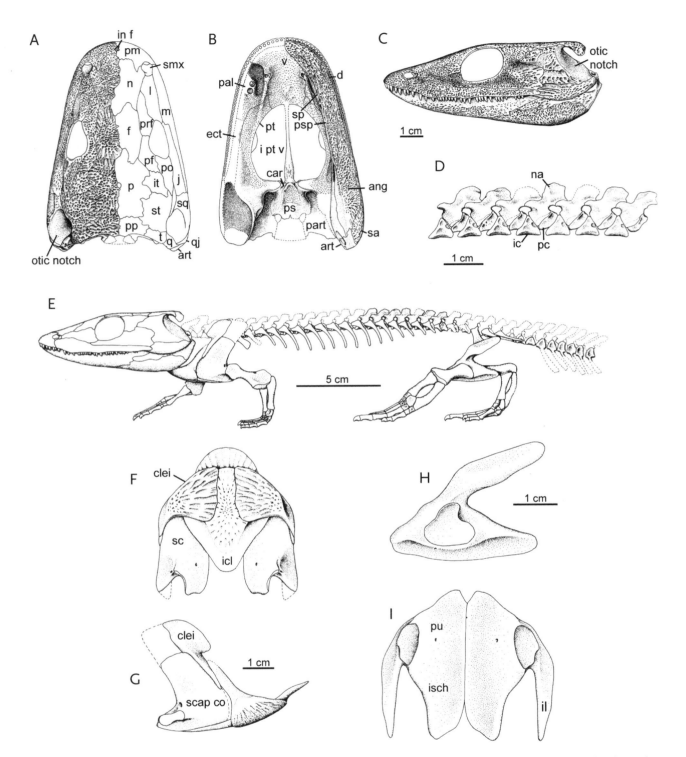

Figure 4.37. Dendrerpeton acadianum, a temnospondyl from the Westphalian A of Joggins, Nova Scotia. *A–C,* Skull in dorsal, palatal, and lateral views. *D,* Anterior trunk vertebrae. *E,* Reconstruction of skeleton. *F–G,* Pectoral girdle in ventral and lateral views. *H, I,* Pelvic girdle in lateral and ventral views. From Holmes et al., 1998.

+ trunk length of about 12 cm, but highly ossified, with long, slender limbs.

The skull alone shows how different the early amniotes were from any other tetrapods from the Carboniferous. There is no trace of a squamosal notch, and the jaw articu-

lation is just behind the back of the skull table. Only a delicate pitting is evident in the dermal bones of the skull, rather than the conspicuous sculpturing of most Paleozoic tetrapods. The intertemporal bone is lost, but all the other paired bones of most Carboniferous amphibians are retained. The

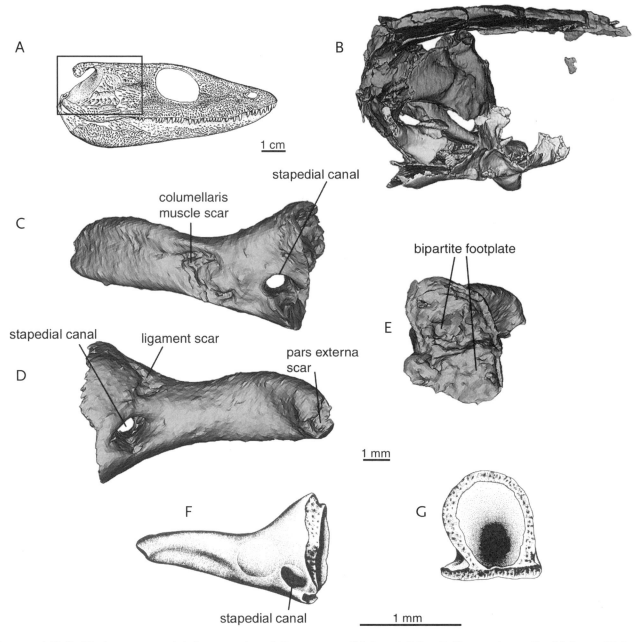

Figure 4.38. A, Skull of *Dendrerpeton;* rectangle indicates area shown in X-ray (*B*). *B*, Interior of skull viewed laterally. *C–E*, Stapes in posterior, anterior, and medial views. *A–E*, from Robinson et al., 2005. *F* and *G*, Stapes of *Doleserpeton* from Lombard and Bolt, 1988.

parietal is relatively wider than in any labyrinthodonts, with the supratemporal reduced to a small splint lateral to the tabular. The postparietals are largely limited to the occipital surface.

A striking feature of the dentition is the presence of canine teeth in a comparable position to those of humans and other carnivorous mammals. The palate is distinct from that of all labyrinthodonts in the absence of large fangs on the lateral bones and from all other Paleozoic tetrapods in the configuration of the pterygoid. Amniotes are unique among Paleozoic tetrapods in the presence of a transverse flange of

the pterygoid, just lateral to the movable basicranial articulation, which bears a row of teeth (Fig. 4.41). Except for the loss of these teeth, the transverse flange has the same configuration as that of primitive living lizards, where the margin extends ventrally to provide a surface for the origin of the pterygoideus muscle, which is one of the major jaw adductors. Neither the flange nor this muscle is known in primitive living amphibians.

The occiput, better known in the Sydney genus *Paleothyris*, exhibits a unique combination of features diagnostic of early amniotes. The occipital condyle, unlike that of microsaurs, is hemispherical, so that it serves as the ball of a ball-and-socket joint linking the skull with the first cervical vertebra. The configuration of this joint allowed controlled movements in all planes, in contrast to being restricted to vertical hinging in a transverse plane, as in microsaurs and advanced nectrideans. A large supraoccipital links the otic capsules with the postparietal bones. The stem of the stapes is oriented ventrally toward the jaw articulation, as in other early amniotes, and would not have served as part of an impedance matching system for the detection of airborne vibrations, a capacity probably not evolved among amniotes until the early Mesozoic. The lower jaws of early amniotes differ from those of Paleozoic amphibians in the reduction of the number of coronoid bones from three to two, and the number of splenials to one.

As in all primitive amniotes, the vertebral column consists of cylindrical pleurocentra, fused to the neural arches in mature specimens, with small, intervening crescentic intercentra. The disarticulated vertebral column of *Hylonomus* is estimated to consist of approximately 25 presacral vertebrae, while *Paleothyris* has a count of 32. The atlas-axis complex resembles that of the embolomere *Proterogyrinus* except for the fusion of the axis intercentrum with the atlas pleurocentrum and of the axis arch with its pleurocentrum. The structure of the atlas-axis complex of early amniotes evolved ultimately from that of the multipartite pattern of early labyrinthodonts, but is highly divergent from that of all the orders grouped among the lepospondyls.

The ribs are long and cylindrical throughout the trunk, curving well around the flanks as in modern lizards. This is suggestive of costal respiration, as in amniotes (including mammals), in contrast with the buccal respiration of most amphibians.

Very little of the scapulocoracoid is known in *Hylonomus*, but that of *Paleothyris* is ossified as a single unit and extends far dorsally. The cleithrum is not known in either genus, but is present in later amniotes. The clavicle retains a fairly wide ventral blade. The interclavicle is widely expanded anteriorly and has a long, narrow posterior stem.

The humerus is about three-quarters the length of the skull, with a long shaft between proximal and distal extremities set at right angles to one another. There is a large entepicondylar foramen posteriorly, and a medially tapering supinator process anteriorly. The hemispherical radial condyle extends directly ventrally. The lower portion of the forelimb is better seen in *Paleothyris,* in which the ulna and radius are short, slender elements. The carpals are fully articulated and consist of the intermedium and radiale in the proximal row. Two centrale and the ulnare, together with a neomorphic pisiform bone, make up the second row. There are five distal carpals. As in most other early amniotes, the phalangeal count is 2, 3, 4, 5, 3.

The three elements of the pelvis are closely integrated in both genera. There is a single posteriodorsally oriented iliac blade, but with a widened area anteriorly that may represent a remnant of the primitive dorsal process, although it is in the same plane as the rest of the ilium. The femur is long and narrow in both genera, but more so in *Paleothyris*. The lower limb and foot are particularly well preserved in *Hylonomus* (Fig. 4.40). In contrast with nearly all other Paleozoic tetrapods, those of amniotes have fused the tibiale, proximal centrale, and intermedium of primitive tetrapods into a single bone, the astragalus. The fibulare remains as a single element, but is now termed the calcaneum. *Hylonomus* appears to have only a single distal centrale, but a second small bone is found medial to it in *Paleothyris* (Fig. 4.42).

Incomplete skeletons found at Joggins have been suggested as belonging to additional amniote genera. One is represented by humeri larger than those of the type of *Hylonomus,* which might indicate the presence of a second group of early amniotes related to the ancestry of mammals. Several skeletons of mammal-like reptiles (or synapsids) have been described from the younger tree stump locality of Sydney, Nova Scotia (Reisz, 1972), demonstrating an early stage in the radiation of amniotes, which will be further discussed in Chapter 7.

Joggins and the locality near Sydney, Nova Scotia, are unique in the preservation of diverse tetrapod remains within the stumps of upright lycopods. Individual, incomplete skeletons have been described from two other localities in Nova Scotia, but nowhere else in the world. This area seems to be unique in the combination of specific geological processes leading to the repeated burial of lycopod trees in an upright position, their subsequent exposure on a new land surface, and the particular pattern of decay of the lycopod genus *Sigillaria*.

The presence of a great diversity of terrestrial vertebrates, and especially of the amniote *Hylonomus,* led to the recognition of the Joggins locality as a UNESCO World Heritage site in July 2008.

COAL SWAMP DEPOSITS

Continuing the description of distinct groups of Carboniferous tetrapods in the sequence of their appearance in the fossil record, we move from the highly terrestrial genera common to the tree stumps of Joggins to a series of shallow water deposits in which were preserved a sequence of faunas ranging in age from the Westphalian A (just prior to Joggins) into the Westphalian D of Europe and North America.

The localities of Jarrow (Westphalian A of the Irish Republic), Linton, Ohio, and Nýřany, Czech Republic (both from the Westphalian D), are all thought to represent deposition in small, oxbow lakes. All are coal deposits that accumulated over very short periods of time, probably lasting no more than a few thousand years, and may individually have been no larger than an acre or so in extent (Behrensmeyer et al., 1992). These are clearly but a tiny sample of the faunal and ecological diversity of this time period. This seems strange in light of the huge areas in Europe and North

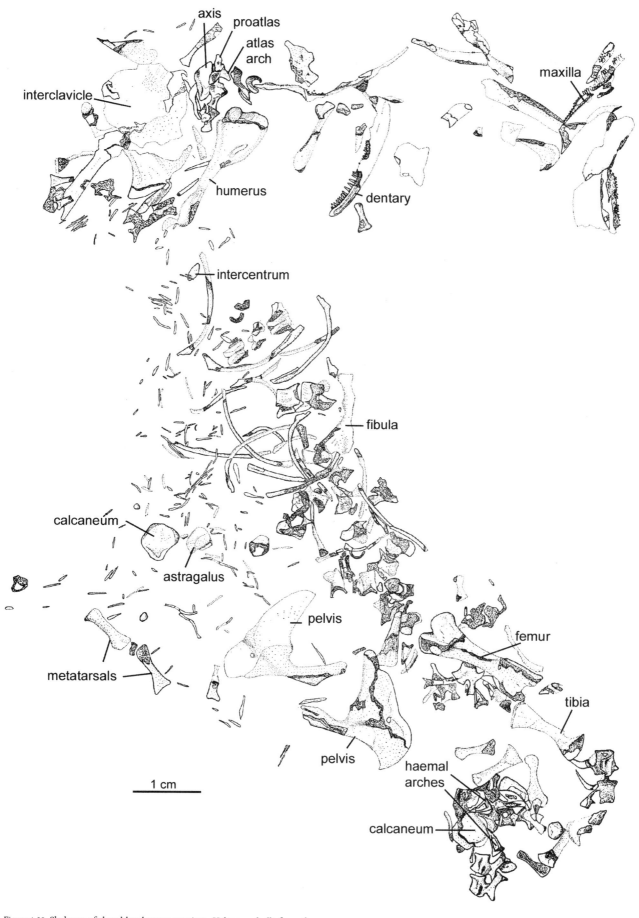

Figure 4.39. Skeleton of the oldest known amniote, *Hylonomus lyelli,* from the tree stump fauna of Joggins, Nova Scotia. From R. L. Carroll, 1964c.

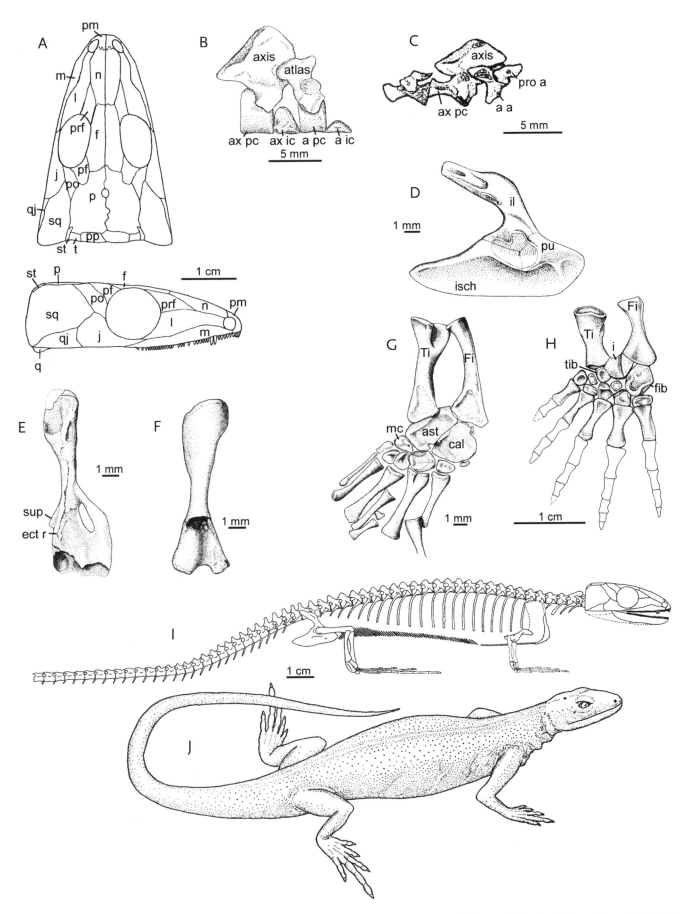

Figure 4.40. *Hylonomus lyelli. A,* Reconstruction of skull in dorsal and lateral views. *B,* Atlas-axis of the anthracosaur *Gephyrostegus. C,* Comparable view of the atlas-axis complex of *Hylonomus. D,* Lateral view of *Hylonomus* pelvis. *E,* Humerus of *Hylonomus* in dorsal view; the distal end is represented by a natural cast, showing the radial condyle. *F,* Dorsal view of femur. *G,* Lower rear limb of *Hylonomus* showing the distinct astragalus and calcaneum of amniotes. *H,* Comparable bones of the primitive "reptiliamorph" *Westlothiana,* showing the larger number of proximal tarsals. *I,* Reconstruction of the skeleton of *Hylonomus. J, Hylonomus,* as it may have appeared in life. *A, B, G,* and *I* from R. L. Carroll, 1964c; *C* and *J* from R. L. Carroll, 1970c; *D, E, F,* and *H* from Smithson et al., 1994.

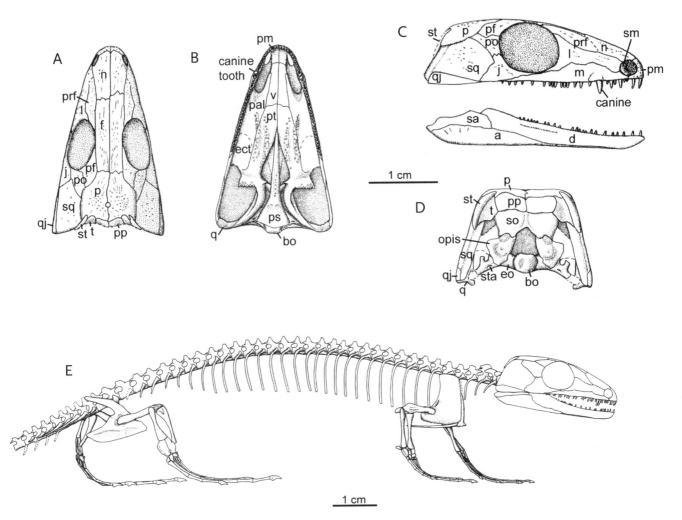

Figure 4.41. *Paleothyris acadiana*, an amniote from the Westphalian C-D tree stump locality near Sydney, Nova Scotia. *A–D*, Skull in dorsal, palatal, lateral, and occipital views, restored on the basis of three skulls of slightly different sizes. *E*, Reconstruction of skeleton. From R. L. Carroll, 1969b.

America from which Carboniferous coal deposits have been mined over the past two centuries, but can be explained by the fact that most coal deposits have a low pH because of the acids produced by decaying vegetation. This normally results in dissolving the bones. Only rarely is enough calcium carbonate dissolved in the water to buffer the acid. In fact, many of the skeletons from Jarrow occur as "ghosts" in which the prior presence of skeletons is indicated by a raised outline on the surface of the coal, but there is no bone underneath. This is comparable to the preservation of mummified bodies of humans lacking the skeleton in some bogs in northern Europe where they have been buried over the past thousand or more years.

Another locality, at the base of the Westphalian D, which also has some elements of these coal swamp faunas is Mazon Creek in Illinois, just south of Chicago. It was not a coal swamp, but rather represented a fluvial or lacustrine to lagoonal or estuarine environment.

Among the most common fossils of these deposits are the remains of another group that has not been discovered earlier in the Carboniferous, the nectrideans, recently reviewed by Kathleen Bossy and Angela Milner (1998), but they also include labyrinthodonts, microsaurs, aïstopods, and a few remains of amniotes.

Nectrideans

Nectrideans are among the most readily characterized of the small Paleozoic tetrapods (Figs. 4.43–4.45; Plate 8). The primitive genera broadly resemble newts, with short bodies, small limbs, and a long, laterally compressed tail. Most appear habitually aquatic. The oldest nectrideans, belonging to two families, are known from the Jarrow locality in Ireland, where they occur with primitive temnospondyls and the aïstopod *Ophiderpeton brownriggi*. They are already well differentiated from one another, suggesting a considerable period of prior evolution, but without evidence from the fossil record. A single specimen of *Dendrerpeton* also occurs at Jarrow (A. R. Milner, 1980a).

The most primitive of the Jarrow nectrideans is *Urocordylus*, initially described by Wright and Huxley (1866) and more recently discussed by Bossy and Milner (1998). *Urocordylus* clearly represents a distinct group, with nothing similar

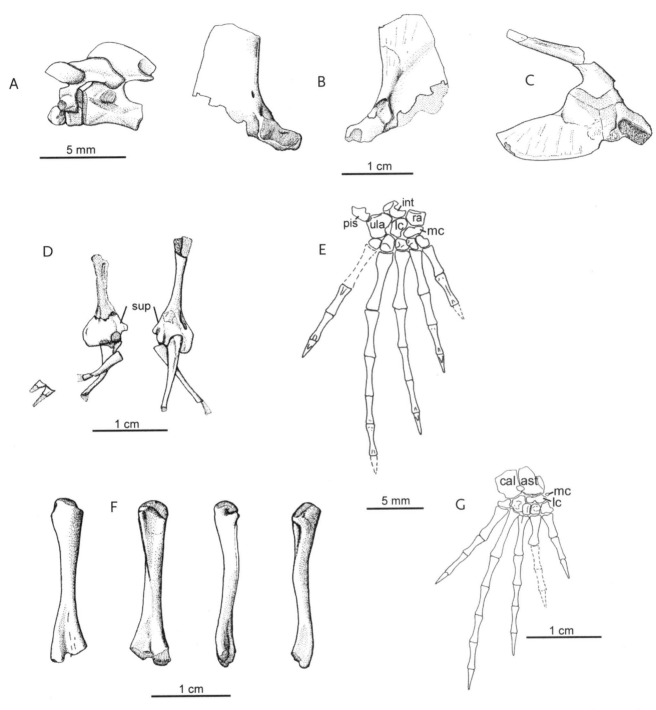

Figure 4.42. Postcranial element of *Paleothyris*. *A*, Atlas-axis complex; note fusion of axis intercentrum and pleurocentrum. *B*, Scapulocoracoid in lateral and medial views. *C*, Pelvis in medial view. *D*, Humeri in ventral and dorsal views, and adjacent ulna and radius. *E*, Reconstruction of carpus and manus, in ventral view. *F*, Femur in dorsal, ventral, anterior, and posterior views. *G*, Reconstruction of tarsus and pes in dorsal view. From R. L. Carroll, 1969b.

known from the earlier Carboniferous. The most obvious feature is the extremely long tail, with very similar haemal and neural arches expanded distally to form an effective structure for sculling. There are only about 19 trunk vertebrae in this genus, and no nectridean has more than 26. The limbs are not long, but neither are they seriously reduced. In contrast with microsaurs, *Urocordylus* retains five toes in the manus, although this is reduced to four in all other nectrideans. On the basis of later urocordylids, the phalangeal formula is reconstructed as 2, 3, 4, 3, 2. Unlike early microsaurs, nectrideans have only a single, dorsoposteriorly angled iliac blade. The pubis and ischium are loosely attached. The femur is stout and the tibia and fibula about two-thirds its length. No tarsals are known in this genus, but the pes shows the

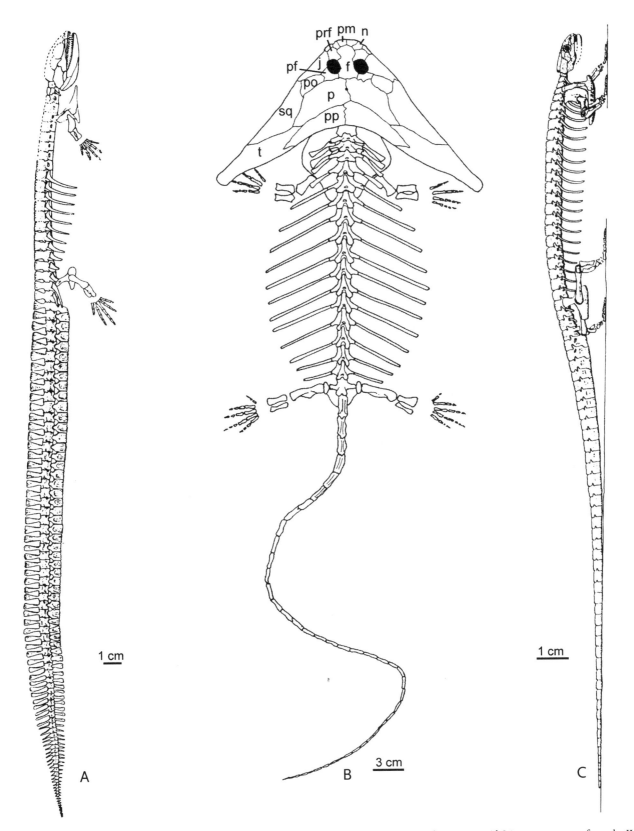

Figure 4.43. Skeletal reconstructions of the three families of nectrideans. *A,* The urocordylid *Urocordylus wandesfordii* from the Westphalian A of the Republic of Ireland. *B,* The diplocaulid *Diplocaulus magnicornis,* from the Lower Permian of Texas. *C,* The scincosaurid *Scincosaurus crassus,* from the Westphalian D of the Czech Republic. From R. L. Carroll et al., 1998.

complete phalangeal count of 2, 3, 4, 4, 2. Some later genera have only four toes.

The skull is not well known in *Urocordylus,* but it is in later members of the same family. In *Urocordylus,* it is relatively short, with the eyes near the middle of its length. The orbit does not extend to the maxilla, but it does in some other members of the family. In most nectridean genera the prefrontal extends into the margin of the external nares, as seen in the earliest microsaurs and also colosteids, but this characteristic probably evolved separately in nectrideans, in which the narial opening is farther posterior to the end of the snout than it is in other early tetrapods. No nectridean has a squamosal notch and the jaw articulation is typically, but not always, at the level of the occiput. In these features they resemble early microsaurs. Where the back of the skull table is preserved, however, it is clearly distinguished by the retention of the supratemporal, which extends posteriorly from the postorbital to the tabular as a narrow splint of bone. The tabular is shorter than the medial edge of the postparietal, but reaches the parietal laterally, a combination of features that distinguishes it from both anthracosaurs and other early labyrinthodonts. The absence of an intertemporal distinguishes all nectrideans from primitive labyrinthodonts. The family Urocordylidae is unique among nectrideans in having a very loose attachment of the cheek and the skull table, which in fact extends forward to the antorbital overlap between the lacrimal and prefrontal (which itself may explain the anterior extension of the prefrontal) (Fig. 4.44). Exactly why urochordylid nectrideans should have such a kinetic skull is more difficult to explain. Bossy and Milner (1998) argued that the dorsal movement of the snout allowed reorientation of the tips of the long recurved premaxillary teeth to a position suitable for impaling prey. Neither of the two other nectridean families have such mobility between the skull bones.

Despite the almost certainly aquatic habits of urocordylids, none show lateral line canal grooves on the skull, but they are present in a second family, the Diplocaulidae.

Most nectrideans, including *Urocordylus,* have an interpterygoid vacuity, as do temnospondyls and many microsaurs, but whether or not this feature is homologous can only be determined from knowledge of more primitive members of each of these clades. They also have a movable basicranial articulation, a feature common to most primitive tetrapods. As in microsaurs, the marginal bones of the palate carry denticles and small teeth, but not the fang and pit pairs common to labyrinthodonts. In most genera the marginal teeth are small and all lack labyrinthine infolding. The stapes is not known in any nectrideans.

The nature of the occipital condyle is not well known in most of the coal swamp genera, but later, more three-dimensionally preserved specimens show two conspicuous condyles arising from the exoccipitals, with the basioccipital recessed. The articulating surface of the atlas is not as laterally expanded as that of most microsaurs, but it would have had the same functional properties of facilitating dorsoven-

tral flexion of the skull on the trunk, but limiting rotation or lateral bending. The braincase is typically poorly ossified, revealing no taxonomically significant features.

The lower jaws are also distinctive in nectrideans, with only one of the original three coronoids and one of the two splenials being retained. Urocordylids have two openings in the medial surface, a large posterior fenestration vaguely reminiscent of that of advanced embolomeres and a smaller one anteriorly. The very short jaws of the diplocaulids have only very small medial openings.

In contrast with microsaurs, no nectrideans possess intercentra, and the pleurocentra are always fused to the neural arches. The haemal arches are not located between the pleurocentra as in all other groups of early tetrapods, but are fused midway in the length of the centra, directly beneath the neural spines. Nectrideans have a second set of articulating facets, the apophyses, above the zygapophyses in both the trunk and the tail. Upper Carboniferous aïstopods also have accessory articulating surfaces between the neural arches in the trunk region, but they are not present in the earliest genus, *Lethiscus,* and so were probably not a primitive characteristic of that group, but evolved separately in the two orders.

Nectrideans retain widely expanded clavicular blades and a large interclavicle, both with conspicuous sculpturing in most genera. The posterior portion of the interclavicle may be rounded or pointed, but without the clearly defined stem of anthracosaurs or microsaurs. The scapulocoracoid is frequently poorly ossified. The humerus of *Urochordylus* retains the primitive configuration of early labyrinthodonts, with very little torsion, but lacks an entepicondylar foramen. The ulna and radius are slender, but may be nearly as long as the humerus. The carpals typically lack ossification. The proportions of the hind limb resemble those of the front.

Urocordylus belongs to the most diverse family of nectrideans, the Urocordylidae, which extends to the end of the Carboniferous in Europe and into the Lower Permian in North America. A second family, the Diplocaulidae, accompanies *Urochordylus* in Jarrow, represented by *Keraterpeton.* Diplocaulids are clearly distinguished from other nectrideans by the broad and flattened skulls (Figs. 4.43, 4.45). The skull table is better integrated with the cheek and the supratemporal bone is lost. The tabular bones extend well behind the postparietals in *Keraterpeton,* and in later genera the entire cheek region is drawn out posteriorly, far behind the level of the quadrate and occipital condyle. The quadrate is level with or anterior to the occiput. The palate of *Keraterpeton* is closed, which might be the primitive condition for nectrideans, but becomes more open in later diplocaulids. The basicranial articulation is immobile. In contrast with urochordylids, lateral line canals are present. No genera are known to have more than 18 presacral vertebrae. In early members of this family the dorsal margin of the cleithrum is greatly expanded posteriorly. *Diceratosaurus* is unique in having a second pair of ventral bones behind the clavicle blades termed accessory

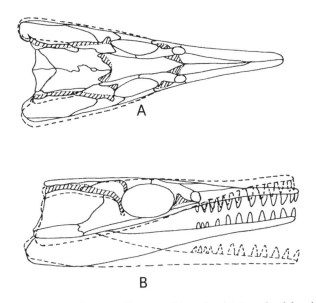

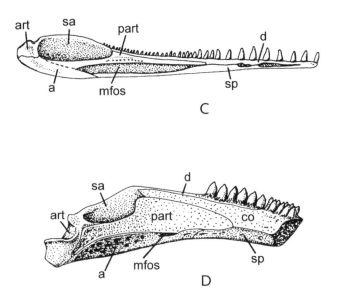

Figure 4.44. Nectridean cranial kinesis and jaws. *A* and *B*, Lateral and dorsal views of the skull of the urocordylid *Sauropleura pectinata* showing areas of extensive overlap of bone along kinetic joints (cross-hatched) and extent of movement of snout relative to the back of the skull (dashed lines) as the jaws are opened and closed. The cheeks appear capable of lateral movement as well. *C,* Medial view of the lower jaw of *Sauropleura pectinata. D,* Medial view of the lower jaw of the diplocaulid *Diploceraspis.* Note great differences in the size of the mandibular fossa and the loss of the postsplenial and coronoids. From R. L. Carroll et al., 1998.

dermal elements. *Diplocaulus* is extremely common in the upper part of the Lower Permian of the Texas red beds, with its crescentic head reaching more than 20 cm in width. Three partial skulls and three presacral vertebrae of *Diplocaulus* have also been found in the lower part of the Upper Permian of Morocco (Dutuit, 1988).

The third nectridean family, the Scincosauridae, has very different body proportions with a very small head to trunk ratio (Fig. 4.43C). The tail is very long, but lacks the elongate neural and haemal arches of urochordylids. It is known from only two genera. *Scincosaurus* is known primarily from Nýřany, but with one specimen from the Stephanian of France. *Sauravus* is also known from France, where it extends into the Lower Permian. *Scincosaurus,* the best-known member of this family, is enigmatic in combining primitive and derived features. It is derived in lacking both the postparietal and the supratemporal, but unlike all other nectrideans, retains an entepicondylar foramen. Also, the carpals and tarsal are well ossified, as would be expected of a more terrestrial way of life. This way of life might have been more primitive than the obligatorily aquatic nature of urochordylids and diplocaulids.

Most specimens of *Scincosaurus* came from Nýřany, a clearly aquatic locality, but it is presumed that they had come back to the water to breed, rather than being permanent residents of the lake. Since it is one of the later members of this order, it is probably not representative of a primitive stock, unless it was already highly terrestrial and so less likely to be preserved in the more common aquatic localities. Whatever the taxonomic position of *Scincosaurus* within the nectrideans, it provides no evidence for the affinities of this order with other amphibian orders.

Coal Swamp Microsaurs

While microsaurs were the most taxonomically diverse group in the fully terrestrial locality of Joggins, Nova Scotia, numerous genera are also known from the coal swamp localities of Linton and Nýřany, although they have not been confirmed from Jarrow. *Tuditanus,* known only at Linton, is thought to be related to *Asaphestera* from Joggins. It has a relatively short trunk and well-developed limbs, suggestive of a terrestrial way of life (Fig. 4.46). It is known from only two skeletons, and so may have spent most of its time outside the lake.

As a microsaur, *Tuditanus* is interesting in having evolved an amniote-like astragalus by the fusion, indicated by traces of the lines of attachment, between three proximal elements of the tarsus. The foot has an amniote-like phalangeal count of 2, 3, 4, 5, 4, but the terminal phalanges are uniquely hoe-shaped, with their tips expanded laterally and bent ventrally. The hand has a phalangeal count of 2, 3, 4, 3 (similar to that of basal temnospondyls). The ilium retains the bifurcate dorsal process seen in earlier microsaurs.

Three other microsaurs, all grouped in the more derived suborder, Microbrachomorpha, are known from Linton or Nýřany. Only *Odonterpeton* is known from Linton. *Microbrachis* and *Hyloplesion* are known from Nýřany. All these genera were adapted to a more aquatic way of life, with elongation of the trunk and reduction of the limbs and number of digits. Lateral line canal grooves are well developed around the orbits, on the cheek, and on the lateral surface of the lower jaw in *Microbrachis,* which has a conspicuous pattern of sculpturing (Fig. 4.47). *Microbrachis* retains gill rakers in the larger specimens, but does not exhibit external gills, even in the smallest individuals. Palpebral cups are present in the

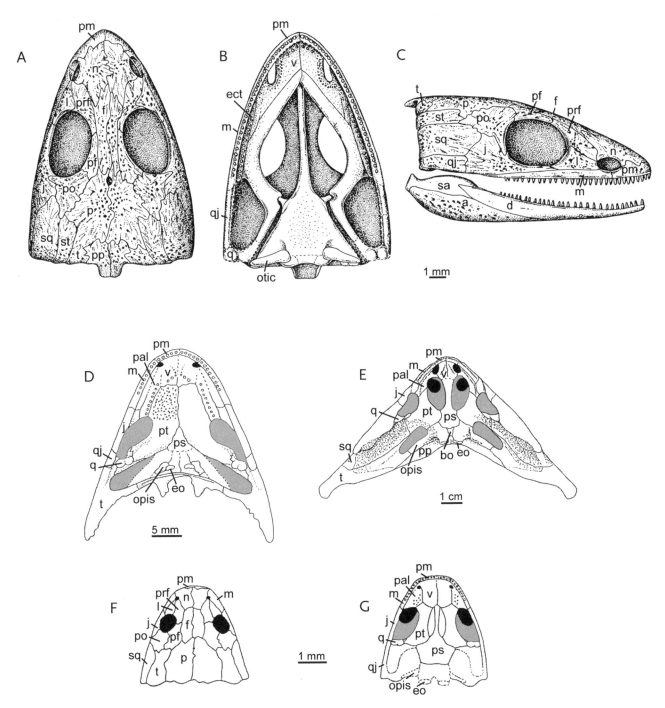

Figure 4.45. Nectridean skulls. A–C, Dorsal, palatal, and lateral views of the urocordylid *Ptyonius marshii*, Westphalian D of Linton, Ohio. D, Palate of the European diplocaulid *Batrachiderpeton reticulatum*, from the Westphalian B of England. E, The palate of *Diplocaulus* from the Lower Permian of North America. F and G, Dorsal and palatal views of the skull of *Scincosaurus crassus* from the Westphalian D of the Czech Republic. All from R. L. Carroll et al., 1998.

eyelid. *Hyloplesion* lacks the sculpturing as well as the lateral line canals, which might have been supported by soft tissue superficial to the skull bones. The lower jaws retain the three coronoids and two splenial of primitive tetrapods.

Microbrachis has 38 vertebrae, but unlike the Goreville microsaurs and long-bodied tuditanomorphs, there is no

trace of intercentra. The iliac blade has lost the terminal bifurcation. The interclavicle is fimbriated. The carpus is without ossification but the tarsus ossifies the intermedium. The phalangeal count of the manus is 2, 3, 3, that of the pes is 2, 3, 4, 4, 3.

The skull of *Hyloplesion* differs in the loose attachment of

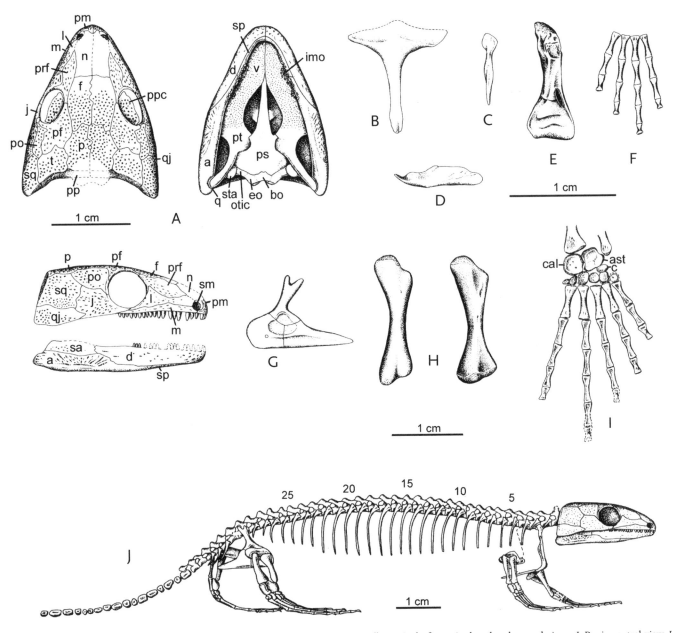

Figure 4.46. *Tuditanus punctulatus*, a microsaur from the Westphalian D coal swamp locality of Linton, Ohio. *A*, Skull in dorsal, palatal, and lateral views. *B*, Interclavicle. *C*, Cleithrum. *D*, Blade of clavicle. *E*, Humerus, with entepicondylar foramen. *F*, Manus, with unique hoe-shaped unguals. *G*, Pelvic girdle. *H*, Right femur in dorsal and ventral views. *I*, Pes in ventral view. *J*, Reconstruction of skeleton. The configuration of distal caudal vertebrae suggests regrowth after autotomy. From R. L. Carroll and Gaskill, 1978; R. L. Carroll et al., 1998.

the tabular, which is smaller than in other microsaurs and fits into a notch in the parietal (Fig. 4.48). One anterior maxillary tooth is enlarged as a canine. There are only 30 presacral vertebrae and both the carpals and tarsals are fully ossified. The scapulocoracoid is well developed in *Hyloplesion*, but much less well in *Microbrachis*. *Microbrachis* retains an entepicondylar foramen, but it is lost in *Hyloplesion*. *Hyloplesion* has three digits on the manus, with a phalangeal count of 2, 3, 3, but the middle carpus in the distal row has small lateral processes that might be interpreted as evidence for the prior existence of five distal carpals. The pes has a phalangeal count of 2, 3, 4, 5, 2+.

Odonterpeton is known from only the anterior part of the body of a single specimen from Linton, Ohio (Fig. 4.49). In relationship to its very small size, the skull is highly modified, with the tabular totally lost, the postparietals fused medially, and the pineal opening at the junction of the frontals and parietals. The carpals are not ossified and the manus has a phalangeal count of 2, 4, 3. This genus is unique among microsaurs in the absence of lateral expansion of the anterior articulating surface of the atlas.

Coal Swamp Aïstopods

We first encountered the extremely long-bodied and limbless aïstopods near the very base of the Carboniferous. Specimens were also noted at the East Kirkton and Jarrow

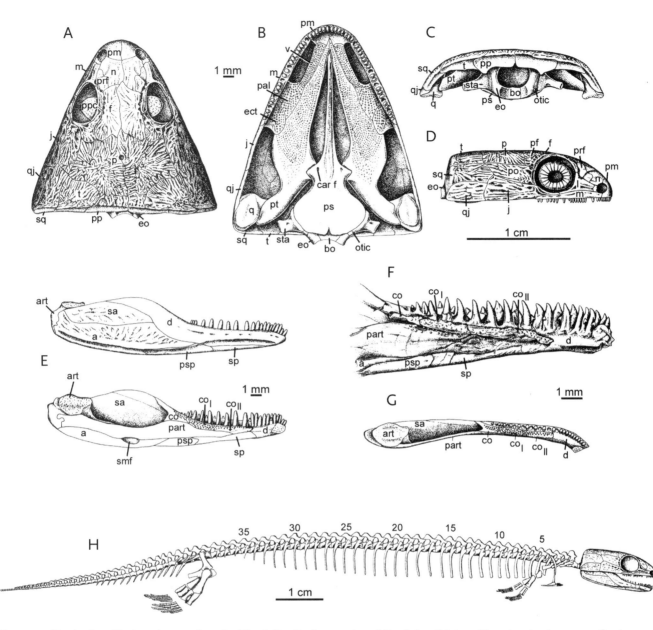

Figure 4.47. Microbrachis pelikani, a microsaur from the Westphalian D of Nýřany, Czech Republic. *A*, Skull in dorsal view. *B*, Palate of a slightly larger specimen. *C*, Occiput. *D*, Lateral view. *E*, Lower jaw in lateral and medial views. *F*, Detailed medial view of lower jaw showing crowns of replacement teeth. *G*, Dorsal view of lower jaw showing coronoid dentition. *H*, Reconstruction of skeleton. From R. L. Carroll and Gaskill, 1978.

localities, but numerous well-preserved specimens are not common in deposits earlier than Linton, Nýřany, and Mazon Creek in the Westphalian D (Anderson, 2002, 2003a, b; R. L. Carroll, 1998a). These specimens provide detailed knowledge of some of the most highly specialized amphibians that ever lived. All are immediately distinguishable on the basis of an exceedingly long vertebral column and the almost complete absence of the appendicular skeleton. Aïstopods are also noteworthy in exhibiting the greatest amount of skeletal change during their evolution of any of the Carboniferous orders (Figs. 4.50–4.54).

Above the level of *Lethiscus*, aïstopods have long been divided into two families, the Ophiderpetontidae and the Phlegethontiidae. The ophiderpetontids, first recognized at Jarrow, are characterized by the anterior position of the orbits, the covering of the cheek region by dermal ossicles, and the presence of osteoderms in the trunk. They are not known past the Carboniferous. The phlegethontiids are distinguished by the extensive loss of dermal bone dorsal to the braincase, the position of the orbits near the mid-length of the skull, the loss of osteoderms, and the greater length of the vertebral column. They extend from the Upper Carboniferous into the Lower Permian.

OPHIDERPETON AND OESTOCEPHALUS. It is now recognized that *Lethiscus* (Fig 4.8), the East Kirkton species *Ophiderpeton kirktonense,* and the Jarrow *Ophiderpeton brownriggi* represent a single lineage, and should not be separated into separate families. *Lethiscus* remains poorly known, but

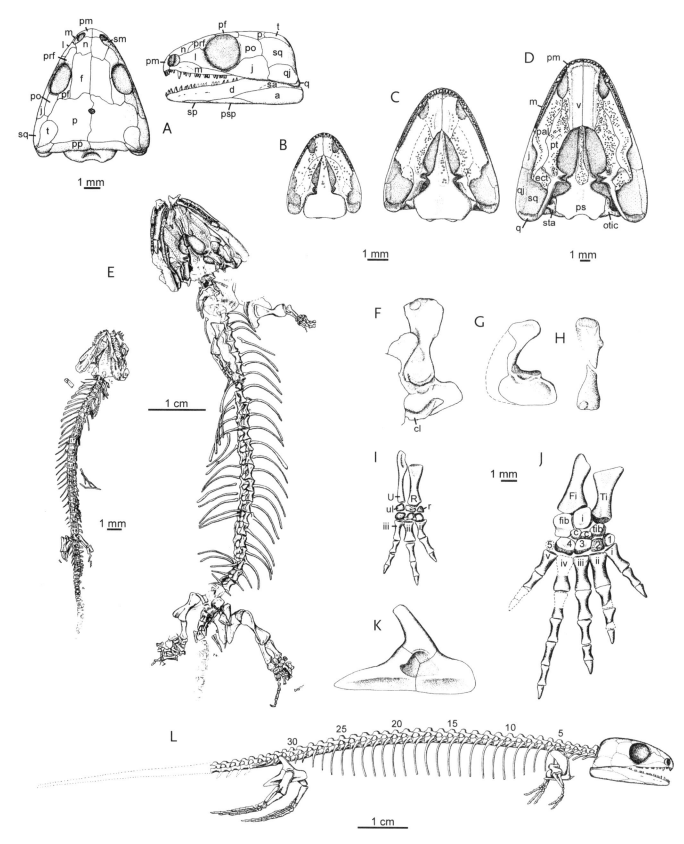

Figure 4.48. *Hyloplesion longicostatus*, a microbrachomorph microsaur from the Westphalian D of Nýřany, Czech Republic. *A*, Dorsal and lateral views of skull. *B–D*, Growth stages in the development of the palate. Note the progressive posterior extension of the jaw articulation. *E*, Skeletons of a juvenile and an adult in ventral view. Note the fully cylindrical caudal centra in the tiny juvenile. *F–H*, Shoulder girdle and humerus. *I*, Lower forelimb. *J*, Lower rear limb. *K*, Pelvic girdle; note loss of anterodorsal iliac process. *L*, Reconstruction of skeleton. From R. L. Carroll and Gaskill, 1978.

most of its observable features resemble those of *Ophiderpeton*. These genera are succeeded in the Westphalian D by *Oestocephalus*, which is known from a single species, *amphiuminus*, ranging from the eastern European locality of Nýřany in the Czech Republic to Linton and Mazon Creek in central North America.

Oestocephalus is one of the best-known aïstopods. The skull is similar to that of the Lower Carboniferous genera in the anterior position of the eyes and the large opening in the cheek, covered by osteoderms. Like all lepospondyls it lacks an intertemporal bone and can be distinguished from *Lethiscus* and *Ophiderpeton* by the absence of the postorbital, which was reduced in size and excluded from the orbital margin in the transition between these genera. *Ophiderpeton kirktonense* and *Oestocephalus* definitely lack a direct connection between the back of the maxilla and the quadrate, and this was apparently the case for *Lethiscus* as well. This area is best known in *Oestocephalus*, where the great reduction in the lateral bones of the cheek exposes the palate in lateral and ventral views. It can be seen that the bones of the palate are incorporated into a single structural element that resembles the palatoquadrate in osteolepiform fish. In genera such as *Eusthenopteron*, a dorsal layer of bone, consisting of the epipterygoid and quadrate, is closely integrated with the pterygoid, vomer, palatine, and ectopterygoid, which form the palate in tetrapods. The epipterygoid remains a very large bone in anthracosaurs, where it extends posteriorly to the quadrate, but is much reduced in most other tetrapods. In *Oestocephalus* and other aïstopods, the area of the palatoquadrate appears to retain the structural role it had in *Eusthenopteron*, linking the quadrate, epipterygoid, and dermal bones of the palate into a single unit. Although it is possible that this condition was re-evolved in aïstopods, it is more parsimonious to assume that it was retained from the condition in *Eusthenopteron* and early tetrapods such as anthracosaurs. The elaboration of the palatoquadrate would appear to distinguish aïstopods from all other lepospondyls. The vomer and palatine are tentatively identified as distinct bones in *Lethiscus* and *Oestocephalus*, but the ectopterygoid has not been reported in this order.

The integration of the palatal bones within a single structural element would have provided strong support for the quadrate to compensate for the great reduction of the original bones of the cheek. The opening of the cheek region in early aïstopods may have been associated with a major change in the distribution of the jaw musculature, part of which, in contrast with all other Paleozoic tetrapods, appears to have inserted on the lateral surface of the lower jaw, as in mammals. This is suggested by the presence of a recessed area on the dorsolateral surface of the surangular bone in both *Oestocephalus* and *Lethiscus*.

The skull roof becomes progressively narrower from *Lethiscus* through *Odonterpeton kirktonense* to *Oestocephalus*. The rear of the skull table maintains the primitive pattern, with a relatively large postparietal, a narrow supratemporal,

and a small tabular at the corner of the skull table that does not reach the parietal.

The jaw articulation is somewhat behind the end of the skull table in *Odonterpeton*. In *Oestocephalus*, it is level with the occiput in small specimens (Fig. 4.51), but with increased size, it becomes displaced posteriorly, giving a progressively greater gape. This is further accentuated in *Coloraderpeton*. This would seem to recapitulate the pattern of growth seen in labyrinthodonts. The lower jaw in *Oestocephalus* has apparently lost all the coronoids and one of the two splenials common to early labyrinthodonts and microsaurs. These elements are also reduced within nectrideans.

The dorsal and lateral margins of the occiput in *Oestocephalus* are formed by flanges of the postparietal, tabular, squamosal, and quadratojugal, but with no clearly defined post-temporal fenestra. What is of much greater significance is the clearly unique configuration of the articulating surface for the atlas vertebra. As seen in *Oestocephalus*, *Phlegethontia*, *Sillerpeton*, and apparently *Odonterpeton*, the occipital articulation is in the shape of a conical recess, comparable with that of the anterior surface of the atlas centrum. These surfaces were in contact only at their edges. This may be interpreted as an abbreviation of the condition seen in *Eusthenopteron*, in which the notochord extends well into the back of the braincase. In aïstopods, the configuration of the space between the occiput and the first cervical is exactly like that between each of the trunk vertebrae, in which the notochord expanded from its constricted configuration in the mid-length of the centra. As is the case for all of the trunk centra, the anterior articulating surface of the atlas is slightly smaller than that of the ring into which it fits. The configuration of the articulation between the occiput and the atlas centrum in aïstopods is unique among land vertebrates and would have allowed limited movement in all planes. *Oestocephalus*, and probably other aïstopods, have a wide, median proatlas that articulates with the atlas arch, although the occiput has no specialized area for its articulation.

The trunk vertebrae are distinguished in having long, amphicoelous, cylindrical centra, fused without a suture to the neural arch. All lack both intercentra in the trunk region and also their caudal homologues, the haemal arches. They appear hourglass-shaped in longitudinal section. Foramina for the spinal nerves pierce the neural arches of all the trunk vertebrae, except in *Lethiscus*, where they are only known to occur in the posterior trunk region. The spinal nerves pass between the neural arches of most other Paleozoic tetrapods. The only exception occurs in two Permian nectrideans, *Crossotelus* and *Sauropleura bairdi*. They may have been present in Carboniferous genera, but most coal swamp fossils are so flattened that they would be difficult to recognize.

Another vertebral specialization, known in *Odonterpeton kirktonense* and *Oestocephalus* but not in *Phlegethontia*, are accessory vertebral articulations above the zygapophyses. Accessory vertebral articulations are also known in all nectrideans, but this character must have evolved separately in the

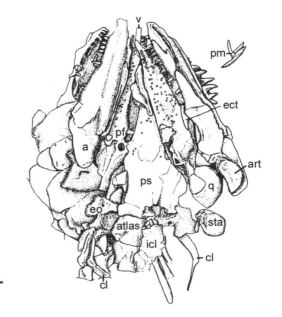

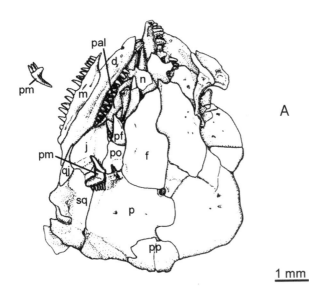

A

1 mm

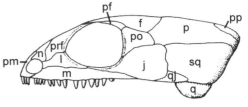

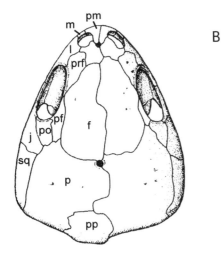

B

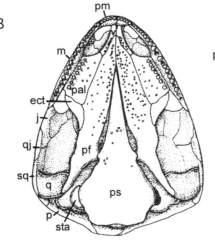

C

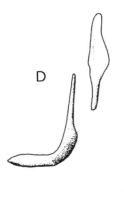

D

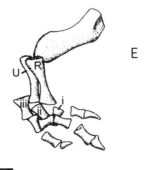

E

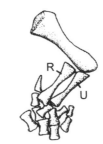

F

1 mm

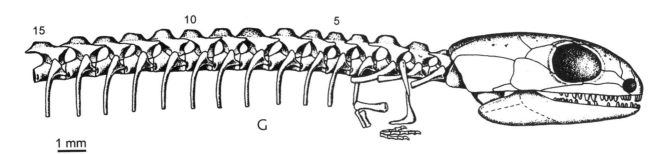

G

1 mm

two groups, since they are not evident in the most primitive aïstopod *Lethiscus*. Caudal vertebrae are recognized by the presence of paired, longitudinal ridges on the ventral surface of the centra.

According to Anderson (2002), all aïstopods have only a single point of attachment between the vertebra and the rib, via the transverse process. The ribs of *Lethiscus* and *Ophiderpeton* have fairly simple, cylindrical shafts, but those of *Oestocephalus* have anterior and posterior processes near the head, giving them a unique *K* shape (Fig. 4.52).

Lethiscus has 78 trunk vertebrae in articulation, all apparently bearing ribs, but the total length of the trunk and the extent of the tail cannot be determined. None of the specimens of *Odonterpeton* are sufficiently complete to provide a vertebral count. *Oestocephalus* has approximately 110 vertebrae, the last 40 of which are rib-free caudals.

PHLEGETHONTIA. *Phlegethontia* occurs together with *Oestocephalus* at Nýřany, Linton, and Mazon Creek, but differs considerably in both its cranial morphology and the number of vertebrae (Fig. 4.53). *Phlegethontia* keeps the large temporal openings of more primitive aïstopods, but the orbits have shifted from a far anterior position to near the middle of the skull's length, and the jaw articulation has moved well anterior to the occiput. *Oestocephalus* retained most of the dermal bones of the skull seen in more primitive Paleozoic amphibians, losing only the intertemporal and postorbital. In contrast, nearly all of the posterior bones of the skull roof are lost in *Phlegethontia,* including the parietals, postparietals, supratemporals, and tabulars. The squamosal is retained, but not the quadratojugal. The loss of the superficial bones of the skull table reveals the highly ossified underlying braincase, appearing as a single unit without sutures. Dorsally, it is distinguished by a conspicuous sagittal crest, extending forward to the pineal opening, where it is clasped by paired processes of the median frontal bone. Posteriorly, the sagittal crest is crossed by a transverse nuchal crest. The occipital surface also consists entirely of fused endochondral bones. Because of the small size of the skull, the foramen magnum appears huge, compared with the circular area of articulation with the vertebral column. The stapes are saucer-shaped, without a stapedial stem or foramen.

The squamosal, postfrontal, jugal, maxilla, prefrontal, lacrimal, nasal, and premaxilla form an open lattice outlining the lateral and anterior margins of the skull. The parasphenoid becomes fused to the base of the braincase. The other bones of the palate include a very elongate pterygoid (continuous with the quadrate) and the epipterygoid, resting against the braincase. The entire palatoquadrate complex articulates with a prominent basicranial articulation (Fig. 4.54). *Phlegeth-*

ontia is also distinguished by a series of narrow, curved bones that are stacked into the orbital opening. They are thought to have strengthened the eyelid. The lower jaws are reduced to two areas of ossification, one bone carrying the dentition overlaps a second that forms the jaw articulation.

Phlegethontia longissima, known from Linton, Nýřany, and Mazon Creek, has 200 to 210 vertebrae and *Phlegethontia linearis,* known only from Linton, Ohio, has 230 to 250, compared with only 140 in *Oestocephalus*. *P. linearis* is also distinguished by the relative shortness of the centra of the anterior cervical vertebrae and their longer neural spines.

The differences between *Phlegethontia* and *Oestocephalus* have long been thought sufficient to place them in separate families, the Phlegethontiidae and the Oestocephalidae. However, Jason Anderson's (2003a) recent recognition of a new genus, *Pseudophlegethontia,* exhibiting intermediate anatomical characteristics, makes it difficult to maintain this taxonomic distinction (Fig. 4.53). A single specimen of a small, immature individual from Mazon Creek shares almost equally the characteristics of *Oestocephalus* and *Phlegethontia*.

A simple quantitative comparison can be made on the basis of number of vertebrae. *Oestocephalus, Pseudophlegethontia,* and *Phlegethontia* all have about 64 precaudal vertebrae, but differ in the number of caudals that do not bear ribs. *Oestocephalus* has 38, *Pseudophlegethontia* 52, and *Phlegethontia* between 100 and 160.

Pseudophlegethontia resembles *Phlegethontia* and differs from more primitive aïstopods in the following derived features: the quadrate condyles are anterior to the otic capsules and the eyes are near the mid-length of the skull. Accessory articulating surfaces on the vertebrae are lost, as are the osteoderms, and the size of the gastralia is reduced.

On the other hand, *Pseudophlegethontia* retains all of the bones at the back of the skull present in *Oestocephalus,* as well as the *K*-shaped ribs common to *Ophiderpeton* and *Oestocephalus*. Posterior processes on the ribs continue to the 57th vertebra in *Pseudophlegethontia,* but they are limited to the first six in *Phlegethontia*. The lower jaw has moved part of the way along the continuum of bone reduction, with the posterior portion having distinct angular, articular, and surangular bones but the anterior portion has apparently lost the remaining splenial, leaving only the dentary.

The last surviving aïstopod is *Sillerpeton permianum,* from the Lower Permian of Fort Sill, Oklahoma (Fig. 4.54). It is known only from the braincase, which further accentuates the features of *Phlegethontia*.

RELATIONSHIPS. The seven aïstopod genera have been separated into as many as four families: Lethiscidae, Ophiderpetontidae, Phlegethontiidae, and Pseudophlegethontiidae.

Figure 4.49. (*opposite*) The highly derived microbrachomorph *Odonterpeton triangularis* from the Westphalian D of Linton, Ohio. *A,* Specimen drawing of skull in dorsal and palatal views. *B,* Reconstructions of skull in dorsal, palatal, and lateral views. Note loss of tabular and medial fusion of postparietals. *C,* Atlas vertebra in ventral view; note lack of expansion of articulating surface for skull, in contrast with all other microsaurs. *D,* Cleithrum and clavicle. *E,* Left and right forelimbs. *F,* Reconstruction of manus. *G,* Reconstruction of anterior portion of the skeleton. The posterior portion remains unknown. From R. L. Carroll and Gaskill, 1978.

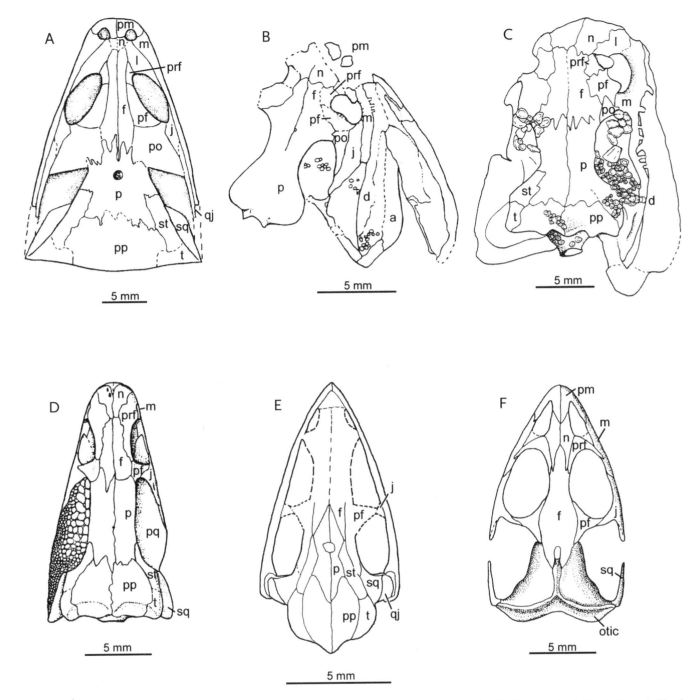

Figure 4.50. Skulls of aïstopods. *A, Lethiscus stocki,* the oldest known aïstopod from the mid-Viséan Wardie shale, near Edinburgh. *B, Ophiderpeton kirktonense* from the late Viséan of East Kirkton. *C, Ophiderpeton brownriggi,* from the Westphalian A of Jarrow, Republic of Ireland. *D, Oestocephalus amphi-* *uminus* from the Westphalian D of Mazon Creek, Illinois. *E, Pseudophlegethontia turnbullorum* from the Westphalian D of Mazon Creek. *F, Phlegethontia linearis longissima* from the Westphalian D of Linton, Ohio. *A–F* from R. L. Carroll, 1998c.

Figure 4.51. (*opposite*) "Ophiderpetontid" aïstopods. *A,* Interclavicle of *Oesto-cephalus nanum* from the Westphalian B of the Newsham locality in Northern England. *B–F,* Dorsal, palatal, lateral, and occipital views of the skull, and medial view of the lower jaw of a juvenile specimen of *Oestocephalus am-phiuminus* from the Westphalian D of Mazon Creek, Illinois. Note the great extent of the palatoquadrate and the extensive fenestration of the cheek. The configuration of the articulating surface between the occiput and the first cervical is common to all aïstopods and unique among tetrapods. *G* and *H,* Dorsal and lateral views of a large skull of *Oestocephalus* from the Westphalian D of Linton, Ohio. Note the posterior extension of the jaw articulation with growth, and the ossicles that cover the temporal fenestration. *I,* Lateral view of cheek, showing the great extent of the palatoquadrate and the emargination of the dorsolateral surface of the surangular to accommodate insertion of the lateral jaw musculature. *J* and *K,* Dorsal and lateral views of *Coloraderpeton brilli,* from the Upper Pennsylvanian (Missourian) beds of Fremont County, Colorado. Note the extreme posterior extension of the jaw suspension. *A,* from Boyd, 1982; *B–I* from R. L. Carroll, 1998a; *J* and *K,* from Anderson, 2003b.

A

1 mm

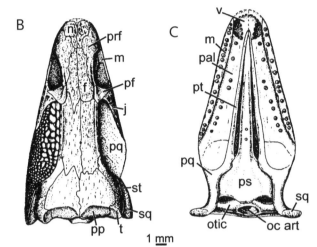

B

n
prf
m
pf
j
pq
st
sq
pp   t

C

v
m
pal
pt
pq
ps
sq
otic   oc art

1 mm

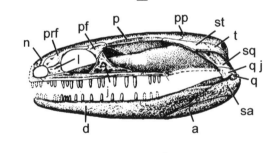

D

n
prf
pf
p
pp
st
l
t
sq
q j
q
sa
d
a

E

sq
t
ptf
pp
qj
q
otic   oc art

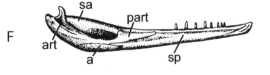

F

sa
part
art
a
sp

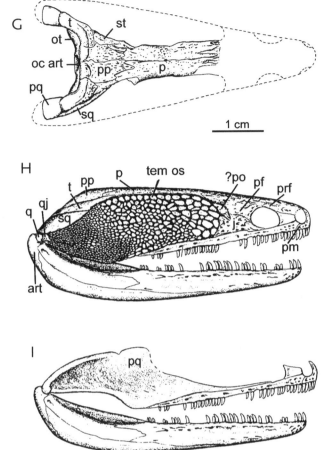

G

t
st
ot
oc art
pp
p
pq
sq

1 cm

H

t
pp
p
tem os
?po
pf
prf
qj
q
sq
pm
l
art

I

pq

J

prf
l
m
pf
f
pf
j
p
pq   m
pp   sq
st
t   qj

5 mm

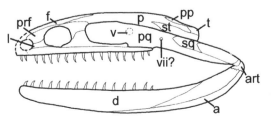

K

prf
f
l
p
pp
t
v
pq
st
sq
vii?
art
d
a

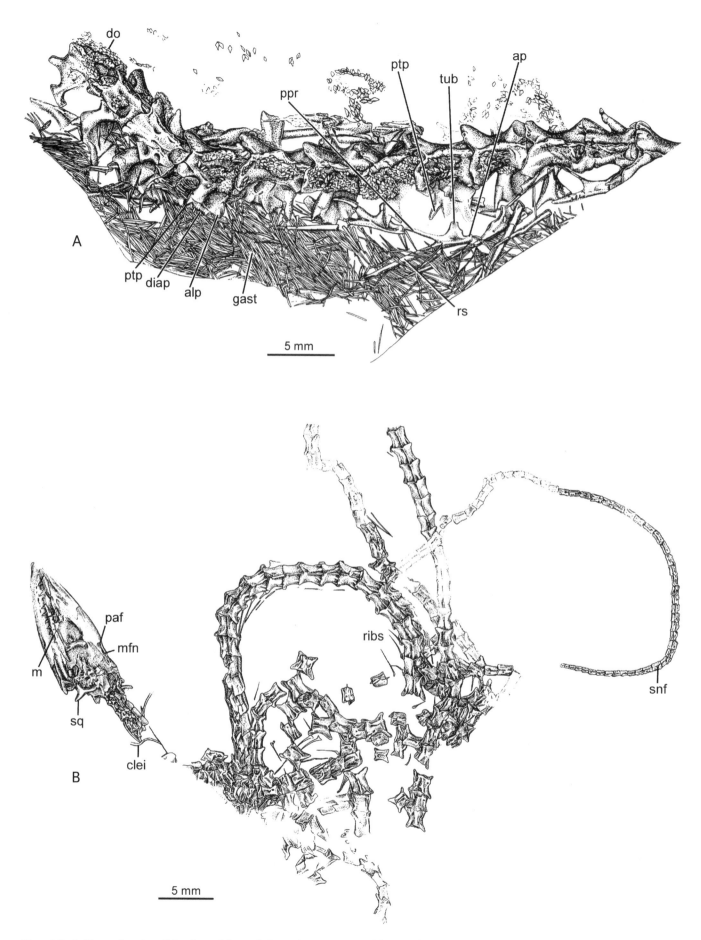

Figure 4.52. A, The trunk region of the aïstopod *Oestocephalus amphiuminus* showing the *K*-shaped rib heads, the transverse processes that articulate with the ribs, the elongate ventral gastralia, and the dorsal osteoderms. The front of the animal is to the right. B, *Phlegethontia linearis* from the Westphalian B of Linton, Ohio. A from Anderson, 2003b; B from Anderson, 2002.

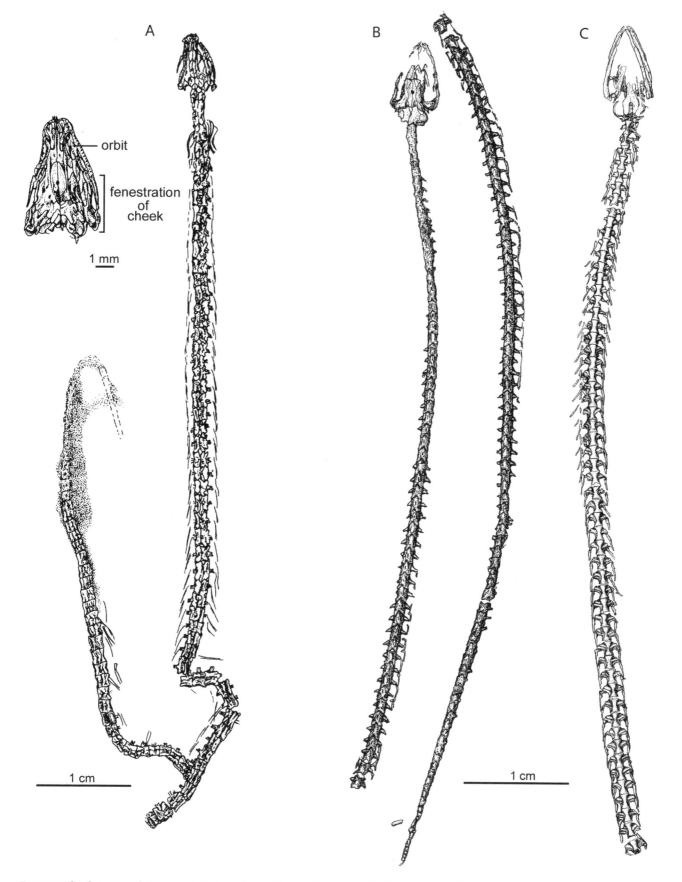

A

orbit

fenestration
of
cheek

1 mm

1 cm

B

C

1 cm

Figure 4.53. The elongation of aïstopods. *A, Oestocephalus amphiuminus* from the Westphalian D of the Czech Republic. From R. L. Carroll, 1998a. *B–C, Pseudophlegethontia turnbullorum* from the Westphalian D of Mazon Creek, Illinois. *B,* Anterior and posterior segments of the entire skeleton in dorsal views. *C,* Ventral view of the anterior portion of the skeleton. *B* and *C* from Anderson, 2003a.

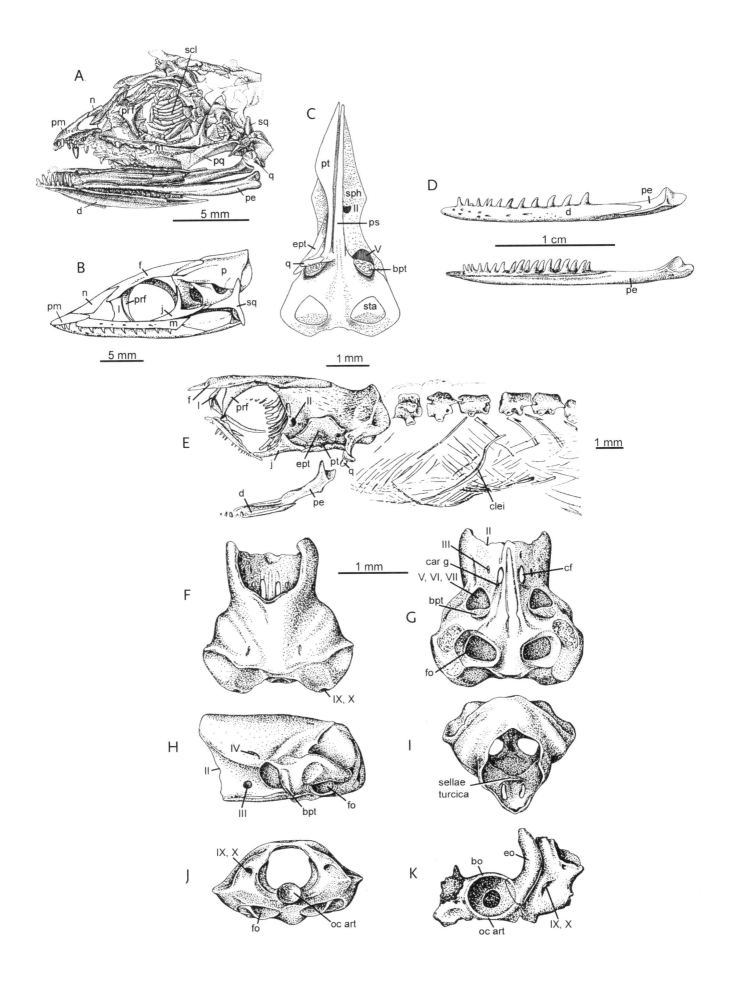

However, as more has become known of these groups, it has become evident that there are no significant gaps allowing taxonomic distinction above the level of genera. One might place them in two groups, based on the dichotomy between ophiderpetontids and phlegethontiids, but this would leave their antecedents as a paraphyletic clade (a lineage that does not included all of its descendants). There are, in fact, no objective ways to divide aïstopods into distinct families. No later amphibians are plausible descendants of this assemblage.

In reference to the increasing number of vertebrae, Anderson (2003a) notes that body elongation in advanced aïstopods proceeds largely thorough addition of vertebrae in the caudal region, in contrast with that in snakes, in which elongation in most lineages occurs within the trunk. There is no evidence that vertebral number increases with growth in an individual species. Retention of the cleithrum indicates that forelimb position is maintained, in contrast with snakes (Cohn and Tickle, 1999). "Aïstopods became elongated via repeated somitomere subdivision throughout the column, primarily in the caudal region, without a change in the expression domains of regulatory genes specifying flank identity" (Anderson, 2003a, 85).

LYSOROPHIDS

This leaves just one more group of tetrapods that first appeared during the Carboniferous, the Lysorophia (Figs. 4.55 and 4.56). A further aquatic locality, which seems to include specimens from two adjacent bodies of water, one quite deep and the other much more shallow, occurs in the Westphalian B of Newsham England. Boyd (1984) argued that the larger genera from this locality, the baphetid *Megalocephalus,* the anthracosaur *Eogyrinus,* and the diplocaulid nectridean *Batrachiderpeton,* normally lived in the deeper water body. The smaller genera, the aïstopod *Ophiderpeton* and a urocordylid, represented by only one specimen each, are thought to have been washed in from an adjacent, shallower body of water. A further species from this shallow water community can be identified as the oldest known member of yet another major lineage of Carboniferous tetrapods, the lysorophids. This group is also known from Linton, Ohio, Mazon Creek, and the Permian red beds of Texas.

Even a small number of vertebrae and associated ribs from Newsham are enough to recognize this specimen as a lysorophid. The vertebral structure is unique among all Carboniferous tetrapods. Only in this group do the neural arches remain paired even in the most mature individuals. The gap

between the halves of the neural arches may have been occupied by a supraneural ligament, which apparently extended the length of the column in the primitive labyrinthodonts *Acanthostega* and *Greererpeton.* The structure in the Newsham specimen is essentially identical with those of *Brachydectes newberryi,* from Linton, Ohio, and *Brachydectes elongatus* from the Lower Permian of Texas, both of which are known from a great number of complete skeletons, described in detail by Carl Wellstead (1991). One specimen of this group has also been tentatively identified from Jarrow, but no specimens have been recognized from the environmentally similar deposits in the Czech Republic. Boyd (1984) suggested that the absence of lysorophids from Nýřany might have been the result of an environmental barrier separating eastern Europe from Great Britain.

In addition to the unique characters of the individual vertebrae, lysorophids are also distinguished by their long vertebral column, from 69 presacrals in *B. newberryi* to 97 in *B. elongatus.* In contrast, the skulls of this group are tiny relative to their body length, spanning the length of only four vertebrae in *B. elongatus.*

The skulls of the Lower Permian species are extremely well known from hundreds of specimens. Many of these are found together in numerous sites, apparently as a result of being preserved in aestivation burrows after long periods without rain. The skulls are highly fenestrate, with a large gap between the anterior margin of the orbit and the anteriorly angled jaw suspension. No other family in the history of amphibians has such a configuration, although proteid salamanders may come close, but not in details.

The configuration of the back of the skull, key to the recognition of most Paleozoic groups, again displays a unique pattern. The intertemporal and supratemporal are lost and the tabular extends far ventrally to support the suspensorium, consisting of the squamosal and quadrate (but without the quadratojugal). The postparietal shares the upper occipital surface with an extensive supraoccipital. Much of the lower occiput is formed by the exoccipitals, which take the lateral supporting role of the opisthotic in some other early tetrapods. The exoccipitals and basioccipital form a broad concave recess for the articulating surface of the atlas, somewhat as in microsaurs. The prootic and opisthotic are not closely integrated, indicating the extensively cartilaginous nature of the otic capsule. The stapes has a large round footplate and a short stem without a stapedial foramen. In keeping with the small size of the skull, there is a huge opening for the Xth nerve between the exoccipital and the opis-

Figure 4.54. (opposite) Dermal skull and braincase of "phlegethontiids." A, Lateral view of the skull of *Phlegethontia longissima,* Westphalian D of Nýřany in the Czech Republic showing the sclerotic plates that protected the eye. B, Reconstruction of the skull of *Phlegethontia longissima* in lateral view. C, Ventral view of the skull of *Phlegethontia longissima,* from the Westphalian D of Mazon Creek, Illinois. D, Lateral and medial views of the lower jaw of *Phlegethontia longissima.* It consists of only two bones, the dentary and the posterior element (pe), which represent the fusion of all the more pos-

terior bones present in primitive aïstopods. E, Lateral view of the braincase and anterior vertebrae of *Phlegethontia longissima* from the Westphalian D of Mazon Creek, Illinois. Note sinuous cleithrum, the only known element of the appendicular skeleton. F–J, Dorsal, ventral, left lateral, anterior, and posterior views of *Sillerpeton permianum,* from the Lower Permian of Fort Sill, Oklahoma. K, Posterior view of occiput of *Phlegethontia phanerhapha* from the Swisshelm Mountains, Cochise County, Arizona. From R. L. Carroll, 1998a.

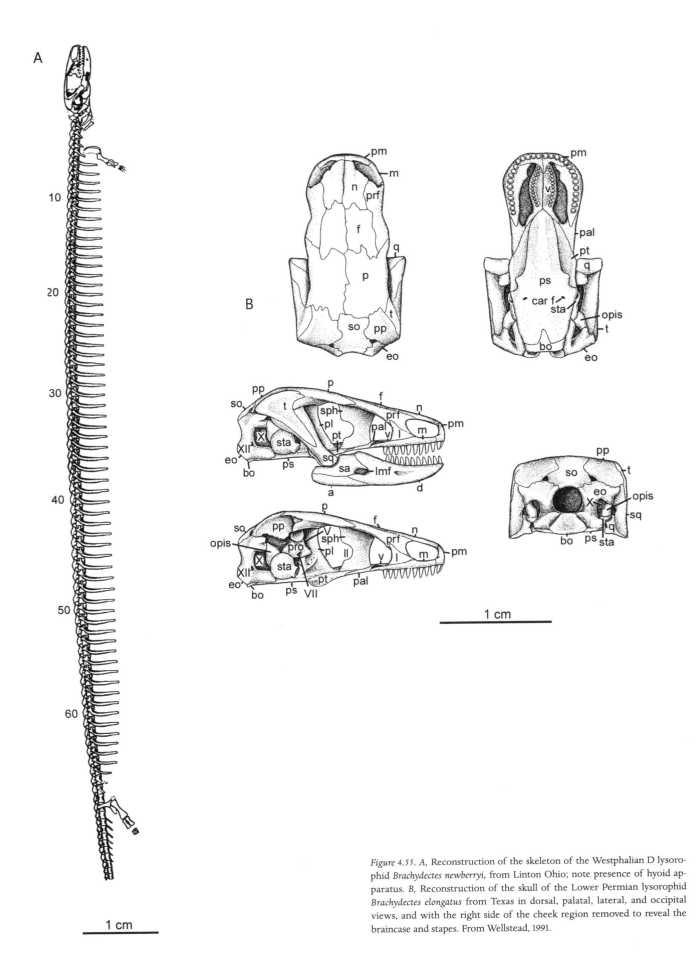

A

10

20

30

40

50

60

1 cm

B

pm
m
n
prf
f
p
q
t
so
pp
eo

pm
v
pal
pt
q
ps
car f
sta
opis
t
bo
eo

pp
p
f
so
t
sph
n
pl
prf
pt
pal
pm
v
m
l
X
sta
XII
eo
bo
ps
sq
sa
lmf
a
d

p
pp
f
so
sph
n
opis
V
prf
pro
pl
ll
pm
XII
X
sta
v
m
l
eo
bo
ps
VII
pt
pal

pp
so
t
eo
X
opis
sq
q
bo
ps
sta

1 cm

Figure 4.55. A, Reconstruction of the skeleton of the Westphalian D lysorophid *Brachydectes newberryi*, from Linton Ohio; note presence of hyoid apparatus. B, Reconstruction of the skull of the Lower Permian lysorophid *Brachydectes elongatus* from Texas in dorsal, palatal, lateral, and occipital views, and with the right side of the cheek region removed to reveal the braincase and stapes. From Wellstead, 1991.

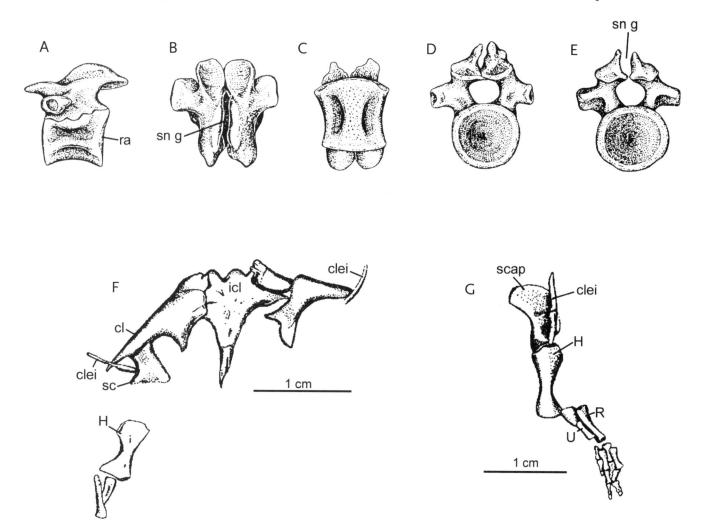

Figure 4.56. Lysorophids. *A–E,* Vertebrae of *Brachydectes elongatus* from the Lower Permian of Texas in lateral, dorsal, ventral, anterior, and posterior views. *F,* Shoulder girdle and forelimb of *Pleuropteryx clavatus* from the West-phalian D of Linton, Ohio. *G,* Pectoral girdle and forelimb of *Brachydectes newberryi* from the Westphalian D of Linton, Ohio. From Wellstead, 1998.

thotic. The anterolateral portion of the braincase is formed by two paired vertical plates, a posterior pleurosphenoid and an anterior sphenethmoid, between which there is a wide gap.

The jaw articulation is far anterior in position, close to the middle of the skull's length. There is no ectopterygoid. The vomers each have a row of teeth, more or less parallel to those of the maxilla. The parasphenoid is very wide and appears to underlap most of the length of the pterygoid. The hyoid apparatus is preserved in many specimens. Its high degree of ossification can be attributed to its use for gape-and-suck feeding in adult lysorophids. Its particular configuration cannot be matched in any other Paleozoic amphibians (Fig. 4.12H).

In addition to the pairing of the neural arches, lysorophids are also distinguished from most lepospondyls in their remaining only loosely attached in the adults. The trunk centra are holospondylous, that is, fully cylindrical and notochordal, with no trace of intercentra. However, as in adelogyrinids and microsaurs, intercentral haemal arches are present in the tail. Elongate, cylindrical ribs are present throughout the trunk in *Brachydectes,* but those of another, less well-known

Carboniferous genus, *Pleuroptyx,* have long posterior flanges. *Pleuroptyx* may also represent the most primitive condition for the pectoral girdle, with extensive clavicles and interclavicle (Fig. 4.56F). The interclavicle has a long, pointed posterior stem and a bifurcate anterior margin. The cleithrum is represented by a narrow stem. The scapulocoracoid is formed by a rounded rectangular plate lacking foramina and a distinct glenoid. The ilium, ischium, and pubis of *Brachydectes* form a simple triangular plate with a short, dorsally directed iliac blade at the apex, but do not co-ossify.

The limbs in all species are small, especially in *B. elongatus. B. newberryi* and *Pleuroptyx* have four digits in the manus, with the former having a phalangeal count of 3, 3, 3, 2. The pes of this species has a formula of 2, 3, 3, 3, 2. No lysorophids are known to possess either dorsal or ventral scales.

Lysorophids may have had the longest time of any Paleozoic amphibians to diverge from possible antecedents among the Early Carboniferous tetrapods, but they are no more divergent than the Early Carboniferous aïstopod, *Lethiscus.* Their relationships remain similarly difficult to establish.

5

Adaptation, Radiation, and Relationships

INCOMPLETE KNOWLEDGE OF THE FOSSIL RECORD

WE HAVE BEEN FOLLOWING the sequence of appearance of a series of amphibian lineages throughout the Carboniferous. Most are clearly distinct from one another, with few obvious similarities indicative of a specific pattern of interrelationships. This is a common problem in paleontology, in which gaps in the fossil record hinder the discovery of intermediate forms. It is an especially acute problem during the early radiation of the many metazoan phyla that arose during the Cambrian explosion (Chapter 1) and the rise of the many orders of placental mammals (Rose and Archibald, 2005), as well as the early stages in the evolution of land vertebrates.

In general, there is a progressive improvement in our knowledge of the history of life over geological time as a result of the increased likelihood of the preservation and recovery of sediments that contain fossils. But there are also biological factors that result in more rapid evolution during periods of major structural change and adaptive radiation. The faster the changes, the lower the likelihood of recovering a sequence of intermediate forms from the fossil record, in contrast with the more common occurrence of members of well-established groups. This is clearly the case during the initial adaptive change between obligatorily aquatic choanate fish and the first facultatively terrestrial amphibians, which must have occurred over a period of less than 15 million years.

On either side of this transition, there were contradictory forces of selection for many aspects of anatomy, physiology, and behavior. Fish such as *Panderichthyes*, *Elpistostega*, and *Tiktaalik*, living in shallow coastal waters and rivers, demonstrate a somewhat intermediate way of life, in which occasional and partial emergence from the water would have led to selection for adaptation to counter the force of gravity. The bones of the paired fins were enlarged and extended distally. The joints between the proximal and distal portions of the limbs were modified so that the body could be more effectively moved along the substrate, and the progressive loss of the posterior bones of the skull roof and the bony operculum enabled the head to be elevated and angled laterally relative to the trunk.

However, a much greater gap in morphology separates *Panderichthyes, Elpistostega,* and *Tiktaalik* from the Upper Devonian tetrapods, *Acanthostega, Ichthyostega,* and *Tulerpeton,* in which both the pectoral and pelvic girdles were much modified for attachment and movement of the limbs. The limbs themselves attained the structural and mechanical features common to all later tetrapods that enabled them to support their bodies against the force of gravity. As yet, no adequately known fossils provide evidence of intermediates between these stages or document the specific time frame during which these changes occurred. All we know is that a minimum time of approximately 10 million years separates the latest fossil record of *Panderichthys, Elpistostega,* and *Tiktaalik* from the earliest known tetrapods.

Such a gap in our knowledge may be attributed to various factors. It could have resulted from very rapid evolution, driven by extremely strong and highly directionally selective forces acting on relatively small populations of a relatively few successive species that had little chance of leaving an informative fossil record. On the other hand, our ignorance of this stage in the origin of tetrapods may result from abiological factors, such as the scarcity of fossil-bearing beds that survived to the present or to our failure to discover such deposits.

The problem of an incomplete fossil record continues into the uppermost Devonian. Although *Acanthostega, Ichthyostega,* and *Tulerpeton* are certainly tetrapods, in the sense of having four limbs and enlarged pectoral and pelvic girdles capable of support and movement on land, only the still incompletely known *Tulerpeton* has skeletal features indicative of specific affinities with any of the later lineages. The other two genera were clearly divergent in the loss or fusion of bones of the skull table that retained a more primitive configuration in the Early Carboniferous amphibians.

There is an even longer gap in our knowledge of the fossil record at the beginning of the Carboniferous, lasting approximately 30 million years, during which only one of the fourteen distinct lineages present in the later Carboniferous is adequately known (Fig. 4.1). This gap is almost certainly a result of the rarity of known fossiliferous horizons during this time, for there is some evidence of greater diversity provided by isolated bones indicating the presence of three or four lineages in the oldest of Carboniferous localities, Horton Bluff.

In contrast, by the Upper Carboniferous, there is a great diversity of distinctive tetrapod groups, including the ancestors of the major living amphibian and amniote lineages. Despite our ignorance of the earlier fossil record, we know that the later Carboniferous tetrapods represent an extensive radiation of land vertebrates into a wide spectrum of semiaquatic and terrestrial habitats. This was roughly equivalent to the radiation of approximately eighteen orders of placental mammals following the extinction of dinosaurs at the end

of the Mesozoic. Suddenly, with the origin of effective limbs and girdles, it was possible for early amphibians to invade a great many previously unoccupied adaptive zones. These included not only more fully terrestrial environments, farther from the sea coast and higher in elevation than the localities of the earliest tetrapods, but also rivers, streams, swamps, and lakes that were isolated from the predators and competitors living in their original marine environment.

Surprisingly, the first amphibians to appear in the fossil record after the very primitive whatcheerid labyrinthodonts were the most derived of all Carboniferous tetrapods, the limbless aïstopods (Fig. 4.8). This suggests that all the other, less highly specialized forms probably diverged long prior to their first appearance in the fossil record. While the labyrinthodonts retained many features of the earliest known tetrapods, the amniotes and the various lepospondyl groups were all highly divergent both from early tetrapods and from each other, to a degree nearly equivalent with that separating the earliest tetrapods from *Tiktaalik.*

ESTABLISHING RELATIONSHIPS

Our very incomplete knowledge of the fossil record for the first 30 million years of the Carboniferous has made it extremely difficult to establish relationships among the many lineages that dominated the remainder of the Paleozoic. Why is this important? We know that all modern tetrapods evolved from among Paleozoic antecedents, but without knowing the specific nature of these relationships we cannot achieve a formal classification of living land vertebrates. We do not know whether modern frogs, salamanders, and caecilians shared a common ancestry from a single group of Paleozoic amphibians or evolved from two or three separate lineages that had already diverged within the Carboniferous. Neither do we know from which of the Paleozoic groups amniotes evolved, or whether they are more closely related to frogs, salamanders, or caecilians. Without determining the specific relationships of modern land vertebrates to particular lineages of Paleozoic amphibians we cannot understand the patterns and modes of evolution leading to their origins.

PAUP

How can we find answers to these questions? The common approach of most biologists today is to undertake a phylogenetic analysis making use of computer programs such as PAUP (Phylogenetic Analysis Using Parsimony [Swofford, 2001]). This is achieved by assembling a database of many characters (which may include anatomical and/or molecular attributes) of all the taxonomic groups in question. To be useful in classification, each character must be divisible into two or more character states that differ among the taxa under study. The computer program assembles this data in the form of a tree, or cladogram, with many branches, each terminating in a particular taxon. The position of the branches is determined on the basis of the most parsimonious arrange-

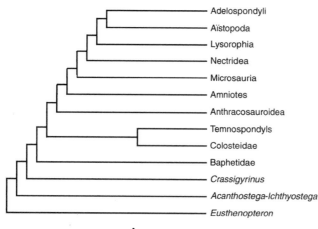

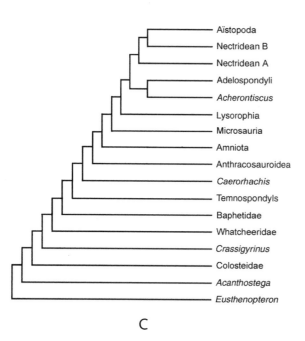

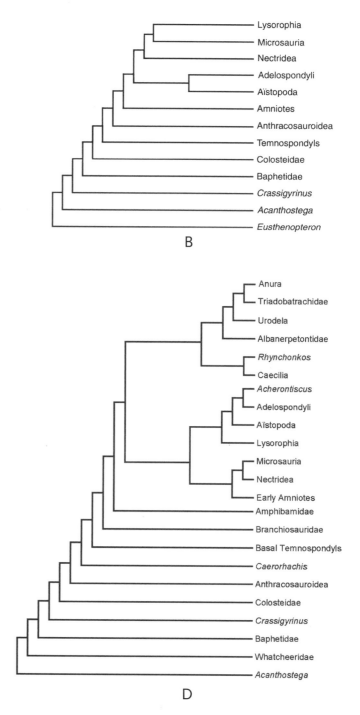

Figure 5.1. Cladistic analyses of Carboniferous tetrapods. Simplified cladograms showing the variable patterns of relationships among major Paleozoic groups, depending on the data assembled by different authors. *A*, from

R. L. Carroll, 1995; *B*, modified from Laurin and Reisz, 1997; Ill *C*, based on data from Ruta et al. 2003; *D*, R. L. Carroll, 2007.

ment of changes in the character states recognized among the various taxa. That is, for the entire tree or cladogram, the pattern that is sought is one produced by the smallest number of character changes. This approach is based on the assumption that evolution proceeds in a parsimonious manner (Sober, 1994; Albert, 2005).

Since the means for establishing phylogenetic trees by this method are essentially constant, it is assumed that given the same taxonomic problem, all investigators should achieve

similar results, and thus that this method is repeatable and objective. In practice, this is not the case, since there are no specific guidelines as to either the nature or number of taxa or characters that should be included in databases. Hence, every analysis run on any particular assemblage will differ from investigator to investigator, and for any one investigator the results will differ from one study to the next as new data is added. A few of the phylogenetic analyses of Paleozoic tetrapods are illustrated in Figures 5.1–5.3. The original

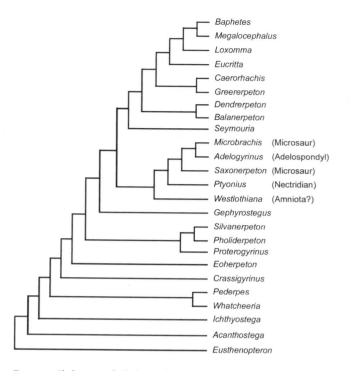

Figure 5.2. Cladogram of Clack (2002b) based on data specifically selected to establish the relationships of *Pederpes* with other early labyrinthodonts.

cladograms have been simplified to emphasize the individual groups discussed in Chapter 4. Most originally included additional major taxa and/or additional taxa within each of the larger clades.

With these simplifications, the relative position of the major Paleozoic lineages comes into sharper focus. All the cladograms have a common geometry. A number of taxa at the base, including *Eusthenopteron* or other osteolepiform fish, *Acanthostega* and/or *Ichthyostega*, whatcheerids, *Crassigyrinus,* and baphetids, appear as a series of separate lineages, without indication of specific relationships between one another. Above the level of *Ichthyostega* and *Acanthostega,* the relative sequence of the clades becomes increasingly variable, with no consensus for any group. For example, counting from the bottom, the position for whatcheerids differs from 3 to 5, adelospondyls from 9 to 14, and aïstopods from 9 to 16.

An even more disparate cladogram was published by Clack and Finney in 2005 (Fig. 5.2). In addition to the larger taxa included in the previously cited cladograms, five genera, long attributed to widely separated groups, were united in a single assemblage between taxa otherwise placed within the Anthracosauria. This assemblage includes *Westlothiana* (a putative sister-taxon of amniotes), *Ptyonius* (a nectridean), *Saxonerpeton* (a tuditanomorph microsaur), *Adelogryinus* (an adelospondyl), and *Microbrachis* (a microbrachomorph microsaur). This grouping can be explained by the choice of different characters than had been used in other analyses. Because this analysis was done primarily for establishing the position of *Pederpes* among the more primitive amphibians, Clack intentionally emphasized characters that could be recognized in that genus, rather than those present in much

more derived tetrapods. A similar grouping of highly divergent taxa into an unresolved polytomy occurred in her paper with Per Ahlberg (Ahlberg and Clack, 1998), which concentrated on the jaws of primitive tetrapods and their immediate relatives among osteolepiform fish. In that case, *Balanerpeton* (a basal temnospondyl), *Gephyrostegus* (an anthracosaur), *Eocaptorhinus* (an amniote), and *Discosauriscus* (a seymouriamorph anthracosaur discussed in Chapter 6) appear as if evolving from a single common ancestor. In both these cases, the choice of characters was appropriate for the particular study, since it was not focused on more highly derived taxa. On the other hand, the cladograms of both Laurin and Reisz (1997) and Ruta et al. (2002) (Fig. 5.3) were concerned with the interrelationships of the modern amphibian orders, and yet were based on very few characters of these groups and so were unable to resolve their specific affinities with any of the Paleozoic taxa. These examples clearly demonstrate that the nature of characters chosen for analysis can greatly influence the pattern of the resulting cladogram.

In addition to the inconsistencies in the sequence of major lineages shown in Figure 5.1A–C, very few are shown as having specific affinities with one another, as illustrated by dichotomous branching. Rather, most appear as a succession of progressively more derived taxa, showing affinities only with the nebulous stem of all subsequent taxa. This presumably reflects the overall achievement of more derived characters over time, but tells us little, if anything, of the specific relationships among the individual clades. Aside from the two most terminal taxa, which are inherently illustrated as if they were sister-taxa (pairs of dichotomously branching lineages), the only pairings of clearly defined clades are between temnospondyls and colosteids (Fig. 5.1A), adelogyrinids and aïstopods (Fig. 5.1B), and adelospondylids and *Acherontiscus* (Fig. 5.1C). Figure 5.1D, based on a newly assembled character list, shows two clusters including lepospondyls and amniotes as an out-group of nectrideans + microsaurs.

Both the inconsistency of the sequence of divergence from the axis of the cladogram and the rarity and inconsistency of the recognition of sister-taxa suggest serious problems in the application of PAUP to establishing relationships among the major groups of early Paleozoic tetrapods. One of these is certainly the inadequacy of the fossil record, but the other may result from the particular pattern of evolution that occurred at the base of this assemblage.

Limits to the Applicability of PAUP

INCOMPLETENESS OF THE FOSSIL RECORD. The primary lesson to be learned from these cladograms is that phylogenetic analyses using PAUP have failed to provide consistent or informative answers to the probable interrelationships of the major clades of Paleozoic tetrapods. Why might this be the case? The answer is that establishing interrelationships among this assemblage is exceptionally difficult, no matter what means are used. The most obvious problem is the rarity of informative fossils over a period of more than

Figure 5.3. Cladogram modified from Ruta et al., 2003, showing that the genera *within* each of the major Paleozoic groups corresponds closely to those recognized by other scientists studying early tetrapods, but the inter-relationships *among* the major groups, listed on the right, differ significantly from those supported by cladograms generated by other authors, shown in Figures 5.1 and 5.2.

30 million years during which early tetrapods underwent their initial radiation. The most primitive known members of all clades were already highly divergent from one another, both morphologically and presumably adaptively, when they first appeared in the fossil record. This makes it very difficult to determine whether character changes (for example, loss or fusion of particular bones, increase or decrease in the number of trunk vertebrae) had evolved within the immediate common ancestors of pairs of lineages and so support their close relationships, or independently, in lineages without an immediate common ancestry.

Our current knowledge of the fossil record probably represents less than 1% of the actual diversity among Carboniferous amphibians in terms of numbers of species, anatomical patterns, ways of adaptation to diverse environments, and geographic distribution. Only a single species is known from Gondwanaland (the land mass including the antecedents of the modern continents of the Southern Hemisphere). All the adelospondyls have come from a single lake basin in Scotland and the Acherontiscidae, Caerorhachidae, and *Casineria* are known from single specimens (all from Scotland). Until near the end of the Carboniferous, nearly all of the individual localities were limited to less than a few square miles of natural outcrops, coal mines, or quarries. Only Joggins and East Kirkton provide informative samples of animals well adapted to a fully terrestrial way of life.

There is no solution to this fundamental problem except the discovery of informative fossils from numerous localities and horizons between the Upper Devonian and the end of the Lower Carboniferous. Comparable problems also exist at the base of the radiation of placental mammals, which as yet preclude understanding of the sequence of divergence of the major placental orders (Archibald, 2005) (Fig. 5.4). These exceptional but extremely important examples imply limits to the applicability of PAUP where there are long gaps in the fossil record and major anatomical disparities among the few adequately known specimens.

PATTERNS AND RATES OF EVOLUTION. In addition to the inadequacies of the fossil record, the problem of recognizing specific interrelationships among the major lineages of Paleozoic tetrapods may result from exceptional patterns and rates of evolution. From the time of Darwin, evolutionary biologists have generally assumed that patterns and rates of morphological change have been relatively constant throughout most of the history of multicellular organisms.

However, as the fossil record has become better known, it has become apparent that there were some periods of more rapid change, leading to the diversification of many clearly distinct lineages, such as the Cambrian explosion (Gould, 1989) and the radiation of the many orders of mammals subsequent to the sudden extinction of dinosaurs (K. Rose and Archibald, 2005). The influence of greatly differing evolutionary rates has not been considered a serious problem in using PAUP as a means of establishing relationships, but this

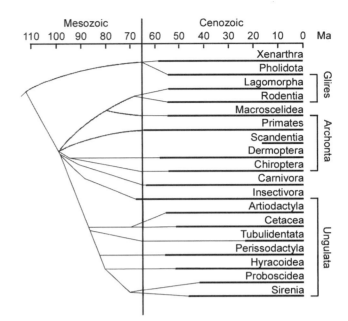

Figure 5.4. Phylogeny of the orders of placental mammals showing approximate duration (bars) of each based on the fossil record. Thinner lines show their putative interrelationships and the relative time of divergence within the late Mesozoic. Note the relatively few lineages for which definitive sister-group relationships can be established. From Archibald, 2005.

may be reflected in the high degree of inconsistency in these analyses of early tetrapods.

It is striking that the cladogram of Ruta et al. (2002) shows a grouping of genera *within* each of the long-recognized families and orders that is broadly consistent with that which has long been accepted by other paleontologists (Fig. 5.3), and yet the phylogenetic positions of the major taxa are very different in all the cladograms. In fact, our concept of the nature of each one of the individual major groups has remained relatively constant for more than a century, with the recognition of the clear distinction *between* the major groups of labyrinthodonts and lepospondyls going back to the summaries of the Miall committee (1875) and Zittel (1890).

Clearly, the degree of similarity among genera *within* each larger taxonomic group is much greater than that *between* the known fossils of the major taxa, both anatomically and adaptively, making objective comparison very difficult. This degree of difference is such that use of the same means of classification is not equally effective in establishing plausible relationships at the two levels. This problem is especially evident in the placement of lineages or individual genera such as the highly derived aïstopods, adelogyrinids, and *Casineria*, which must have evolved very rapidly to have appeared so early in the Lower Carboniferous. Clearly, some means of analysis other than PAUP is necessary to investigate the patterns of divergence and interrelationships of these larger clades.

An Alternative Approach

The initial divergence of the major tetrapod lineages clearly involved rapid and profound changes in the anatomy and behavior of the derived groups that make it difficult to compare characters of the individual lineages with one another and their putative ancestors. Many of the differences in the skull, vertebrae, and limbs were divergent, which would give no direct evidence of relationships without a fairly continuous series of intermediate forms. Even more difficult to deal with are the occurrences of similar changes in groups that might be indicative of close relationships but could be the result of convergent evolution. For example, increase in the number of trunk vertebrae and reduction or loss of limbs occurred during the evolution of aïstopods, adelospondylids, and lysorophids, and are coded as the same characters in all three groups. This results in their appearing as close relatives in most cladograms. However, the similarity of these traits is in strong contrast with the very different cranial anatomy and the configuration of the individual vertebrae in these three groups. These differences strongly suggest that increase in vertebral number and reduction of the limbs resulted from convergence, as each of the three groups independently adapted to an sinusoidal mode of locomotion. It is certain that very similar attributes evolved independently among modern snakes, amphisbaenids, and many lizard genera (Greer, 1991).

Relatively few specific changes may actually reflect common ancestry between any of the major groups, in contrast with the numbers that resulted from divergence or convergence. This makes it unlikely that they would be recognized by PAUP, which is based on the assumption that evolution proceeds in a parsimonious manner, with only a small number of convergent events, and no weight being given to uniquely divergent characters. Rather than trying to apply PAUP to establish the probable pattern of interrelationships among the major groups of Paleozoic tetrapods, we might look at an earlier method of phylogenetic analysis that concentrated on individual characters, determining the monophyly of major groups, and the specific identification of sister-group relationships between taxa. This brings us back to the methodology and nomenclature summarized by Willi Hennig in his 1966 publication *Phylogenetic Systematics*. Hennig concentrated on the recognition of individual derived traits rather than overall similarity, which had been the practice of previous systematists. He designated two categories of derived characters: those that were unique to a single group (of whatever taxonomic rank), which he termed autapomorphies, and those that were uniquely shared by two groups, which he called synapomorphies. It was Hennig who coined the term sister-groups to apply to taxa joined by synapomorphies.

Critical to the recognition of synapomorphies is knowledge of the polarity (that is, direction) of evolutionary change. This he achieved by out-group comparison, based on prior knowledge of more general evolutionary relationships. In the case of the polarity of character change among Paleozoic amphibians, this would be based on the nature of characters in their immediate out-group—advanced choanate fish such as *Eusthenopteron*. Ironically, the polarity of

character state change is not determined prior to the input of data in PAUP (unless specifically designated by the person inputting the data), but is determined by the results of the parsimony analysis.

Hennig did not depend on overall parsimony for establishing relationships, but rather concentrated on recognition of the congruence of character states among a small number of traits in a limited number of taxa. In looking for the congruence of traits, he recognized that the probability of their homology depended on the similarities of multiple other traits. This is in clear contradiction to the procedures of PAUP, which treat all traits as independent of one another.

In looking for individual traits that might unite the various groups of Carboniferous tetrapods, it is necessary to recognize the large number of derived traits (autapomorphies) that distinguish each of the individual clades, and to realize that the few similarities that they happen to share with highly divergent groups are not necessarily homologous, unless they are relatively complex and/or are not encountered elsewhere in the assemblage. In contrast with the views of Hennig, who argued that only derived characters were significant in establishing affinities, some of the most critical similarities that appear to unite Carboniferous tetrapod groups are primitive characters (plesiomorphies) that support common ancestry near the base of tetrapod radiation, such as retention of an extensive epipterygoid, an intertemporal bone, and archaic jaw structure.

The complexity of character change and the combination of primitive and derived traits in particular groups are illustrated by the diversity of occipital structures and vertebral configurations shown by individual taxa (Figs. 5.5 and 5.6). Some of the changes are indicative of broad aspects of adaptation to the terrestrial environment, such as support and movement of the head on the trunk and support of the trunk by the vertebral column, but others, such as the particular configuration of the vertebral elements, show unique specialization within each clade.

THE UNCERTAINTIES OF RELATIONSHIPS

In searching for shared characters, both plesiomorphies and synapomorphies, the groups long characterized as labyrinthodonts and lepospondyls are treated in succession because anatomical comparisons are facilitated by the general similarity in size within each of these groups. However, this does not preclude various lepospondyl groups from having evolved separately from more than one clade of labyrinthodonts. On the other hand, a pattern of relationships among labyrinthodonts can be established without knowledge of lepospondyls, since all lepospondyl groups are highly derived relative to all known labyrinthodont clades.

Labyrinthodonts

UPPER DEVONIAN TETRAPODS. *Acanthostega* and *Ichthyostega* exhibit a great many primitive character states relative to those of later tetrapods, including multiplicity of digits, retention of a preopercular bone, and a fully notochordal attachment of the braincase and trunk. Hence they provide the primary basis for determining the polarity of most character state changes among subsequent Carboniferous taxa. However, the close integration of the cheek and skull table and the loss of the intertemporal bone and fangs on the ectopterygoid in both genera, the medial fusion of the postparietal in *Ichthyostega,* and the highly derived configuration of the tabular in *Acanthostega* indicate early divergence. This precludes their having a specific sister-group relationship with any of the known Carboniferous tetrapods, which must have diverged prior to the appearance of these specializations.

In contrast, the less well-known *Tulerpeton* retains the primitive state for these particular characters, and is not known to possess any other, uniquely divergent, characters. The common presence of primitive traits does not in itself demonstrate sister-group relationships, but neither does it preclude it. On the other hand, the derived nature of the humerus of *Tulerpeton* and most Carboniferous amphibians, including the reduction or loss of accessory foramina, are features that support affinities with later labyrinthodonts, most specifically the anthracosaurs. The great antiquity of anthracosaurs is otherwise suggested by the presence of both humeri and femora at Horton Bluff that greatly resemble those of the early anthracosaur *Eoherpeton* (Fig. 4.3). Unfortunately, the absence of any adequately known anthracosaurs until the end of the Viséan precludes their specific phylogenetic placement.

WHATCHEERIDAE. The whatcheeridae are the oldest known lineage of Carboniferous labyrinthodonts for which nearly all the skeleton is known. *Pederpes* and *Whatcheeria* are clearly distinguished by their highly expanded orbits, their very high cheek, and its sharp angle with the skull table. The sutural connection between the skull table and cheek distinguishes them from *Tulerpeton* and anthracosaurs, and is not comparable with that of *Ichthyostega* or *Acanthostega*. The reduction of ossification of the pectoral and pelvic girdles in whatcheerids suggests secondary adaptation to life in the water, but in a different manner than that of the well-known Upper Devonian genera. Whatcheerids show no recognizable synapomorphies with any one of the other individual groups of Carboniferous tetrapods, but the retention of an intertemporal bone and an ectopterygoid fang suggests divergence prior to the lineages represented by *Acanthostega* and *Ichthyostega.*

COLOSTEIDS. Colosteids are the next labyrinthodont group to appear in the fossil record. Like the whatcheerids, the relatively low degree of ossification of the limbs suggests an aquatic way of life, but they have adapted to life in the water in a much different way than the previous taxon. Instead of the high skull of whatcheerids, theirs is very low and broad. They reduced and eventually eliminated the squamosal notch, but retained some degree of mobility between

the table and cheek. The vertebral column is much elongated, and the limbs progressively reduced.

In view of the many differences in the appendicular skeleton, it seems very unlikely that the one significant derived character of the skull that they exhibit, an initial stage in the formation of an interpterygoid vacuity, is homologous with that of temnospondyls. Although the vertebrae retain the pattern of primitive tetrapods in the paired nature of the pleurocentra, those in most of the column have extended ventrally and medially, to approach the configuration in primitive anthracosaurs rather than that of early temnospondyls (Fig. 4.13). The ilium resembles that of temnospondyls in the loss of the anterior process, but this is also lost in several other groups.

One strikingly primitive feature of colosteids that is clearly expressed in *Greererpeton* is the retention of a very extensively ossified epipterygoid, as in their osteolepiform predecessors. Again, no specific sister-group relationship can be proposed. However, an early divergence may be reflected in the discovery of a similar small, flat, L-shaped humerus at Horton Bluff (Fig. 4.3).

ANTHRACOSAURS. The term anthracosaur (coal reptile) was initially used for the large embolomeres from the Carboniferous of Great Britain, but has since been expanded to include the gephyrostegids and the seymouriamorphs, which will be discussed in the following chapter. All these groups share the synapomorphy of the parietal extending distally to reach the tabular. No other labyrinthodont group shares this character, although it is expressed in nectrideans. Otherwise, anthracosaurs are distinguished by the dorsal attachment of the otic capsule to the tabular and postparietal.

Anthracosaurs may have been one of the most ancient of labyrinthodont lineages, if the similarity of the phalangeal count of *Tulerpeton* and the geometry of the humerus of this genus and isolated bones found at Horton Bluff are homologous with those of primitive embolomeres from the late Viséan. An early time of divergence is also supported by the retention of an anocleithrum in some genera, presence of an intertemporal bone, a degree of mobility between the skull table and cheek, and, in some primitive genera, an extensive epipterygoid. All retained an anterior dorsal process of the ilium. Their vertebrae are distinguished from more primitive labyrinthodonts by the ventral fusion of the primitively paired pleurocentra and the dorsal extension of both the intercentra and the pleurocentra.

CRASSIGYRINUS. The most striking characteristics of *Crassigyrinus* can be attributed to a unique mode of aquatic feeding and locomotion, with reduced ossification throughout the postcranial skeleton and drastic reduction of the forelimb. *Crassigyrinus* has long been associated with anthracosaurs because of the apparent mobility between the skull table and cheek and the connection between the tabular and opisthotic, but the skull retains the contact between the parietal and supratemporal to the exclusion of the tabular. The anterior portion of the palate resembles that of *Eusthe-*

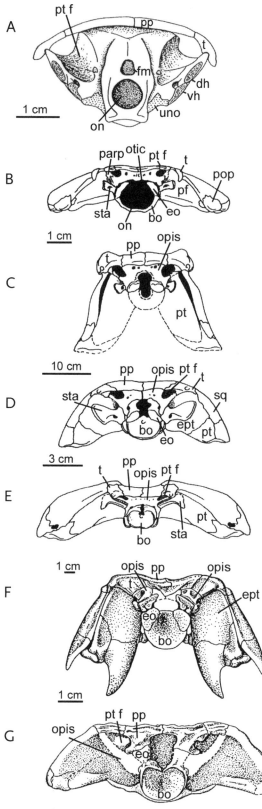

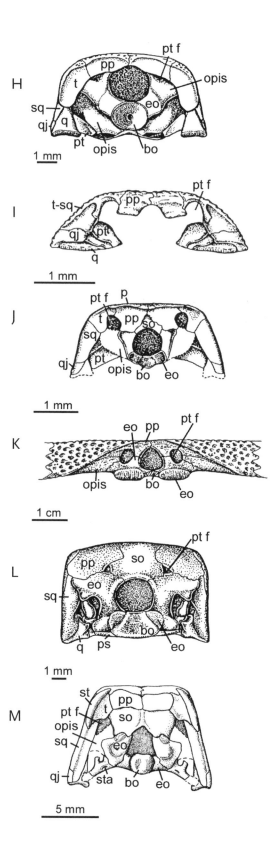

nopteron, with huge fangs on the vomers, suggesting a very early divergence from other tetrapods, as does the primitive nature of the lower jaw (Ahlberg and Clack, 1998).

Any affinities of *Crassigyrinus* must lie well back in the Carboniferous, if not in the Upper Devonian, although it is surprising that earlier remains have not been discovered, considering the large size and conspicuous nature of this genus.

BAPHETIDS. Opinions regarding the affinities of baphetids have oscillated between anthracosaurs and temnospondyls, yet they are extremely divergent from both when they first appear in the fossil record. No known group has a comparably shaped orbit. All unquestioned baphetids differ from anthracosaurs in the absence of a tabular-parietal contact, the lack of the interpterygoid vacuities of temnospondyls, and the retention of a very large epipterygoid. The postcranial skeleton is known in only a very few specimens, but shows the limbs were relatively small, and none shows the length of the vertebral column. All were presumably permanently aquatic, to judge by the persistence of open lateral line canal grooves.

TEMNOSPONDYLS. The most distinctive derived characters of primitive temnospondyls are large interpterygoid vacuities extending to the vomers, a large otic notch, into which is directed a small, dorsolaterally oriented stapes, loss of a distinct anterior iliac blade, a four-toed manus with a phalangeal count of 2, 2, 3, 3, and for the pes, 2, 2, 3, 4, 3

(Table 5.1). Primitive genera lack lateral line canal grooves and have well-ossified limbs, indicating a primarily terrestrial way of life. Primitive features include retention of an intertemporal bone and rhachitomous vertebrae with paired pleurocentra and crescentic intercentra. The skull roof is sculptured by a regular pattern of small pits. Aquatic larvae are known in many species. No antecedents from the Lower Carboniferous can be identified as plausible sister-taxa on the basis of clearly defined synapomorphies.

CAERORHACHIS. The skull of *Caerorhachis* is superficially similar to that of temnospondyls in the uniform pitting of the dermal bones, large fang and pit pairs on all marginal bones of the palate, and a primitive interpterygoid vacuity, not extended to the vomers. The ilium also resembles temnospondyls in the great reduction of the anterior dorsal process of the ilium. On the other hand, the vertebrae resemble those of anthracosaurs, especially gephyrostegids, in the crescentic configuration of both the intercentra and pleurocentra. There are approximately thirty-two presacral vertebrae and the limbs are relatively short, as might be associated with aquatic locomotion, but the skull shows no lateral line canal grooves and the tarsus is well ossified.

Caerorhachis might be a primitive sister-taxon of temnospondyls, but has few convincing synapomorphies with that group. The lower jaw is primitive in retaining the configuration of Upper Devonian and Early Carboniferous tetrapods

Figure 5.5. (opposite) Occiputs of *Eusthenopteron* and representatives of the major clades of Paleozoic tetrapods showing progressive changes over time in the manner of attachment of the skull roof to the braincase and of the configuration of the surface for attachment of the skull to the trunk. Other changes in the shape of the skull are of a more divergent nature, which may be attributed to adaptation to different environments and ways of life. *A, Eusthenopteron,* in which the occiput consists of a single area of ossification, loosely attached to the skull table. The notochord passed through the occiput and extended to the sphenethmoid, without bony connection between the occiput and the first cervical vertebra. The hyomandibular links the lateral commissure with the operculum and the palatoquadrate. *B,* The most primitive condition in tetrapods, illustrated by *Acanthostega.* The notochord continues to pass through the occiput, but is surrounded by a ring of bone that is homologous with the basioccipital and exoccipitals in later tetrapods. The medially fused otic capsules link the skull roof and the bones surrounding the notochord. The posttemporal fossae are reduced to small openings. The area of the lateral commissure loses its ossification and can be recognized as the fenestra ovalis. The hyomandibular is reduced in size and is now recognizable as the stapes. Its footplate fits into the fenestra ovalis and its flattened distal extremity rests against the quadrate ramus of the pterygoid, which may have helped to support the braincase. *C,* The basal Carboniferous labyrinthodont *Pederpes.* The exoccipitals and basioccipital ossify separately and form an effective surface for articulation with the centrum of the atlas vertebra. *D,* The colosteid *Greererpeton.* The otic capsules are paired and attached ventrally to the exoccipital. The widely expanded distal end of the stapes rests against the epipterygoid portion of the palatoquadrate. *E,* The baphetid *Kyrinion.* The otic capsules are fused medially, and the paroccipital processes attach to the tabulars surrounding the posttemporal fossae ventrally. This area of attachment was cartilaginous in more primitive tetrapods. *F,* The early embolomere *Proterogyrinus.* The otic capsules are not fused medially, but they were in the more primitive *Eoherpeton.* They separate the postparietals from the exoccipitals. Embolomeres lack the posttemporal fossae. *G,* The primitive temnospondyl *Neldasaurus.* The postparietals are attached to the exoccipitals, medial to the occipital exposure of the otic cap-

sules. The posttemporal fossae are retained. As a result of their much smaller size, the occiputs of lepospondyls (*H–L*) are much modified to accommodate the relatively much larger foramen magnum and otic capsules, but these changes are strongly divergent from group to group. *H,* Aïstopods, as exemplified by *Oestocephalus,* retained an essentially notochordal attachment between the occiput and the atlas centrum, which were in contact only on their periphery. Limited bending would have been possible in all directions. The posttemporal fossae are almost closed and the otic capsules are primarily lateral to the exoccipitals. *I,* The adelospondyl *Adelospondylus* had postparietal lappets that presumably extended ventrally to the exoccipital, and the basioccipital was recessed to receive a narrow hemispherical extension of the atlas. The posttemporal fossae remained large. *J,* The early microsaur *Asaphestera* resembles *Adelospondylus* in the retention of posttemporal fossae and the postparietals definitely reached the exoccipitals, but the occipital condyle is much wider so the movement relative to the anterior surface of the atlas was restricted to hinging in the transverse plane. The dorsal margin of the foramen magnum was formed by a new bone, the supraoccipital. *K,* Nectrideans, represented by the Lower Permian *Diploceraspis,* expanded the exoccipitals even more, but the basioccipital was recessed so that it did not make contact with the atlas. No nectridean is known to have had a supraoccipital. *L,* The lysorophid *Brachydectes* has a large supraoccipital that extends onto the skull roof between the postparietals, but very small posttemporal fossae. The exoccipitals extend to the squamosals, taking the role of the paroccipital process of the otic capsules in supporting the braincase laterally. *M,* The early amniote *Paleothyris* has a very large supraoccipital linking the postparietals and exoccipitals. The occipital condyle differs from that of all amphibians in having a narrow, hemispherical occipital condyle, consisting of closely integrated exoccipitals and basioccipital that fits into a recess formed by the anterior elements of the atlas like the ball of a ball-and-socket joint, allowing limited rotation and hinging of the skull on the neck in all directions. The relatively massive stapes extends ventrolaterally toward the base of the quadrate. *A, F–L* from R. L. Carroll and Chorn, 1995; *B–E* from Clack, 2003; *M* from R. L. Carroll, 1969a.

(Ahlberg and Clack, 1998). The slope of the tooth-bearing margin of the dentary continues dorsally toward the jaw articulation and there are a series of openings on the medial surface between the elongate prearticular and the very slender postsplenial. The pelvis is extremely primitive in the extension of the unfinished surface of ossification of the acetabulum to the anterior margin of the pubis as in Upper Devonian tetrapods.

Lepospondyls

Labyrinthodonts have been analyzed in sequence, not because they necessarily form a monophyletic assemblage distinct from other Paleozoic tetrapods, but because their generally large size is associated with a great number of comparable cranial and vertebral features in which all are more primitive than any lepospondyls. However, this does not preclude one or more of these lineages from being the sister-taxon of one or more of the lepospondyl clades or of amniotes, which they must have been in the absence of any evidence that lepospondyls or amniotes evolved directly from osteolepiform fish.

While it is difficult to establish the specific sequence of divergence of the major labyrinthodont lineages, they all presumably evolved from among primitive members of a single assemblage of stem tetrapods sometime between the Upper Devonian and the end of the Viséan, to judge by their anatomy and fossil record.

The question of lepospondyl ancestry is much more difficult to resolve. All groups are so disparate from any labyrinthodonts that no convincing arguments have been made for affinities with any of the known clades (Fig. 5.7). The term lepospondyl was already coined in the late 19th century on the assumption that aïstopods, nectrideans, and microsaurs belonged to a single assemblage, to which lysorophids, adelospondylids, and *Acherontiscus* were later appended. However, little evidence is available from the fossil record to support specific sister-group relationships among the individual orders or to particular labyrinthodonts.

One can tabulate a long list of apparent synapomorphies of lepospondyls:

1. Absence of large fang and pit pairs on the marginal bones of the palate
2. Absence of labyrinthine infolding of the dentine
3. Reduction of dermal bones of the skull. All lack the intertemporal, and in many the supratemporal, tabular, and/or postorbital are lost or fused with adjacent bones.
4. Absence of squamosal notch
5. Jaw articulation typically at the level of the back of the skull table or more anterior in position
6. Cylindrical pleurocentra, fully ossified early in development
7. Common fusion of pleurocentrum to neural arch

Figure 5.6. (opposite) Progressive (increased ossification and consolidation of elements) and divergent changes in vertebral structure among early tetrapods. A, The osteolepiform fish *Eusthenopteron* had all the elements of the vertebral column present in early tetrapods (neural arches, dorsal pleurocentra, and ventral intercentra) but all were paired. There are no zygopophyses to link the arches longitudinally and no bony contact with the occiput. The ribs were short and served primarily to link the arches and the intercentra. B, *Acanthostega* illustrates a primitive condition for tetrapods. Paired proatlantes link the first pair of neural arches with the occiput. Prezygapophyses are evident, but not postzygapophyses. All the arches are paired, and all but the first cervical and the sacral intercentra. Pleurocentra are only ossified in the more posterior portion of the trunk. C, Anterior and left lateral views of the atlas, axis, and posterior trunk vertebra of *Crassigyrinus* showing paired nature of neural arches, limited medial fusion of initially paired intercentra, and absence of ossification of pleurocentra. Postzygopophyses are poorly developed. D, Neural arches of trunk vertebrae of *Pederpes*, showing well-formed postzygopophyses. The arches remain paired, as do the pleurocentra; the intercentra are crescentic. E, Anterior view of pleurocentra of *Whatcheeria*, which are fused dorsally. F, Vertebrae of the colosteid *Greererpeton*. Lateral views of atlas-axis complex and mid-trunk vertebrae; anterior view of mid-trunk vertebrae shows that the pleurocentra met ventrally. A paired proatlas links the occiput and the paired atlas arches; the axis and more posterior neural spines are fused at the midline in mature individuals, but paired in juveniles. G, The rhachitomous pattern as seen in the temnospondyl *Eryops*, in which the pleurocentra remain small and paired. H, The primitive condition for anthracosaurs, as seen in gephyrostegids, in which the intercentra and pleurocentra are both crescentic rather than cylindrical. I, The primitive embolomere *Proterogyrinus*. Compete atlas-axis complex and two trunk vertebrae show the crescentic nature of the intercentra but nearly cylindrical pleurocentra. J, The fully embolomerous condition of *Eogyrinus*, in which both intercentra and pleurocentra are cylindrical. K, The enigmatic *Caerorhachis*, a genus originally described as a temnospondyl, but vertebrae resemble those of the anthracosaur *Gephyrostegus*. L, Atlas-axis complex of

the early amniote *Paleothyris*. M, Trunk vertebra of a primitive amniote. N, Cylindrical intercentra and pleurocentra of the lepospondyl *Acherontiscus*. O, Anterior and lateral views of the first cervical of the aïstopod *Oestocephalus*, with fused rather than paired proatlantes, and a funnel-shaped anterior articulating surface for articulating with the occipital surface. P, Anterior view of the first cervical of the microsaur *Rhynchonkos*, showing the laterally expanded surfaces for articulated with the paired condyles of the occiput. Q, Lateral view of trunk vertebrae of the microsaur *Llistrophus*, which lacks intercentra, and *Euryodus*, which possesses them. R, Anterior and lateral views of the first cervical of the nectridean *Diploceraspis*. S, Caudal vertebra of the nectridean *Sauropleura scalaris* showing fusion of the haemal arches midway in the centrum, which is characteristic of this group. T, Caudal vertebra of the lysorophid *Brachydectes* in which the haemal arches articulate between the caudal centra. U, Posterior view of the neural arch of *Adelospondylus*, which resembles labyrinthodonts in the considerable length of the transverse processes. V, The lysorophid *Brachydectes*, in which the supraneural canal, common to early labyrinthodonts, remains open dorsally to accommodate the supraneural ligament.

The patterns of the vertebral centra have long been used to distinguish the major groups of Paleozoic labyrinthodonts, but knowledge of additional taxa demonstrates that their configuration is extremely variable and not reliable in establishing sister-group relationships of the major clades. On the other hand, the atlas-axis complex is consistent among primitive labyrinthodonts, temnospondyls, anthracosaurs, and early amniotes. Lepospondyls are consistently distinct in the similarity of the second cervical with more posterior vertebrae.

A, from R. L. Carroll, 1995; B, Coates, 1996; C, Panchen, 1985; D, Clack and Finney, 2005; E, Lombard and Bolt, 1995; F, Godfrey, 1989; G, modified from Moulton, 1974; H, R. L. Carroll, 1970a; I, Holmes, 1984; J, Panchen, 1966; K, Holmes and Carroll, 1977; L, R. L. Carroll, 1969b; M, R. L. Carroll, 1980; N, R. L. Carroll, 1969d; O, P, R. L. Carroll and Chorn, 1995; Q, R. L. Carroll and Gaskill, 1978; R, R. L. Carroll and Chorn, 1995; S, R. L. Carroll et al., 1998. T, V, Wellstead, 1991; U, Andrews and Carroll, 1991.

8. Common loss of intercentra and haemal arches
9. Absence of recognizable larval stage with external gills
10. Common increase in number of vertebrae and length of trunk
11. Several groups reduce or lose much of appendicular skeleton and limbs

This is a substantial list of apparent synapomorphies, but many and perhaps all can be associated with a single additional attribute—small size. Sizes of adult skulls of a spectrum of labyrinthodonts and lepospondyls are plotted in Figure 5.8. While there is some overlap, nearly all adult labyrinthodonts were substantially larger than any lepospondyls.

Table 5.1. Phalangeal Counts of Frogs, Salamanders, and a Variety of Palaeozoic Tetrapods

	Manus	Pes
Most primitive known temnospondyl		
Balanerpeton woodi (A. R. Milner and Sequeira, 1994)	2233	22343
Superfamily Dissorophoidea		
"Branchiosauridae" from Nyřany (pers. observation)	2233	22343
Amphibamus grandiceps (pers. observation)	2233	22343
Eoscopus lackardi (Daly, 1994)	—	22343
Micropholis (Broili and Schröder, 1937)	2233	22343
Micromelerpeton credneri (Boy and Sues, 2000)	2233	22343
Apateon (Royal Ontario Museum, no. 44276)	2233	22343
Apateon pedestris (Boy and Sues, 2000)	2232	22343
Apateon caducus (Boy and Sues, 2000)	2232	22343
Jurassic and Cretaceous salamanders		
Chunerpeton tianyiensis (Gao and Shubin, 2003)	22?2	22343
Karaurus sharovi (Ivachnenko, 1978)	2232	22343
Jeholotriton paradoxus (Wang, 2000)	2232	22332
Valdotriton gracilis (Evans and Milner, 1996)	2232	22342
Mesozoic and Tertiary frogs		
Vieraella herbsti (from Roček, 2000)	2233	—
Notobatrachus degiustoi (from Roček, 2000)	2333	22343
Notobatrachus degiustoi (from Sanchiz, 1998)	2233	22343
Eodiscoglossus santonjae (from Roček, 2000)	2233	22343
Palaeobatrachus grandipes (from Sanchiz, 1998)	2233	22343
The albanerpetontid *Celtedens* (from McGowan, 2002)	2332	23443
"Lepospondyls" (R. L. Carroll et al., 1998)		
Microsauria		
Tuditanus	2343	23454
Batropetes	2332	23341
Microbrachis	233	23443
Hyloplesion	233	2345?
Odonterpeton	243	—
Lysorophia and Nectridea		
Brachydectes	3?32	23332
Urocordylus	23??2	23442
Ptyonius	2343	2343
Sauropleura scalaris	2343	23442
Keraterpeton	2333	23333
Diceratosaurus	2343	233?3
Other Early Tetrapods (R. L. Carroll and Holmes, 2006)		
Embolomeres	23454	23455
Seymouria	23453	23454
Limnoscelis	23453	23454
Early amniotes	23453	23453

The importance of size can be investigated through the study of larval and juvenile stages of larger labyrinthodonts. In specimens within the size range of most adult lepospondyls, the marginal teeth do not yet show labyrinthodont infolding and distinct fang and pit pairs are not yet developed. Posterior jaw extension only occurs late in growth, in animals with an adult size larger than most lepospondyls. The nature of these features in lepospondyls can all be attributed to truncation of development, resulting in retention of juvenile characteristics into the adult stage.

On the other hand, development of the vertebrae in lepospondyls is much accelerated, with a high degree of ossification of the centra being achieved in animals much smaller than labyrinthodonts in which they long remain cartilaginous and do not ossify until after the neural arches.

Early ossification, in fact, may be the means by which lepospondyls remain small. In modern mammals and lizards, growth is controlled by the timing of ossification of the epiphyseal plates near the ends of the limb bones. Their presence in juvenile animals permits continuing growth, but when they ossify, the areas of growth are lost. Epiphyseal plates are not evident in Paleozoic tetrapods, but growth probably occurred at the extremities of the unossified areas of articulation at the ends of the limb bones. Hormonal changes could result in their earlier ossification. The simplification of the vertebrae, from multipartite elements to single cylinders, may also be associated with small size. Among modern salamanders, which are not only small, but have much larger than normal cell size, an individual cell develops into the osteoblast that results in the formation of an entire centrum (Hanken, 1993). This would make it difficult, if not impossible, to form the multipartite centra of typical labyrinthodonts. Precocial ossification of the skull may also limit the number of dermal bones, also seen in miniaturized salamanders (Fig. 5.9).

The common phenomenon of trunk and caudal elongation in lepospondyls can also be attributed to size reduction. As discussed by Carl Gans (1975), trunk elongation and limb reduction can both be associated with small size and life close to the ground in which sinusoidal bending of the trunk and tail can be as effective or more so than limb locomotion. Greer (1991) tabulated evolution of trunk elongation and limb reduction as occurring convergently sixty-two times in fifty-three lineages of modern lizards. This naturally raises questions regarding the homology of limb loss in animals as otherwise distinct from one another as aïstopods, *Acherontiscus,* and adelospondylids.

If most of the unique characteristics of lepospondyls can be attributed to their small size, the next question is, what was the selective advantage of size reduction? All of the early tetrapods and their immediate antecedents among osteolepiform fish were relatively large animals, a meter or more in length. Their large size would have enabled them to take advantage of the radiant heat of the sun to raise and maintain a high body temperature and metabolic rate, as was dis-

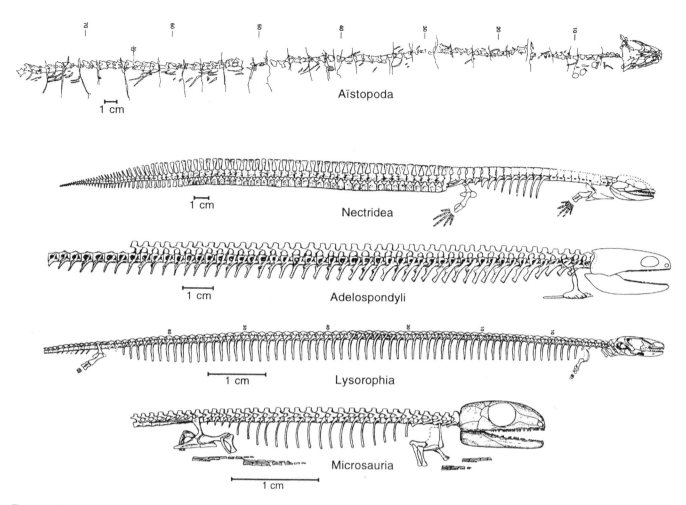

Aïstopoda

1 cm

Nectridea

1 cm

Adelospondyli

1 cm

Lysorophia

1 cm

Microsauria

1 cm

Figure 5.7. Reconstructions of the oldest adequately known genera of the five major lineages of lepospondyls. All have highly distinct cranial anatomy, detailed differences of vertebral structure, and different degrees of expression of the appendicular skeleton. None share convincing synapomorphies with one another or with any of the known labyrinthodonts. All are separated by at least 30 million years from the initial radiation of tetrapods in the uppermost Devonian. The Aïstopoda are represented by *Lethiscus* from the mid-Viséan of Scotland. The Nectridea are represented by *Urocordylus*, Westphalian D of Ireland. The Adelospondyli by *Palaeomolgophis* from the Viséan of Scotland. The lysorophid is *Brachydectes newberryi*, Westphalian D, Ohio. The microsaur is *Utaherpeton*, Lower Pennsylvanian of Utah. From R. L. Carroll, 1999.

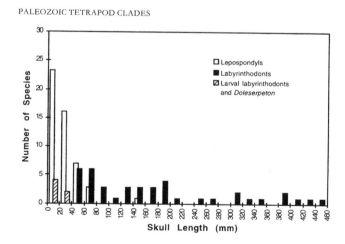

PALEOZOIC TETRAPOD CLADES

□ Lepospondyls
■ Labyrinthodonts
☒ Larval labyrinthodonts and *Doleserpeton*

Figure 5.8. Size distribution of cranial length in Paleozoic amphibians. Open bars: lepospondyls. Solid bars: adult labyrinthodonts. Hatched bars: larval, possibly neotenic branchiosaurs, and *Doleserpeton*. From R. L. Carroll 1999.

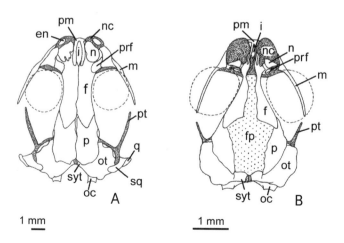

1 mm   1 mm

Figure 5.9. Skulls of two closely related salamanders showing striking differences in the pattern of the dermal bones to accommodate the great relative increase in size of sense organs as a result of the smaller absolute size of the skull. A, *Pseudoeurycea goebeli*. B, The much smaller animal *Thorius narisovalis*. From R. L. Carroll, 1997.

cussed in Chapter 3. The larger the body volume, the lower the surface to volume ratio, the longer large labyrinthodonts would retain a high body temperature after basking in the sun. This would be an advantage to large adults, living on the water-land interface and depending on a diet of fish. Extrapolating from what is known of the stomach contents of *Eusthenopteron*, it is probable that adults of the earliest land vertebrates relied primarily on fish. In fact, there were few, if any, large-sized terrestrial prey for early land vertebrates to feed on in the Late Devonian and Early Carboniferous, except other amphibians. On the other hand, insects were already radiating extensively during the Carboniferous, and small, non-flying arthropods would have been common (Gray and Schear, 1992).

There would also have been large numbers of small arthropods breeding in shallow water streams, ponds, and swamps. They would have provided a copious and diverse diet for larvae and juveniles of early amphibians, which almost certainly hatched out and began to grow in the water, just as do the larvae of modern amphibians. The advantage of a steady aquatic food supply for the smaller-sized young would have resulted in selection for remaining small and/or staying in the water. At least in some circumstances, this would have led to protracted larval or juvenile stages and to reduction in the size of the adults. This would also have resulted in the evolution of animals that were anatomically more effective in feeding on small prey.

The inherent patterns of skull growth in primitive tetrapods, which can be traced back to *Eusthenopteron* (Schultze, 1984), make it particularly effective for juveniles to feed on small prey. The jaw is short and the back of the cheek nearly vertical. The fibers of the muscles occupying the adductor chamber were essentially vertical, so that the greatest force for grasping and manipulating the prey occurred when the jaws were nearly closed, as would be the case if they fed on small prey such as juvenile fish, invertebrates, and their larvae. With growth, the jaw articulation extended posteriorly, providing a longer gape, suitable for larger prey, but also changing the orientation of the jaw muscles so their greatest force (at a right angle to the orientation of the muscle fibers) occurred when the jaws were wide open (Olson, 1961) (Fig. 5.10). By maintaining their small skull and body size, lepospondyls could perpetuate their juvenile feeding habits into adult life, in the water or on land. The fact that this habit was not inherited from a single common ancestry is indicated by the many different morphologies of the adult skull and manner of locomotion in the various groups illustrated in Chapter 4.

The earliest known members of the lepospondyl orders were even more divergent than those of adequately known labyinthodonts, precluding current understanding of either their specific ancestry or affinities with one another.

AÏSTOPODS. Aïstopods were the second adequately known lineage of tetrapods to appear in the Carboniferous (after the whatcheerids) and yet were already the most divergent, and share no specific derived characters in common with other lepospondyls, aside from those that can be associated with their small size. The retention of an essentially notochordal articulation between the occiput and the first cervical is perhaps the most clearly expressed primitive character state, but the integration of the epipterygoid with the dermal elements of the palate also appears to be a perpetuation of an osteolepiform complex, suggesting, as does their early appearance in the fossil record, a very early time of divergence.

ADELOSPONDYLS. Like aïstopods, adelospondyls greatly extended the length of their vertebral column and lost all trace of limbs, the endochondral elements of the pectoral girdle, as well as the pelvis. However, the anatomy of the skull and the configuration of the individual vertebrae are very different, questioning the homology of appendicular skeleton reduction. Instead of the conspicuous fenestration of the cheek region in all aïstopods, the skull is fully roofed in adelospondyls, and several of the bones retained by aïstopods have been lost or fused into the extensive tabular-squamosal element. The reduction and loss of the postorbital bone certainly occurred separately within the two groups. Adelospondyls appear to have evolved a bony connection between the occiput and the first cervical, with a median extension of the atlas fitting into a recess in the basioccipital to form a ball-and-socket joint, supplanting the notochordal attachment of basal tetrapods. Although both have lost the scapulocoracoid, adelospondyls have retained all the original dermal bones of the shoulder girdle, in contrast with the absence of all but the cleithrum in aïstopods. Instead of a solid attachment of the neural arches and the centra, these elements do not co-ossify in adelospondyls.

Despite the fully cylindrical pleurocentra, total absence of intercentra, and lack of evidence for haemal arches, the neural arches of adelospondyls are strikingly similar to those of the early labyrinthodonts *Greererpeton* and *Proterogyrinus* in having long transverse processes, at the base of which are broad areas for articulation with the centrum (Fig. 5.6U). In contrast with the remainder of the skeleton, these primitive features connect them ultimately with the base of the labyrinthodont radiation, and clearly distinguish them from other lepospondyls.

The extension of ventral lappets of the postparietals over the occipital surface toward the area of the otic capsules, medial to the post-temporal fenestrae, somewhat resembles the pattern in early temnospondyls as well as other lepospondyls (Fig. 5.5I), but this may only reflect comparable adjustment to the reduced size of the occiput. No adelospondyls are known to have had a supraoccipital bone, as does occur in basal microsaurs and lysorophids.

ACHERONTISCUS. *Acherontiscus* is immediately distinguishable from aïstopods and adelogyrinids by the retention of both pleurocentra and intercentra and their broadly embolomerous configuration (Fig. 4.29). The skull, however, appears superficially similar to that of a particular family of

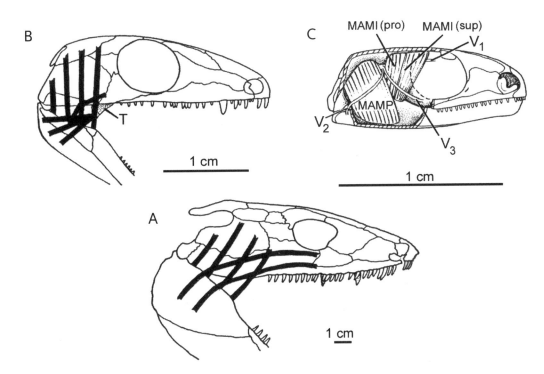

Figure 5.10. Orientation of major jaw muscles in the primitive labyrinthodont *Palaeoherpeton*, A; The early amniote *Paleothyris*, B, and the microsaur *Hapsidopareion*, C. In the labyrinthodont, the jaw articulation is well behind the occiput and the posterior wall of the chamber for the adductor jaw muscles is angled posteriorly, so that the muscles are oriented at a right angle to the jaw when it is wide open. This would enable the jaw muscles to exert their greatest force when the jaw was open, so that they could hold large prey. In the early amniote and the microsaur, with much smaller skulls, the jaw articulation is at the level of the occipital condyle, and the jaw muscles would have been oriented nearly vertically when the jaw was nearly closed, enabling them to capture and hold small prey. A and B from R. L. Carroll, 1969a; C from R. L. Carroll, 1988a.

advanced microsaurs (the gymnarthrids), but with no derived similarities with either of the earlier appearing lepospondyls. Like aïstopods and adelogyrinids, it has lost all trace of limbs, but like adelogyrinids it has retained elements of the dermal shoulder girdle. In contrast, its hyoid apparatus is unique among early tetrapods. The single incomplete skeleton provides no evidence of specific affinities with any other tetrapods, but suggests derivation from an as yet unknown lineage of basal amphibians.

MICROSAURS. Microsaurs retain the most characteristics that can be compared with labyrinthodonts. Early members have no temporal fenestration, and the configuration of the skull table indicates the primitive incorporation of the supratemporal and intertemporal into a single unit termed the tabular. Large teeth are present on all the marginal bones of the palate of primitive genera, reflecting the pattern seen in primitive labyrinthodonts although without the adjacent pit pairs.

Small, crescentic intercentra are present in some primitive and elongate species, and haemal arches are retained in the tail. These are primitive features among lepospondyls, clearly separating microsaurs from adelogyrinids, aïstopods, and lysorophids, as does their primitively small number of vertebrae and the absence of any reduction of the appendicular skeleton. In contrast, their fully cylindrical pleurocentra, strongly attached to the neural arches, are not encountered in any basal Carboniferous labyrinthodonts. Other characters clearly distinguishing microsaurs from all labyrinthonts but implying affinities with other lepospondyl clades include the unipartite atlas, its broad articulation with the occiput, and similarity of the second cervical with all more posterior trunk vertebrae. Almost certainly a matter of convergence, the microsaurs share the presence of a supraoccipital bone in the occiput with such divergent groups as amniotes and lysorophids. However, the small size of the epipterygoid, shared with the earliest temnospondyls, might be of more taxonomic significance.

Microsaurs retained all the primitive elements of the pectoral and pelvic girdle, and the early genera kept the anterior dorsal process of the ilium common to Devonian tetrapods, whatcheerids, baphetids, and anthracosaurs, but not temnospondyls or other lepospondyls. No genera are known to have possessed more than four digits in the manus, but the Lower Permian genus *Pantylus* has five distal carpals, suggesting the presence of a fifth digit sometime in their past. The phalangeal count of the manus is close to that of temnospondyls and that of the pes is similar to that of anthracosaurs. The configuration of the humerus with a long shaft and the ends twisted at a 90 degree angle to one another resembles that of early temnospondyls, but may be attributed to relatively small size and a more terrestrial way of life rather than common ancestry. No uniquely derived characters are shared between microsaurs and any particular lineage of early labyrinthodonts that convincingly support a common ancestry, and nothing links them to any other Paleozoic lepospondyls.

NECTRIDEANS. The skull of early nectrideans is the most primitive of any lepospondyl in having lost only a single bone of the primitive complement, the intertemporal. It also retains the primitive mobility between the skull table and the cheek. However, the skull is derived in the posterior extension of the parietal to reach the tabular. The combination of these characters was used by Keith Thomson and Kathy Bossy (1970) to support specific affinities between nectrideans and anthracosaurs. The presence of a five-toed manus in the most primitive genus and a phalangeal count of the pes also resemble those of anthracosaurs. The presence of a five-toed manus is primitive, but it is difficult to accept the contact between the parietal and the tabular as being homologous with that of anthracosaurs in view of the extensive differences in the rest of the skull and the entire postcranial skeleton.

The loss of trunk intercentra, the solid fusion of the fully cylindrical trunk pleurocentra with the neural arches, and the fusion of the haemal arch midway in the length of the caudal centra clearly distinguish nectrideans from all labyrinthodonts, and preclude specific comparison with any known groups. Their retention of nearly all elements of the appendicular skeleton makes it unlikely that they had close affinities with the earlier occurring aïstopods, adelospondyls, or *Acherontiscus*, and the persistence of a very short trunk region and a solidly roofed skull make affinities with lysorophids extremely improbable.

LYSOROPHIDS. As the last group to appear in the fossil record, lysorophids may have had the longest period of time to diverge from among more primitive tetrapods. One character that may be a relict from basal tetrapods is the failure of the neural arches to fuse medially, allowing the intervertebral ligament, assumed to have passed through a foramen in the neural arch in early tetrapods, to extend the length of the spinal column. Relative to aïstopods and adelospondyls, lysorophids are primitive in the retention of articulated haemal arches and most of the elements of the appendicular skeleton. However, the number of trunk vertebrae has increased to 69 in the oldest known species, the pleurocentra are fully cylindrical, and the trunk intercentra have been lost.

The skull is highly derived relative to all other Paleozoic tetrapods in the loss of the postfrontal, postorbital, and jugal, as well as converting the squamosal into an elongate, anteriorly directed jaw suspension, supported by the tabular. The length of the jaw is greatly reduced. None of the major changes in the cranial anatomy can be considered as synapomorphies with any other Paleozoic tetrapods.

AMNIOTES

Amniotes are unique among Paleozoic tetrapods in having an unquestioned link to living descendants. Sir William

Dawson identified *Hylonomus* as a reptile because of its overall similarity with living lizards, even without knowledge of any intervening fossils. He immediately recognized its distinction from the contemporary temnospondyl, *Dendrerpeton,* which he linked to modern urodeles. On the other hand, it has remained difficult to establish specific affinities between *Hylonomus* and any other early Paleozoic lineages. It has long been assumed that amniotes arose from anthracosaurs on the basis of the pattern of the bones at the back of the skull table and the presence of cylindrical pleurocentra and crescentic intercentra. However, Laurin and Reisz (1997) hypothesized a sister-group relationship with lepospondyls. In the cladogram of Ruta et al. (2002) amniotes appeared between anthracosaurs and microsaurs.

In fact, all early amniotes have a pattern of the atlas and axis vertebrae that precludes close affinities with any lepospondyls, which have reduced the atlas to a single center of ossification and in which the axis has lost its distinction from other anterior vertebrae. Amniotes maintain the same basic pattern of the atlas and axis that was initiated in basal tetrapods, such as *Acanthostega,* and was further elaborated among anthracosaurs (Fig. 5.6B, I). As in anthracosaurs, the parietal extends back to the tabular. Early amniotes lose only one skull bone from the complement in anthracosaurs, the intertemporal. However, they differ significantly in the absence of a squamosal notch. None of the lepospondyls have the phalangeal count of the manus of anthracosaurs and amniotes—2, 3, 4, 5, 3 (4).

Among anthracosaurs, investigations into the ancestry of amniotes were initially focused on *Seymouria,* from the Lower Permian, although it appears long after the first occurrence of amniotes. In fact, *Seymouria* is highly divergent in the nature of the middle ear and shared few definitive synapomorphies with the earliest amniotes then known. In 1969–70, I argued for affinities with gephyrostegids, but they too were far too late in time and had only vague similarities with the oldest amniotes, notably an initial stage in the formation of the astragalus from the tibiale, intermedium, and proximal centrale.

Discovery of *Westlothiana* provided the first evidence of an animal older than the oldest known amniote that had some characters in common with that group. It provided additional support for affinities between amniotes and anthracosaurs, especially the configuration of the skull table, the nature of the trunk vertebrae, and the loss of the anterior dorsal process of the ilium. But it also had characters that were very improbable in a close sister-taxon of amniotes, including the very reduced limbs and the presence of approximately 36 presacral vertebrae.

More recently, the discovery of the middle Viséan genus *Casineria* (Fig. 4.9), which resembles early amniotes in its small size and the phalangeal count of the manus, suggests a more plausible ancestor. The vertebrae remain similar to those of primitive anthracosaurs, but without a skull neither affinities with amniotes nor anthracosaurs can be strongly argued. Phylogenetic analysis by Paton et al. (1999) placed it in an unresolved polytomy with *Westlothiana* and several early amniotes, but this particular result is somewhat dubious since the data matrix consisted primarily of cranial features (78 out of a total of 111), which could not be judged in the headless skeleton.

If additional material of *Casineria* more fully demonstrates affinities with amniotes, it could pull the origin of this clade back to the Early Carboniferous, prior to the temnospondyls and microsaurs, both of which have been suggested as including the ancestry of the modern amphibian orders. In any case, definitive amniotes certainly appeared in the fossil record long before the occurrence of any of the modern amphibians. We must look much later in the Paleozoic to establish the probable affinities of frogs, salamanders, and caecilians among the known assemblage of late Carboniferous tetrapods.

CONCLUSIONS

The most important lesson to be learned from this chapter is that what we now know of the many groups of Carboniferous amphibians is insufficient to determine their specific ancestry or interrelationships. No means of manipulating the currently known data will provide a reliable phylogeny or understanding of their modes of evolution. The answers to these questions will only come from the discovery and detailed analyses of fossils from currently unknown localities and horizons throughout the world that represent a full range of the environments in which early tetrapods lived and the spectrum of their adaptive evolution and geographical distribution. This work awaits the next generation of paleontologists.

6

The Zenith of Amphibian Diversity

LATE CARBONIFEROUS AND PERMIAN LOCALITIES

THE PRIMARY RADIATION OF terrestrial vertebrates was concentrated in the latest Devonian and Early Carboniferous, but it is only in the latter part of the Upper Carboniferous and the Permian when we can appreciate the dominance of amphibians throughout the world. Our knowledge of this diversity results not only from the more extensive radiation of already existing lineages, but also from the presence of much more extensive fossil deposits in many parts of the world. These encompass large areas of swamp and lake deposits in the Upper Carboniferous of Great Britain and throughout continental Europe and eastern North America, as well as aquatic, deltaic, and fully terrestrial deposits in the southwestern United States, western Europe, Russia, eastern Asia, and Africa.

SEYMOURIAMORPHS

Some of this later radiation was already alluded to in Chapter 4, with reference to Lower Permian aïstopods, embolomeres, nectrideans, and lysorophids, all of which occurred in the deltaic and flood plane deposits of north-central Texas and Oklahoma. In contrast, the greatest Late Carboniferous and Permian radiations were to be seen among the temnospondyls and microsaurs. We will begin, however, with possible relatives of the anthracosaurs for which there is little if any evidence in the Carboniferous. These belong to an enigmatic group termed the seymouriamorphs, named for the town of Seymour, in north-central Texas. This is in the midst of the Texas-Oklahoma red beds—the most extensive terrestrial deposits representing the first 25 million years of the Lower Permian. These beds accumulated in an enormous delta complex from a river system flowing toward the coast, much like the Mississippi delta today, which extended far into the ocean. Animals and plants representing a wide range of local environments were preserved, including osteolepiform fish, freshwater sharks, obligatorily aquatic amphibians such as diplocaulid nectrideans, bog-dwelling embolomeres, and a host of more terrestrial amphibians and early reptiles. From these deposits one can collect at least scattered remains of hundreds of animals in a single day.

For the paleontologist, if not the farmer, it is a great advantage that this area is now semiarid, with relatively sparse vegetation and subject to continuous erosion to expose the fossils. Compared with the few isolated deposits in the Lower Carboniferous, this is a fossil hunter's paradise and accounts for much of our knowledge of Lower Permian tetrapods.

Seymouria is known from many nearly complete skeletons from Texas and New Mexico and also recently discovered remains from Germany (Berman et al., 2000) from which every bone in the body can be described (Fig. 6.1). *Seymouria* was clearly a terrestrial animal, with massive limbs, highly ossified cylindrical pleurocentra, and the body held well off the ground. In addition to the proportionately large-sized limbs, the carpals and tarsals were highly ossified, indicating more effective adaptation to terrestrial locomotion than any other anthracosaur. This is particularly true of the species *Seymouria sanjuanensis,* which is known from both the Bromacker locality in central Germany and in New Mexico, both of which are considered to be among the most terrestrial, upland environments of deposition of any Lower Permian tetrapods. In contrast, the Texas *Seymouria baylorensis* is less well ossified and occurs within a fauna of more aquatic tetrapods.

The high degree of terrestriality implied by the postcranial skeleton of *Seymouria* was initially thought to support relationships with early reptiles, but the discovery of larval stages in the very similar European and Asian discosauriscids (Klembara and Bartík, 2000; Klembara and Ruta, 2004a, b; Klembara and Ruta, 2005a, b) demonstrated conclusively that seymouriamorphs were not biologically amniotes. Three genera of larval seymouriamorphs are recognized: *Discosauriscus* from the Lower Permian of the Czech Republic, France, Poland, Germany, and Russia; *Ariekanerpeton,* also placed in the Discosauriscidae, from Tadzhikistan; and *Utegenia* from the ?Upper Carboniferous–Lower Permian of Kazakhstan.

Cranial features, such as the retention of an intertemporal bone and a large tabular in contact with the parietal, imply affinities with earlier anthracosaurs, but *Seymouria* is primitive among anthracosaurs in retaining a broad vomer with vomerine fangs, in common with gephyrostegids rather than embolomeres, and the occiput retains post-temporal fenestrae (Fig. 6.2). These features indicate divergence prior to the origin of embolomeres. *Discosauriscus* (Fig. 6.3A) is unique in having an anocleithrum, a primitive feature also reported in the embolomere *Pholiderpeton.*

Surprisingly, *Seymouria* has what appears to be an impedance matching middle ear. The squamosal is large and deeply embayed, as in temnospondyls. The stapes is very small and light, suggesting that it could have responded to airborne vibrations. However, the fenestra ovalis is unique in being at the end of a long tube formed by the prootic and a lateral extension of the parasphenoid.

Seymouriamorphs also have a derived structure of the vertebrae, with much reduced intercentra. The pleurocentra in adults are fused to extremely wide and dorsally expanded (or swollen) neural arches with nearly flat zygopophyses (Fig. 6.2E–G). Such a vertebral structure is also present in a number of animals that have been thought to be close to the ancestry of amniotes: *Limnoscelis, Diadectes,* and *Tseajia.* The specific affinities of these genera are still being debated, but the peculiar structure of their vertebrae is almost certainly associated with their large size rather than close affinities with seymouriamorphs. The common phalangeal counts of 2, 3, 4, 5, 3 and 2, 3, 4, 5, (3)4 may be a synapomorphy of embolomeres, seymouriamorphs, gephyrostegids, and amniotes, but also resemble those of *Tulerpeton,* perhaps indicating a common ancestry at the level of the Upper Devonian.

It is hence surprising that no seymouriamorph specimens have been reported prior to the very end of the Carboniferous. *Seymouria* itself is only known from the Lower Permian of North America and Europe, but the occurrence of the discosauriscid *Ariekanerpeton* in Tadzhikistan and *Utegenia* from the Upper Carboniferous or Lower Permian of Kazakhstan raises the possibility that this group had migrated from northern China and ultimately from Gondwana, as refugees from the continental ice sheets that covered that area in the Late Carboniferous and Early Permian (A. R. Milner, 1996). Movement of these genera into Europe and North America was only possible when the Kazakhstan plate finally sutured with Euramerica (Fig. 6.4).

The last known seymouriamorph is *Kotlassia,* an apparently neotenic form from the Upper Permian of Russia, and other members of the family Karpinskiosauridae (Fig. 6.3C).

CHRONIOSUCHIA

Another, enigmatic group tentatively associated with anthracosaurs includes a number of species from the Upper Permian and Lower Triassic of Russia termed the Chroniosuchia, recently reviewed by Novikov et al. (2000) (Fig. 6.3D). They are grouped with anthracosaurs because of the sutural contact between the tabular and parietal, but lack the intertemporal bone and have a large foramen between the orbit and external naris that is unique among labyrinthodonts. The orbits are relatively large and intersect the frontal bones. The palate is closed, as in all earlier anthracosaurs, but the palatal teeth are arranged in an almost continuous row. The Chroniosuchia are also distinguished by medial armor plates that either lie above the neural spines or are sutured to them, in the manner of dissorophids (discussed later in this chapter). The presence of armor suggests that they were

Figure 6.1. Two nearly complete skeletons of *Seymouria sanjuanensis* from the Bromacker locality in Germany, thought to represent one of the most terrestrial depositional environments known for the Lower Permian. From Berman et al., 2000.

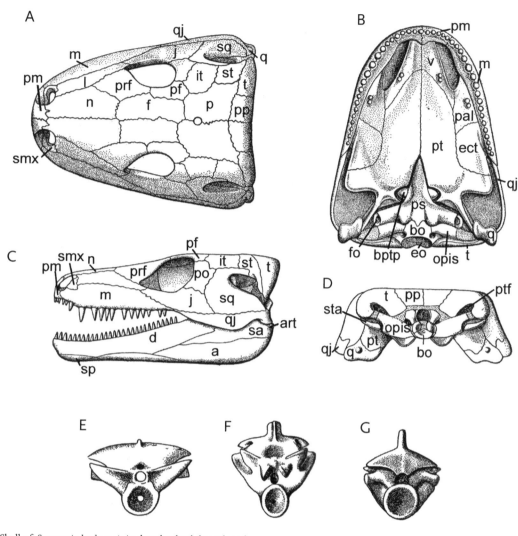

Figure 6.2. A–D, Skull of *Seymouria baylorensis* in dorsal palatal, lateral, and occipital views. *E–G,* Anterior views of the trunk vertebrae of *Seymouria, Diadectes,* and *Limnoscelis.* Modified from Romer, 1956.

terrestrial and there is little evidence of lateral line canal grooves.

MICROSAURS

Microsaurs had achieved considerable diversity within the upright tree fauna of Joggins by the Westphalian A and radiated into the coal swamp environment by the Westphalian D. The long-bodied gymnarthrids, with their expanded teeth, continued into the Lower Permian, and were particularly common and varied within the fissure fillings of the Fort Sill locality in Oklahoma (Fig. 6.5A, B). The large Joggins genus *Trachystegos* (Fig. 4.36) may have given rise to the common Lower Permian *Pantylus,* with its massive skull, extremely large teeth, and short trunk (Fig. 6.5C). Two much longer-bodied genera, *Pelodosotis* and *Micraroter,* were the lon-

gest microsaurs known, the latter reaching more than 70 cm. Their flattened snouts imply a burrowing habitus. *Micraroter* was unique among all amphibians in having three pairs of sacral ribs (R. L. Carroll and Gaskill, 1978).

Another family of elongate microsaurs, numerous at one Lower Permian locality in Oklahoma, is the Goniorhynchidae, represented by the single genus *Rhynchonkos* (Fig. 6.5F). It is exceptional among all lepospondyls in being a possible antecedent of one of the living amphibian groups, discussed in Chapter 12.

While all these microsaurs can be classified among the Tuditanomorpha, two other Lower Permian genera, *Batropetes* and *Quasicaecilia,* were apparently derivatives of the Microbrachomorpha, judging from the great lateral expansion of the parietal and reduction or loss of the tabular. *Batropetes* (Figs. 6.6 and 6.7) is known from numerous specimens from

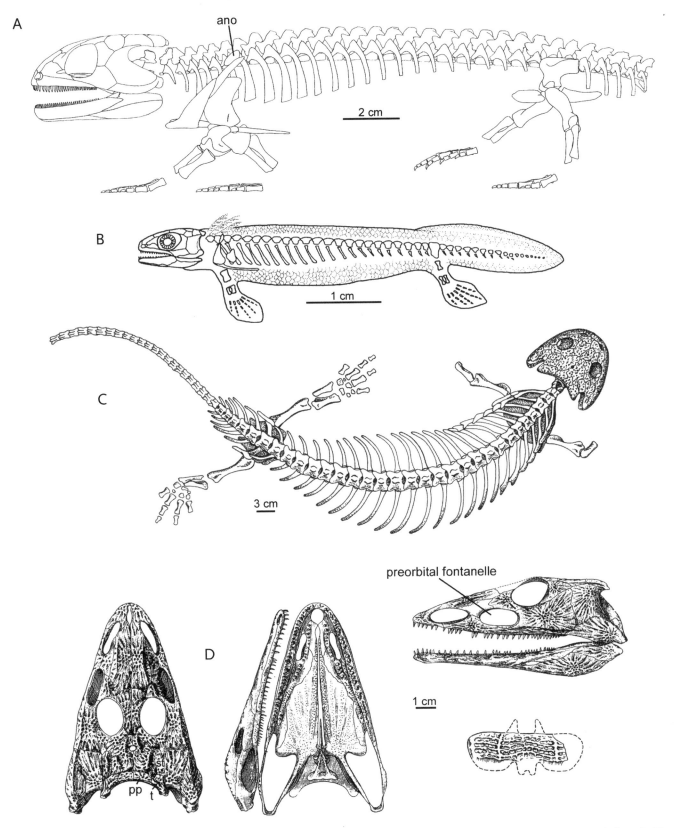

Figure 6.3. A, A juvenile specimen of the seymouriamorph *Discosauriscus austriacus* from the Lower Permian of the Czech Republic. Note the retention of the anocleithrum (ano), a relict of embolomeres and Upper Devonian tetrapods. Modified from Klembara and Bartík, 2000. B, Reconstruction of the larva of *Ariekanerpeton*, a seymouriamorph from Tadzhikistan. Based on Laurin, 2000. C, *Kotlassia*, a neotenic seymouriamorph from the Upper Permian of Russia. Modified from Bystrow, 1944. D, Skull and dorsal armor of *Chroniosuchus*, a possible anthracosaur from the Upper Permian of European Russia. Based on Novikov et al., 2000.

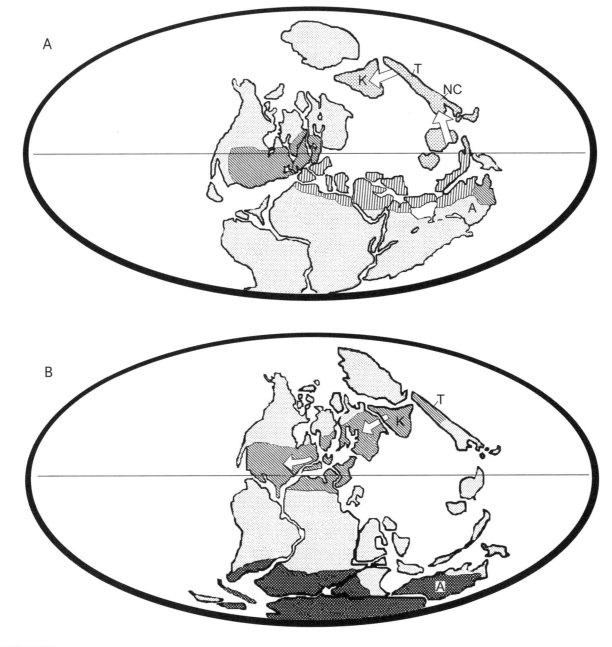

 Late Paleozoic ice cap

Inferred range of Carboniferous seymouriamorphs

 Known range of Carboniferous tetrapods

Inferred equatorial range of Carboniferous tetrapods

Figure 6.4. A, Early Carboniferous world map showing distribution of tetrapods. Early tetrapods may originally have been widely distributed along the margins of the Tethys Sea, on the continents of Pangaea in the north and Gondwanaland in the south. Large water gaps initially separated the early tetrapods in Australia from those in the north. *B,* Position of land masses in the Early Permian. Continental drift was responsible for the movement of large land masses from close to Australia to positions north and west, so that they eventually came in contact with what was to become the Eurasian land mass. This made it possible for the seymouriamorphs to spread into western Europe and North America. A, Australia. K, Kazakhstan. NC, North China. T, Tarim. Modified from Milner, 1996.

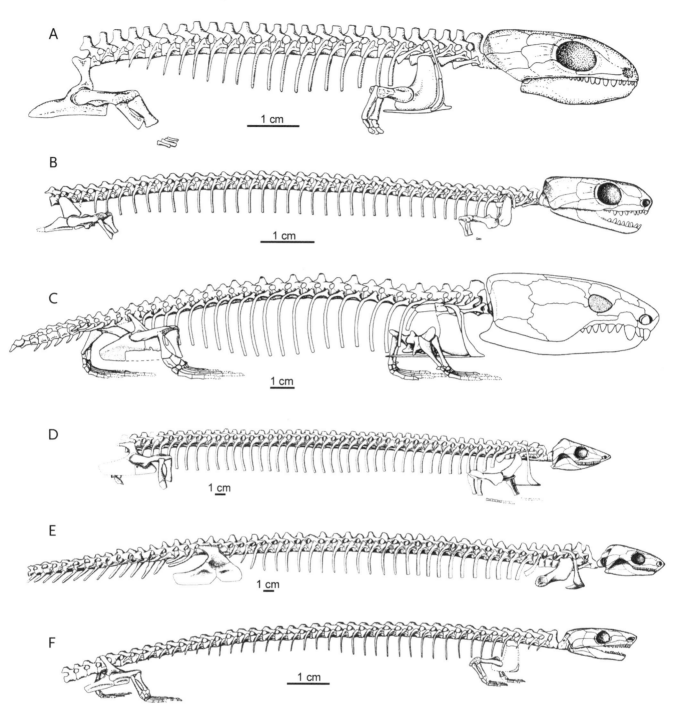

Figure 6.5. Skeletons of microsaurs. *A* and *B,* The gymnarthrids *Sparodus* from the Upper Carboniferous of the Czech Republic and *Cardiocephalus* from the Lower Permian of Fort Sill, Oklahoma, showing the changes in body proportions from a large head and relatively short vertebral column, to a small head at the end of a very long trunk region. This may be associated with adaptation to burrowing. Intercentra are present throughout the column in *Cardiocephalus,* but are not known in *Sparodus. C,* The Lower Permian

Pantylus, with the largest skull of any microsaur. It has a crushing dentition similar to that of *Trachystegos* from the mid-Carboniferous of Joggins, Nova Scotia. *D* and *E,* The large and very elongate ostodolepids *Pelodosotis* and *Micraroter* from the Lower Permian of Texas. *F,* The elongate and short-limbed *Rhynchonkos,* a goniorhynchid from the Lower Permian of Oklahoma, which resembles the most primitive known caecilian in many features. From R. L. Carroll, 2000.

Germany, showing a short trunk, with only 19 presacral vertebrae, stout limbs, and the proximal bones of the tarsus fused together to form a structure comparable to the astragalus in amniotes (R. L. Carroll, 1991). The skull is very small, with relatively large orbits, and the jaw articulation far forward. *Quasicaecilia* is known from only a single skull from

the Lower Permian of Texas, showing an extremely high degree of specialization associated with its tiny size (Fig. 6.7C). Much of the skull is dominated by the great dorsal and ventral expansion of the massively ossified braincase.

Despite their great diversity and locally high population size in the Lower Permian of both Europe and North America, no microsaurs have yet been described from other continents, nor are they known from later in the Paleozoic.

TEMNOSPONDYLS
Primitive Temnospondyls

The most numerous, widespread, and anatomically diverse of late Paleozoic and early Mesozoic amphibians were the temnospondyls (Fig. 6.8). They were the last of the major labyrinthodont groups to arise in the Carboniferous, but their oldest known representatives were already clearly distinct, with no obvious antecedents among earlier tetrapods. *Balanerpeton* and *Dendrerpeton* were common in the East Kirkton and Joggins localities, which were dominated by tetrapods indicative of adaptation to a terrestrial way of life. These genera lacked lateral line canals and had large limbs with fully ossified carpals and tarsals.

These early temnospondyls ranged from small to medium size, compared with other labyrinthodonts, with adult skulls ranging from approximately 5 to 10 cm in length. The position of the orbits near the middle of the skull length was probably primitive for temnospondyls as a group, since they had that position in the early members of many other Carboniferous lineages. Primitive aspects of their anatomy provide a good model for establishing the polarity of character state changes among other temnospondyls, and so assist in determining their interrelationships.

Other, incomplete remains of larger labyrinthodonts have been reported from both the Greer and East Kirkton localities (A. R. Milner and Sequeira, 1994). They have been suggested as being related to later large temnospondyls, implying a somewhat earlier radiation leading to genera such as *Capetus* from the Upper Carboniferous (Sequeira and Milner, 1993) and *Edops* from the Lower Permian of Texas, with a skull length of almost 70 cm (Fig. 6.9A, B).

A further lineage of primitive temnospondyls is represented by the family Cochleosauridae, which extended from the Westphalian A of Jarrow, Ireland (Sequeira, 1996), to the very end of the Permian. It is known from eastern North America, western Europe, and Niger in northern Africa (Fig. 6.10). All except the earliest is judged to be aquatic by the elongation of the snout, flattening of the skull, and relatively small limbs. However, only the latest genus, *Nigerpeton*, has lateral line canal grooves (Steyer et al., 2006).

Cochleosaurids, *Capetus*, and *Edops* have been grouped as a primitive superfamily, the Edopoidea, recognized by their small interpterygoid vacuities (not extending forward to the vomers), retention of intertemporal bones, and absence of a distinct posterior process of the premaxilla. All have the orbits in a more posterior position than in *Balanerpeton* and *Dendrerpeton*. Another genus, *Saharastega,* from the terminal Permian of the Sahara (Damiani et al., 2006), has been assigned to the Edopoidea, but is highly derived in the far anterior position of the jaw articulation, the sutural attachment of the palate and the base of the braincase, the exceptionally large "horns" of the tabular, and the great reduction or loss of palatal tusks (Fig. 6.10E). If an edopoid, it is certainly the most highly divergent member of this group, but it has no other more probable taxonomic affinities.

ERYOPOIDS, TRIMERORHACHOIDS, AND ARCHEGOSAUROIDS

A large number of other temnospondyl lineages, advanced above the level of edopoids, radiated in the Late Carboniferous. Many were initially placed in the superfamily Eryopoidea, but this group has been progressively limited as more and more genera and families have been placed in separate taxonomic groups. The family Eryopidae is reduced to only three genera: *Eryops* (Figs. 6.9C and 6.11A), whose remains are extremely common in the Texas red beds; *Onchiodon,* from the Lower Permian of central Europe; and *Clamorosaurus,* from the Upper Permian of Russia. These are all large animals, a meter or more in length. *Eryops* was presumably the dominant terrestrial predator among the Late Carboniferous and Early Permian amphibians of North America, although it may have been equally effective in the water. The smaller and much less common *Parioxys* (Fig. 6.9D) was even more highly adapted for a terrestrial way of life, with cylindrical intercentra, two pairs of sacral ribs, and elongate limbs (R. L. Carroll, 1964a). It is placed in a separate family within the Eryopoidea.

Other temnospondyl lineages are known from the Late Carboniferous, Permian, and Triassic, but clearly had separate evolutionary histories from the edopoids and eryopoids. Two superfamilies were adapted from an early stage to a strictly aquatic way of life as indicated by the retention of lateral line canal grooves into the adults. The earliest genus that might represent an ancestral trimerorhachoid is *Eugyrinus* (Fig. 6.12A), from the Westphalian A of Lancashire, England (A. R. Milner, 1980a). It is known primarily from the skull, but both the dorsal and palatal surfaces are visible. It is less than 2 cm in length, which implies that it was immature. This is supported by the presence of lateral line canal grooves, as well as unossified vertebral centra and the fimbriated anterior edge of the interclavicle. Poorly preserved but equally small skulls of *Dendrerpeton* are known, but do not show lateral line canal grooves, indicating that the *Eugyrinus* specimen was larval, or at least less developed at this size. However, the retention of lateral line canals in a skull with fairly well integrated sutures implies that it may have been paedomorphic or neotenic, and so was a putative antecedent of temnospondyls from the later Carboniferous that were certainly aquatic as adults.

As in other primitive temnospondyls, *Eugyrinus* retained an intertemporal bone and had a distinct otic notch and large palatal vacuities. These characters are reported in most of the Upper Carboniferous and Lower Permian trimerorhachoids. It also shares with them a retroarticular process on the lower jaw. Within the trimerorhachoids, the more primitive saurerpetontids are represented by the Upper Carboniferous

Figure 6.6. Ventral view of the skeleton of the microbrachomorph microsaur *Batropetes* from the Lower Permian near Dresden. From R. L. Carroll, 1991. Note the minor fault that crosses the stone diagonally separating the pelvis and the trail from the trunk. White bar = 1 cm.

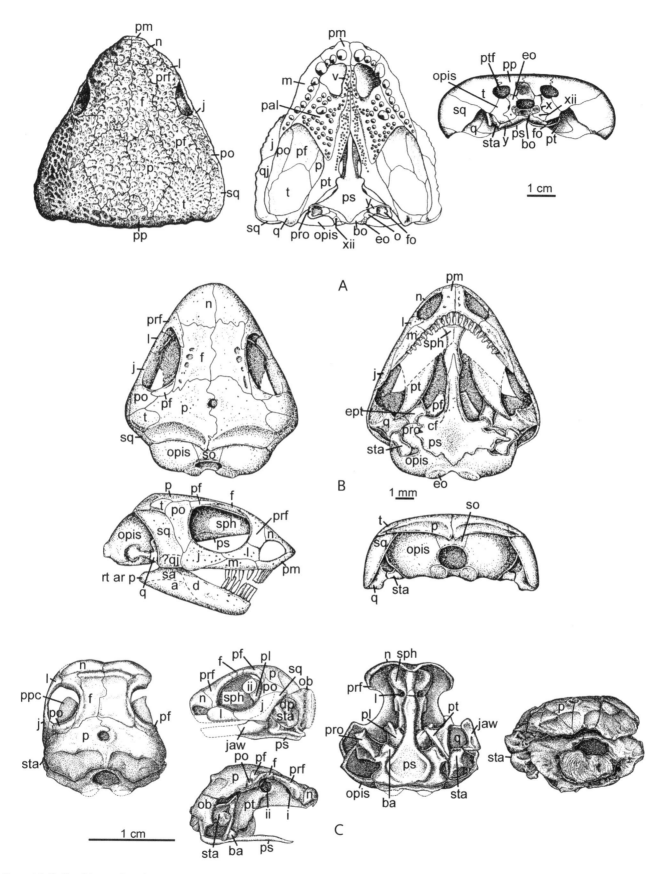

Figure 6.7. Skulls of Lower Permian microsaurs. *A, Pantylus. B, Batropetes.*
C, The tiny skull of *Quasicaecilia* from the Lower Permian of Texas in dorsal, lateral, palatal, and occipital views, and a lateral view of the skull with

the cheek removed to show the extent of the braincase. From R. L. Carroll,
1998b. Bone y, near the fenestra ovalis, is in the position of the operculum in
frogs and salamanders.

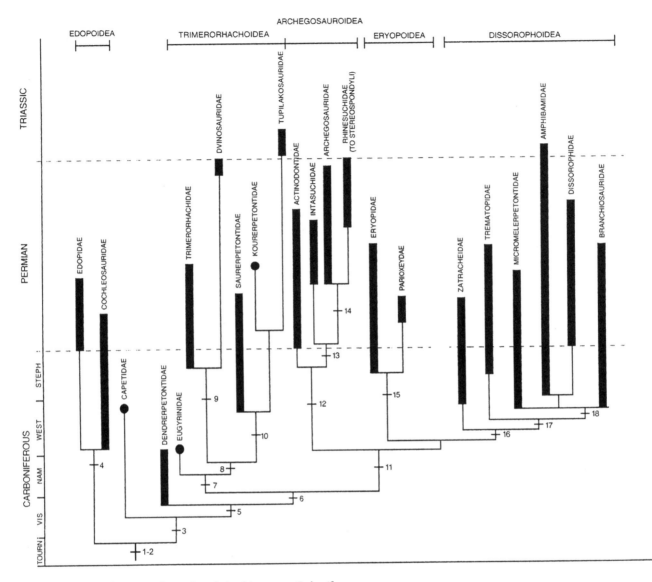

Figure 6.8. Geological ranges and tentative relationships among Carboniferous and Permian temnospondyls. Slightly modified from Holmes, 2000. Numbers indicate synapomorphies of the major clades.

and Lower Permian *Isodectes* (previously termed *Saurerpeton*), which lost the otic notch; also, the palatine bone was exposed beneath the orbit, separating the jugal and lacrimal (Fig. 6.12*B*, *C*). Sequeira (1998) has shown that the same species of *Isodectes* survived from the Westphalian D into the Lower Permian, a period of approximately 35 million years. Exposure of the palatine bone in the orbital margin was also expressed in some dissorophoids, but this was almost certainly a result of the small size of the skull and the relatively large orbit, rather than reflecting close relationship.

The more advanced trimerorhachids, *Neldasaurus* and the very common *Trimerorhachis*, are present in the Lower Permian red beds of Texas. The otic notch is maintained in

Trimerorhachis, but both genera are derived in the presence of the anterior palatine fenestrae to receive large parasymphysial tusks, as in cochleosaurids. This is logically related to the very low profile of the snout. The increasing width to depth ratio of the occiput among large, secondarily aquatic temnospondyls is associated with spreading apart of the articulating surfaces of the exoccipitals so that the movement of the skull relative to the vertebral column is constrained to hinging in the vertical plane. The orbits face primarily dorsally, as in most aquatic temnospondyls. The size and degree of ossification of the limbs was progressively reduced within these groups. A well-ossified hyoid apparatus is known in the adults of *Trimerorhachis* and other large trimerorhachoids.

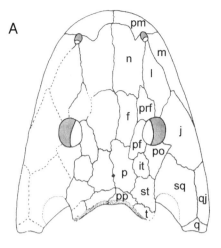

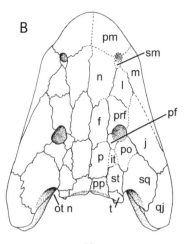

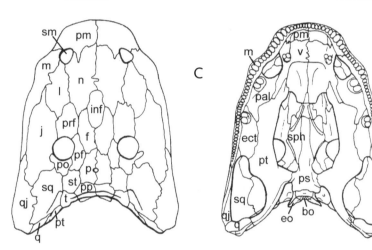

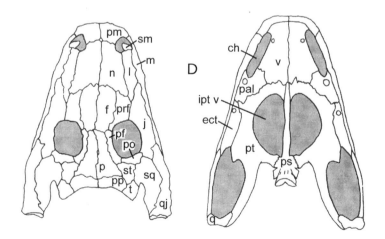

Figure 6.9. Skulls of representative edopoids and eryopoids. A, Capetus, from the Upper Carboniferous. B, Edops, from the Lower Permian of Texas. C, Dorsal and palatal views of Eryops, from the Upper Carboniferous and Lower Permian of North America. D, Dorsal and palatal views of Parioxys. From Holmes, 2000.

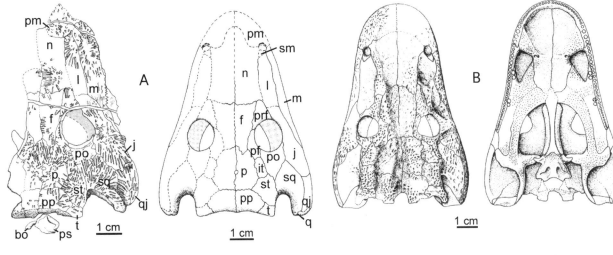

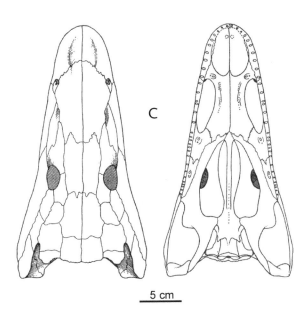

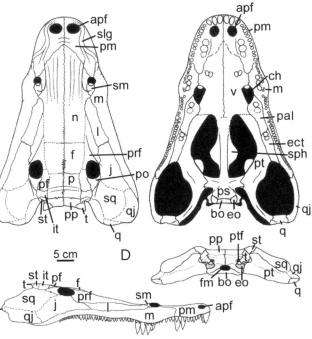

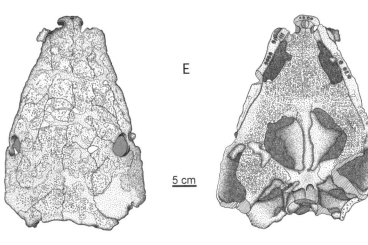

Figure 6.10. Edopoids. A–D, Cochleosaurids. A, Putative cochleosaurid, *Procochleosaurus jarrowensis,* from the Westphalian A of Ireland. Based on Sequeira, 1996. B, *Cochleosaurus florensis,* from the lycopod stump fauna of Florence, Nova Scotia. Modified from Godfrey and Holmes, 1995. C, *Chenoprosopus milleri,* Lower Permian of New Mexico. Modified from Langston,

1953. D, *Nigerpeton ricqlesi,* from the Upper Permian of Niger. Based on Steyer et al., 2006. Note very low profile of skull and large palatal and marginal teeth. E, Dorsal and palatal views of the putative edopoid *Saharastega moradiensis* from the Upper Permian of Niger. Derived from Damiani et al., 2006. Skull roof is too badly crushed for sutures to be identified.

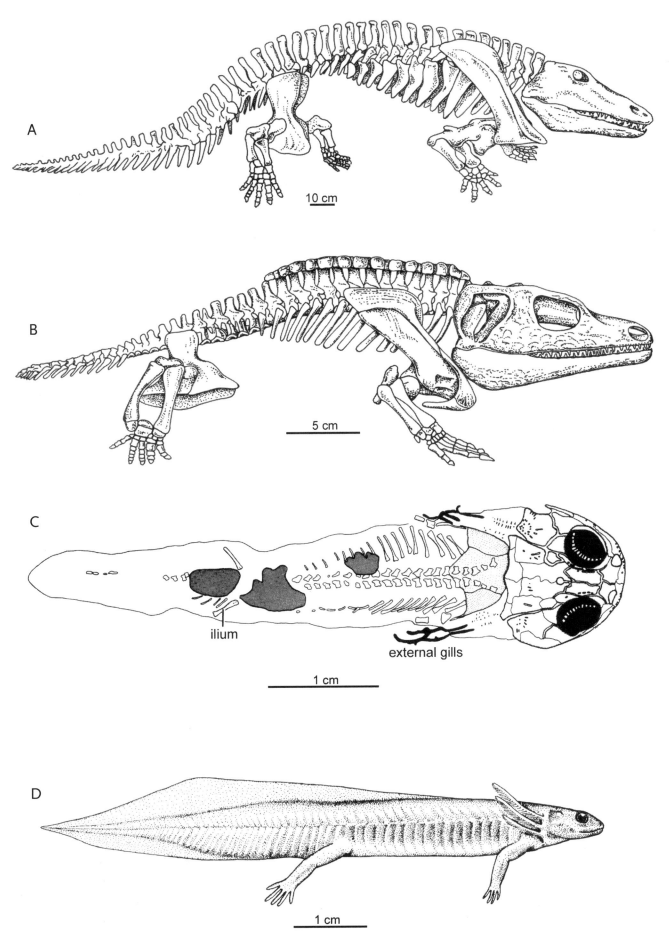

Figure 6.11. The diverse size and body forms of temnospondyls. *A*, The eryopoid *Eryops*. Modified from W. K. Gregory, 1951. *B*, The dissorophid *Cacops*. From Williston, 1910. *C*, Larva of the primitive trimerorhachoid *Isodectes* (*Saurerpeton*) from the Upper Carboniferous of Mazon Creek, Illinois. Modified from A. R. Milner, 1982. *D*, Restoration of the larva of a branchiosaurid, probably *Apateon*, from the Lower Permian of Germany. From A. R. Milner, 1982.

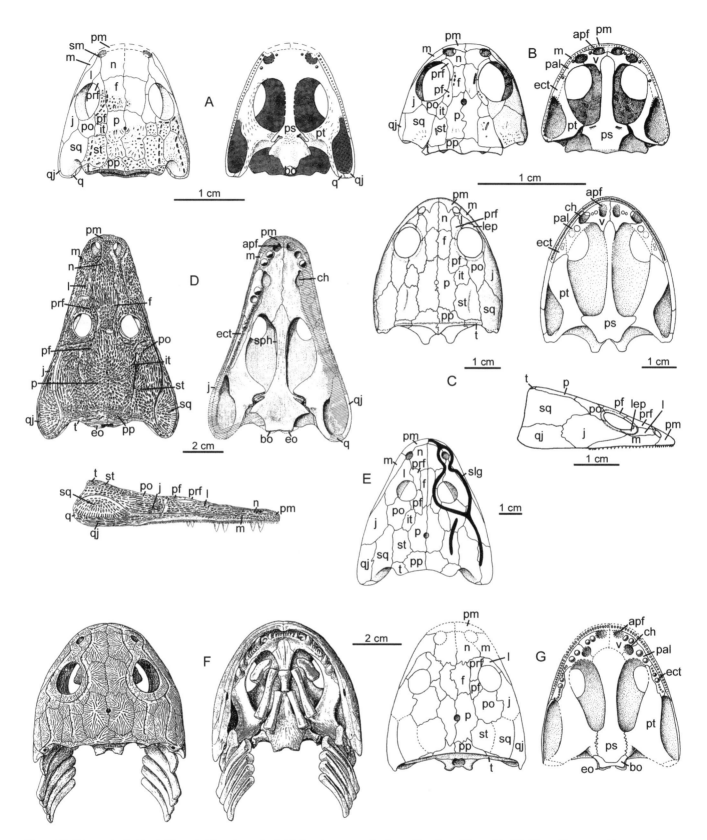

Figure 6.12. Trimerorhachoids. *A,* The primitive trimerorhachoid *Eugyrinus* from the Middle Carboniferous of Great Britain. Skull in dorsal and ventral views. From A. R. Milner, 1980a. *B,* The sauropleurid *Isodectes obtusus,* based on a juvenile specimen from the Westphalian D, of Mazon Creek. From A. R. Milner, 1982. *C, Isodectes obtusus* adult from the Arroyo Formation of Texas. From Sequeira, 1998. *D,* The trimerorhachoid *Neldasaurus,* from the Lower Permian of Texas. Modified from Chase, 1965. *E, Trimerorhachis,* from the Lower Permian of Texas. Modified from Case, 1935. Note conspicuous lateral line canals. *F,* The neotenic *Dvinosaurus* from the Upper Permian of Russia in dorsal and ventral views. Note high degree of ossification of branchial arches in the adult. Modified from Bystrow, 1938. *Tupilakosaurus* from the Lower Triassic of Greenland and Russia in dorsal and ventral views. From Shishkin, 1961.

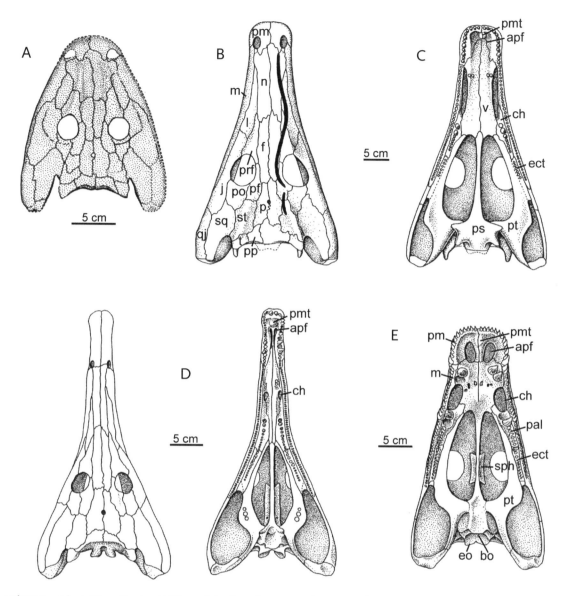

Figure 6.13. Archegosauroids. *A,* The actinodontid *Sclerocephalus* from the Lower Permian of Germany. Modified from Boy, 1988. Differentiated from eryopoids in the separation of the lacrimal bone from the narial opening by the sutural contact of the maxilla and nasal. *B* and *C,* The archegosaurid *Archegosaurus,* which is extremely common in the Lower Permian of Europe. From Holmes, 2000. *D* and *E, Platyoposaurus* and *Konzhukovia* from the Upper Permian of Russia. Adapted from Gubin, 1991.

Trimerorhachoids extended to the end of the Permian in Russia, where they are represented by the neotenic genus *Dvinosaurus* (Novikov et al., 2000) (Fig. 6. 12F). *Tupilakosaurus,* from the Lower Triassic of Greenland and Russia, may also belong to this superfamily, but the shorter snouts of these genera suggest origins from more primitive saurerpetontids such as *Isodectes.* A. R. Milner and Sequeira (2004) argue that fragmentary remains of *Slaugenhopia,* from the uppermost Lower Permian of Texas, may provide a link between these genera.

A second superfamily of larger and longer-snouted aquatic temnospondyls, the Archegosauroidea, may be traced to genera of more modest proportions once placed among the eryopoids, including the German, Lower Permian *Sclerocephalus,* distinguished from eryopoids by the separation of the lacrimal bone from both the naris and the orbital margin (Fig. 6.13A). The much longer-snouted *Archegosaurus* is known from hundreds of animals from central Europe, represented by very extensive growth series (Witzmann, 2006). More specialized members of this group extended into the Upper Permian in Russia.

THE DISSOROPHOIDEA

Most of the remainder of Upper Carboniferous and Lower Permian temnospondyls are thought to have belonged to a single, albeit heterogeneous assemblage, the Dissorophoidea, which is first known from the Westphalian D coal swamp deposits of North America and central Europe. By the Lower Permian, six families are clearly distinguishable from one another: Amphibamidae, Branchiosauridae, Dissorophidae, Micromelerpetontidae, Trematopidae, and Zatracheidae. Most can be traced to more primitive Upper Carboniferous fossils that converge toward a broadly similar morphology as they are traced back toward the late Westphalian, but their specific interrelationships and pattern of divergence remain uncertain (Fig. 6.8).

The earliest specimens come from the ironstone nodules of the Mazon Creek fauna, exposed near Chicago, Illinois (A. R. Milner, 1982). The specimens are largely preserved as natural casts, but when molded with latex or silicon yield extremely informative replicas of the original bone surface (Fig. 6.14). These fossils, in common with many other early dissorophoids, are of small size, with skull lengths of adults of 2 cm or less, and yet show no sign of lateral line canals. Smaller specimens, showing external gills, have skulls only about 5 mm in length.

Among earlier temnospondyls, skull proportions are most similar to those of Balanerpeton (Fig. 4.26) although the orbits and the otic notches are even more prominent. The ventral margin of the orbit is partially formed by the palatine bone, bridging a gap between the jugal and lacrimal to accommodate the eye. There is evidence of two dissorophoid families at Mazon Creek. Four more appear in the later Carboniferous or Lower Permian. All are initially characterized by their small size, similarity of the general configuration of the skull, loss of the intertemporal, and reduction or absence of lateral line canal grooves in the skull.

Amphibamidae

The oldest and plausibly most primitive dissorophoid is *Amphibamus grandiceps* from the Westphalian D of the Mazon Creek locality in Illinois (Fig. 6.14) (J. T. Gregory, 1950; A. R. Milner, 1982). It is represented by many specimens, ranging from larvae exhibiting external gills to metamorphosed individuals in which much of the skeleton is well ossified. Although the carpals, tarsals, and ends of the limb bones remained cartilaginous, this species is striking in its rapidity to reach a near adult condition at very small size, with the skull less than 2 cm in length. This may be a specialization of species that emerged at the base of the family Amphibamidae, but it might be the heritage of a highly terrestrial lineage stemming from conservative early temnospondyls such as *Balanerpeton*.

The small size of the skull is emphasized by the relatively large size of the orbital openings, which intersect the palatine bone ventrally and almost reach the frontals dorsally, and the great extent of the interpterygoid vacuities. The quadrate bone, usually late to ossify in early amphibians, is fully formed, with a dorsal process comparable to the structure that supports the tympanic annulus in frogs. The palate is covered with a shagreen of tiny denticles, but it lacks the palatal fangs that occur in adult labyrinthodonts. This is almost certainly the result of its small size, comparable to the condition in lepospondyls, in which these teeth never develop the labyrinthodont configuration.

The trunk region is about twice the length of the skull and the limb proportions are approximately comparable to those of *Balanerpeton*, taking into consideration the incomplete ossification of the ends of the limb bones resulting from the immaturity of the specimens. The reconstruction gives the impression of an agile, terrestrial animal.

Several larval specimens can be attributed to *Amphibamus grandiceps*, which are among the oldest adequately known of any labyrinthodonts. The skulls of the largest larvae are sufficiently well known to assign them to this species, while the specific identity of smaller specimens is less certain. Some retain calcium reserves in the endolymphatic sacs of the inner ear (Fig. 6.14D). All the dermal bones of the skull appear to ossify nearly simultaneously in the smallest known specimens.

A second species, *Amphibamus lyelli* (also termed *Platyrhinops*), comes from the slightly later Linton locality in Ohio. Adult specimens have about twice the cranial length of *A. grandiceps*, and do have small fangs on the vomers, palatines, and ectopterygoids, as in most later dissorophoids, but like most members of this superfamily, maintain an unfused area of articulation between the base of the braincase and the pterygoid (Fig. 6.15B). Very similar fossils from the Nýřany locality in the Czech Republic have been referred to as *Amphibamus calliprepes* by R. L. Carroll (1964b). The size and general configuration of the skull in *Amphibamus* make it a plausible ancestor for most, if not all other dissorophoids, most of which appear initially slightly later in the fossil record.

A second genus, *Eoscopus*, occurs at the very end of the Carboniferous in eastern Kansas and a third, *Tersomius*, in the Early Permian of Texas and Oklahoma (R. L. Carroll, 1964b; Daly, 1994). The skull-trunk length of *Eoscopus* is about 10 cm, with the body proportions similar to those of *Amphibamus*. The skulls differ in the retention of the union between the jugal and lacrimal beneath the orbit in adults of *Tersomius texensis*, but not in *Eoscopus*, in which the palatine enters the margin of the orbit. Despite its relatively small size, the skeleton of *Eoscopus* was fully ossified, including all the carpals and tarsals, indicating that it was a highly agile, fully terrestrial vertebrate. It may have had a role similar to that of modern insectivorous lizards such as small iguanids.

The skull of *Tersomius* (Fig. 6.15D) is striking in the presence of palpebral cups that form a bony protection within the eyelid, giving it a "bug-eyed" appearance. It also has irregular bony plates between the lower jaws. Like all other terrestrial dissorophoids, *Tersomius* has a very large otic notch, implying the presence of an extensive tympanum for the detection of airborne vibrations. Fossils of *Tersomius* are extremely common, with more than 50 specimens found in Oklahoma sites.

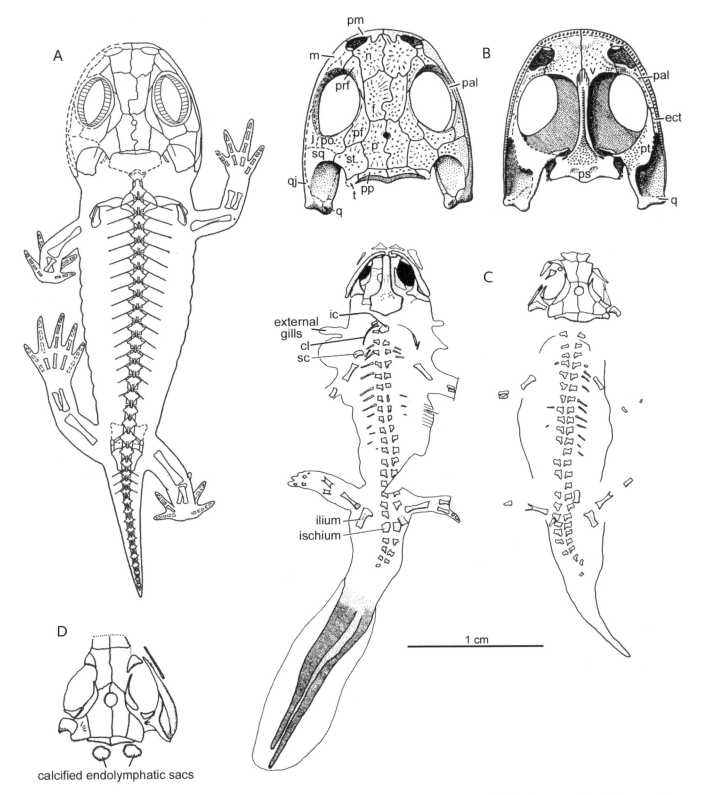

Figure 6.14. Amphibamid dissorophoids. *A, Amphibamus grandiceps* skeleton from the Westphalian D of Mazon Creek, Illinois. Modified from J. T. Gregory, 1950. *B,* Dorsal and palatal views of the skull of *Amphibamus grandiceps.* *C,* Ventral and dorsal views of a larval specimen of *Amphibamus grandiceps.* *D,* Skull of larva showing calcified endolymphatic sacs. *B–D* from A. R. Milner, 1982.

A further genus that is thought to be closely related to early amphibamids is *Doleserpeton*, known from a single Lower Permian locality near Fort Sill, Oklahoma (Fig. 6.16C, D). It is represented by hundreds of disarticulated skeletons as well as several well-preserved skulls. *Doleserpeton* resembles *Amphibamus grandiceps* in its small size and absence of palatal fangs, but the high degree of ossification of the entire skeleton indicates that the larger specimens were of

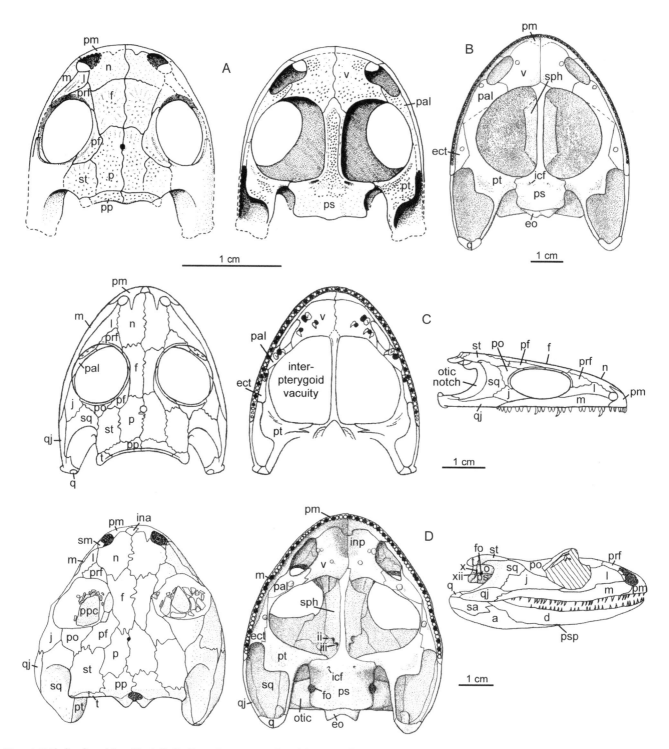

Figure 6.15. Skulls of amphibamids. *A,* Skull of juvenile specimen of *Amphibamus lyelli* (also termed *Platyrhinops*) from the Westphalian D of Linton, Ohio. Based on A. R. Milner, 1982. *B,* Skull of adult specimen of *Amphibamus lyelli,* from Linton, Ohio. From R. L. Carroll, 1964b. *C,* Skull of *Escopus lockardi* from the Upper Pennsylvania of Kansas. From Daly, 1994. *D,* Skull of *Tersomius texana* from the Lower Permian of Texas. From R. L. Carroll, 1964b.

adults. The orbits are very large and intersect the frontals and palatine bones, and the interpterygoid vacuities are surrounded by only a narrow border of palatal elements. What is of particular importance is the presence of a unique configuration of the marginal teeth, in which their bases (also termed pedicles) are not co-ossified with the crowns. These are termed pedicellate teeth. Teeth with this configuration have long been considered a diagnostic character of frogs, salamanders, and caecilians (Parsons and Williams, 1963).

The vertebrae are also derived relative to those of other

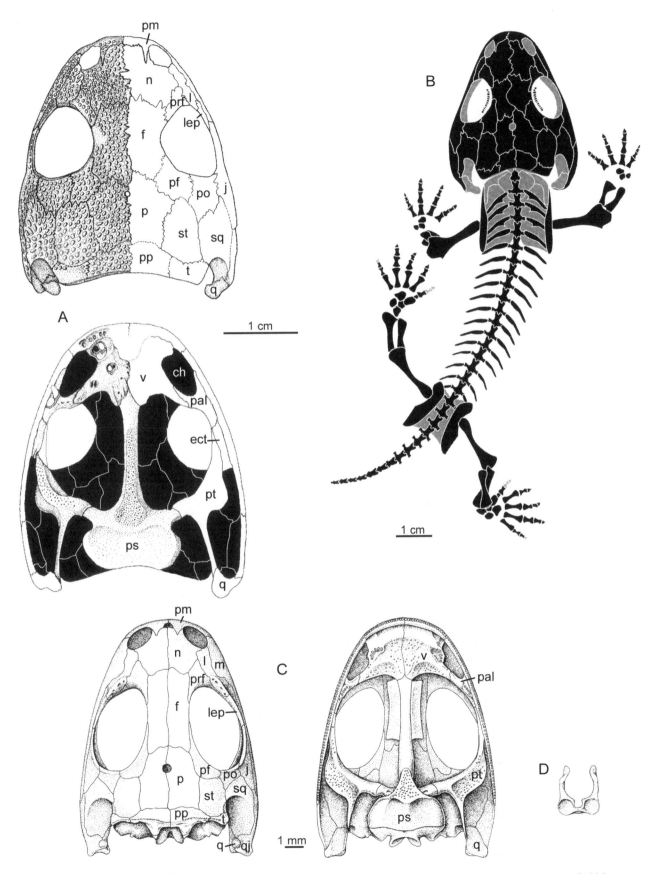

Figure 6.16. Advanced amphibamids. *A* and *B*, Skull and reconstruction of the skeleton of the latest known amphibamid, *Micropholis*, from the Lower Triassic of South Africa. From Schoch and Rubidge, 2005. *C* and *D*, Skull and atlas vertebra of *Doliserpeton* from the Lower Permian of Oklahoma; a genus with many characters resembling frogs. Redrawn from Bolt, 1969.

Paleozoic temnospondyls in the ontogenetic fusion of the originally separate elements. The adult atlas is formed of a single bone, as in modern amphibians, and the pleurocentra of the trunk vertebrae are cylindrical and fused to the neural arches in the adult. The intercentra are reduced to small crescents. From its initial description (Bolt, 1969), *Doleserpeton* has been hypothesized as being close to the ancestry of one or more of the modern amphibian orders. This subject will be discussed in more detail in Chapters 9 and 10.

Amphibamids have not been found in North America or Europe in beds later than the Early Permian, when both areas of deposition became drier and most of the Paleozoic lineages appear to have died out. Surprisingly, a single genus, *Micropholis*, indicates the persistence of the family into the Lower Triassic of South Africa (Schoch and Rubidge, 2005). Superficially, the skull is similar to that of *Tersomius*, although somewhat more slender and with a pustular ornamentation (Fig. 6.16*A*, *B*). Palpebral cups are not present, but the palatine enters the orbital margin. There is an irregular increase in the length of the skull relative to the trunk throughout the history of amphibamids, from 49% in *Amphibamus grandiceps* to 56% in *Micropholis*. *Micropholis* is the last known of the dissorophoids, and in fact of all fully terrestrial temnospondyls.

Dissorophidae

The genera now included in the Amphibamidae were long grouped with a number of more derived temnospondyls in the family Dissorophidae. Eleanor Daly (1994) divided up these taxa, with the genera *Aspidosaurus, Broiliellus, Dissorophus,* and *Cacops* retaining the original family name. These genera are clearly distinguished from amphibamids in the possession of bony plates either lying above the neural spines or fused to them, or having the neural spines extending far above the neural arches (Fig. 6.17). Dissorophids are known primarily from the Lower Permian of Texas, but extend to the Upper Permian in Russia (R. L. Carroll, 1964b; DeMar, 1966). The most primitive genus, *Broiliellus,* has the size and general cranial anatomy of *Tersomius,* but with clearly defined ridges outlining the skull table, extending medially around the orbits, and anteriorly from the prefrontals and nasals to the external nares (Fig. 6.17*B*). As in other dissorophids, dermal plates, located above the vertebral column, extend posteriorly from the occiput. In *Broiliellus brevis* they are narrow, limited to a single segment, and are not attached to the neural spines. In the slightly later species, *Broiliellus texensis,* the armor plates are wider, but remain unattached, and one per segment (Fig. 6.17 *C, D*). *Dissorophus angustus* (Fig. 6.17*A*) is contemporary with *B. brevis,* but differs in having two narrow plates per segment, with external and internal plates alternating with one another and extending posteriorly over the first 17 vertebrae. The vertebrae of the somewhat later genus, *Cacops* (Figs. 6.11*B*, 6.17*G*), could have evolved by the fusion of the lower row of plates of an animal resembling *Dissorophus angustus* to the ends of the neural spines. *Cacops* and *Dissorophus* separately evolved a closure of the otic notch

by down-growth of the tabular. *Dissorophus multicinctus* had two rows of wider plates, with the first several fused into a longer shield and the inner row extending ventral processes between the neural spines (Fig. 6.17*E, F*). The genera *Aspidosaurus* (Fig. 6.17*H*) and *Conjunctio* differed in the fusion of a single row of narrow plates with the dorsal extremities of the neural spines.

Even more spectacular modifications of the neural spines evolved in another genus of dissorophid affinities, *Platyhystrix rugosus.* Isolated pieces of laterally compressed and ornamented neural spines have long been recognized from Lower Permian and even Upper Carboniferous sites in North America, but rarely with even fragments of the skull or appendicular skeleton. Lewis and Vaughn (1965) described the most complete sequence of neural spines, showing a pattern broadly resembling the "sail" of edaphosaur pelycosaurs, from the area of the atlas arch to the 15th dorsal vertebra (Fig. 6.17 and Plate 9). Subsequently, Berman et al. (1981) prepared a skull, originally collected by David Baldwin in 1881 from the Lower Permian Cutler Formation in New Mexico, which showed an extremely rugose texture, closely associated with spines initially attributed to *Aspidosaurus.* The lateral edges of the skull table are ridged as in *Broiliellus* and *Dissorophus.* Comparable neural spines designated *Astreptorhachis ohioensis,* from the Upper Pennsylvanian Conemaugh Group of Ohio, support the earlier divergence of this group, near the time of emergence of the amphibamids.

Dissorophids are not known in North America or western Europe past the Lower Permian, when the red beds of the southwestern United States became very dry and the localities in Europe greatly restricted. However, their persistence during this period is indicated by the presence of *Kamacops* and *Tatusaurus* in the Upper Permian of Russia (Gubin, 1980). Both were large and had flattened skulls with dorsally facing orbits, suggesting they had reverted to aquatic habits. *Kamacops* resembles *Cacops* in the posterior closure of the otic notch.

Branchiosaurs

The term branchiosaur was coined by Anton Fritsch (1876) to describe very small amphibians from the Upper Carboniferous (Westphalian D) of Nýřany, in the Czech Republic, which had external gills broadly comparable to those of modern salamanders. It was first thought that they represented a separate group of early tetrapods, subsequently termed the Phyllospondyli because of the unusual leaf-like shape of the vertebrae. Later, Romer (1939) suggested that they were the larvae of larger amphibians found in the coal swamp deposits.

It was not until the work of Jurgen Boy (1972, 1974, 1995; summarized by Boy and Sues, 2000) that it was recognized that most of the species identified as branchiosaurs were members of two distinct families, the Branchiosauridae and the Micromelerpetontidae, nearly all of which were neotenic (spending their entire life as aquatic, larva-like forms). Larvae

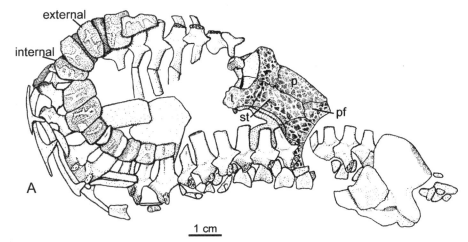

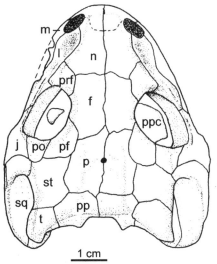

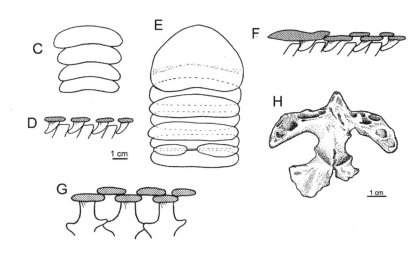

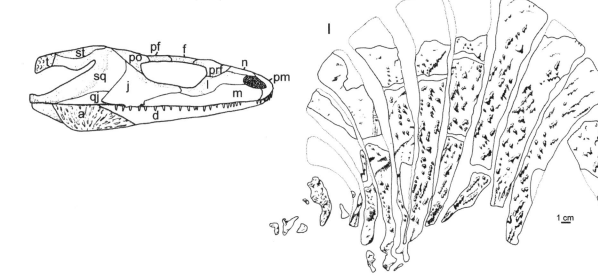

Figure 6.17. Skulls and armor of dissorophids. *A,* Skull and axial skeleton of *Dissorophus angustus* showing ridges on the margin of the skull and the presence of external and internal armor plates. *B,* Dorsal and lateral views of the skull of *Broiliellus brevis. C* and *D,* Dorsal and lateral views of the armor of *Broiliellus texensis. E* and *F,* Dorsal and lateral views of anterior armor plates of *Dissorophus multicinctus. G,* Lateral view of the armor of *Cacops,* in which the internal plates are fused to the ends of the neural spines. *H,* A vertebra of *Aspidosaurus* in anterior view, showing the lateral elaboration of the neural spine. *I,* Greatly extended and ornamented neural spines of *Platyhystrix;* anterior is to the left. *A, B,* and *H* from R. L. Carroll, 1964b; *C–G* from Holmes, 2000; *I* from Lewis and Vaughn, 1965. All from the Lower Permian of the southwestern United States.

and young of other families of Paleozoic amphibians have also been recognized, but they are clearly distinct from branchiosaurs. The closest relatives of both families of branchiosaurs seem to lie among the dissorophoids, and more specifically the small, unspecialized early amphibamids.

Branchiosaurs provide the best evidence of the modes of development of Paleozoic labyrinthodonts (Figs. 6.18–6.20). They include many species, widely distributed in central Europe and locally extremely abundant, making it possible to trace their development from hatching to maturity, although very few actually metamorphosed into terrestrial adults. Amphibians are the only vertebrates for which long sequences of early development can be studied on the basis of the fossil record. Not only can the patterns and sequences of development be described, but this information can be used to test the hypothesis that ontogeny recapitulates phylogeny. From branchiosaurs, it is possible to determine to what extent distinct patterns of development may influence particular modes of evolutionary change.

MICROMELERPETONIDAE. Two families of branchiosaurs have been recognized on the basis of significant and consistent differences in anatomy and in the sequence of early development. The more primitive of these families is the Micromelerpetontidae, of which the earliest fossils belong to the genus *Limnogyrinus* from the Nýřany locality (Fritsch, 1883). *Branchierpeton* is known from the latest Carboniferous and earliest Permian and *Micromelerpeton* only from the earliest Permian (Boy and Sues, 2000). The skulls of the adults are comparable to those of amphibamids in the large size of the orbits, intersecting the frontals in the later species and exposing the palatine bone ventrally. They also have very large interpterygoid vacuities. However, even at small size, they developed fangs on the marginal bones of the palate. As in most other labyrinthodonts for which larvae are known, they have four rows of bony, tooth-bearing plates that were attached to the ceratobranchial elements of the hyoid apparatus (Fig. 6.18B). These are thought to have been used for filtering out small particles of food that were captured by gape-and-suck feeding. The quadrate was not ossified.

The skull is small, relative to the trunk, ranging from 33 to 37% of its length. The presacral vertebrae number from 25 to 29 and the limbs are relatively short. The vertebral centra are only ossified in the largest specimens of *Micromelerpeton*, which also has a mosaic of bones in the orbit resembling the palpebral cup of *Tersomius*. Only *Limnogyrinus* has dermal ossicles covering the interpterygoid vacuities, as in *Balanerpeton* and *Amphibamus grandiceps*. *Micromelerpeton* provides an informative example of the sequence of ossification throughout the skeleton among primitive labyrinthodonts. Nearly all the dermal bones ossify simultaneously, at a very small skull size, prior to the appearance of the limbs (Fig. 6.19).

BRANCHIOSAURIDAE. A. R. Milner (1982) tentatively identified two specimens from the Westphalian D of Mazon Creek, just slightly older than the Nýřany deposit, as belonging to the second branchiosaur family, the Branchiosauridae. Surprisingly, these are the only early branchiosaurs yet described from North America; almost all the others are restricted to central Europe. The Mazon Creek specimens are distinguished from the larvae of *Amphibamus* by ossification of anterior elements of the hyobranchial apparatus and the presence of internal carotid foramina on the basal plate of the parasphenoid. Milner tentatively assigned this species to *Branchiosaurus*, but without designating a particular species.

Boy and Sues (2000) recognized *Branchiosaurus salamandroides* from Nýřany as the oldest European branchiosaurid. Later branchiosaurids, *Apateon* and *Melanerpeton*, are known from the Carboniferous-Permian transition and *Schoenfelderpeton* from the earliest Permian. The skulls of *Branchiosaurus* range from 11 to 15 mm in length, and those of *Schoenfelderpeton* from 22 to 35 mm. The number of presacral vertebrae declines from 24 to 19 between the Late Carboniferous and the Permian genera.

Branchiosaurids are distinguished from micromelerpetontids by a number of highly significant characters. In contrast with the larvae of micromelerpetontids and all other labyrinthodonts and their fish ancestors, there are six rather than four rows of gill rakers associated with the hyoid apparatus (Fig. 6.18D). Instead of arising from thin bony plates, they were attached directly to the four pairs of ceratobranchials, with their tips facing across the gaps between them like the teeth of a zipper. In addition to filtering food particles, comparison with modern salamander larvae indicates that they probably served to close the gaps between the external gill slits so as to maintain the vacuum within the mouth and pharynx during suction feeding (Lauder and Schaffer, 1985).

The most striking difference between the larvae of branchiosaurids and those of all other amphibians from the Paleozoic is the sequence and pattern of ossification of the skull. In micromelerpetontids, all the dermal bones ossified essentially simultaneously in the smallest specimens that can be identified. This pattern is common to all other Paleozoic amphibians and their fish ancestors, as exemplified by *Eusthenopteron* (Fig. 6.19E). Among branchiosaurids, the later genera show a unique, sequential ossification of the skull bones, most fully described in *Apateon caducus* and *Apateon pedestris* (Schoch and Carroll, 2003; Schoch 2004) (Fig. 6.20). The first bones to ossify were the tooth-bearing elements of the upper and lower jaws. This enabled the larvae to use them for feeding when they were very small, and yet maintain a very flexible skull for gape-and-suck feeding. The bones forming the main axis of the skull roof ossified next, and only later the bones surrounding the eyes and forming the back of the skull table. In some of the largest specimens of *Apateon* the marginal teeth change from very long and slender to much thicker in diameter and the crowns become separated from the base, as in the pedicellate teeth of modern salamanders.

It is uncertain as to how widespread sequential ossification of the skull bones was in branchiosaurids. In the apparently oldest members of this family, from Mazon Creek, skulls about 6 mm in width appear to have a fully ossified skull roof, with the margins of the orbits clearly defined. This also seems to be the case for the fossils of *Branchiosaurus salamandroides* from Nýřany described by Fritsch (1883) and undescribed material from this locality in the Natural History Museum in Vienna. Since the branchiosaurids from Nýřany are among the smallest members of this family, it may be that even the smallest known fossils are too large to show the earliest stages of development. On the other hand, it is possible that the capacity for sequential ossification of the skull bones evolved within the family, and was not yet expressed in the earliest known members.

Nearly all branchiosaurid species are represented by gilled larvae, but almost none are known from terrestrial adults. This implies that most were neotenic, spending their entire life in the water. However, all had body proportions more like terrestrial amphibamids than the micromelerpetontids and had a relatively long and stable phase in late larval life, undergoing growth, but without apparent morphological change.

Only one branchiosaurid species, *Apateon gracilis,* has been described as undergoing a clearly defined metamorphosis involving a rapid and dramatic change to an anatomical pattern consistent with life on land (Schoch and Fröbisch, 2006). *Apateon gracilis* is one of the smallest branchiosaurids, with the skull of most specimens ranging from 6 to 12 mm and the largest reaching only 18 mm. Yet, the metamorphosed individuals show rapid ossification of the endochondral bones of the braincase, palatoquadrate, and shoulder girdle that remained cartilaginous throughout development in all other branchiosaurs. In addition, the larval hyobranchial apparatus was not ossified and branchial denticles were lost. Their overall adult appearance resembles the amphibamids *Amphibamus* and *Doleserpeton.*

The highly distinctive features of the latest Carboniferous and Early Permian branchiosaurids, especially *Apateon,* may be due in part to their common occurrence in high-altitude lakes in Germany and France (Boy and Sues, 2000). Based on geological reconstructions, branchiosaurs lived in relatively deep lakes along the northern margin of the tropical belt in intermontane basins of the Variscian mountains at elevations of up to 2000 m or more above sea level. Temperatures were low due to the high elevation. These localities probably had a greater degree of seasonality than the typical coal swamp deposits of Mazon, Linton, and Nýřany. In similar environments today, amphibians face a critical choice. As animals that can live either in the water or on land, branchiosaurs may have had some range of physiological response enabling them to stay for a longer period of time in the water if conditions on land were unsuitable for survival, or to metamorphose at an earlier or later time. Among living salamanders, some genera always undergo metamorphosis, others spend all of their life in the water, while a few may or may not undergo metamorphosis, depending on the conditions (Whiteman, 1994). This choice may also have been open to branchiosaurids. The importance of branchiosaurids

in understanding the ancestry of salamanders is discussed in Chapter 11.

Zatracheidae

Advanced zatracheids, from the Lower Permian of Texas and New Mexico, have the most striking cranial anatomy of any Permo-Carboniferous temnospondyls, with very flat skulls reaching more than 12 cm in length, small orbits, and spines extending from the back of the skull table and cheek (Fig. 6.21). The spines presumably provided some protection from aquatic predators such as pleuracanth sharks, whose teeth are common in the same deposits. The failure to protect the more posterior portion of the body may explain the almost total absence of postcranial remains of the most specialized genus, *Zatrachys* (Langston, 1953).

More primitive zatracheids have been described from the Lower Permian of Germany, represented by *Acanthostomatops* (initially termed *Acanthostoma*), which already shows the elaboration of spines and the presence of an internarial foramen on the dorsal surface of the snout that may have held a glandular structure. Postcranial remains show relatively short limbs. Steen (1930) described the intercentra as being paired, but Langston (1953) described vertebrae with a typical rhachitomous structure in *Zatrachys,* with paired pleurocentra dorsally and crescentic intercentra ventrally. More recently, Werneburg (1998) described a larval specimen of *Acanthostomatops* in which the pleurocentra were fused ventrally but open dorsally, somewhat as in *Caerorhachis,* described in Chapter 4. This indicates some degree of variation in vertebral development in this family.

The oldest zatracheid, *Stegops,* from the Westphalian D of Linton, Ohio (Romer, 1930; Steen, 1930), demonstrates their presence in the Upper Carboniferous and supports an origin from the base of the dissorophoid complex. The palate resembles that of early amphibamids, with extremely large interpterygoid vacuities although the pterygoids still extend to the vomers. The orbits are smaller than those of amphibamids, with the prefrontal and postfrontal in broad contact. Ridges, apparently formed above the course of the lateral line canals, have a pattern closely resembling that of dissorophids such as *Broiliellus.*

Trematopidae

A further family, the Trematopidae, is also assumed to have evolved from the base of the dissorophoid complex (Fig. 6.22). It is known primarily from the Upper Pennsylvanian and Lower Permian of the southwestern United States. All trematopsids have more elongate skulls than the amphibamids and dissorophids, with the eyes somewhat posterior to the middle of the skull length. More striking is the great elongation of the external nares, suggesting the presence of a gland in the more posterior portion.

Trematopsids have an enlarged tooth near the posterior end of the premaxilla and at about the position of the sixth tooth from the front of the maxilla. *Acheloma* also has a very

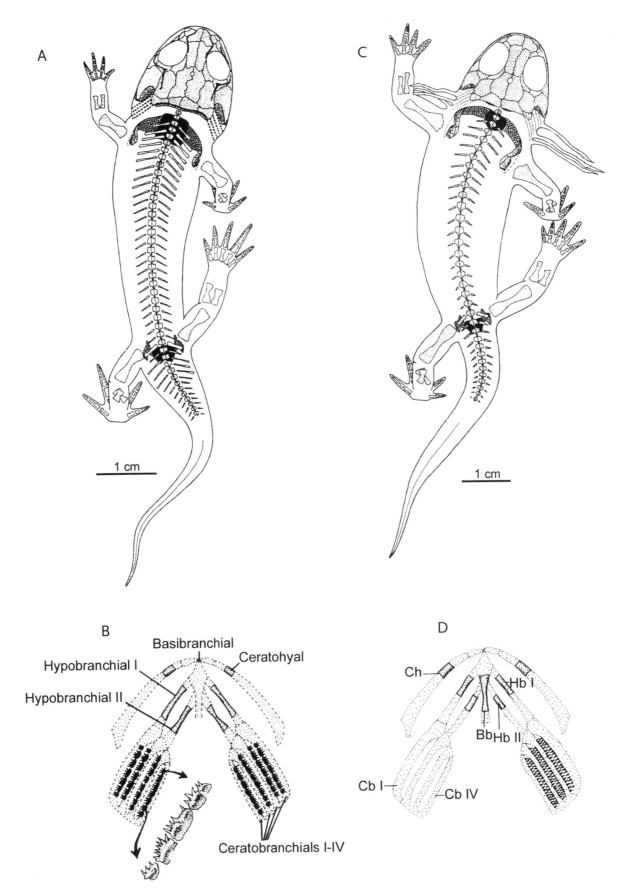

A

C

B

Basibranchial

Hypobranchial I

Hypobranchial II

Ceratohyal

Ceratobranchials I-IV

D

Ch

Hb I

Bb Hb II

Cb I

Cb IV

Figure 6.18. Branchiosaurs. *A,* Skeleton of the micromelerpetontid *Micromel-erpeton. B,* Reconstruction of the hyoid apparatus of *Micromelerpeton. C,* Skel- eton of the branchiosaurid *Branchiosaurus. D,* Reconstruction of the hyoid apparatus of *Branchiosaurus.* From Boy, 1972; Boy and Sues, 2000.

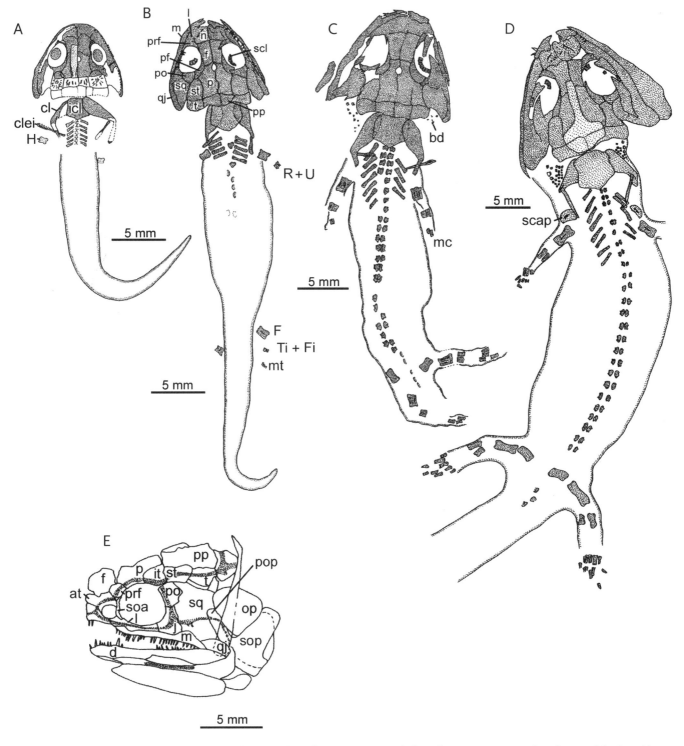

Figure 6.19. Growth series of *Micromelerpeton* showing the very early ossification of the dermal bones of the skull and dermal shoulder girdle and the later ossification of the endochondral bones of the postcranial skeleton. *A* and *D* show the skull roof in ventral view; *C* and *B* in dorsal view. From Witzmann and Pfretzschner, 2003. *E,* Extent of ossification of the dermal bones of the skull of a very small specimen of the fish *Eusthenopteron* that is close to the ancestry of amphibians. From Schultze, 1984.

long and narrow otic notch that was closed posteriorly, although not in the same way as that of the dissorophid *Cacops.* Dilkes (1990) described a separate Lower Permian genus, *Phonerpeton,* that has a more normally shaped otic notch, like that of the amphibamids (Fig. 6.22C).

Anconastes, from the Late Pennsylvanian of New Mexico

(Berman et al., 1987), and *Actiobates* (Fig. 6.22B), from slightly older deposits in Kansas, show the earlier appearance of enlarged premaxillary and maxillary teeth and the elongation of the external nostril, characteristic of later trematopsids. The latter character is also evident in *Mordex* from the Nýřany deposit in the Czech Republic (Fig. 6.22A), which is

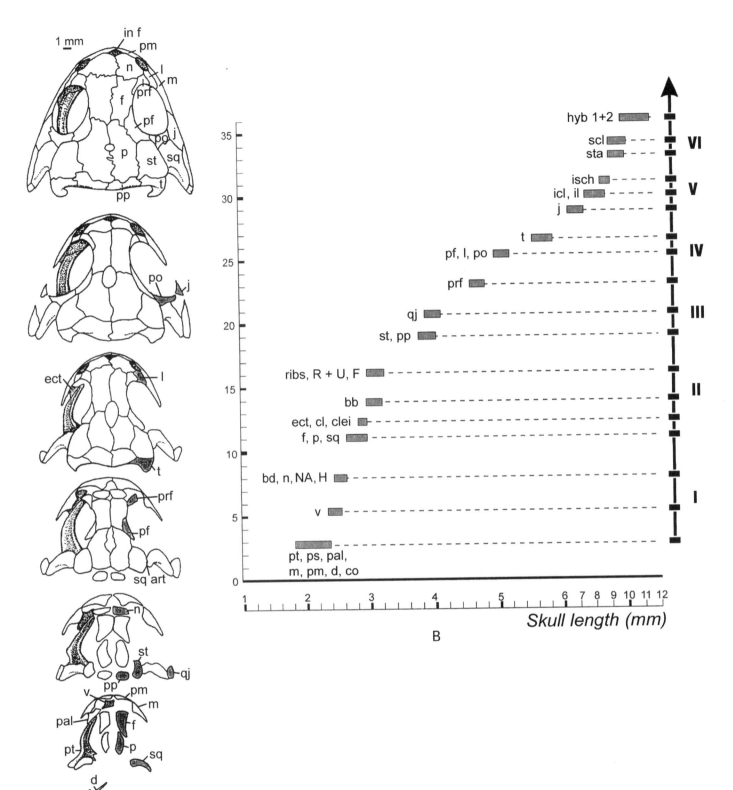

Figure 6.20. Sequence of ossification of the branchiosaurid *Apateon*. *A*, Sequence of skulls from an individual locality showing the progressive ossification of bones. From Schoch and Carroll, 2003. *B*, Stages of ossification of bones of the skull and postcranial skeleton compiled from a great number of specimens. This sequence does not correspond exactly with that shown by the individual specimens illustrated in *A*. From Schoch, 2004.

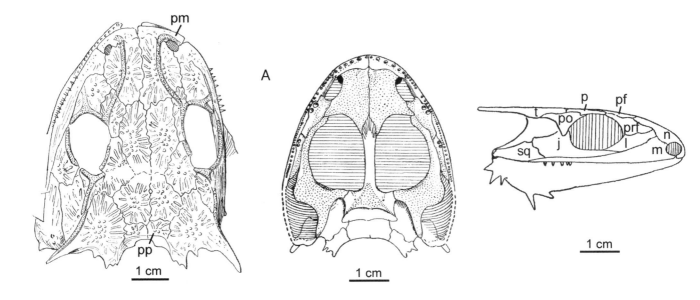

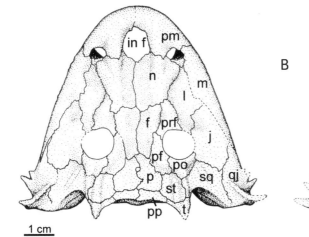

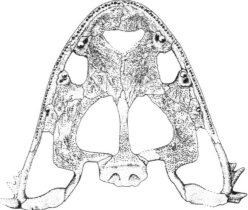

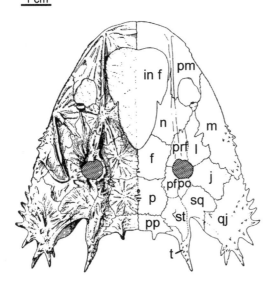

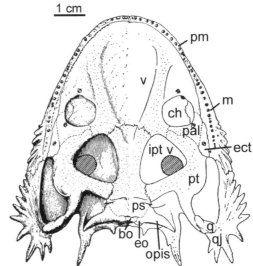

Figure 6.21. Representatives of the family Zatracheidae. *A,* Dorsal, ventral, and lateral views of *Stegops* from the Upper Carboniferous of Linton, Ohio. Modified from Steen, 1930. *B,* Dorsal and palatal views of *Acanthostomatops* from the Lower Permian of Germany. From Boy, 1989. *C, Zatrachys,* from the Lower Permian of New Mexico. Modified from Langston, 1953.

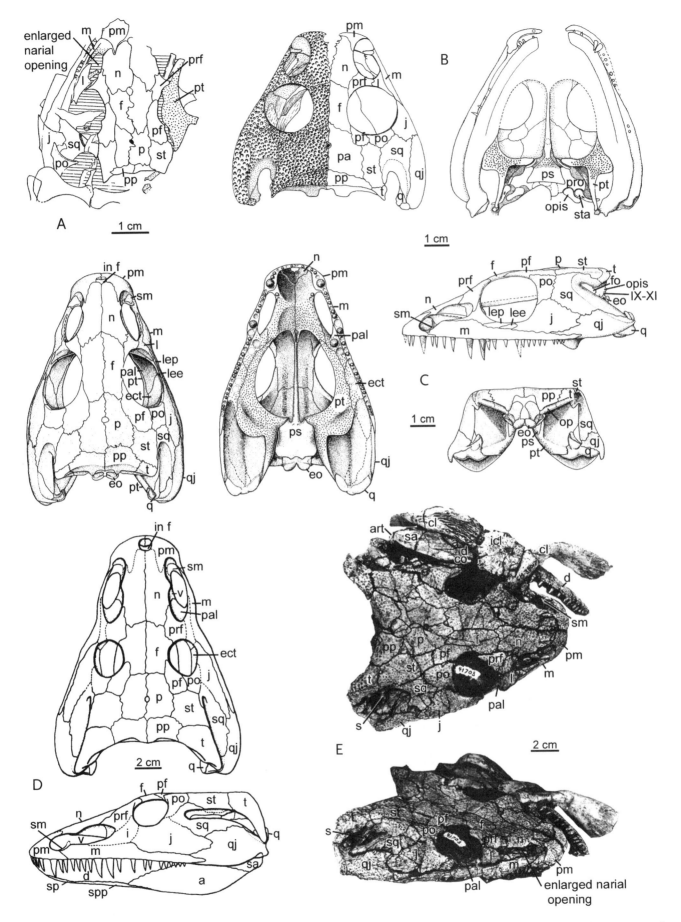

Figure 6.22. Trematopsids. *A, Mordex,* from the Upper Carboniferous of Nýřany, Czech Republic. Modified from Steen, 1930. *B, Actiobates,* from the Upper Carboniferous of Kansas in dorsal and palatal views. Modified from Eaton, 1973. *C, Phonerpeton,* from the Lower Permian of Texas. From Dilkes, 1990. *D, Acheloma cumminsi,* from the Lower Permian of Texas. From Dilkes and Reisz, 1987. *E, Ecolsonia* (whose specific relationships among the Disso-rophoidea are in question). From Berman et al., 1985.

the only possible European representative of this family. *Anconastes* is unique among trematopsids in having very small sculptured osteoderms that formed a dense, non-overlapping armor-like covering of the trunk region.

Ecolsonia

One last, enigmatic dissorophoid requires special mention. *Ecolsonia* (Fig. 6.22E), from the Early Permian of New Mexico, shows a complex of characters shared with amphibamids, dissorophids, and trematopsids (Vaughn, 1969; Berman et al., 1985). Superficially, the skull (despite its larger size) is similar to that of the contemporary *Tersomius texana* in the uniform pitting of the surface, anterior position of the orbits, and presence of many small marginal teeth. It lacks the ridges on the skull table and around the eyes characteristic of dissorophids and has extensive interpterygoid vacuities. However, it has elongate external nares, nearly identical to those of trematopsids, and a posterior closure of the otic notch like that of the dissorophids *Dissorophus* and *Cacops*. It lacks the dermal plates associated with the vertebrae in that family, but has tiny sculptured osteoderms covering the body, as in the trematopsid *Anconastes*. Uniquely, the dorsal extremities of the neural spines are irregularly expanded, not in the manner of dissorophids, but suggestive of areas for the attachment of ligaments to help support the vertebral column. Vaughn (1969) argued that *Ecolsonia* may represent an offshoot from the same basal stock as trematopsids and dissorophids, while Berman and his colleagues (1985) referred to it as a dissorophid, when that family was understood to include both amphibamids and the armored dissorophids.

The entire assemblage of dissorophoids shows a much wider range of anatomical and presumably adaptive modes than any other group of temnospondyls. This may be related to the small size of the ancestral species, which, as in lepospondyls, appears to enable greater changes in early developmental stages. Or there may have been genetic modifications that enabled changes to occur in the timing of bone ossification and other aspects of development, as is particularly evident among the branchiosaurids.

LATE PERMIAN AND TRIASSIC TEMNOSPONDYLS

The greatest number and diversity of primarily terrestrial temnospondyls have been described from the Lower Permian of North America and western Europe. During the early part of the middle Permian, sediments from the fossil deposits in the southwestern United States show evidence of progressively drier conditions and increasingly fewer fossils. Gaps in the fossil record also occur in western Europe and southern Africa. Only northern Africa, from which have come edopoids, and European Russia, from which are known a wide range of temnospondyls, provide us with evidence of

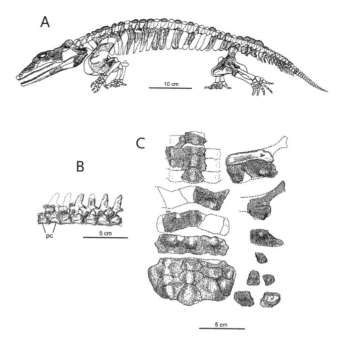

Figure 6.23. *Peltobatrachus pustulatus* from the Upper Permian of Tanzania. A, Skeleton as restored. B, Section of vertebral column showing large size of pleurocentra and absence of intercentra. C, Dermal ossification running down the midline, in association with the ribs, and along the tail. From Panchen, 1959.

the continuation of terrestrial lineages that first appeared in the Late Carboniferous.

Peltobatrachus

Peltobatrachus pustulatus, from the Upper Permian of East Africa, is one of the last large temnospondyls that was specialized for a terrestrial way of life (Fig. 6.23) In its large size and robust limbs it resembled *Eryops,* but is clearly distinct in the presence of complex dermal armor and the structure of the vertebrae. Alec Panchen (1959) argued that it might be related to one of the many groups of large aquatic stereospondyls, the plagiosaurs that will be described in Chapter 8, because of the elaboration of the pleurocentra rather than the intercentra as was the common pattern among earlier Permian temnospondyls. However, the remainder of the anatomy appears as an extension of the pattern of adaptation to a terrestrial way of life seen in the Lower Permian, with no adaptive similarities with stereospondyls. Bands of large scutes, vaguely reminiscent of turtles rather than of dissorophids, covered the central portion of the trunk, extended over the tail, and were attached to the ribs. The skull had a pustular ornamentation, like that of the dissorophid *Micropholis* and plagiosaurid stereospondyls.

Late Russian Temnospondyls

More than 40 genera, belonging to 20 families of temnospondyls, have been described from the earliest Late Permian into the Middle Triassic of Russia, primarily from east of the Ural mountains (Shishkin et al., 2000). The vast majority were aquatic, either clearly relicts of Lower Permian families common to North America and Europe, or elements of a second radiation grouped together as the stereospondyls. Eryopoids were represented by *Clamorosaurus*. Members of the closely related endemic family Intrasuchidae were distinguished by their narrower skulls and loss of ectopterygoid teeth.

The latest Permian beds include a single skull of the dissorophid *Kamacops* and the possible branchiosaurid *Tungussogyrinus*. The most diverse amphibian assemblages were the fully aquatic archeosauroids, including three genera of the long-snouted, gavial-like archegosaurids and three genera of the endemic, alligator-like Melosauridae. The trimerorhachoid *Dvinosaurus* survived to the very end of the Permian, with only *Tupilakosaurus* surviving into the Lower Triassic. The many groups of obligatorily aquatic stereospondyls that continued into the Lower Cretaceous will be discussed in Chapter 8.

THE ECLIPSE OF TERRESTRIAL AMPHIBIANS

In contrast with North America and Europe, where deposits bearing terrestrial vertebrates became increasingly unproductive in the later Permian, those in Russia and southern Africa have yielded a vast number of fossils, not of terrestrial amphibians, but of the ancestors of later reptiles and mammals. In the otherwise richly fossiliferous beds of eastern Russia, we see the progressive decline of terrestrial amphibians throughout the Late Permian, with none surviving into the Triassic. To understand the almost complete extinction of terrestrial amphibians by the end of the Paleozoic, it is necessary to consider the origin of what were to become the most successful of all descendants of the original tetrapods radiation in the Early Carboniferous, the amniotes.

PLATES

PLATE 1. Upper Devonian fish common to the Miguasha Formation, Gaspe, Canada, including sarcopterygians close to the ancestry of tetrapods. From top to bottom: the advanced air-breathing fish *Panderichthyes*, a shallow water fish that could have basked at the interface between water and land; adult and young of *Eusthenopteron*, a strictly aquatic choanate fish, but with many skeletal features of early amphibians; a ray-finned palaeoniscoid; two placoderms; and three ostracoderms, living close to the bottom. Illustration by Tonino Terenzi.

PLATE 2. Early tetrapods from the Upper Devonian of East Greenland, basking *Ichthyostega* and more aquatic *Acanthostega*. Illustration by Tonino Terenzi.

PLATE 3. Amphibians discovered at Horton Bluff, Lower Carboniferous of Nova Scotia, Canada. At the top, a possible descendant of *Ichthyostega*, represented by similar shoulder girdles; to the left, an elongate, short-limbed amphibian, represented by a small, poorly ossified humerus, resembling that of the colosteid, *Greererpeton*; on the sand bar, a large amphibian resembling later anthracosaurs. In the water, a huge rhizodontid sarcopterygian fish, very common at the locality; and an acanthodian. No bony remains have yet been found of the very large amphibian that left trackways at Horton. Illustration by Tonino Terenzi.

PLATE 4. Amphibians discovered at the Cowdenbeath Locality, late Lower Carboniferous of Scotland. The fauna consisted primarily of deep water, aquatic species. On the far bank, an anthracosaur, modeled after *Eoherpeton* and *Proterogyrinus*; venturing into the water, a colosteid, such as *Greererpeton*. On the right, the anterior portion of a baphetid; below, the gigantic *Crassigyrinus*, with tiny forelimbs. Illustration by Tonino Terenzi.

PLATE 5. The primarily terrestrial amphibians whose remains are represented in the East Kirkton Locality, late Lower Carboniferous of Scotland. Animals are identified in the accompanying outline: A, a large temnospondyl, known only from disarticulated bones. B, the anthracosaur *Eldeceeon*. C, the enigmatic *Eucritta*. D, the anthracosaur *Silvanerpeton*. E, the aïstopod *Ophiderpeton*. F, a eurypterid G, a scorpion. H, *Westlothiana*, an elongate animal that may be related to early

amniotes. I, the oldest known temnospondyl *Balanerpeton*. Illustration by Tonino Terenzi.

PLATE 6. The diversity of microsaurs. A, *Asaphestera*, from the Upper Carboniferous of Joggins, Nova Scotia. B, *Tuditanus*, from the Upper Carboniferous of Linton, Ohio. C, *Saxonerpeton*, from the Lower Permian, near Dresden, Germany. D, *Pantylus*, from the Lower Permian of Texas. E, *Rhynchonkos*, from the Lower Permian of Oklahoma. F, *Cardiocephalus*, from the Lower Permian of Oklahoma. G, *Pelodosotis*, from the Lower Permian of Texas. H, *Trihecaton*, from the Upper Carboniferous of Colorado. I, *Microbrachis*, from the Upper Carboniferous of the Czech Republic. J, *Hyloplesion*, from the Upper Carboniferous of the Czech Republic. K, *Brachystelechus*, from the Lower Permian, near Dresden, Germany. L, *Odonterpeton*, from the Upper Carboniferous of Linton, Ohio. Note that the largest known microsaur, *Pelodosotis*, is reproduced at one-half the size of all other microsaurs. Color adaptation by Tonino Terenzi of illustrations in R. L. Carroll and Gaskill, 1978.

PLATE 7. Amphibians and reptiles from the early Upper Carboniferous locality of Joggins, Nova Scotia. All the vertebrates from this locality were highly terrestrial as adults, as indicated by their presence in the upright tree stumps of the giant lycopod *Sigillaria*. From the top, basking on a *Sigillaria* log, the medium-sized embolomere *Calligenethlon*. Center, the very common temnospondyl, *Dendrerpeton*. With red stripes, the early amniote *Hylonomus*. Other amniotes and microsaurs, known from only incomplete remains, are represented by the smaller figures. Illustration by Tonino Terenzi.

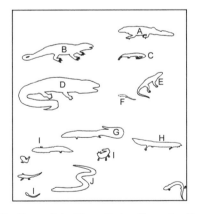

PLATE 8. The fauna of the Upper Carboniferous locality of Nýřany, Czech Republic. Amphibians and reptiles are identified in the accompanying outline. A, the advanced anthracosaur *Solenodonsaurus*, whose bones are also shown deep in the water. B, the anthracosaur *Gephyrostegus*. C, the microsaur *Sparodus*. D, the large, semi-aquatic temnospondyl *Cochleosaurus*. E, the primitive amniote *Brouffia*. F, the terrestrial nectridean *Scincosaurus*. G, an aquatic nectridean, *Sauropleura*. H, the obligatorily aquatic microsaur *Microbrachus*. I, the branchiosaur *Branchiosaurus*. J, the aïstopod *Ophiderpeton*. Illustration by Tonino Terenzi.

PLATE 9. Fauna from the Lower Permian of north-central Texas. Amphibians and reptiles are identified in the accompanying outline. *A,* the carnivorous pelycosaur *Dimetrodon. B,* the giant microsaur *Pelodosotis. C,* the herbivorous pelycosaur *Edaphosaurus. D.* The large-headed microsaur *Panylus. E,* the elongate microsaur *Goniorhynchus. F,* the sail-backed dissorophoid *Platyhystrix. G,* the large and extremely common temnospondyl *Eryops. H,* the very elongate and small-limbed microsaur *Cardiocephalus. I,* very elongate and tiny-limbed lysorophids. *J,* the nectridean *Diplocaulus,* with a head shaped like a boomerang. *K,* lungfish, which, like the lysorophids, are found aestivating in burrows. Illustration by Tonino Terenzi.

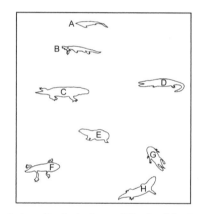

PLATE 10. A shore line in the Lower Triassic of the Southern Hemisphere, dominated by stereospondyl amphibians. Amphibians identified in accompanying diagram: *A,* ichthyosaurs, which will dominate the seas in the Later Mesozoic. *B,* the highly aquatic trematosaurs. *C,* a capitosaur. *D,* a rhinosuchid, a family that survived through the end-Permian extinction. *E,* a primitive brachiopoid. *F, Micropholis,* an amphibamid dissorophoid that survived into the Lower Triassic. *G,* an ancestor of the plagiosaurs. *H, Lydekkerina,* a moderate-sized temnospondyl close to the base of stereospondyls. All these genera, or their close relatives, were discovered in the Southern Hemisphere, at a time when South America, Africa, India, Madagascar, and Australia were joined in a single land mass. Illustration by Tonino Terenzi.

PLATE 11. Diversity of frogs. *A,* reconstruction of the Lower Triassic antecedent of frogs, *Triadobatrachus,* from Madagascar. *B–K,* representatives of the major living families. *B,* Ascaphidae: *Ascaphus truei,* from northwestern North America. *C,* Discoglossidae, *Discoglossus pictus,* Europe. *D,* Pipidae, *Pipa pipa,* South America. *E,* Rhinophrynidae, *Rhinophrynus dorsalis,* Mexico. *F,* Megophryidae, *Megophrys nasuta,* from Asia. *G,* Leptodactylidae, *Eleutherodactylus augusti,* Mexico. *H,* Myobatrachidae, *Notaden bennettii,* Australia. *I,* Hylidae, *Agalychnis callidryas. J,* Ranidae, *Rana palustris,* North America. *K,* Bufonidae,

Atelopus varius, Central America. Images *B–K,* courtesy of Dr. David Kirshner in Cogger and Zweifel, 1998.

PLATE 12. Tadpoles of the pipoid frog *Shomronella jordanica* from the Lower Cretaceous of Israel. Photograph provided by Ariel Chipman.

PLATE 13. Diversity of modern salamanders. *A,* the salamandrid *Salamandra salamandra. B,* the hynobiid *Batrachuperus pinchonii. C,* the plethodontid *Eurycea bislineata. D,* the ambystomatid *Ambystoma tigrinum. E,* the dicamptodontid *Dicamptodon ensatus. F,* the cryptobranchid *Cryptobranchus alleganiensis. G,* the sirenid *Pseudobranchus striatus. H,* the proteid *Necturus maculosus. I,* the amphiumid *Amphiuma means.* Assembly by Tonino Terenzi.

PLATE 14. Comparative views of the sequence of ossification of the dermal skull bones of the uppermost Carboniferous branchiosaurid *Apateon caducus* (on the left), and the extant salamandrid *Notophthalmus viridescens* (on the right). A_1–E_1, succession of growth stages illustrated in specimens from the Geological and Palaeontological Institute of Mainz; specimen numbers 1478, 1460, 14654, 1280, and 1530. A_2–D_2, developmental sequence of cleared and stained specimens of the salamandrid *Notophthalmus viridescens* (Redpath Museum, McGill University, specimen numbers 5007, and 5009–5011). Bone is red and, where viewed on edge, black; cartilage is blue. Note the very early appearance of the squamosal, when most of the surrounding skull is still cartilaginous. The maxilla, however, ossifies long after the premaxilla. From R. L. Carroll, 2007.

PLATE 15. Reconstruction of the Jehol fauna in northern China, from which have come a wealth of Middle Jurassic salamanders and occasional frogs. Animals are identified in the accompanying outline: *A,* an iguandontid dinosaur. *B,* a pterosaur. *C,* putative cryptobranchid. *D, Jeholotriton. E,* unidentified salamander larva. *F,* Unnamed terrestrial salamander. *G,* primitive frog with either a vestigial tail or an ascaphid-like copulatory organ. *H,* a discoglossid frog. *I,* frog eggs and larvae. Illustration by Tonino Terenzi.

PLATE 16. Diversity of modern caecilian families. *A, Ichthyophis kohtaoensis,* a member of the primitive family Ichthyophidae. *B, Dermophis mexicanus,* an advanced caecilian, in the family Caeciliidae. *C, Schistometopum thomense,* a viviparous caeciliid. *D, Typhlonectes natans,* a member of the primarily aquatic Typhlonectidae. Illustrations reproduced from Nussbaum, 1998.

PLATE I

PLATE 2

PLATE 3

PLATE 4

PLATE 5

PLATE 6

1 cm

A

C

B

D

E

F

1 cm

G

H

I

J

K

L

PLATE 7

PLATE 8

PLATE 9

PLATE 10

PLATE II

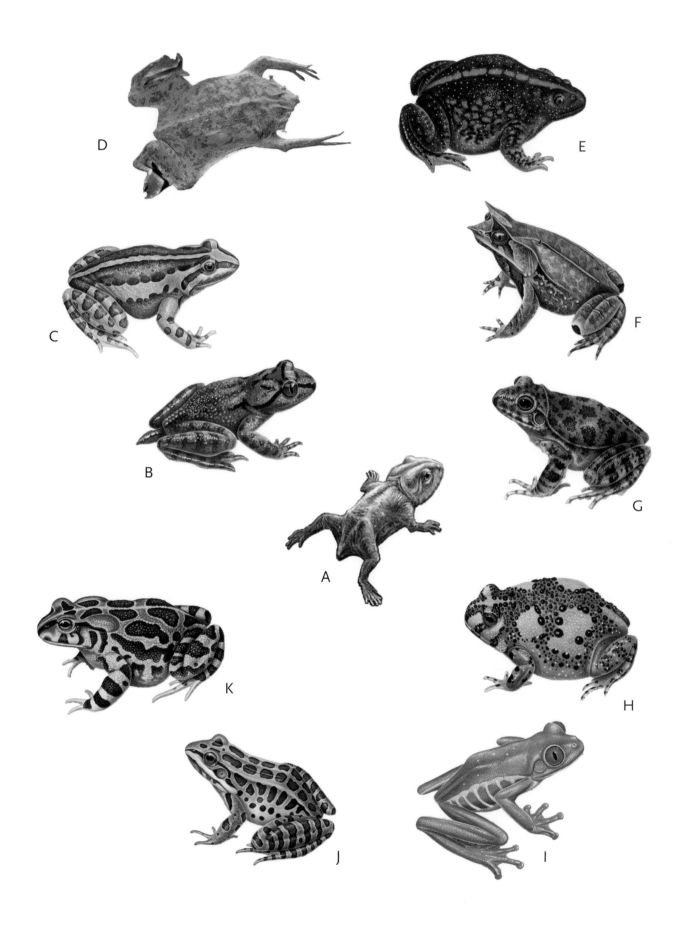

PLATE 12

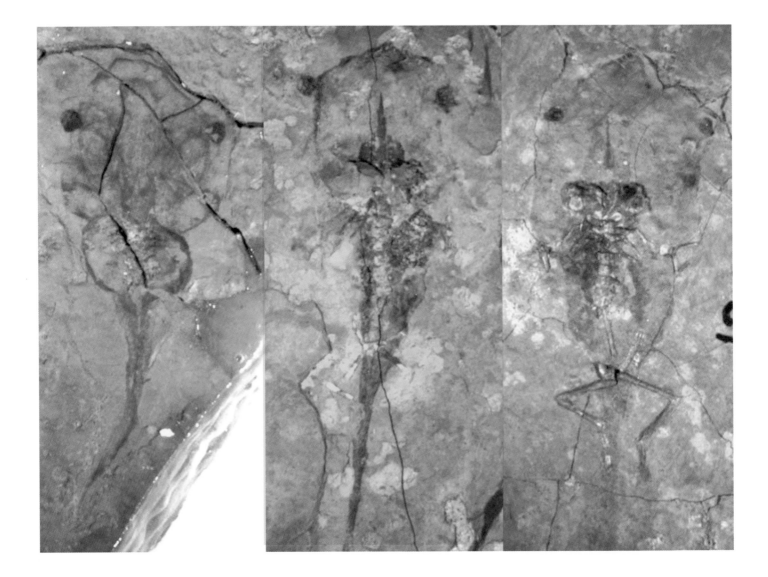

PLATE 13

PLATE 14

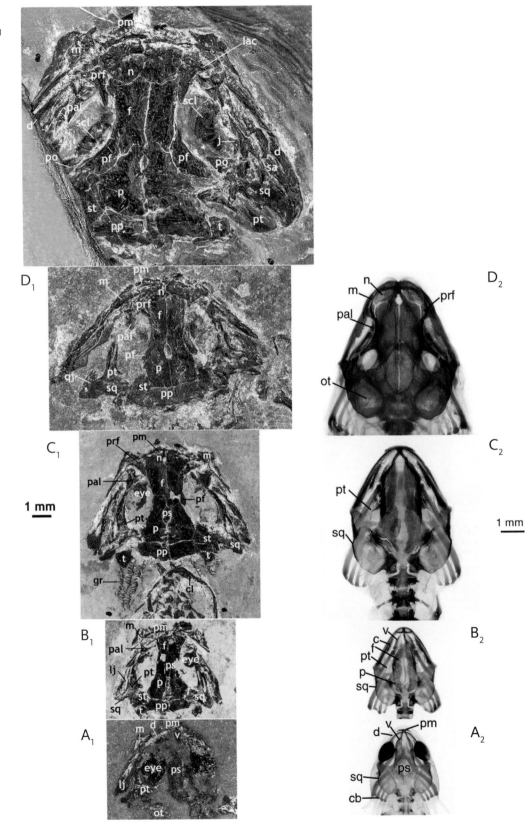

PLATE 15

PLATE 16

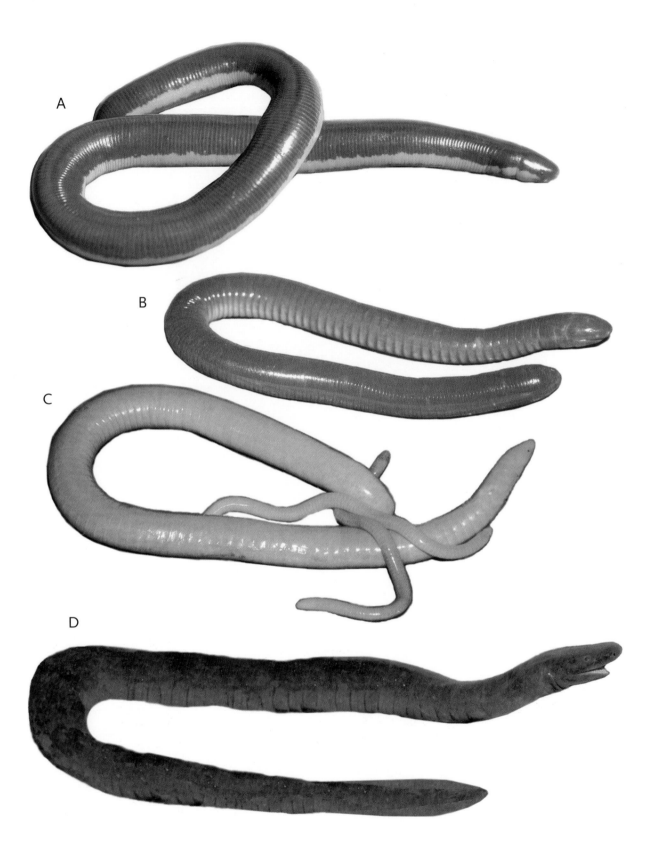

The Origin of Amniotes

Escape from the Water

FROM THE UPPER DEVONIAN, through the Carboniferous, and for much of the Permian, amphibians had radiated extensively and spread into a host of shallow water and terrestrial environments throughout the world. But by the end of the Permian most of the non-aquatic families had become extinct or reduced to a few isolated species. Their demise can be attributed primarily to the success of one of their own descendants that had diverged near the base of the Carboniferous, the amniotes, which would eventually give rise to the modern reptiles, birds, and mammals.

It may seem odd to devote an entire chapter in a book on amphibians to the origin of amniotes, but this is necessary to explain the profound influence of their evolutionary history on amphibians and their role in the origin of all more advanced land vertebrates.

THE ANCESTRY OF AMNIOTES

The earliest known amniotes, *Hylonomus* and *Paleothyris,* from the upright trees of Joggins, Nova Scotia, and the nearby Florence locality, are represented by nearly complete skeletons (Figs. 4.39–4.41 and 7.1–7.3). These genera were already clearly distinct from all other Carboniferous tetrapods. This can be partially attributed to the almost 40 million-year gap since the initial amphibian radiation in the Early Carboniferous, but also reflects the distinct environment in which they lived and their probable mode of reproduction.

In order to determine the nature of the evolutionary changes leading toward amniotes, it is necessary to establish their phylogenetic relationships. Unfortunately, we have very little knowledge of plausible ancestors among the earlier Carboniferous tetrapods, although *Casineria* and *Westlothiana* (discussed in Chapter 4) are possible sister-taxa of this lineage. On the other hand, our very complete knowledge of early amniotes near the time of their initial radiation and the very similar skeletal anatomy

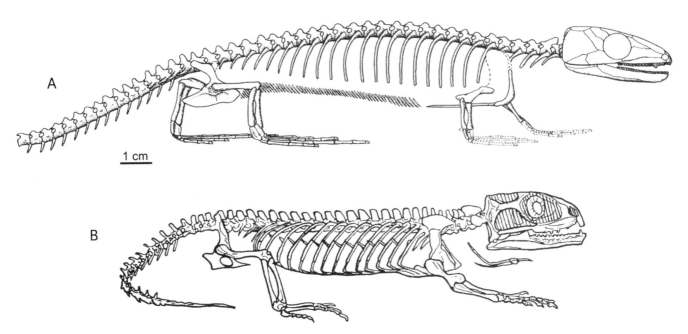

1 cm

Figure 7.1. A, Reconstruction of the skeleton of *Hylonomus lyelli,* the oldest known amniote, from the Westphalian A of Joggins, Nova Scotia. From Carroll, 1969c. *B,* Skeleton of *Sphenodon,* one of the most primitive living reptiles. Modified from W. K. Gregory, 1951.

of the early genera clearly demonstrate the polarity of their character state changes relative to other early tetrapods, and document their highly distinctive nature.

The following combination of characters distinguishes all early amniotes. Most are autapomorphies—characters unique to this group relative to primitive tetrapods (indicated by A), while others are primitive features (P), indicating divergence from very early land vertebrates. Only a few appear to be synapomorphies (shared derived characters—S) possessed in common with specific antecedents among earlier tetrapods. This tabulation involves the evaluation of all distinctive characters, without minimizing the importance of primitive characters, as did Hennig (1966), or unique derived characters, as does PAUP.

Distinctive Characters of Early Tetrapods

General
1. Small size, approximately 15 cm from the tip of the snout to the base of the tail (A)
2. High degree of ossification of the skeleton, even in very small individuals (A)
Skull
3. Absence of intertemporal (A)
4. Vertical posterior margin of cheek (no squamosal notch) (A)
5. Narrow supratemporal extending to occiput (A)
6. Tabular in contact with parietal (S)
7. Orbit bordered by frontal dorsally (A)
8. Presence of canine teeth (A)
9. Loss of labyrinthine infolding of teeth (A)
10. Absence of palatine fangs (A)
11. Transverse flange of pterygoid (A)

12. Absence of interpterygoid vacuities (P)
13. Absence of lateral line canal grooves (A)
14. Supraoccipital, a thin plate of bone extending from the postparietals to the opisthotics (A)
15. Integration of articulating surface of exoccipitals and basioccipital into occipital condyle (A)
Vertebrae and Ribs
16. Retention of crescentic intercentra throughout the trunk (P)
17. Retention of haemal arches (P)
18. Retention of paired proatlas (P)
19. Retention of paired atlas arch (P)
20. Large axis arch (P)
21. 25 to 32 presacral vertebrae (P)
22. Two pairs of sacral ribs (A)
Appendicular Skeleton
23. Narrowing of clavicular blade (A)
24. Narrowing of stem of interclavicle (A)
25. Supinator process on humerus (A)
26. Pisiform bone in carpus (A)
27. Phalangeal count of manus 2, 3, 4, 5, 3 (S)
28. Ilium lacking anterodorsal process (A)
29. Tibiale, intermedium, and proximal centrale fused into astragalus (A)
30. Phalangeal count of pes 2, 3, 4, 5, 3 (S)

The common expression of this distinctive suite of characters among all early amniotes strongly supports not only the monophyletic nature of this group, but also the homology of these characters throughout this assemblage. This is in strong contrast with the many traits observed among other groups of Paleozoic tetrapods for which homology is difficult to establish since there is much less congruence with other characters.

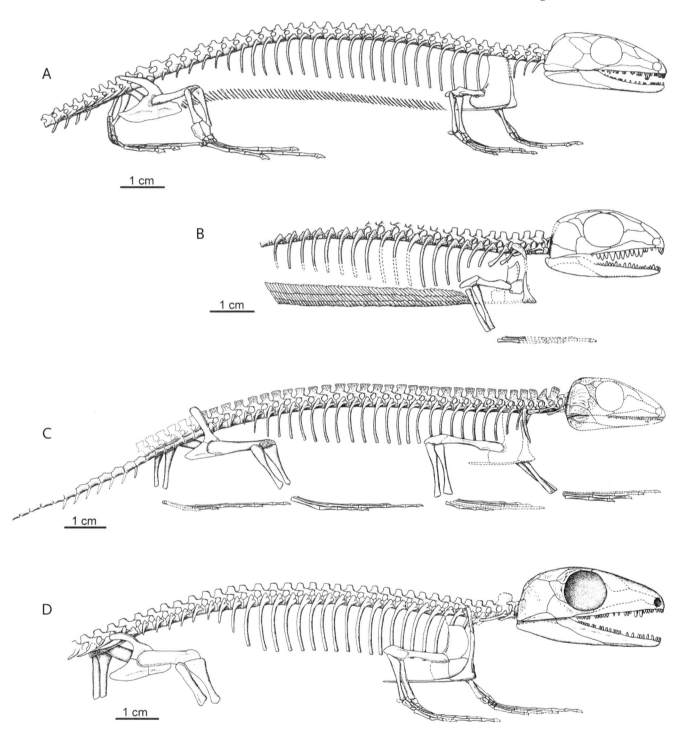

Figure 7.2. Reconstructions of later Carboniferous amniotes from North America and Europe. *A, Paleothyris* from the Westphalian B-D of Florence, Nova Scotia. *B,* The large-toothed *Cephalerpeton,* from Mazon Creek, Illinois. *C, Anthracodromeus,* from Linton, Ohio. *D, Brouffia* from Nýřany, Czech Republic. From R. L. Carroll and Baird, 1972.

For example, many derived groups of tetrapods, including aïstopods, advanced temnospondyls, nectrideans, and microsaurs, lack an intertemporal, but other cranial characters of these groups differ so greatly that homology of intertemporal loss is impossible to establish and so cannot be used to determine affinities among the many groups. Similarly, a pisiform is present in one microsaur and one nectridean, but its occurrence in individual genera among these groups is certainly the result of convergence, as is the consolidation of proximal tarsals into an element resembling the amniote astragalus in the anthracosaur *Gephyrostegus* and the microsaur *Pantylus.*

There are relatively few derived characters that appear to be uniquely shared between amniotes and other Paleozoic clades. These include the very similar phalangeal counts of

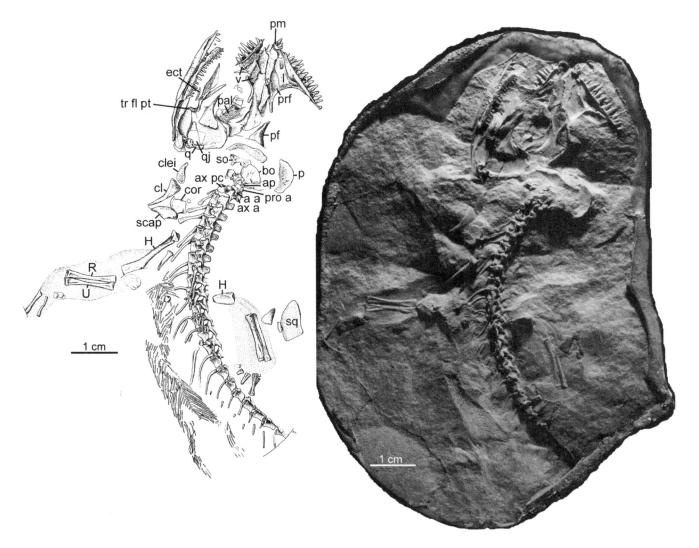

Figure 7.3. Specimen drawing and photograph of the early amniote *Cephaler-peton.* From R. L. Carroll and Baird, 1972.

both the manus and the pes and the contact between the tabular and the parietal, otherwise present in anthracosaurs, in which most other aspects of the skull table appear homologous (Fig. 7.4). These groups also share many primitive characters in common, most significantly the retention of the intercentra and haemal arches throughout the vertebral column. In contrast, these bones are variably reduced or lost in all lepospondyls. Of the Paleozoic tetrapods, only anthracosaurs are plausible sister-taxa of amniotes, but this affiliation may lie at the level of *Tulerpeton,* which indicates a very long ghost lineage (a period of time during which putative ancestral forms are likely to have been present, but for which no fossils are known).

Functional Complexes

Enumeration of the traits present in *Hylonomus* and *Paleothyris* documents the unique anatomy of early amniotes, but does not, in itself, explain why they would rapidly achieve a position of evolutionary dominance. This brings us to a consideration of functional complexes that further emphasize

their distinctive nature. Although little of the soft anatomy is preserved in any of these early amniotes, comparison with primitive living species of similar size and body plans, such as the tuatara *Sphenodon* and iguanid lizards, allows us to hypothesize similar functional complexes associated with feeding and locomotion.

The structure of the palate and the nature of the dentition indicate different modes of feeding than those of the contemporary labyrinthodonts and lepospondyls. The palate is clearly distinct from that of labyrinthodonts in the absence of palatal fangs and in the presence of a downturned portion of the pterygoid termed the transverse flange (Fig. 7.4). The configuration of the chamber containing the jaw muscles also differs in having a vertical rather than a sloping posterior margin as a result of the loss of the squamosal notch.

The configuration of the portion of the skull surrounding the jaw muscles in early amniotes resembles that of modern iguanid lizards, implying a comparable arrangement of these muscles (Fig. 7.5). In the absence of a squamosal notch, the direction of orientation of the main jaw adductors was

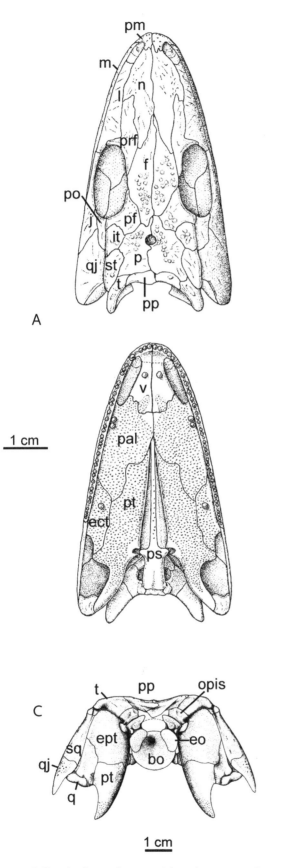

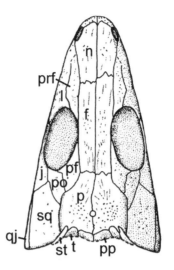

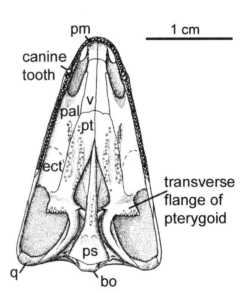

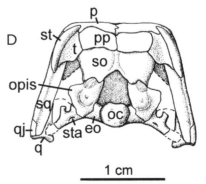

Figure 7.4. A, B, C, Skull roof, palate, and occiput of the anthracosaur *Gephyrostegus* (on the left) and the early amniote *Paleothyris* (on the right). Note loss of both the intertemporal bone and the squamosal notch as well as reduction of the supratemporal, postparietal, and tabular in *Paleothyris*. This genus has also lost the labyrinthine enfolding of the dentine and the fangs on the vomer, palatine, and ectopterygoid and evolved a transverse flange of the pterygoid. In both genera the orbit is bordered by the frontal. This is probably the result of convergence because of small skull size, rather than being indicative of close relationship. From R. L. Carroll, 1969a.

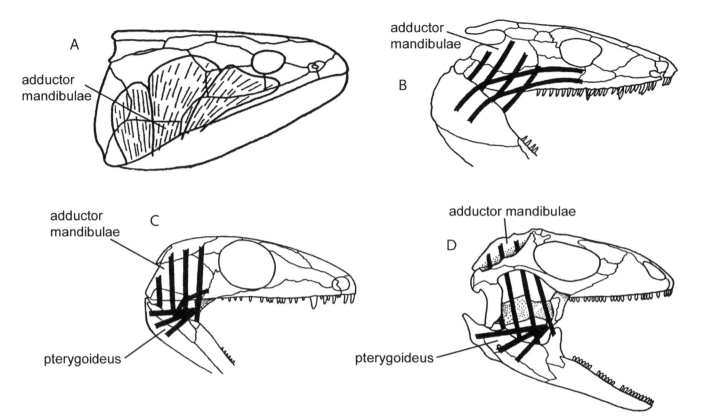

Figure 7.5. Changes in the configuration of the skull and the associated jaw musculature between labyrinthodonts and early amniotes. *A,* The choanate fish *Ectosteorhachis* with the major portion of the jaw musculature restored as consisting of a series of units, with the fibers angling from obliquely posterior to nearly vertical. *B,* The anthracosaur *Palaeogyrinus,* with most of the musculature angling obliquely posterior, which would have resulted in the strongest force for closing the jaw occurring when the jaw was wide open. *C, Paleothryis,* in which the presence of a transverse flange of the pterygoid and the nearly vertical posterior end of the cheek made it possible to divide the jaw-closing muscles into more or less vertical and horizontal units. The pterygoideus could exercise its greatest strength when the jaw was wide open, and the more vertical muscles when the jaw was nearly closed. This would also allow *Paleothyris* to feed on prey of a fairly wide size range. *D,* Known orientation of the jaw muscles in a modern iguanid lizard. From R. L. Carroll, 1969a.

vertical, which results in expression of their maximum force when the mouth was nearly shut. This is most effective for feeding on relatively small prey. On the other hand, the presence of the transverse flange of the pterygoid provides a new surface for the origin of another muscle, the pterygoideus, which has a more horizontal orientation and so is more effective in closing a widely gaping jaw and for holding larger prey. The ventral edge of the transverse flange of the pterygoid in Carboniferous amniotes was covered by denticles or a transverse row of small teeth that would have served to hold the prey between bites by the marginal dentition.

Capture and holding of prey would also have been more effective in amniotes than in any other early tetrapods because of the configuration of the occipital condyle and atlas-axis complex (Fig. 7.6). The hemispherical occipital condyle and surrounding elements of the atlas vertebra formed a closely fitting ball-and-socket joint, enabling controlled movements in all directions. In contrast, the occiput of primitive labyrinthodonts provided only a shallow basin for articulation. In most lepospondyls, the skull was limited to hinging in the vertical plane, with little if any lateral mobility or rotation (Fig. 5.5).

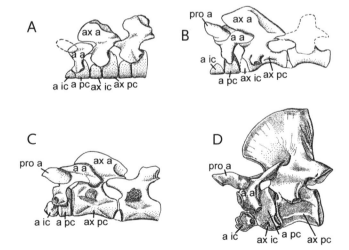

Figure 7.6. Evolution of the cervical vertebrae in early amniotes. *A, Gephyrostegus,* showing the primitive pattern for labyrinthodonts, going back to *Acanthostega. B, Hyonomus lyelli,* retaining all the primitive elements but enlarging the axis pleurocentrum and fusing it with the neural arch. *C,* The slightly later genus *Paleothyris,* in which the atlas pleurocentrum is fully cylindrical and the axis intercentrum is either lost or incorporated into the axis pleurocentrum. *D,* The pelycosaur *Dimetrodon* in which the axis intercentrum is fused to the axis pleurocentrum. From R. L. Carroll, 1969a.

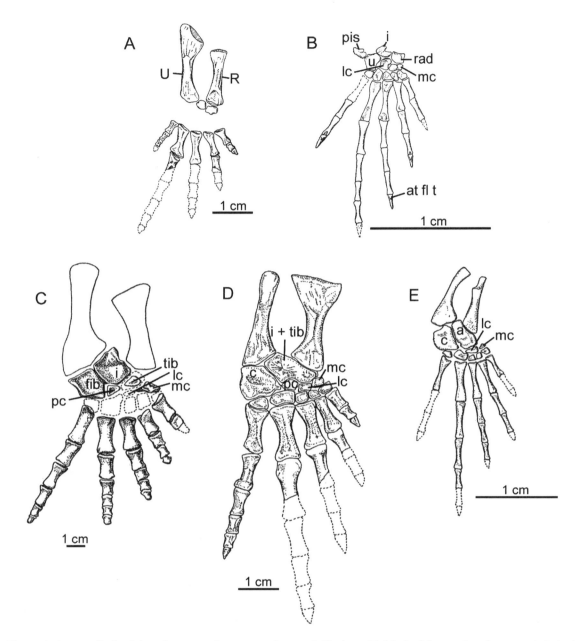

Figure 7.7. Changes in the appendicular skeleton between anthracosaurs and early amniotes. *A,* The forelimb of *Gephyrostegus* showing the low degree of ossification of the carpus. *B,* The lower forelimb of *Paleothyris,* in which all the elements of the carpus are ossified, even in a relatively small specimen, including an element not known in anthracosaurs, the pisiform, which in living amniotes forms a point of attachment for the tendons of muscles along the posterolateral margin of the wrist that pull the lower limb posteriorly. *C,* The lower hind limb of the aquatic anthracosaur *Archeria,* showing the absence of ossification of the distal tarsals and the presence of four proximal tarsals. *D, Gephyrostegus,* in which the entire carpus is ossified, and the intermedium and tibiale have fused with one another. *E, Hylonomus,* in which the astragalus has formed by the fusion of the tibiale, intermedium, and proximal centrale. From R. L. Carroll, 1969a.

The limb joints in early amniotes ossified at a very small size, providing greater agility and control. This is most clearly seen in the ankle joint, in which the many small bones in early tetrapods provided only a very flexible hinge that could not be controlled effectively by the associated muscles (Fig 7.7). Ancestral amniotes (and a few lepospondyls) evolved a more effective joint by the fusion of three proximal bones, the tibiale, intermedium, and centrale, into a single element, the astragalus, while the fibulare enlarged to form the calcaneum, or heel bone. In addition, tubercles evolved on the proximal ventral surface of the terminal phalanges of the hands and feet so that they could be sharply bent to grasp prey. These and the addition of a second pair of sacral ribs brought the early amniotes nearly to the level of primitive living lizards.

Surprisingly, early amniotes had not evolved one of the basic sensory structures of most modern amniotes, an impedance matching middle ear. *Hylonomus* and other Carboniferous amniotes did not have an otic notch or other surface for support of a tympanum and the stapes remained a massive element that would not have been able to transmit airborne vibrations. Such a system did, however, begin to evolve by the very end of the Carboniferous in the lineage that led to mammals.

The Significance of Small Size

What was probably the most important attribute of early amniotes was not a specific aspect of the skeleton but rather their much smaller adult size than any of the early labyrinthodont groups. As discussed in Chapter 5, most of the derived characters of lepospondyls can be attributed to their very small body size. These included changes in the skull, dentition, vertebrae, and appendicular skeleton, all of which can be associated with miniaturization of the skeleton and ossification at a very early stage in development.

Comparable processes can also be cited for the ancestry of amniotes. The great abundance of prey among the larvae and adults of small arthropods by the middle Carboniferous would have served as a selective force for reduction in adult body size, which could have occurred via ossification of the skeleton at progressively earlier stages of development. Growth ceased before the appearance of palatal fangs, the posterior extension of the jaw articulation, and the formation of a squamosal notch. Earlier ossification of the vertebrae may have led to formation of the pleurocentra from a single medial area of ossification and its firm attachment to unpaired neural arches (as in most lepospondyls), but not the complete loss of the intercentra, which occurred in some, but not all, lepospondyls.

The retention of the separate elements of the atlas-axis complex indicates strong counter-selection to maintain effective multidirectional movement of the head on the trunk in early amniotes. This can also be associated with the evolution of a much altered configuration of the back of the skull.

Small body size was also closely associated with the origin of the unique mode of reproduction among amniotes.

Reproduction

The earliest amphibians were able to emerge from aquatic ancestors while maintaining a fishlike mode of reproduction. Judging from the larvae of Carboniferous labyrinthodonts, the earliest tetrapods must have laid their eggs in the water, with the hatchlings emerging as fishlike larvae that gradually developed limbs that made terrestrial locomotion possible. Having an amphibious life history enabled amphibians to leave the water, but required that their eggs be very small so that the embryos could survive on passive exchange of respiratory gases with their environment. The limited nutrients in such small eggs made it impossible to support extended periods of development during which complicated adult structures and functions could be elaborated. In order for the ancestors of modern reptiles, birds, and mammals to achieve a fully terrestrial mode of reproduction, a series of novel structures, the extra-embryonic membranes, had to evolve.

The most important characteristic that unites amniotes—the nature of the egg—cannot be directly determined from the skeleton. However, two lines of evidence suggest that the evolution of the amniote egg had occurred shortly before the appearance of *Hylonomus* in the fossil record. 1. Extra-embryonic membranes develop in a similar way in all living groups of amniotes. It is more parsimonious to assume that their common ancestor had already developed this feature than to propose that it has evolved separately two or three times. 2. The fossils of early amniotes are just slightly larger than the largest of living amphibians that can hatch from non-amniotic eggs laid on land. This suggests that they had only recently achieved a pattern of reproduction approaching that of modern amniotes.

Among modern salamanders, there is a close correlation between the size of the eggs and the size of the adults within three broad categories: those laying eggs in still water, those laying eggs in streams, and those laying eggs on land (Fig. 7.8). Closest comparison to amniotes occurs among plethodontid salamanders, which lay their eggs on land and whose young hatch out as miniatures of the adults. The eggs in this family are the largest in relationship to body size. Although small, the yolk sac contains enough nutrients for the young to develop to maturity before hatching. However, without the capacity for effective exchange of gases between the air and the embryo and retention of water for continuing growth, both afforded by extra-embryonic membranes, a larger size cannot be achieved.

The largest plethodontid salamanders that undergo direct development on land are no more than about 80 mm long as adults, as measured from the snout to the base of the tail. Lizards of this body length lay eggs that are similar in size to those of the largest plethodontids. By analogy, we can argue that the origin of amniotes proceeded via a stage in which non-amniotic eggs were laid on land and development was direct.

Prior to the evolution of extra-embryonic membranes, hatchling size would have been restricted as it is in plethodontids. Larger amphibians, such as most labyrinthodonts, presumably hatched out of similarly small eggs, but had a protracted larval period during which they depended on the water for support, food, and exchange of respiratory gases.

According to Szarski (1968), the first of the extra-embryonic membranes to evolve would have been the allantois. In modern amniotes, it forms as a bladder to retain water that is otherwise lost from the embryo. Solutes within the urine render its contents hypertonic relative to freshwater. If the egg is laid on damp ground, water is drawn into the allantois by osmosis. In modern reptiles, including some lizards and turtles that have permeable eggshells, the water content within the egg may more than double during growth because of absorption into the allantois. With the enlargement of the allantois, the more superficial layers of tissue

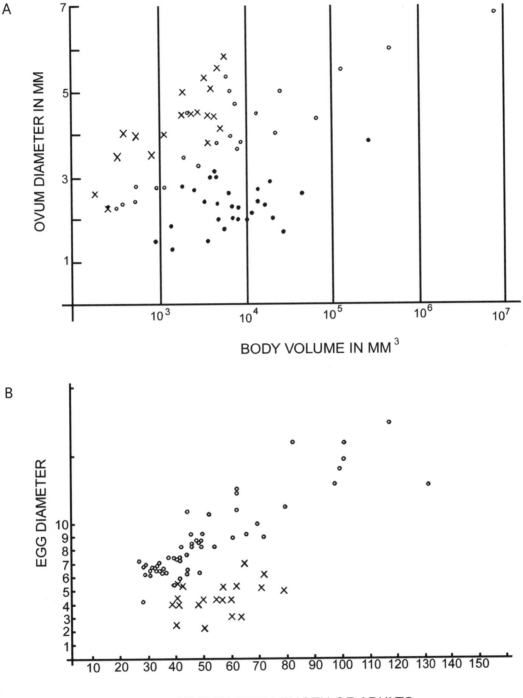

Figure 7.8. A, The relationship between body volume and ovum diameter in salamanders. Solid circles: amphibians reproducing in still water. Open circles: amphibians reproducing in running water. *X*'s: amphibians reproducing on land. *B,* Correlation between mean diameter of eggs and adult body size in gekkos, as an example of small lizards (open circles) and plethodontid salamanders laying their eggs on land (*X*'s). Modern plethodontids laying their eggs on land provide a model of a condition prior to the emergence of amniotic membranes. From R. L. Carroll, 1970b.

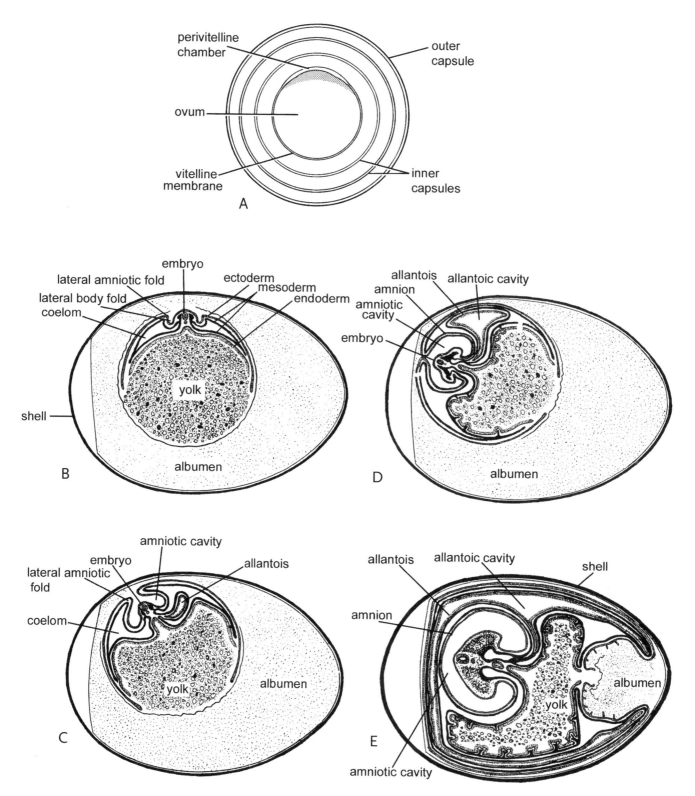

Figure 7.9. A, Generalized amphibian egg showing membranes and capsules. No membranes are comparable to those of amniotes. From Duellman and Trueb, 1986. *B–E*, Stages in the development of the amniotic egg of a chicken, from 2 to 14 days of incubation. Initially, the membranes that will form the external surface of the body (the ectoderm), the lining of the gut (endoderm), and the lining of the body cavities (the mesoderm) spread out over the yolk. Then the allantois extends out from the base of the gut to form the embryonic bladder, which serves to retain water and metabolic wastes.

The allantoic cavity progressively expands, thrusting the overlying membrane, now termed the amnion, over the embryo, enclosing a water-filled chamber (the amniotic cavity) that protects the embryo. The common surface of the allantois and amnion serves as an area for the resorption of atmospheric oxygen and the osmotic pressure that develops in the allantois results in the absorption of water from the ground. Primitive amniotes had a membranous covering, like that of some living turtles and lizards, without a rigid shell. Modified from Patten, 1958.

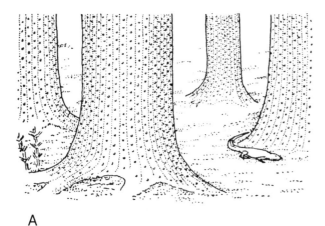

A

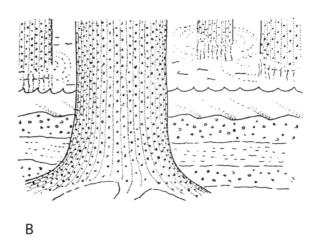

B

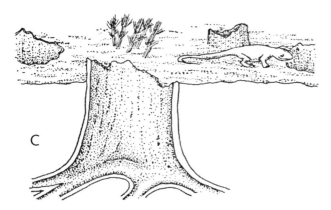

C

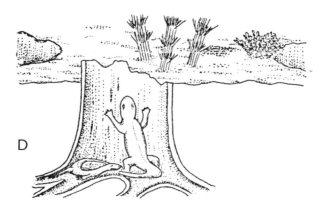

D

Figure 7.10. Sketches of the events leading to the entrapment of terrestrial animals in lycopod stumps at Joggins and Florence, Nova Scotia. *A,* Growth of forest of *Sigillaria* trees. *B,* Trees are buried and killed by sediments as a result of flooding. *C,* Water subsides, center of trees rot out forming deep pits in new land surface. *D,* Amphibians and reptiles fall into pits and become trapped. Additional sediments cover animals. From R. L. Carroll, 1970c.

that formed around the embryo in more primitive vertebrates, the extra-embryonic ectoderm and mesoderm, were forced out from their original position and reflected over the body (Fig. 7.9). These layers, together with the surface of the allantois, serve for gas exchange. A further membrane, the amnion, forms a fluid-filled chamber that surrounds the embryo to provide support and protection.

In order to fertilize eggs laid on land, internal fertilization probably evolved early in amniote evolution. Since copulatory organs are absent in the primitive living genus *Sphenodon* and have apparently evolved separately in the ancestors of lizards and crocodiles, they were almost certainly absent in early amniotes. Presumably they relied on cloacal apposition.

Instead of having two developmental stages, larval and adult, with contradictory selective forces, amniotes could make use of their entire period of early development perfecting characters optimal for life on land. Some members of all three modern amphibian orders are capable of direct development, but all must live in a damp environment to fa-

cilitate cutaneous respiration, even if they spend their entire life on land.

Environment

Another important aspect of early amniotes was the environment in which they lived and were preserved. Their distribution suggests that they were among the most terrestrial of all early tetrapods. The oldest known amniotes have been found within the upright stumps of lycopod trees in Joggins and Florence, Nova Scotia (R. L. Carroll, 1964c, 1969b; Reisz, 1972). Dozens of specimens have been found at both localities, including complete, articulated skeletons. As discussed in Chapter 4, only fully terrestrial vertebrates were found at these sites. The fauna found in the upright stumps must have accumulated from animals living on dry land who fell into the trees after their bases had been buried by flood waters and the tops again exposed in a newly formed forest floor (Fig. 7.10 and Plate 7).

A few early amniotes have also been discovered in the later coal swamp localities of Mazon Creek, Linton, and Nýřany

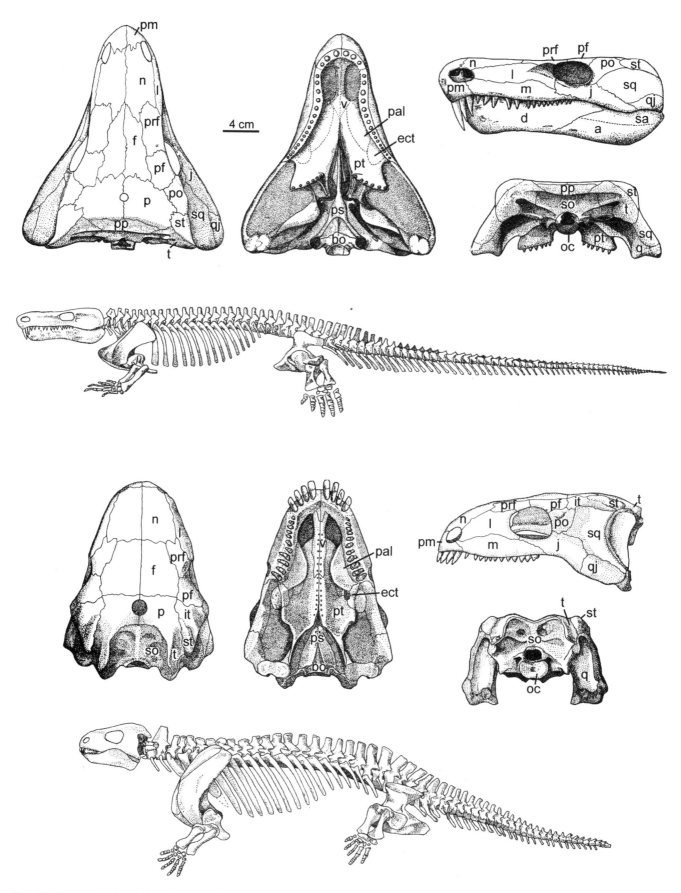

Figure 7.11. Large, primitive relatives of amniotes from the uppermost Carboniferous and Lower Permian. Top, *Limnoscelis,* in which the marginal teeth retain enfolding of the enamel and the bones that make up the astragalus in *Hylonomus* are not yet fused. However, palatal fangs are lost, and the ptery-goid has evolved a transverse flange. Bottom, *Diadectes,* a large herbivore to judge by its massive cheek teeth and nipping incisors. Some species of this genus have not yet consolidated the astragalus. From R. L. Carroll, 1969c.

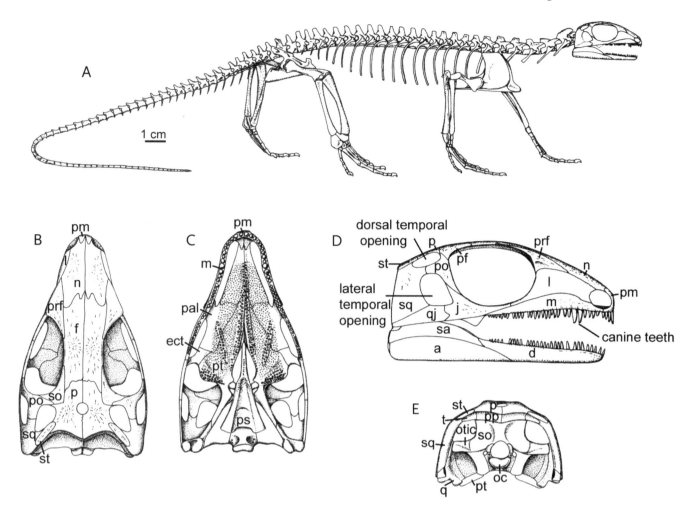

Figure 7.12. The skeleton and skull of the oldest known diapsid reptile, *Petrolacosaurus,* from the Upper Carboniferous of Kansas, distinguished by the presence of both dorsal and lateral temporal openings. Other features, such as the small size of the skull, the position of the orbit midway in its length, and the nature of the dentition, suggest closer affinities with the most primi- tive amniotes rather than with the pelycosaurs. This animal has much longer limbs than either primitive amniotes or synapsids. Animals similar to *Petrolacosaurus* were presumably ancestral to all later diapsid reptiles, including lizards, snakes, crocodiles, dinosaurs, and birds. From Reisz, 1977, 1981.

(Figs. 7.2 and 7.3) but in much smaller numbers than the aquatic and semi-aquatic amphibians. The absence of any amniote fossils prior to Joggins presumably results from our failure to find any fully terrestrial deposits where they may have been preserved.

Early Amniote Radiation

By the time of appearance of *Hylonomus* at Joggins, a major division among the early amniotes had probably already occurred between those that would eventually give rise to modern reptiles, dinosaurs, and birds (collectively termed sauropsids) and those that gave rise to mammals (the synapsids). Nearly all the amniote specimens at Joggins can be attributed to *Hylonomus,* which is structurally close to the base of all amniotes, but a relatively large humerus has been attributed to a pelycosaur, the lineage that eventually gave rise to mammals. Antecedents of both modern reptiles and mammals are definitely present in a slightly later tree stump locality—that at Florence, near Sydney, Nova Scotia—dated as equivalent to the European Westphalian C-D.

A further lineage, of uncertain relationships, the limnoscelids, is also present at Florence. They are represented by a single incomplete skeleton, with an estimated body size at least twice that of *Hylonomus* or *Paleothyris,* but more primitive in retaining labyrinthine enfolding of the teeth and having laterally expanded neural arches resembling those of *Seymouria.* Better-known species from the Lower Permian lack an astragalus, but have a transverse flange on the pterygoid (Fig. 7.11). Limnoscelids might be approaching the amniote condition anatomically, but there is no way to determine their mode of reproduction.

An apparently related group of large animals also thought to be close to early amniotes were the diadectids, relatively common in North America and western Europe, which were unique in having widely expanded cheek teeth that strongly suggest an herbivorous diet (Fig. 7.11) (Berman and Henrici, 2003; Berman et al., 2004). Neither limnoscelids nor diadectids are known to have survived beyond the Lower Permian (Reisz, 2007).

Other lineages of early amniotes retained the relatively

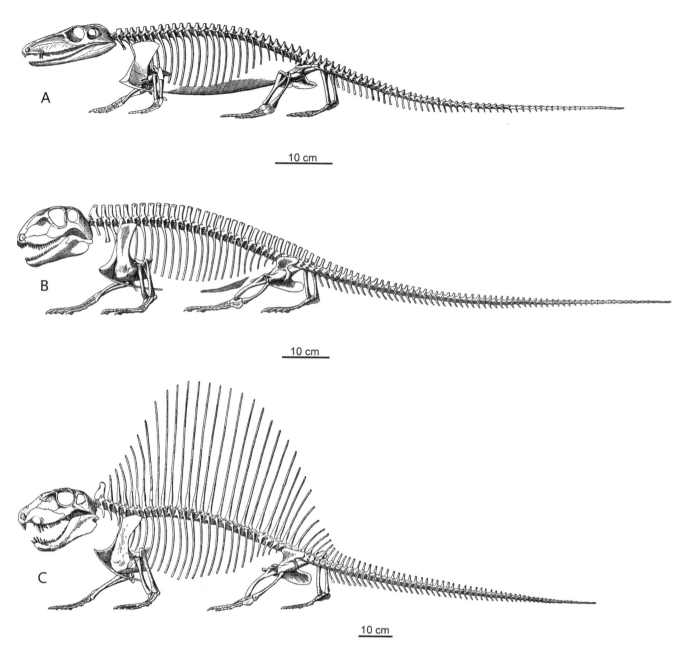

Figure 7.13. Three carnivorous pelycosaurs. A, The primitive genus *Varanosaurus*, which retains the low skull and conspicuous canine teeth of *Hylonomus*, but shows the lateral temporal opening that distinguishes pelycosaurs, later synapsids, and mammals. The snout region is much extended relative to the postorbital portion of the skull. B, *Haptodus*, a primitive sphenodontid pelycosaur in which the neural spines are somewhat elongate. C, *Dimetrodon*, in which they are extended into a sail-like structure that may have been used for temperature regulation. Modified from Romer and Price, 1940.

small size and many of the anatomical features of *Hylonomus* and *Paleothyris* (Clark and Carroll, 1973; Heaton, 1979). The captorhinids were common in North American and Europe in the Early Permian and larger descendants persisted into the Upper Permian of Africa. Many developed multiple tooth rows that may have been used to crush mollusks and/or arthropods.

More divergent was *Petrolacosaurus* (Fig. 7.12), known from the Garnett locality, Late Carboniferous of Kansas (Reisz, 1977, 1981). The size and general configuration of the skull resembled that of *Paleothyris*, but there are two openings behind the eye, the dorsal and lateral temporal fenestrae. These openings may have evolved initially to distribute the forces generated by the temporal musculature, so as to avoid separation of bones at the point where two or three meet in the cheek region. They may also have served to lighten the skull, and eventually to form edges for the attachment of muscles. Their presence serves to characterize a distinct group of amniotes, the diapsids, which came to dominate the earth and its oceans in the Mesozoic (Fig. 1.2). Their descendants can be divided into two major groups—on the one hand, the relatively primitive tuatara, lizards, and snakes, and on the other, the more derived crocodiles, dinosaurs, and birds. Many of the groups of marine reptiles common to the Mesozoic also

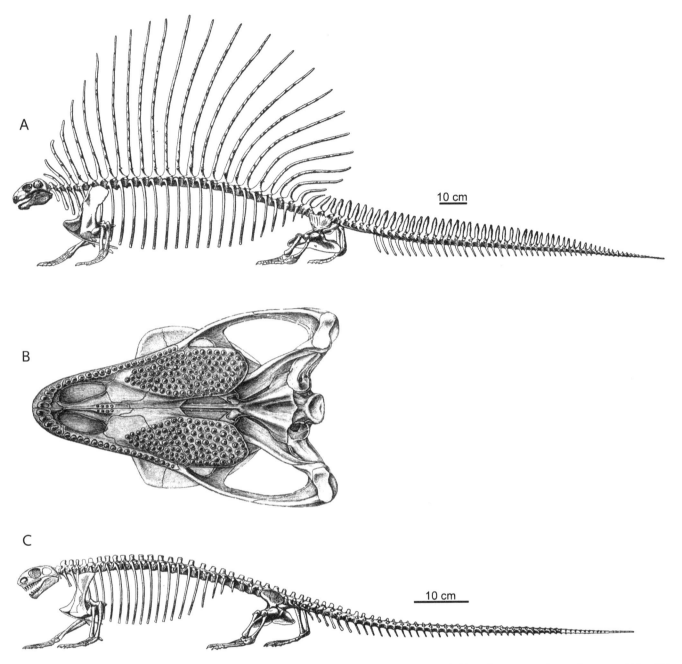

Figure 7.14. Herbivorous pelycosaurs. *A* and *B, Edaphosaurus,* from the Upper Carboniferous and Lower Permian. The long spines are ornamented by short laterally projecting processes. The small skull but massive palatal dentition identifies this genus as a herbivore. *C, Casea* a herbivore whose rib cage is greatly expanded laterally to surround a very large gut within which plant material was presumably broken down by bacterial symbionts, as in living ungulates. Modified from Romer and Price, 1940.

evolved from among the diapsid assemblage. All of these groups impacted the later amphibians in one way or another.

From the human standpoint, the most important derivatives of the early amniotes were the pelycosaurs, which were the first of two successive groups included in the Synapsida—the antecedents of mammals. They were distinguished as early as the Westphalian C-D of Florence, Nova Scotia, by the presence of a single temporal opening, low in the cheek, and the relatively posterior position of the orbit (Reisz, 1972). They were already larger than the contemporary *Paleothyris*, and became the dominant terrestrial carnivores by the end of the Upper Carboniferous when they were common in the Garnett locality in eastern Kansas (Reisz and Berman, 1986; Currie, 1977). This locality was drier than most of those in the Late Carboniferous and Permian, to judge by the presence of the primitive conifer *Walchia* and leaves resembling those of the living ginkgo. Not surprisingly, few amphibians are known from this locality.

Despite the continuing diversity of temnospondyls in the Early Permian, pelycosaurs including *Varanosaurus, Haptodus,* and *Dimetrodon* (Fig. 7.13) had taken over as the dominant terrestrial carnivores and *Edaphosaurus* (Fig. 7.14*A, B*) had evolved to a new trophic level, that of a terrestrial herbivore. This is demonstrated by the elaboration of blunt teeth over most of the surface of the palate and medial portion of the lower jaws. *Casea* (Fig. 7.14C) was another large herbivore from the Lower Permian, but with only a single row of marginal teeth. However, the great lateral expansion of the rib cage suggests that it, like modern cattle, had a much expanded gut and made use of symbiotic bacteria for fermentation of plant material. Aside from the tadpole larvae of frogs, amphibians never evolved the capacity to subsist on plant material.

In the Late Permian and Triassic, a second radiation of advanced synapsids, the therapsids, became dominant throughout the world, with fossils common in South Africa, Russia, North and South America, China, and even Antarctica. By that time, terrestrial amphibians had dwindled to only a few relict lineages, as the advanced therapsids and the antecedents of the dinosaurs, together with a host of smaller, lizard-like reptiles, rose to overwhelming dominance, leaving the aquatic environments as the only relatively safe abode for non-amniotes.

8

Stereospondyls

Escape from the Land

THE END-PERMIAN EXTINCTION

FOLLOWING THE PROGRESSIVE DECLINE OF amphibians toward the end of the Permian, we come to a major milestone in the history of life on earth, the end of the Paleozoic. As early as the mid-19th century, major changes in the earth's biota and its geological features were observed between the Late Permian and the Early Triassic that were used to distinguish the Paleozoic and Mesozoic eras. This is now recognized as a time of massive extinction among nearly all elements of animal and plant life on earth. Numerous books have been written on this subject in recent years: *The Great Paleozoic Crisis: Life and Death in the Permian* (Erwin, 1993), *Mass Extinctions and Their Aftermath* (Hallam and Wignall, 1997), *When Life Nearly Died* (Benton, 2003), *Extinction: How Life on Earth Nearly Ended 250 Million Years Ago* (Erwin, 2006). This subject was further reviewed by MacLeod (2006).

The results of this extinction are shown graphically in Figure 8.1. Many of the major groups of marine organisms suffered an immediate, catastrophic drop in diversity at the Permian-Triassic boundary, 251 million years ago. Only a few became completely extinct, including the rugose and tabulate corals and complex foraminiferans, while others recovered to a degree later in the Triassic. Bottom-dwelling organisms, especially those making up reef communities, were the most strongly affected.

Extinction Rates

Quantitative estimates of the percentage of extinction at different taxonomic levels were tabulated by Benton (2003). Beginning with unicellular organisms, approximately 5000 species of foraminifera living on the seafloor became extinct. All Permian species of chert-forming radiolarians were eliminated, but other radiolarians reappeared in the mid-Triassic. The number of sea urchin species dropped from 20 to 30 down to one or two. Brachiopod superfamilies dropped from 26 to 10. The extinction rate at the family level was 91%, that of genera 95%, and that of species 99%.

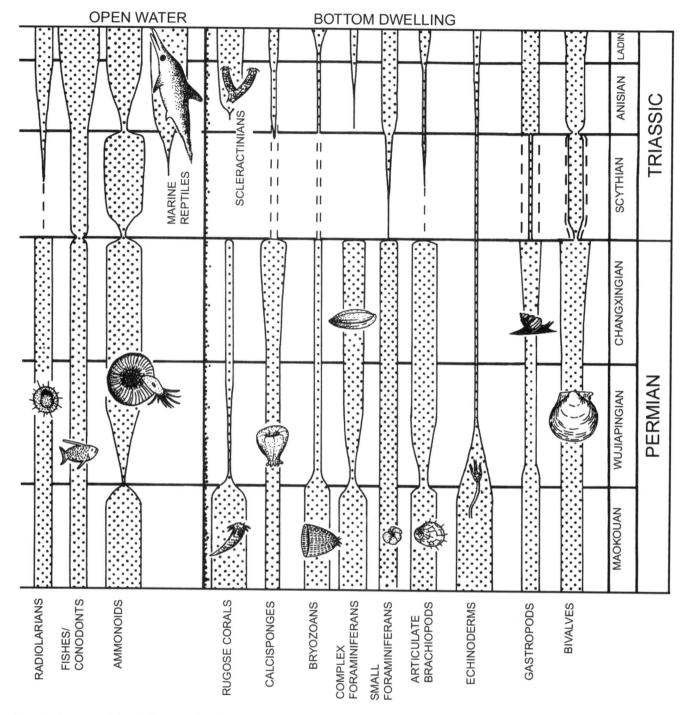

Figure 8.1. Summary of the relative proportion of extinction within major clades across the Permian-Triassic boundary. From Hallam and Wignall, 1997. Width of columns for each group denotes the relative changes of diver-sity within each group and is not intended to indicate differences in diversity between groups. Geological divisions within the Permian are based on the newly described Chinese sections.

On the other hand, the similar-appearing bivalve mollusks lost only 3 families of the 40 that had been present at the end of the Permian, and later radiated extensively. Among the cephalopods, the ammonoids (extinct relatives of the pearly nautilus) lost 20 of the 21 genera and 102 of the 103 species, but recovered in the later Triassic. The trilobites, which had dominated the earlier Paleozoic, were reduced to only one or two species by the end of the Permian, when they too

were lost. Crabs and lobsters, in contrast, were not severely affected, nor were marine vertebrates. Cartilaginous fish, the sharks, were unaffected, and the bony fish lost only 2 of 8 families, which were quickly replaced by a new radiation. Benton et al. (2004) gave an overall extinction rate for families on land and in the sea of 50%. At the species level, he calculated an extinction rate of 80 to 96%. The only benefactor of the end-Permian extinction were fungi, whose spores become extremely common at this time, presumably as a result of the abundance of decaying plant and animal matter.

Geological Events

A great number of changes affected the earth during the Late Permian and Early Triassic. Although the southern continents continued to be closely connected to one another, all were moving away from the South Pole, resulting in melting of the polar ice cap and warming of the entire region. In contrast with the Late Carboniferous and Early Permian, there is no evidence of any continental glaciation during the Triassic. By the end of the Permian, Gondwana and Laurasia had formed a single world continent, Pangaea. This led to the unimpeded flow of ocean currents from tropical to polar areas, producing a relatively uniform warming over all areas of the earth.

Two additional factors that led to global warming were of greater magnitude and occurred at the very end of the Permian. Both resulted in the release of great amounts of CO_2. The most conspicuous source was the enormous extent of basalt flows in eastern Asia, the Siberian Traps, covering more than 2.5 million km². Another source of CO_2 was the weathering of very large Permian coal deposits exposed in southern Africa.

The increased CO_2 led to warming of both the land and the sea. The higher temperatures would have driven oxygen, normally dissolved in the oceans, out of solution, producing anoxia in reef communities and among animals living on the seafloor. Marine vertebrates, making use of surface waters, would have been less severely affected. Previous research had suggested a serious lowering of sea level, but more recent evidence indicates a major transgression at the end of the Permian (Hallam and Wignall, 1997). Any abrupt change in sea level would certainly have been disastrous to the reef community.

Higher temperatures may have also precipitated the mass extinction of the once dominant glossopterus flora of the southern continents (Benton, 2003). This contributed to what Retallack et al. (1996) termed the "coal gap," a period of approximately 7 million years following the end-Permian event from which no major coal seams are known anywhere on earth. The loss of the coal swamp environment, which was the habitat of a wide range of Carboniferous and Lower Permian amphibians, would have certainly been a great blow to their survival into the Triassic.

Another possible cause of extinction that has been investigated, but without finding convincing evidence, is an impact by an extraterrestrial body such as a large meteorite (Benton and Twitchett, 2003). No large craters have been discovered in rocks dating from the time of the Permian-Triassic boundary and no iridium spike has been confirmed. This is in contrast with the end-Cretaceous extinction in which large amounts of this element have been found throughout the world at exactly the Mesozoic-Cenozoic boundary. Such high concentrations of iridium are known only in extraterrestrial objects.

There is no question regarding the great magnitude of extinction at the time of the Permian-Triassic boundary. Much new information has recently been gained from study of the previously poorly known sedimentary sequence in China that represents deposition across this boundary, as well as specification of its date as close to 251 million years ago. This date presumably represents the culmination of geological events that resulted in worldwide changes in the environment of the sea, land, and atmosphere.

However, there remain many questions regarding the relative significance of various environmental changes and their effect on particular groups of organisms. This is complicated by the fact that some major environmental changes were certainly of much longer duration and impacted different groups of organisms in different ways. The diversity of environments affected and the probability of fossilization in different depositional environments must also be considered in determining the proportions of species, genera, or families that may have become extinct during particular periods of time.

We saw in Chapter 5 how difficult it is to determine relationships among Paleozoic amphibians because of the rarity of fossils from the Lower Carboniferous. Yet, we know from isolated bones preserved at the Horton Bluff locality and the occurrence of individual specimens from the following 30 million-year period of Romer's gap that many lineages must have diverged in the earliest Carboniferous, but without being preserved in deposits that have so far been discovered. The factors of scattered distribution, restriction to poorly represented habitats, or subsequent loss by erosion or tectonic activity may have been even greater for animals living in the Upper Permian and Lower Triassic. In fact, the entire duration of the Triassic is almost devoid of fossils of animals that are plausible ancestors of living amphibians, although this geological period was conducive to the presence of a host of other, aquatic amphibians and many terrestrial amniotes.

The problem of the intensity of the end-Permian extinction and its varied effects on different taxonomic groups is shown by different views of the magnitude of extinction in Russia and South Africa, both of which have more or less continuous deposition across the Permian-Triassic boundary and a relatively rich fossil record of other vertebrates. Benton et al. (2004) estimated an 82% extinction rate at the family level over a period of approximately 500,000 years, spanning the Permian-Triassic boundary in the south Urals basin. In contrast, Modesta et al. (2001, 2003) estimated an extinction rate of only 29% at the level of species over this interval in southern Africa among a particular family of amniotes, the procolophonids. Especially among the therapsids, the group that would eventually lead to mammals, there were major gaps within the Lower Triassic, but many of the earlier families reappeared in later beds, indicating the continuation of the lineages, although they were represented by different species or genera. This is clearly apparent above a zone dominated by the genus *Lystrosaurus*, which was deposited in a primarily aquatic environment, in contrast with the terrestrially derived sediments in the beds above and below. In this case, the more terrestrial therapsids may have simply migrated to other habitats and then returned when the prior environmental conditions were reestablished.

THE EMERGENCE OF STEREOSPONDYLS

Whatever the extent of the extinctions, the amphibians that survived into the Triassic would have inhabited a very different world from that of the Late Permian. Geological changes and global warming would have produced very different environments to which the remaining amphibians would have had to adapt. Most of the more terrestrial amphibians, consisting primarily of temnospondyls, had apparently become extinct well before the end of the Permian, as did nearly all of the small, semi-aquatic groups, including most of the lepospondyls and the primarily neotenic branchiosaurs. On the other hand, one trimerorhachoid temno-

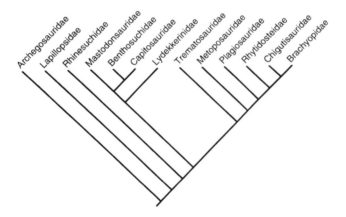

Figure 8.2. Cladogram of relationships of the principal taxa of stereospondyls, as hypothesized by A. Warren, 2000.

spondyl, *Tupilakosaurus,* survived into the Lower Triassic of Greenland and Russia, and a number of archegosauroids persisted to the end of the Permian in Russia and one, *Australerpeton,* in South America.

The descendants of archegosauroids survived to give rise to a new radiation of primarily large aquatic amphibians that dominated the rivers, lakes, and coastal areas throughout the world during the Triassic, with some surviving into the Jurassic and Lower Cretaceous (Fig. 8.2). This assemblage has long been known as the stereospondyls, based on the cylindrical configuration of the vertebral intercentra in some of the more advanced genera (Fig. 8.3). It is now recognized that many members of this group retained crescentic intercentra and some had paired pleurocentra, as is the case in temnospondyls, but they continue to be referred to as stereospondyls.

There remain uncertainties as to the specific ancestry of stereospondyls and the possibility that they arose from more than one group of temnospondyls. A. R. Milner (1990) hypothesized that the groups typically included among the stereospondyls had evolved from four different Upper Permian families. However, Schoch and Milner (2000) and A. Warren (2000) recognized broad similarities between the body form and presumed ways of life between archegosauroids and stereospondyls. The Russian genera *Platyoposaurus* and *Konzhukovia* are close to intermediates, but remain primitive in retaining long pterygoids that suture with the vomers, and either a narrow sutural contact between the pterygoid and the parasphenoid or a mobile joint between these bones.

There is sufficient similarity between the Upper Permian archegosauroids of Russia and the stereospondyls of the Upper Permian and Triassic of the southern continents to support ancestor-descendant relationships. However, there are enough anatomical differences between them, as well as a change in their geographical distribution, to justify recognizing them as distinct groups. This gap may be partially attributed to the scarcity of fossils immediately following the end-Permian extinction, but the presence of two families of

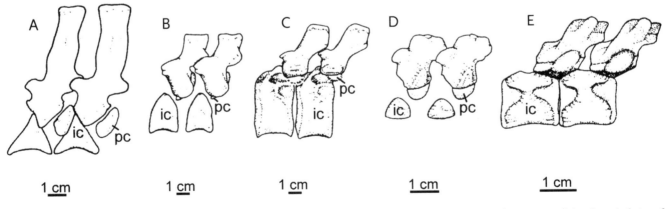

Figure 8.3. Reconstructions of the vertebrae of Mesozoic stereospondyls in lateral view. A, The temnospondyl *Eryops. B, Parotosuchus pronus,* a capitosaur. C, The metoposaur *Metoposaurus. D,* The rhytidosteid *Rewana quadricuneata.*

E, The plagiosaur *Gerrothorax.* It has been suggested that the articulation of the neural arches with two successive cylindrical centra may have contributed to dorsoventral undulation of the trunk. From A. Warren, 2000.

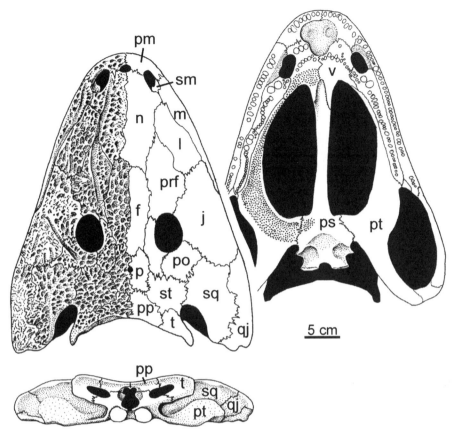

Figure 8.4. The rhinesuchid *Rhineceps nyasaensis,* from the Upper Permian of Malawi, in dorsal, palatal, and occipital views. This genus represents a common pattern for the adults of primitive stereospondyls. From Schoch and Milner, 2000.

stereospondyls in the Late Permian of the Southern Hemisphere and the great diversity of this group at the very base of the Triassic in Australia suggest that they largely survived that event in the southern continents.

Adam Yates and Anne Warren (2000) suggested several alternative scenarios to describe the emergence of a new group of amphibians in the Late Permian. 1. Only a single species of stereospondyl survived the end-Permian extinction and proceeded to radiate spectacularly in the Early Triassic. 2. The Stereospondyli could have been competitively superior to the other temnospondyls and directly caused their extinction by competitive exclusion. 3. The Stereospondyli may have been present as a component of many temnospondyl faunas in the Permian and some aspect of their biology fortuitously enabled them to survive the Late Permian extinction event. 4. The Stereospondyli may have begun their radiation in the Permian but were restricted to a geographical area that was less severely affected by the end-Permian extinction event.

The evidence that is currently available best supports the fourth hypothesis. The only known stereospondyls from the Upper Permian are the rhytidosteid *Trucheosaurus* from New South Wales, Australia, and numerous rhinesuchids in South Africa (Fig. 8.4). A diversity of stereospondyls occurred in the Lower Triassic of Australia: capitosaurids, trematosaurids, rhytidosteids, plagiosaurids, chigutisaurids, and brachyopids. The concentration of early stereospondyls in the southern continents suggests that this may have been a refuge within which they survived the end-Permian catastrophe and went on to spread throughout the rest of the world.

Characteristics of Stereospondyls

The anatomy of the earliest known stereospondyls is clearly distinct from that of individual species of late temnospondyl lineages, suggesting that they were adapting to a different way of life. This is made more conspicuous by the nature of the deposits in which they are found. Stereospondyls occupied all aquatic but non-marine habitats from upland playa lakes to coastal deltas, and some, including the trematosaurs, may have ventured into marine waters. However, few if any have been collected from conditions resembling the coal swamps and bogs from which the great diversity of Carboniferous and Lower Permian amphibians were found.

Several major reviews of the anatomy, diversity, and relationships of stereospondyls have been published in recent years: A. R. Milner (1990); Yates and Warren (2000); A. Warren (2000); and Schoch and Milner (2000). A. Warren's work provides a broad overview, emphasizing a relatively small number of major groups, and includes discussions of their geographic and temporal distribution as well as problems of their ancestry.

The skull of stereospondyls is typically more flattened than in temnospondyls, with the occipital condyles extended laterally, resulting in a hinge-like articulation between the cervical vertebrae and the skull. The posterior portion of the endochondral braincase is usually poorly ossified. The interpterygoid vacuities are very broad and the pterygoids have an extensive sutural attachment with the parasphenoid. There is commonly a broad depression at the front of the palate, between the internal nares. Nearly all have conspicuous lateral line canal grooves. Some had very long retroarticular processes that probably bore muscles attached to the top of the occiput for rapid opening of the lower jaws. Some stereospondyls were perennibranchiate, with the hyoid skeleton ossified into adulthood. The larvae presumably relied upon suction for feeding and ventilation.

The dermal shoulder girdle is greatly expanded, with the interclavicle and clavicle combined into a very large, kite-shaped structure. This has been attributed to their living close to the bottom and requiring protection from a stony substrate. The pubis never ossified, and the ilium and ischium are not co-ossified. The tail fin is supported by neural spines. The limbs are generally small and poorly ossified but with little, if any, loss of elements. The carpals and tarsals are rarely ossified.

STEREOSPONDYL DIVERSITY
Rhinesuchidae

The rhinesuchids are the best-known family of stereospondyls from the Upper Permian, appearing primarily in southern Africa, where they may have been the dominant predators in upland lakes, but with genera in Madagascar, India, and possibly Russia, with some extending into the Lower Triassic. Their skulls approached 60 cm in length, and the total length was more than 2 m. They are generally accepted as being at, or close to, the base of stereospondyls, and near to the ancestry of most, if not all, of the more advanced families. Primitive features include the great anterior extension of the pterygoid, which reaches the vomer, the presence of a conspicuous otic notch, and the lack of contact between the tabular and the squamosal. *Uranocentrodon* had ossified branchial arches, indicating that it was neotenic.

Lydekkerinidae

The skull of *Lydekkerina* from the Early Triassic of South Africa (Fig. 8.5) appears primitive in the position of the orbits near the mid-length of the skull, but this can be attributed to its small size, about 8 cm in length. This is comparable to an immature specimen of the capitosaurid *Parotosuchus* (Fig. 8.6), demonstrating that orbital position is subject to change during growth and may not be indicative of close relationship. Both the short anterior extent of the pterygoid and the broader contact between the parasphenoid and the pterygoid suggest that *Lydekkerina* is advanced relative to rhinesuchids, but evidence from the postcranial skeleton indicates that it was more terrestrial (Pawley and Warren, 2005).

Lapillopsis, from the Lower Triassic of Australia and possibly India, is another small genus, with large, centrally placed orbits and a narrow contact between the parasphenoid and pterygoid, but a pterygoid that reaches only midway in the length of the ectopterygoid. The tabular reaches to the squamosal. This combination of primitive and advanced characters explains why its taxonomic position is not firmly established. The absence of lateral line sulci suggests that it may have been primarily terrestrial.

Capitosauridae

Capitosaurids extended from the beginning of the Triassic until its end. The long-snouted adults superficially resembled crocodiles, but the postcranial skeleton was poorly ossified with a massive dermal shoulder girdle (Fig. 8.6). They were unusual in having the orbits intersect the frontal bone. Capitosaurids are considered to be mid-water predators, but have very small marginal teeth. Conservative classification recognizes only four genera, starting with *Parotosuchus* from Australia and *Kestrosaurus* from South Africa, both Lower Triassic, to the Middle Triassic *Paracyclotosaurs* (Australia) and the Upper Triassic *Cyclotosaurus* (Arizona). They show progressive closure of the otic notch, and have poorly developed lateral line sulci. The largest specimens reached 4 m in length.

Mastodonsauridae

There is general agreement on the close relationship of the Capitosauridae and the Mastodonsauridae. The latter family is known only from the late Early Triassic to the Middle Triassic in Europe. They had massive skulls, up to 125 cm in length, with large orbits and huge palatal tusks. The parasymphyseal teeth of the lower jaw extend through paired openings in the premaxillae (Fig. 8.7).

Figure 8.5. (opposite) Primitive stereospondyls. A, *Lapillopsis,* from the Lower Triassic of Australia, a possible model for an ancestral stereospondyl. Its small size may be responsible for the position of the orbits midway in the length of the skull and their contact with the frontals. On the other hand, it could result from the immaturity of the specimen (see Figure 8.6 for growth stages in another stereospondyl). The narrow extent of the contact between the parasphenoid and the pterygoids is certainly a primitive character, but the failure of the pterygoid to reach the palatine is advanced, as is the contact between the tabular and squamosal, which are separated by the supratemporal in rhinosuchids in relationship to their deeper otic notches. No lateral line canals are evident, suggesting a terrestrial way of life. B, *Lydekkerina,* from the Lower Triassic of South Africa. It too remains primitive in the central position of the orbits, but is larger and has lateral line canals. The pterygoid has a longer line of contact with the parasphenoid, but makes contact with the palatine. Such combinations of primitive and derived features make it difficult to establish the specific affinities of these genera. *Lapillopsis* from Schoch and Milner, 2000; *Lydekkeria* from A. Warren, 2000.

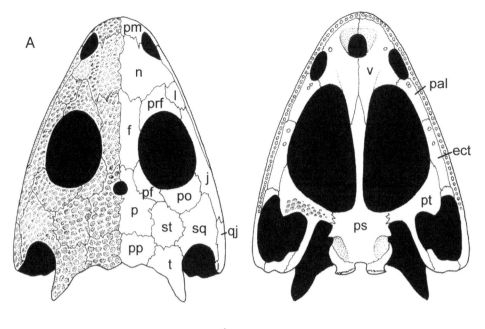

A

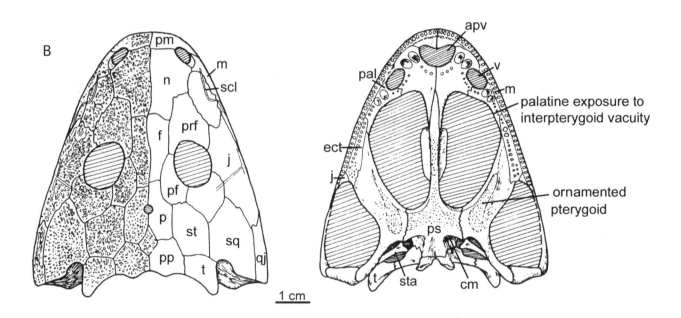

B

1 cm

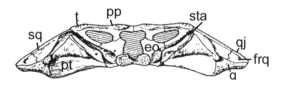

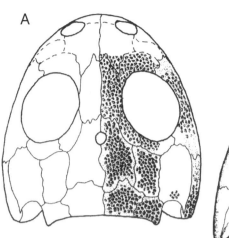

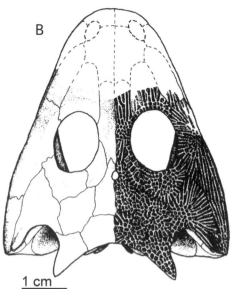

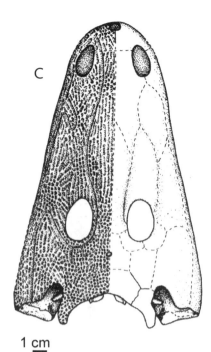

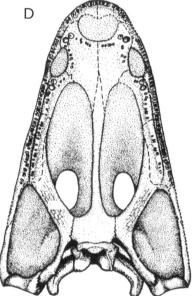

A

1 mm

B

1 cm

C

1 cm

E

10 cm

F

G

H

D

I

10 cm

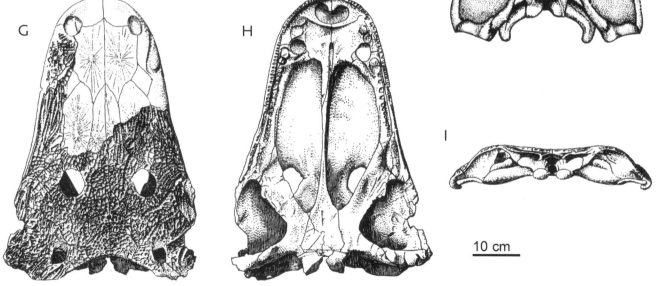

Figure 8.6. The Family Capitosauridae. *A–C,* Growth stages in the skull of *Parotosuchus aliciae* from the Lower Triassic of Australia. *D,* Palate of *Parotosuchus. E* and *F,* Dorsal and lateral views of the skeleton of the Middle Triassic *Paracyclotosaurus. G–I,* skull of *Cyclotosaurus* in dorsal, palatal, and occipital views, from the Upper Triassic. Note circular otic "notches" facing dorsally. From A. Warren, 2000.

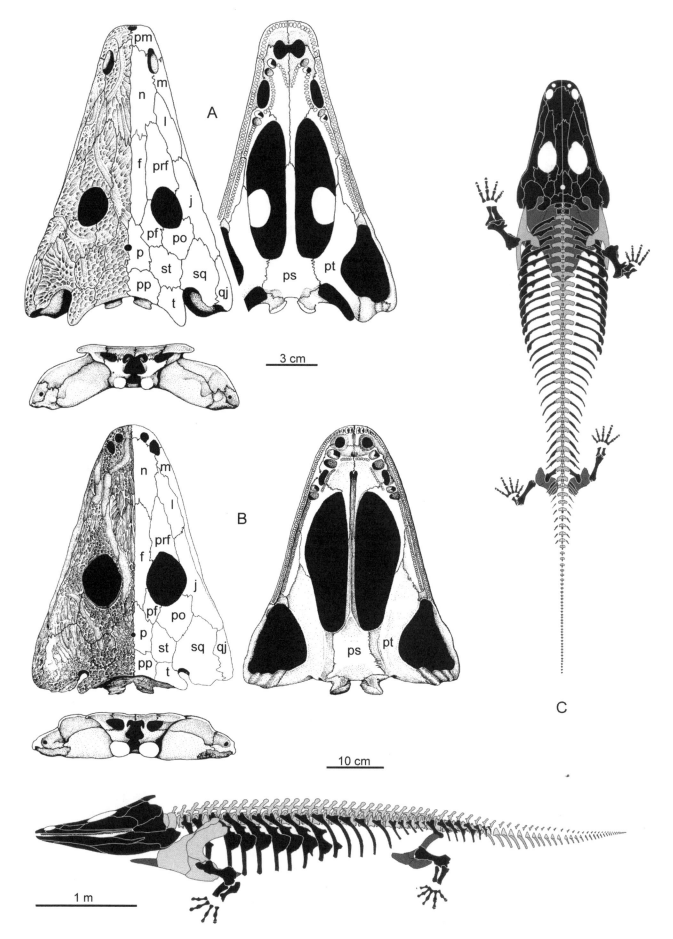

Figure 8.7. A, Dorsal, palatal, and occipital views of the skull of *Benthosuchus sushkini* from the Lower Triassic of eastern European Russia. *B,* Skull of *Mastodonsaurus giganteus. C,* Reconstruction of the skeleton in dorsal and lat-eral views. The family Mastodonsauridae, closely related to the capitosaurs, ranges from the late Early Triassic to the Middle Triassic of Europe. From Schoch and Milner, 2000.

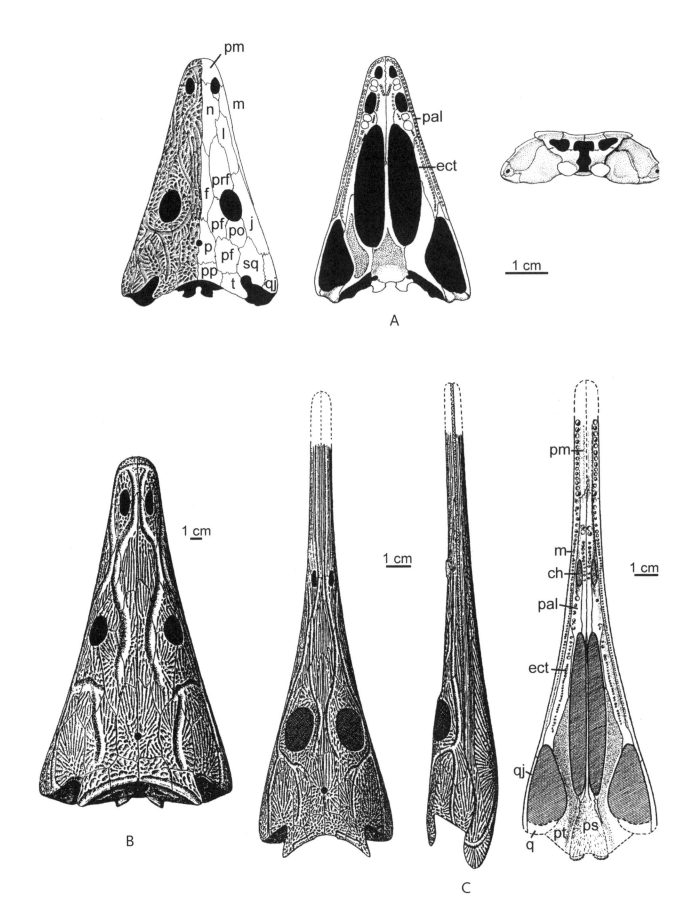

Figure 8.8. A, Dorsal, palatal, and occipital views of the skull of *Thoosuchus yakovlevi* from the Lower Triassic of eastern European Russia. From Schoch and Milner, 2000. It is a close antecedent of trematosaurs. *B,* The primitive, short-snouted trematosaur *Trematosaurus brawni. C,* The advanced trematosaur *Aphaneramma* in dorsal, lateral, and occipital views. From A. Warren, 2000.

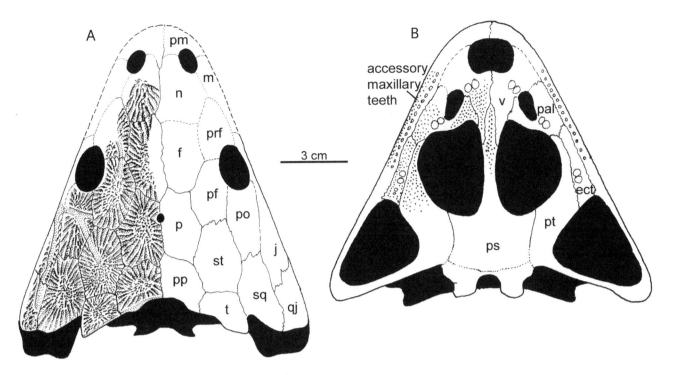

Figure 8.9. Rhytidosteids. *A, Peltostega erici,* Lower Triassic of Madagascar. *B, Deltasaurus kimberleyensis,* Lower Triassic of Australia and Tasmania. From Schoch and Milner, 2000.

Benthosuchidae

Also close to the capitosaurs in general appearance were the benthosuchids from the Lower Triassic of northeastern Europe and northeast Asia and the Middle Triassic of western Europe and Arizona (Fig. 8.7). Their skull proportions are also similar to those of the trematosaur *Thoosuchus.* A closely related genus, *Deltacephalus,* is known from Madagascar and a probable relative occurs in Algeria. The large lateral line canals suggest they were fully aquatic. Their small orbits did not intersect the frontal bones.

Trematosauridae

The derived members of the Trematosauridae are highly distinctive in having extremely elongate snouts, extending well beyond the external nares. The parasphenoid-pterygoid suture extends posteriorly toward the occiput (Fig. 8.8). Their remains are found in marine beds in Spitzbergen and Madagascar and deltaic deposits in western Australia, suggesting that at least some members of this group lived in the ocean. They were the most widespread of all stereospondyls (Fig. 8.16), with a temporal range from the Lower Triassic of India to the Middle Triassic of Nova Scotia. One specimen is known from the early Late Triassic of Europe. The more primitive genera had smaller and more triangular skulls. In the one specimen in which the postcranial skeleton is adequately known, the vertebrae are rhachitomous and the pectoral girdle is narrow, in contrast with most other stereospondyls. The Russian Thoosuchidae appears closely related to primitive trematosaurs, with the antorbital region substantially elongated even in a skull barely 11 cm in length.

Rhytidosteidae

The rhytidosteids spanned the Permian-Triassic boundary (Fig. 8.9). The oldest known is the Australian *Trucheosaurus.* They were rare but cosmopolitan, although not known beyond the Lower Triassic. The skulls were parabolic or triangular in outline, with very small teeth and a highly denticulate palate. Some genera had pustular ornamentation of the dermal skull bones. They reached up to 40 cm in length, but had very poorly ossified skeletons. They are known from fluvial to deltaic deposits and may have been euryhaline. A single deposit in the Lower Triassic of Queensland, Australia, has four genera. Others ranged from Siberia to the Russian Cisurals and Spitzbergen. A newly found rhytidosteid, *Cabralia,* was described by Dias-da-Silva et al. (2006) from the Lower Triassic of Brazil. The retention of a lacrimal suggests relationship with the Indian *Indobrachyops.*

Schoch and Milner (2000) placed the Rhytidosteidae within the infraorder Rhytidostea, including the Lydekkerinidae, Derwentiidae, and Chigutisauridae, united by their generally small to medium size, wide skull, short face, and lateral position of the orbits. The phylogenetic position of rhytidosteids differs from near the base of the stereospondyls (Schoch and Milner, 2000) to near the top of their phylogeny (A. Warren, 2000).

Brachyopidae

Brachyopids were perhaps the rarest of all stereospondyls, with 15 described skulls and only four specimens for which there is postcranial material. *Xenobrachyops,* from the basal Triassic of Australia, is the oldest known member and *Gobios,*

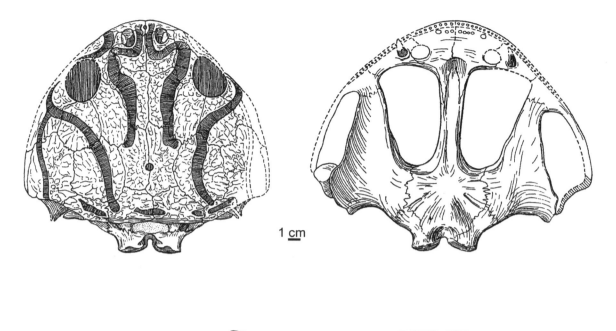

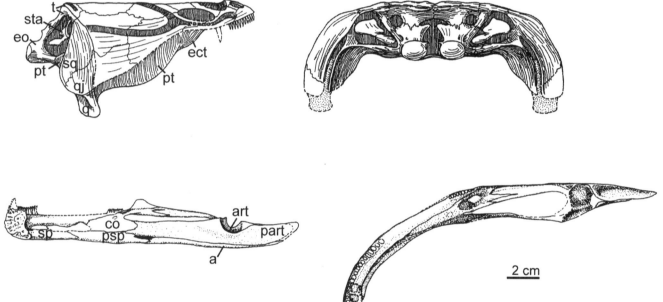

Figure 8.10. Brachyopids. Skull of the brachyopid *Batrachosuchus watsoni* from the Triassic of Queensland in dorsal, palatal, lateral, and occipital views; brachyopid jaw in medial and dorsal views. From A. Warren, 2000.

from the Jurassic of Mongolia, the latest. The early genera had a rhachitomous vertebral pattern but the last had stereospondylous centra. Brachiopods are readily recognized by their short, broad skull, deeply indented sensory canals, and cheeks that extend far ventrally, resulting in a very deep oral cavity that suggests gape-and-suck feeding (Fig. 8.10). This is supported by the very long retroarticular processes that would have enabled the jaws to be opened rapidly. They have no trace of an otic notch. The skull of *Hadrokkosaurus* from the Moenkopi of Arizona reached 30 cm in length, but a jaw fragment from the Upper Triassic or Lower Jurassic of Lesotho, described by Steyer and Damiani (2005), suggests a skull width of about 1.7 m and an estimated body length of

about 7 m. Recently discovered specimens have been found in the Middle Triassic of India and Algeria (Steyer and Damiani, 2005).

Schoch and Milner (2000) did not include either brachyopids or plagiosaurs in their review of stereospondyls. A. R. Milner (1990) allied brachyopids with the Permo-Carboniferous saurerpetontids and plagiosaurids with *Peltobatrachus*.

Chigutisauridae

The chigutisaurids are the most recent group of stereospondyls to be recognized. They extend from the Early Triassic to the Early Cretaceous in Australia, but were most common in the Late Triassic of Argentina and India (Fig. 8.11). The complete skeleton of *Siderops* was described by A. Warren and Hutchinson (1983) from the Early Jurassic of Australia, but the most surprising discovery was the well-preserved jaws of the genus *Koolasuchus* from the Lower Cretaceous (A. Warren et al., 1997) (Fig. 8.12). Chigutisaurids are primitive in retaining an otic notch, rhachitomous vertebrae, and a quite restricted dermal shoulder girdle. The degree of development of the limbs indicates that they were somewhat terrestrial, although lateral line canals are retained. The enlarged, recurved teeth of *Siderops* and *Koolasuchus* suggest that they may have been effective fluvial predators. A. Warren (2000) argues for the common ancestry of chigutisaurids and brachyopids.

Metoposauridae

The metoposaurs are only known from the Late Triassic. They are readily recognized by the anterior position of their eyes and the elongate skull table (Fig. 8.13). The palatal dentition forms a nearly continuous row parallel to the marginal teeth. The intercentra were fully spool-shaped, but remnants of the pleurocentra remain in one genus. The dermal bones of the skull and pectoral girdle were extremely thick. Metoposaurs are known from Canada, western Europe, Madagascar, India, and North Africa as well as the western United States, where their remains number in the hundreds in what appear as desiccated ponds. Surprisingly, they have more completely ossified limbs and girdles than most Upper Triassic stereospondyls. All occurred in what were tropical areas in the Triassic. Schoch and Milner (2000) place the Metoposauridae within the superfamily Trematosauroidea, along with the Benthosuchidae, but in the absence of Lower or Middle Triassic fossils it is difficult to connect them with any other stereospondyl family.

Plagiosauridae

The extreme degree of specialization of the plagiosaurids makes them even more difficult to classify; Schoch and Milner do not consider them as stereospondyls, although their centra are fully cylindrical. Like brachyopids, the cheek in some genera extends ventrally well below the level of the palate. But they differ in having very large orbits, directly over interpterygoid vacuities of similar dimensions (Figs. 8.14 and 8.15). The skull is very short, but the width may be as great as 60 cm. *Gerrothorax* is notable in being perennibranchiate. In contrast with other stereospondyls, plagiosaurids were apparently limited to the northern continental area, Laurasia. The centra are composed of a single unit that articulates with neural arches both anterior and posterior to it. A. Warren (2000) suggests that this may have promoted undulatory locomotion, rather than the lateral bending common to most amphibians. The trunk of both *Gerrothorax* and one species of *Plagiosternum* were covered with dermal bone both dorsally and ventrally. Plagiosaurs ranged from the Early Triassic of eastern Europe to the Middle Triassic of Russia and Germany, and the Late Triassic of Germany, Greenland, Sweden, and possibly Thailand.

GEOGRAPHICAL DISTRIBUTION

The early history of amphibians, discussed in Chapters 3 to 6, is known primarily from the northern continents—North America, Europe, and with some evidence from western Asia—all of which were then united in a single land mass, Laurasia. Almost nothing is known of the fossil record of early amphibians in the southern continents of South America, Africa, Antarctica, and Australia during the Carboniferous and Lower Permian, when they were united in the supercontinent Gondwana. In Chapter 6 (Fig 6.4), we saw that the sudden appearance of seymouriamorphs in Europe and North America could be attributed to northern and western movement of continental plates that provided at least temporary links between Australia, China, and other portions of eastern Asia with Russia. The northern movement of the southern continents continued through the later Permian and Early Triassic until all continents were linked into a more or less contiguous supercontinent, Pangaea. At that time, the Arabian peninsula, India, and Madagascar were all closely connected to Africa and South America within the Southern Hemisphere (Fig. 8.16).

With all the major land masses of the world united, land animals and plants had the potential to spread to all areas of the earth. The Late Permian and Early Triassic also coincided with other geological factors that enabled the preservation and later recovery of fossils from nearly all parts of the world. This is the first time in the earth's history for which we have knowledge of fossil tetrapods from all the major continental areas.

However, other factors, including climate, both local and worldwide, ecology, and the ways of life of individual taxa, certainly influenced their particular patterns of adaptation and distribution. Trematosaurs, which were clearly aquatic, are thought to have been capable of living in a near shore marine environment, which would explain their extremely wide distribution despite their relatively short temporal span. Rhytidosteids, also with a very short temporal span, were common to Australia and Brazil, but also spread as far as western Russia, Siberia, and Spitzbergen.

TEMPORAL DISTRIBUTION

The temporal distribution of stereospondyls is shown in Figure 8.17. The apparent ancestors of stereospondyls, the archegosauroids, extend from the Upper Carboniferous to the very end of the Permian, but are not known from the Triassic. They overlapped in time with the Rhinesuchidae, known primarily from the Upper Permian, and the rhytidosteids, which also crossed the Permian-Triassic boundary. Most other families appear at the base of the Triassic, implying their initial divergence in the Late Permian. They may

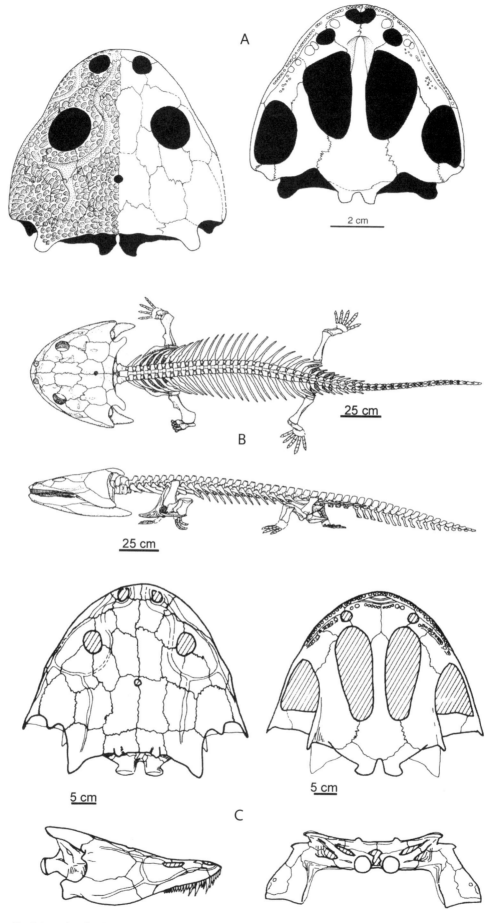

Figure 8.11. Chigutisaurids. *A,* Dorsal and palatal views of *Keratobrachyops australis* from the Lower Triassic of Australia. *B,* Reconstructions of the skeleton of *Siderops* in dorsal and lateral views, from the Lower Jurassic of Aus-tralia. *C,* The skull of *Compsocerops cosgriffi* from the Upper Triassic of India. *A* from Schoch and Milner, 2000; *B–C* from A. Warren, 2000.

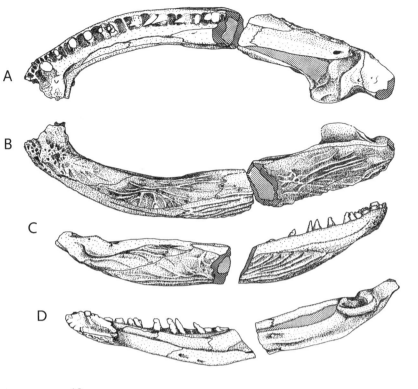

Figure 8.12. The latest surviving stereospondyl. *A,* Dorsal; *B,* Ventral; *C,* Lateral; and *D,* Medial views of the lower jaw of the chigutisaurid *Koolasuchus* *cleelandi* from the late Lower Cretaceous of Australia. From A. Warren et al., 1997.

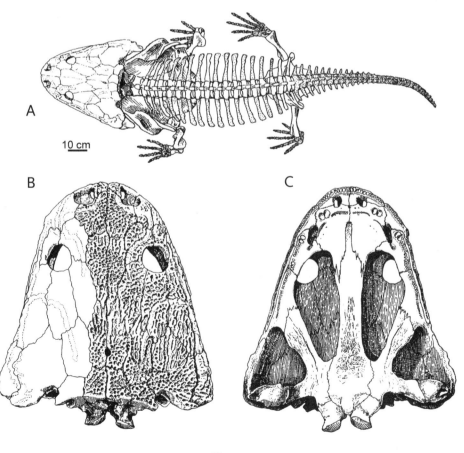

Figure 8.13. Dorsal view of skeleton and dorsal and palatal views of the skull of the metoposaurid *Buettneria,* locally very common in the early Late Triassic of the southwestern United States. Note far anterior position of orbits. From A. Warren, 2000.

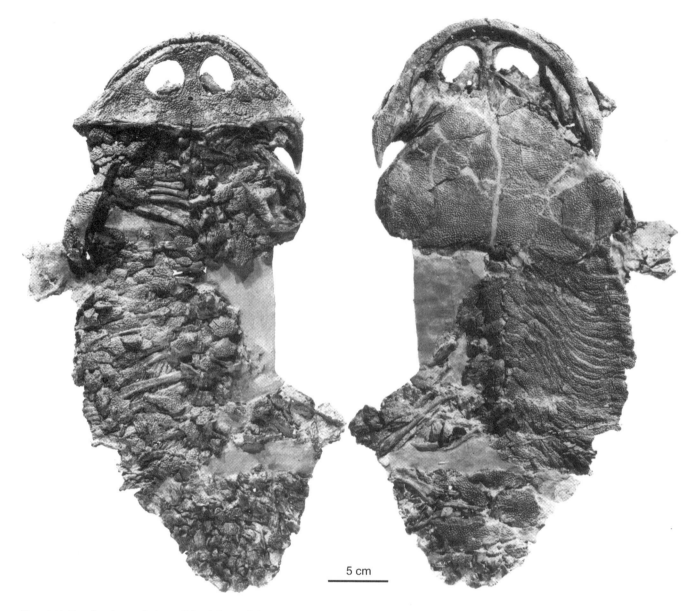

5 cm

Figure 8.14. Dorsal and ventral views of the skeleton of the Upper Triassic plagiosaurid *Gerrothorax pustuloglomeratus.* Note heavy armor plating, both dorsally and ventrally. From Hellrung, 2003.

have evolved in other geographical areas or in environments not represented in the known fossil localities. Such ghost lineages argue against the ubiquity of the end-Permian extinction, and suggest that more diverse faunas may eventually be discovered in the Southern Hemisphere. Eight families of stereospondyls have been recognized by A. Warren (2000) during the short duration of the Lower Triassic, seven of which continue into the Middle Triassic. One additional family, the Metoposauridae, is only known in the Upper Triassic, but only four other families persisted to the end of that period. However, the brachyopids extended into the Upper Jurassic and the chigutisaurids survived into the Lower Cretaceous. The progressive extinction of stereospondyls may be attributed to predation and/or competition with a diversity of freshwater and marine amniotes, culminating in the

phytosaurs, ichthyosaurs, and crocodiles that reverted to an aquatic way of life in the Upper Triassic and Lower Jurassic (R. L. Carroll, 1988a).

OTHER TRIASSIC AMPHIBIANS

Nearly all Triassic amphibians may be grouped as stereospondyls. The exceptions include three groups referred to in Chapter 6: the Chroniosuchia, which presumably evolved from among Permian anthracosaurs; *Micropholis,* clearly a descendant of the amphibamid dissorophoids; and the trimerorhachoid *Tupilakosaurus* from the Lower Triassic of Greenland and Russia. These animals were obviously relicts of lineages common in the late Paleozoic.

On the other hand, there are also a small number of speci-

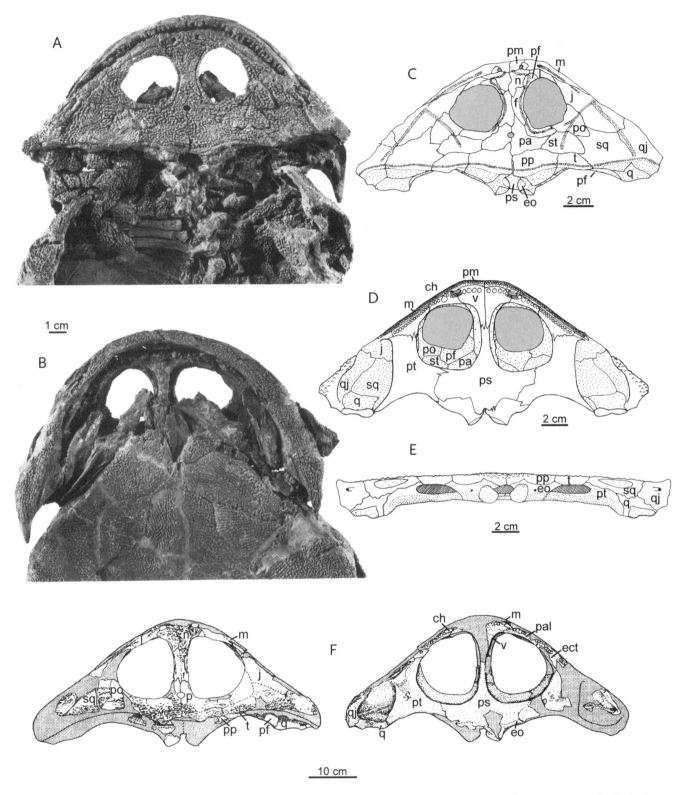

Figure 8.15. A and B, Dorsal and ventral views of the anterior portion of the skeleton of the Upper Triassic plagiosaurid *Gerrothorax pustuloglomeratus*. C–E, Reconstruction of the skull of *Gerrothorax pustuloglomeratus* in dorsal, palatal, and occipital views. From Hellrung, 2003. F, Dorsal and palatal views of the skull of *Plagiosternum granulosum* from the Middle Triassic of Russia and Germany. From A. Warren, 2000.

mens from the Triassic that point toward modern amphibians. Three localities are known in the Lower Triassic. Gao et al. (2004) have reported a new locality in Gansu Province, far western China, which contains a wide variety of vertebrates, including sharks and bony fish, microsaur-like lepospondyls, and branchiosaurid-like temnospondyls. The presence of a eucynodont, an advanced mammal-like reptile not known prior to the Lower Triassic, provides a maximum limit for

Figure 8.16. Geographical distribution of stereospondyls mapped on the position of the continents during the Triassic, based on data from A. Warren, 2000, and Schoch and Milner, 2000. This does not differ substantially from the continental positions in the Upper Permian. During the entire duration of stereospondyls, there were no continuous marine barriers between the continents. 1. Archegosauridae and Melosauridae; 2. *Lapillopsis;* 3. Rhinesuchidae; 4. Lydekkerinidae; 5. Capitosauridae; 6. Mastodonsauridae; 7. Benthosuchidae; 8. Trematosauridae; 9. Rhytidosteidae; 10. Brachyopidae; 11. Chigutisauridae; 12. Metoposauridae; 13. Plagiosauridae.

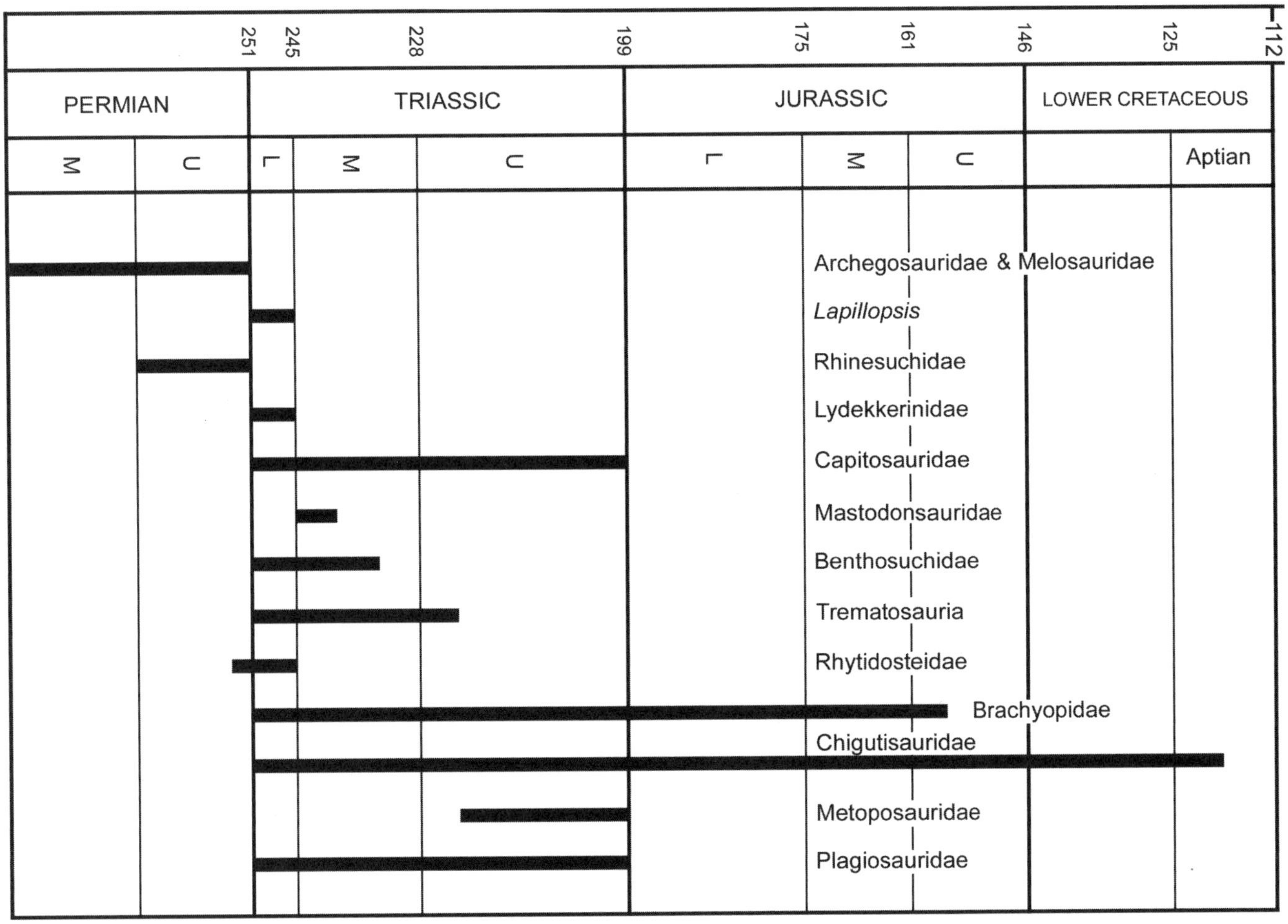

	PERMIAN		TRIASSIC			JURASSIC			LOWER CRETACEOUS	

Figure 8.17. Temporal ranges of fossil record of major lineages of stereo-spondyls, based on data from A. Warren, 2000, and Schoch and Milner, 2000.

Geological time scale from Gradstein et al., 2004. Time in millions of years. Data from several sources cited in text.

the date of this locality. Better-known are the genera *Triadobatrachus* from Madagascar and *Czatkobatrachus* from Poland, both of which represent the antecedents of frogs (Roček and Rage, 2000a; Evans and Borsuk-Białynicka, 1998). *Triadobatrachus* is represented by a single specimen from a deposit otherwise dominated by fish. *Czatkobatrachus* is known only from disarticulated bones found in a fissure filling. A single fossil designated *Triassurus*, from the Upper Triassic of Uzbekistan, was described as a salamander by Ivachnenko (1978). Richard Estes (1981) considered it to be a larval temnospondyl, but A. R. Milner (2000) cited a number of salamander-like characters. These genera will be considered in more detail in later chapters.

9

The Enigma of Modern Amphibian Origins

THE PROBLEM

IT MAY SEEM STRANGE THAT only passing reference has yet been made in this book to modern amphibians—the frogs, salamanders, and caecilians. The reason is simple: not a single fossil that was obviously a member of any of these groups has yet been found in deposits prior to the Early Jurassic (Fig. 9.1). As a result, both the interrelationships among the three living orders and their affinities with earlier amphibians remain subject to controversy. This is one of the greatest problems in the phylogeny of terrestrial vertebrates. Hypotheses range from a common ancestry of all three groups to separate origins of frogs, salamanders, and caecilians from different lineages of Paleozoic amphibians. The proposed nature of their Paleozoic ancestors also differs greatly from author to author.

Fossils of frogs, salamanders, and caecilians are known from the Middle and Lower Jurassic, going back nearly 200 million years, but all are basically similar to their living descendants and provide little more evidence as to their ancestry than do the modern species. During the preceding Triassic, lasting another 50 million years, there is only a single specimen that provides clear evidence of the prior history of any of the living orders.

From what we have seen of the stereospondyls, it is clear that none are plausible ancestors of the modern amphibians. While some of the smaller Paleozoic amphibians vaguely resemble salamanders, and several groups of lepospondyls have reduced or lost their limbs in common with caecilians, none are sufficiently similar to any of the modern orders, as adults, that they can be considered as being closely related.

In addition to the long temporal gap between modern and Paleozoic amphibians and the manifest differences in their anatomy and probable ways of life, establishing relationships between them is made difficult by the fact that their study has been largely conducted by two different groups of scientists. Herpetologists have concentrated primarily on the soft anatomy, physiology, development, behavior,

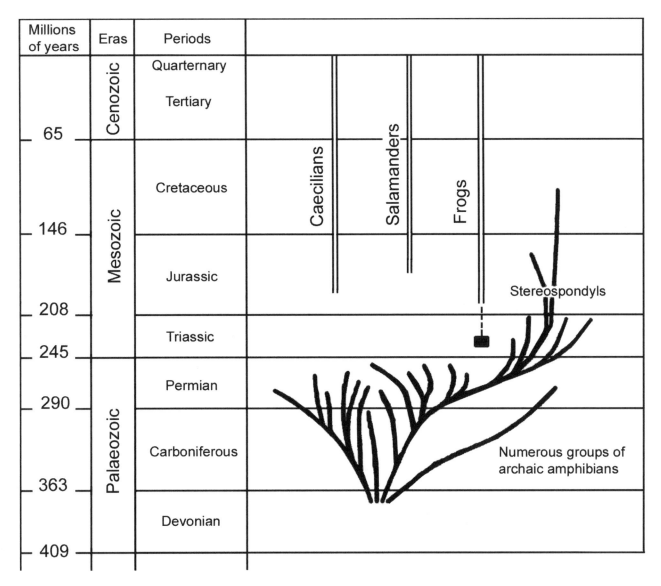

Millions of years	Eras	Periods			
	Cenozoic	Quarternary			
		Tertiary			
65					
	Mesozoic	Cretaceous	Caecilians	Salamanders	Frogs
146					
		Jurassic			
208					Stereospondyls
245		Triassic			
	Palaeozoic	Permian			
290					
		Carboniferous			Numerous groups of archaic amphibians
363					
		Devonian			
409					

Figure 9.1. Temporal ranges of modern and archaic amphibians. All three orders of living amphibians are known from the Jurassic, in nearly their modern body form. Archaic amphibians began their radiation in the Upper Devonian, with the stereospondyls extending into the Mesozoic. The black square is a plausible frog ancestor from the Lower Triassic, *Triadobatrachus*. From R. L. Carroll, 2001.

and ecology of the more than 6000 living species of frogs, salamanders, and caecilians. None of these attributes can be interpreted effectively from fossil remains. Hardly surprising, paleontologists have focused primarily on the bony anatomy as a means of establishing relationships between Paleozoic and modern amphibians.

Because frogs, salamanders, and caecilians share a number of similarities in soft anatomy and physiology, as well as their amphibious lifestyle, it has been natural for herpetologists to think of them as a single phylogenetic unit, to which the term Lissamphibia has been applied. This implies a single common ancestry, separate from that of amniotes. Paleontologists, who are aware of a vast diversity of Paleozoic amphibians, none of which are obvious antecedents of frogs, salamanders, or caecilians, have suggested that they

may have evolved from two or three separate groups of ancestral amphibians. However, no consensus has yet been reached.

The most extreme views of a divergent ancestry of modern amphibians were postulated by Holmgren (1933, 1949) and Jarvik (1942, 1960). Both suggested that frogs and amniotes had evolved from different groups of fish than those that gave rise to salamanders. Holmgren argued, on the basis of the polarity of development of the bones of the hands and feet, that salamanders had evolved from lungfish, and that all other tetrapods had evolved from choanate fish such as *Eusthenopteron*. In contrast, Jarvik proposed, on the basis of the anatomy of the snout region, that salamanders had a separate ancestry from another group of sarcopterygian fish, the porolepiforms, as opposed to the common ancestry of frogs and amniotes from fish similar to *Eusthenopteron*.

The hypothesis that the terrestrial limbs of salamanders had evolved separately from those of frogs and amniotes seemed so improbable that paleontologists and herpetologists put forward various alternative patterns of relationship based on a common ancestry of all tetrapods (J. T. Gregory et al., 1956; Eaton, 1959; Szarski, 1962; Reig, 1964). The most convincing of these arguments was that of the herpetologists Parsons and Williams (1962, 1963). In going through the extensive collections at Harvard, they recognized that the teeth of nearly all specimens of frogs, salamanders, and caecilians had a unique structure that was not shared with any other tetrapods, living or fossil. Instead of the tooth being a solid structure, the base (or pedicel) was separated from the crown by a zone of fibrous tissue that was considerably softer than the dentine and enamel of the rest of the tooth (Fig. 9.2).

In museum specimens, this commonly resulted in breakage, with the smaller crown separating from the base. Such teeth are referred to as pedicellate. No Paleozoic amphibians were then known to have such a structure. No obvious function for the loss of tooth crowns has been demonstrated, although it has been suggested that it may be advantageous for tongue-feeding animals so that the ends of the teeth do not get stuck in the prey and hinder swallowing. The greatly similar structure of this peculiar type of tooth in all three orders and the difficulty of imagining its function suggest that it was very unlikely to have evolved independently in three different lineages, and so appeared as strong evidence for the monophyletic origin of the lissamphibians.

Parsons and Williams then searched for other derived characters shared by frogs, salamanders, and caecilians. They tabulated the following list of lissamphibian synapomorphies. The characteristics indicated by an asterisk can be

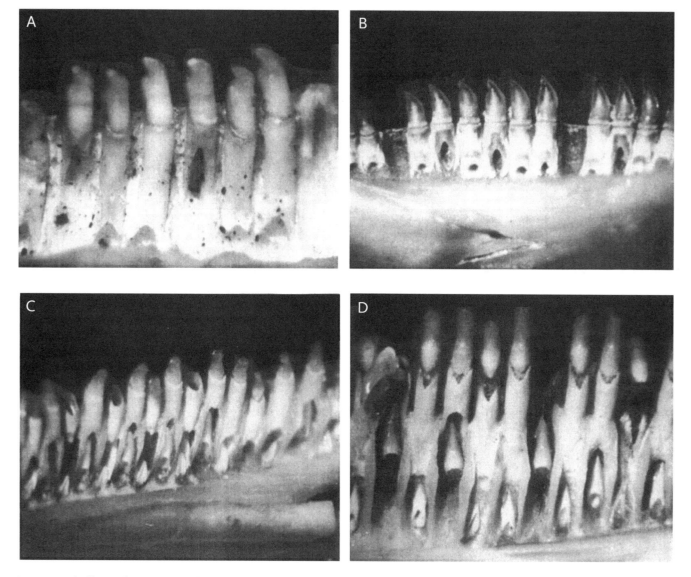

Figure 9.2. Pedicellate teeth, as represented by the three modern amphibian orders. *A,* The salamander *Amphiuma means. B,* The caecilian *Gymnopis mexi-* *canus. C,* The frog *Rana occipitalis. D,* The leptodactylid frog *Calyptocephalus gayi.* From Parsons and Williams, 1962.

recognized in fossils and so are important in making comparisons with Paleozoic species.

Those That Clearly and Uniquely Link the Three Modern Orders

1. Pedicellate teeth*
2. The operculum-plectrum complex*
3. The papilla amphibiorum
4. Green rods in the eye
5. Fat bodies
6. The structure of the skin glands
7. Cutaneous respiration and its attendant specializations

Characters That the Three Modern Orders Share But That Are Known to Have Been Acquired or Approached by Other Tetrapod Groups

1. The peculiar fenestration of the posterolateral skull roof*
2. The loss of the posterior series of skull bones*
3. The advanced type of palate*
4. The unusual presence of only 10 cranial nerves*
5. Two occipital condyles*
6. The characteristic atlas*

Characters That May Be Indicative of Some Special Relationships between the Three Modern Orders But That Show Enough Differences between the Orders to Be Ambiguous

1. Reduction of ossification and simplification of bone structure*
2. Pattern of the nasal organ
3. Simplification of brain structure

Parsons and Williams searched the paleontological literature looking for similar structural complexes in Carboniferous and Permian amphibians, but without success. They also provided an illustration of what the skull of a common ancestor might look like (Fig. 9.3), but no Paleozoic species matches this image. Parsons and Williams did not quote the work of Hennig in their papers, but they followed his precepts in concentrating on unique shared derived characters rather than general similarities in trying to establish the relationships of the lissamphibians.

Paleontologists continue to search for probable ancestors of each of the groups, but no consensus has been reached as to whether they had a single common ancestor (A. R. Milner, 1988, 1993), or if salamanders and caecilians were more closely related to one another than either are to frogs (R. L. Carroll et al., 1999), or if frogs and salamanders are more closely related to one another than either are to caecilians (R. L. Carroll, 2000, 2007; Schoch and Carroll, 2003). However, no one has suggested that frogs and caecilians are more closely related to one another than either are to salamanders.

With the development of PAUP, it became possible to analyze very large datasets, including all the major groups

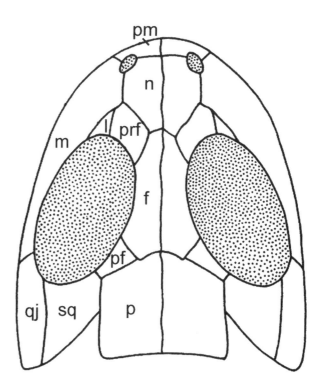

Figure 9.3. Diagrammatic representation of the skull of a putative lissamphibian ancestor from the Paleozoic as proposed by Parsons and Williams, 1963.

of fossil and living amphibians and large numbers of characters. The results, however, have been very different from one author to another, as was the case with the interrelationships of the Carboniferous amphibians. The cladograms of Laurin and Reisz (1997) and Ruta et al. (2003) both united the lissamphibians as a monophyletic assemblage, but identified very different Paleozoic groups as sister-taxa (Figs. 5.3 and 9.4). Ruta and his colleagues hypothesized an ancestry of all modern amphibians from temnospondyl labyrinthodonts, while Laurin and Reisz proposed the lysorophids, a group with a very long vertebral column and greatly reduced limbs (Fig. 9.5).

In view of the fact that frogs and salamanders were distinguished from each other as early as the time of the ancient Greeks, it is surprising that they appear as having diverged from a single node in these cladograms. It is as if they had no distinguishing characters. Even a three-year-old child should be able to differentiate frogs, salamanders, and caecilians. Why was this not possible using PAUP?

In fact, none of the many unique derived features that distinguish each of the three groups of living amphibians from one another were included in these databases. This is because such characters are not thought to be useful in determining sister-group relationships among terminal taxa. Not including derived characters unique to each of the modern groups would be justified if these characters had evolved *within* the crown groups, *after* their phylogenetic divergence from one another, in which case they could not be recognized among

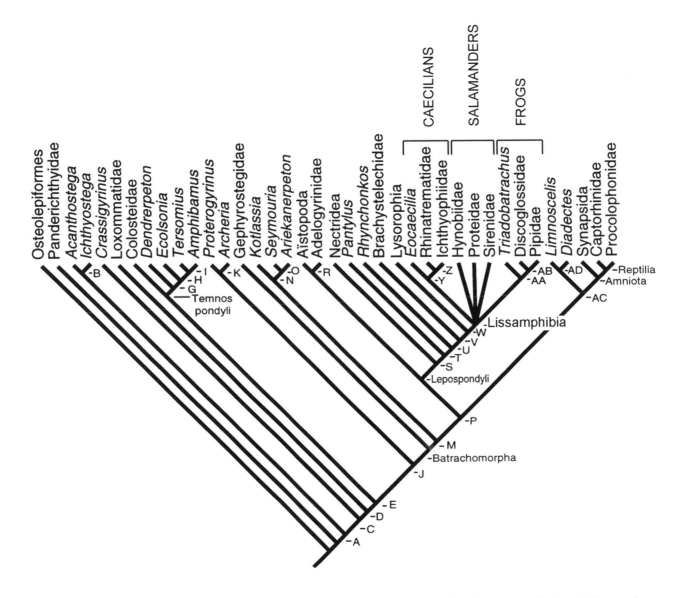

Figure 9.4. Cladogram resulting from a phylogenetic analysis of Laurin and Reisz (1997), showing the Lissamphibia as an unresolved polytomy. Their closest sister-taxon is the Paleozoic Lysorophia. See also Figure 5.3 for a comparable cladogram of Ruta et al. (2003).

any other amphibian lineages (Fig. 9.6A). On the other hand, failure to include distinctive derived features of each of the crown groups in investigations of their ancestry inherently biases the analysis against recognition of affinities with Paleozoic taxa that might share these derived character states (Fig. 9.6B). Even if it were true that frogs, salamanders, and caecilians had an immediate common ancestry from a single antecedent stock, there would still be many problems of determining how and when the key characters of the living orders had evolved and what was the nature of the immediate common ancestor. Neither of these questions is even considered in the analyses of Laurin and Reisz or Ruta and his colleagues.

In contrast with the precepts of Hennig (1966), primitive character states may also be useful in classification if they are unique to one or more of the groups under consideration. For example, primitive characters that are shared by frogs and salamanders but exist in a derived state in caecilians suggest that frogs and salamanders diverged from the lineage leading to caecilians prior to the emergence of these traits in the latter group. In summary, one must consider all characters that can be studied in modern species and interpreted from fossils in order to establish their patterns of evolution, probable relationships, and, where possible, the nature of their ultimate common ancestor(s), without initial bias as to whether these characters are primitive or derived, unique or shared. Clearly, this is an ongoing study as more information becomes available from both modern and fossil species. Much new information has been gained in the past five years, but gaps still remain in the earliest portion of the fossil record of frogs,

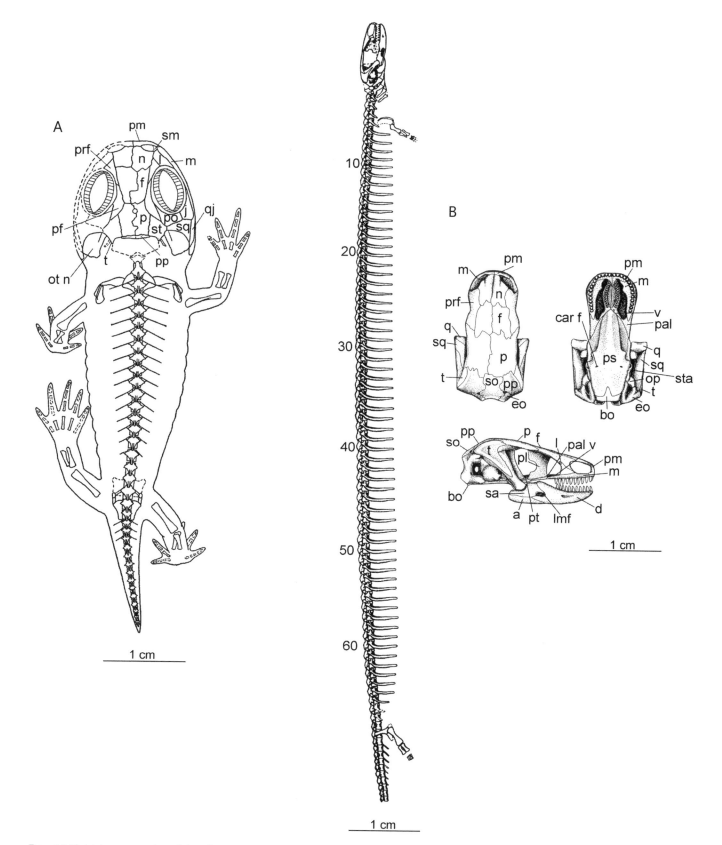

Figure 9.5. Skeletal reconstructions of the Paleozoic sister-taxa of lissamphibians proposed by Ruta et al. (2003) and Laurin and Reisz (1997). *A*, A temnospondyl labyrinthodont. *B*, A lysorophid lepospondyl.

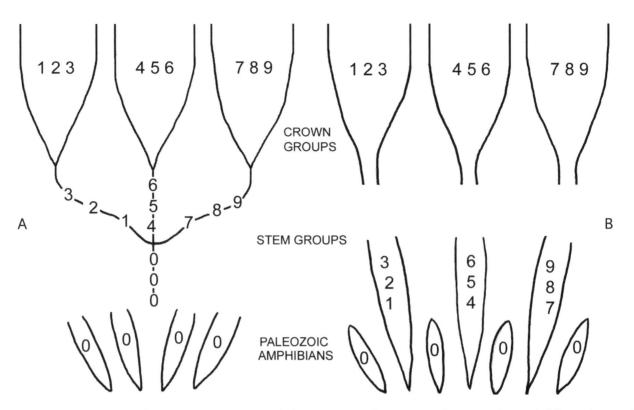

Figure 9.6. Use of distinctive features of crown groups. *A,* Situation in which unique derived characters (1–9) evolved *within* crown groups, or their immediate stem taxa, and so are of no value in establishing affinities among the crown groups or with putative antecedents. *B,* Situation in which distinctive features of crown groups had evolved among antecedent taxa and so can be recognized as synapomorphies uniting them with different plesiomorphic lineages. Numbers within crown groups indicate traits observed in all members. Sequences of numbers within stem taxa imply temporal succession of acquisition of traits. 0 = absence of derived traits.

salamanders, and caecilians that preclude a definitive answer to the current questions.

METHODOLOGY

If the prior applications of PAUP have failed to resolve the specific patterns of interrelationship among the modern amphibian orders or to find their plausible ancestry among one or more Paleozoic lineages, what other means of analysis might be followed? We can begin our study of the ancestry of modern amphibians by considering the wide range of characters common to each of the living orders, including their general way of life, body form, and skeletal and soft anatomy. Next, we can examine the skeletons of the oldest known fossils of each order and attempt to reconstruct particular aspects of their behavior and soft anatomy. These animals provide the best-known evidence of skeletal structures that distinguished frogs, salamanders, and caecilians from each other, but few that support a common ancestry. Lower and Middle Jurassic fossils of both frogs and salamanders can be placed close to or within modern families. No earlier fossils are known that are unquestionably members of the crown groups.

With this information in hand, we can search for evidence of comparable traits in the various groups of Triassic and Paleozoic amphibians discussed in the preceding chapters. Finally, we can integrate this information into a reconstruction of the sequence of changes that presumably occurred between the relevant Paleozoic genera and their descendants in the Jurassic. Since there is a very extensive fossil record of many groups of Carboniferous and Permian amphibians, the search for relationships among the antecedents of modern amphibians is much more hopeful than it was for establishing patterns of relationships among the early Carboniferous tetrapod clades (discussed in Chapter 5) for which the earlier fossil record is so very incomplete.

THE ANATOMY AND WAYS OF LIFE OF THE MODERN AMPHIBIAN ORDERS
Distinctive Features of Lissamphibians

Lissamphibians are unquestionably distinct relative to all other modern tetrapods. Among their most significant features are those associated with reproduction. None have the extra-embryonic membranes present in amniotes that provide protection and an effective means of exchange of respiratory gases and nutrients. Most have an aquatic larval stage with external gills but initially lack limbs and undergo some degree of metamorphosis into a terrestrial adult. All probably had the capacity for gape-and-suck feeding at some stage in their life. These are clearly primitive traits that indicate retention of amphibious features common to basal tetrapods that were lost in the lineage leading to amniotes. However, these primitive characteristics provide no evidence for establishing specific relationships, except to indicate that they did not share close affinities with ancestral amniotes.

Another distinctive feature of all modern amphibians is their reliance on cutaneous respiration. This almost certainly evolved subsequent to their divergence from the most primitive tetrapods, which were very large and so had too small a surface to volume ratio for this means of gas exchange to be practical. A further problem among early tetrapods was their extensive covering of overlapping scales that would have limited gas exchange. We would expect to find the ancestry of lissamphibians from among relatively small predecessors. However, since many later Paleozoic amphibians were sufficiently small that cutaneous respiration would have been practical, and some had lost most or all of their scales, this practice may have arisen many times. The necessity for cutaneous respiration in all modern amphibians also requires that they remain in fairly damp habitats, and this was probably the case for their Paleozoic ancestors.

Frogs and salamanders are distinguished from all amniotes by the presence of a second ear ossicle in addition to the stapes, the operculum, which is associated with detection of ground-borne vibrations. A possibly comparable structure is also present in the earliest known fossil caecilian, but its homology is difficult to establish. Members of all three orders have mentomeckelian bones, paired elements at the symphysis of the lower jaws that ossify from Meckel's cartilage. Pedicellate teeth are known as far back as the Lower Jurassic in frogs and caecilians and are present in most modern salamanders, but they have recently been discovered in more primitive amphibians from the late Paleozoic.

Lissamphibians have long been thought to have a unique sensory structure in the inner ear, the papilla amphibiorum. However, this has recently been demonstrated as developing during the ontogeny of caecilians by division of the macula neglecta (a second sensory tissue occurring in other groups of vertebrates) rather than being a distinct tissue, limited to lissamphibians (Fritzsch and Wake, 1988). The presence of only 10 cranial nerves has been cited as a common feature, but primitive salamanders retain 12.

All three modern orders have paired occipital condyles that articulate with widely separated surfaces of the atlas, in contrast with the more nearly circular occipital condyles of most Paleozoic temnospondyls and early amniotes. As adults, all living lissamphibians have unipartite vertebrae, with the neural arches firmly fused to cylindrical pleurocentra, but lack trunk intercentra. Primitively, the centra are amphicoelous, that is, both ends are deeply concave to accommodate persistent notochordal tissue. These characteristics give them a general resemblance to the vertebrae of lepospondyls rather than those of labyrinthodonts, with their multipartite centra and loosely attached neural arches.

DISTINCTIVE FEATURES OF FROGS, SALAMANDERS, AND CAECILIANS

In addition to the characters that are found in all lissamphibian orders, there are many more that distinguish frogs, salamanders, and caecilians from each other. These features do not necessarily preclude their having had an ultimate common ancestor, but they do indicate very different ways of life and the independent origin of a host of unique anatomical, reproductive, developmental, and behavioral attributes, at least some of which might find their origin in different groups of Paleozoic antecedents.

If you look at either the general body forms or the skeletons of frogs, salamanders, and caecilians, you are immediately struck by how different they are from one another (Fig. 9.7). Frogs and caecilians are the most divergent, in both cases associated with the manner of locomotion. Almost every aspect of the postcranial skeleton of frogs is associated with jumping, as is nearly all of the musculature. Caecilians, in contrast, have modified their bodies entirely for burrowing, from their extremely compact skull to the great elongation of the trunk, with up to 285 vertebrae. In living families, the limbs are totally absent, without even a trace of embryonic precursors (D. B. Wake, 2003). The frog tadpole is also drastically different from the larvae of salamanders and caecilians, as is the rate and magnitude of anuran metamorphosis. No traces of these specializations are seen in salamanders or caecilians.

In contrast with frogs and caecilians, most salamanders are conservative in their body form and limb proportions. Most retain features in common with many Paleozoic genera that enable locomotion in both aquatic and terrestrial environments. On the other hand, the modern salamander families are much more divergent from one another than those of frogs or caecilians in their overall appearance and ways of life. Salamanders, like caecilians, lack the impedance matching middle ears of frogs. Of particular interest, several specific aspects of development distinguish salamanders from all other groups of terrestrial vertebrates.

Even the oldest known members of the modern amphibian orders, from the Lower or Middle Jurassic, are very distinct from one another in both anatomy and their presumed ways of life, but clearly resemble their living descendants in most features of their skeletons. Only Lower Jurassic caecilians are noticeably more primitive in having a much shorter vertebral column and retaining rudiments of the limbs. We will now look in more detail at the anatomy of each of the modern orders in an effort to find characteristics that may help us recognize their specific ancestry among the diverse groups of Paleozoic amphibians. Emphasis will be placed on structural and functional complexes rather than on the attributes of individual bones as was the case of the analyses of Laurin and Reisz and Ruta and his colleagues.

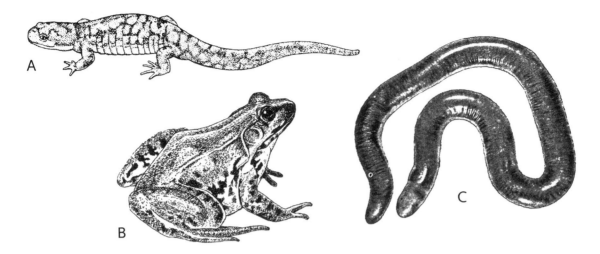

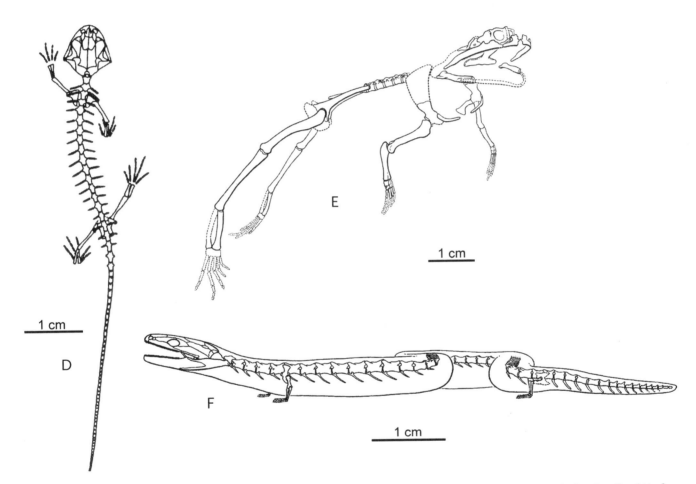

Figure 9.7. Modern frogs, salamanders, and caecilians and skeletons of their early Mesozoic antecedents. *A,* The salamander *Ambystoma. B,* The frog *Rana. C,* A caecilian. *D,* The Lower Cretaceous salamander *Valdotriton,* from Evans and Milner, 1996. *E,* The Lower Jurassic frog *Prosalirus bitis,* from Shubin and Jenkins, 1995. *F,* The Lower Jurassic caecilian *Eocaecilia micropodia,* from Jenkins et al., 2007.

Frogs

SKULL AND CRANIAL SENSE ORGANS. The skulls of frogs are immediately distinguished by their highly fenestrate nature, with very large openings dorsally for the eyes and extremely large interpterygoid vacuities in the palatal surface (Fig. 9.8). It has been speculated that the large size of these openings is related to feeding, since it has long been observed that frogs systematically pull their eyes ventrally, toward their palate, during the process of swallowing prey. This use of the eyes was recently documented by Robert Levine and his colleagues (2004), who compared the time required for swallowing by frogs in which the nerve supply-

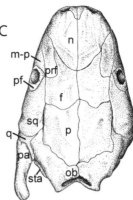
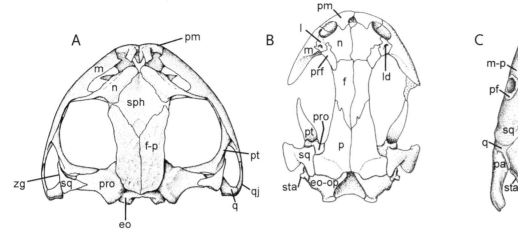
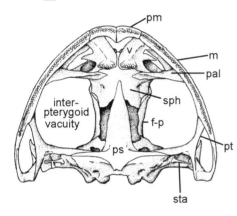
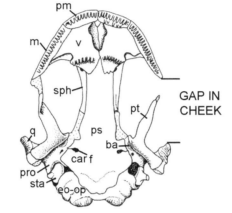
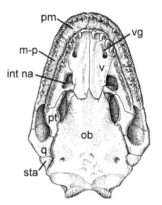
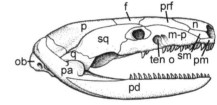
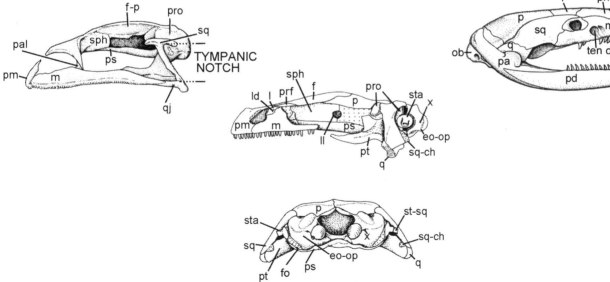

A

1 mm

1 mm

1 mm

inter-
pterygoid
vacuity

GAP IN
CHEEK

TYMPANIC
NOTCH

Figure 9.8. Skulls of modern amphibian orders in various views. *A,* Dorsal, palatal, and lateral views of the hylid frog *Gastrotheca walkerii*. From Du-ellman and Trueb, 1986. *B,* Dorsal, palatal, and lateral views of the hynobiid *Batrachuperus sinensis* and occipital view of *Hynobius naevius*. From R. L. Carroll and Holmes, 1980. *C,* Dorsal, palatal, and lateral views of the conservative caecilian *Ichthyophis glutinosus*. From Jenkins et al., 2007.

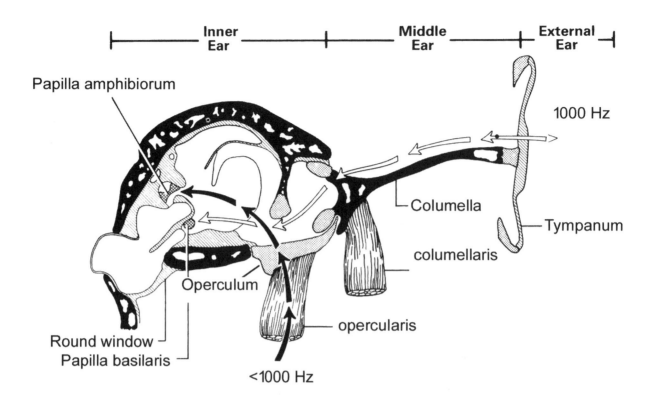

Figure 9.9. Diagram of an anuran auditory system. Solid arrows indicate transmission pathway of frequencies of less than 1000 Hertz via the opercular-papilla amphibiorum complex; open arrows indicate transmission pathway of frequencies of more than 1000 Hertz via the tympanic-columellar-papilla basilaris complex. From Duellman and Trueb, 1986.

ing the retractor bulbi muscle, which attaches to the back of the eye, was severed versus that of frogs in which this nerve was intact. While the frogs could still swallow, the number of swallows increased by 74% in those that could not retract their eyes.

Uniquely, the frontal and parietal bones of frogs are fused, forming paired frontoparietals. These bones provide a longitudinal supporting axis of the skull, beneath which rests the braincase. Frogs have lost all of the circumorbital bones of primitive tetrapods and those that made up the back of the skull table. In contrast with salamanders, the quadratojugal remains as a link between the maxilla and the quadrate. The otic capsule forms a lateral strut connecting the cheek region with the braincase.

Nearly all frogs have a complex middle ear structure including a tympanum that is attached to a cartilaginous tympanic annulus supported by the squamosal bone at the back of the cheek region (Fig. 9.9). The large size of the tympanum provides an extensive surface for reception of airborne vibrations. These are mechanically amplified and transmitted to the inner ear by the stapes. No other living amphibian group has such an ear structure. In contrast with amniotes, in which the footplate of the stapes acts as a piston moving at right angles to the membrane separating the middle and inner ear, that of frogs hinges with the ventral margin of the fenestra ovalis and compresses the fluid in the inner ear with

a rocking motion. The tympanic membrane is much larger than the fenestra ovalis and this difference results in matching the acoustic impedance of air to the higher impedance of the fluid in the inner ear (Duellman and Trueb, 1986).

The inner ear of frogs is specialized to respond to high-frequency vibrations that may approach 5000 Hertz, but are more commonly below 4000 Hz. The normal range for humans is between 800 and 2500 Hz. Hearing in frogs appears to serve a much more limited function than that of mammals. However, most studies have concentrated on its use for recognition and attraction of members of the opposite sex within the same species. Each frog species has a highly specific call, in terms of frequency, pattern, and duration. The sensory structures of the inner ear, the amphibian and basilar papilla, are configured so that they respond preferentially to the calls of the same species. Hence, even if several different species occupying a single area are calling at the same time, relatively few incorrect matings occur. This enables many species to cohabit a particular region and breed around a single pond. The capacity for frogs to produce a great diversity of calls and so isolate adjacent species may be one of the reasons for the great proliferation of species, exceeding 5000, in comparison with the few hundred species of salamanders and caecilians.

In common with salamanders, frogs also have a second ear ossicle, the operculum, an oval disc that fits into the fenestra

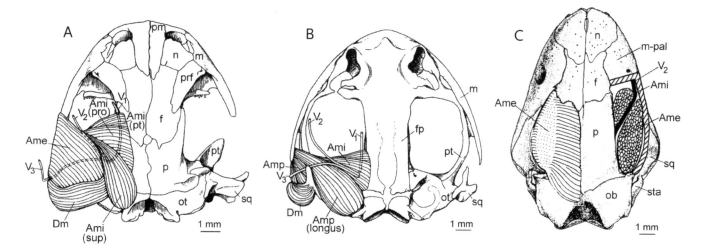

Figure 9.10. Dorsal view of the skulls of primitive members of each of the modern amphibian orders showing the distribution of the major adductor jaw muscles and their associated nerves. *A,* The salamander *Ambystoma macu-* *latus. B,* The very primitive frog *Ascaphus truei. C,* The caecilian *Epicrionops petersi.* From Jenkins et al., 2007.

ovalis adjacent to the footplate of the stapes. It is connected to the shoulder girdle by the opercularis muscle. The operculum serves to pick up lower-frequency vibrations—less than 1000 Hertz, and typically in the range of 200 to 300 Hertz. These too can be used for mate recognition and for sensing vibrations transmitted through the ground. Both systems may also serve for recognition of predators or prey species, but this has been little studied.

The presence of a large middle ear cavity at the back of the cheek considerably restricts the area of the jaw musculature in frogs, which may account for the different distribution of its components relative to those in salamanders and caecilians. Three major elements of the musculature that closes the mouth are recognized in all terrestrial vertebrates by their position relative to branches of the Vth, or trigeminal, nerve (Fig. 9.10). The adductor mandibulae internus lies lateral to branch V_1 and rostral to branch V_2, the adductor mandibulae externus lies between V_2 and V_3, and the adductor mandibulae posterior is behind V_3.

Different portions of the adductor mandibulae extend over the back of the braincase in the three groups of living amphibians: the longus branch of the adductor mandibulae posterior in frogs, the superficial head of the adductor mandibular internus in salamanders, and the adductor mandibulae externus in caecilians. Spread of these muscles over the braincase could only occur after the dermal bones of the back of the skull roof were lost. The posterior extension of different portions of the adductor musculature in frogs, salamanders, and caecilians may be interpreted as indicating that this occurred independently in each of these groups (R. L. Carroll and Holmes, 1980).

The lower jaw has lost its dentition in all frogs and both the upper and lower teeth were lost in the toad family Bufonidae. The anterior portion of the Meckelian cartilage in the lower jaw is ossified to form the mentomeckelian bones, which are associated with tongue protrusion.

STRUCTURE OF HYOID AND TONGUE PROTRUSION. The structures associated with feeding and respiration are unique for both the larvae and adults of frogs compared with those of all other tetrapods (Figs. 4.12 and 9.11). Most salamanders and caecilians maintain features similar to those of their fish ancestors in both the structure and the function of the jaws and hyoid apparatus in the aquatic larvae, and to a lesser extent in the adults. As you may recall from Chapter 4, the hyoid apparatus of the earliest known amphibians was composed of a series of elongate, paired elements, the ceratobranchials, attached to a series of medial structures, the basibranchials. Feeding and respiration are achieved by movements of the medial elements that occur in the sagittal plane (Fig. 9.15). Primitively, adults of salamanders and caecilians can feed effectively in the water as well as on land.

Among anurans, only the divergent pipids routinely feed in the water as adults; all other frogs feed on land. The structure and function of the apparatus associated with aquatic feeding in the larvae and terrestrial feeding of the adults have both diverged radically in frogs. Looking first at the adults, the hyoid apparatus is much simpler than that of salamanders (Fig. 9.11E). Instead of a multipartite structure composed of rod-shaped bones inherited from their fish ancestors, the hyoid consists of a large, primarily cartilaginous plate with three bony elements. In the primitive living frog *Leiopelma* and the Lower Jurassic *Vierella,* there is a Y-shaped parahyoid bone near the middle of the cartilaginous plate. From the posterior margin of the plate of all frogs, there extend two paired rods, the posteromedial processes, which are directed posteriodorsally. These appear at metamorphosis and are not homologous with any elements present in the tadpole, or in salamanders or caecilians. Their function is to support the larynx, which is responsible for vocalization. Frogs are the only modern amphibians that use vocalization for species recognition and attraction of mates.

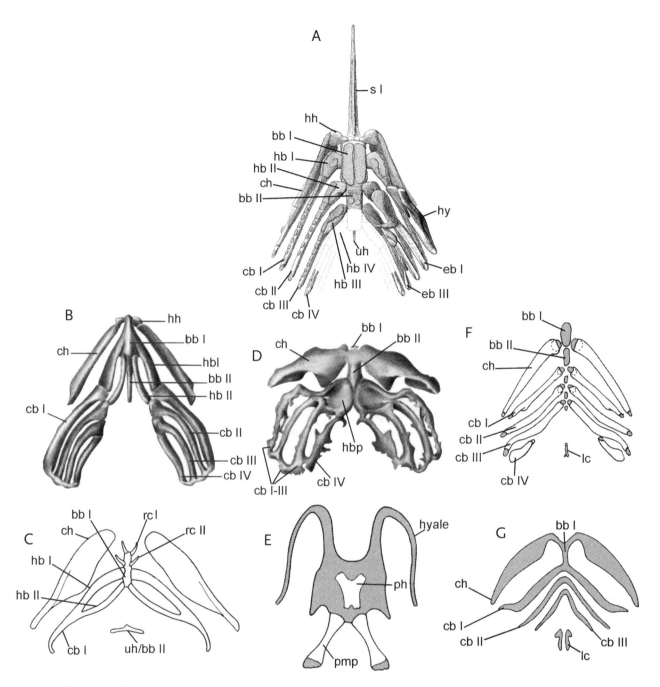

Figure 9.11. Hyoid apparatus of *Eusthenopteron*, frogs, salamanders, and caecilians. *A,* Dorsal view of the hyoid apparatus of *Eusthenopteron. B* and *C,* Ventral views of hyobranchial apparatus of larval and adult *Salamandra salamandra. D,* Ventral view of hyobranchial apparatus of larval *Rana temporaria.*

E, Hyoid plate of adult *Leiopelma hochstetteri,* one of the most primitive living frogs. *F* and *G,* Hyobranchial apparatus of larval and adult individuals of the primitive caecilian *Epicrionops.* From R. L. Carroll, 2007.

Both frogs and salamanders have protrusable tongues, but their structure and function are so different as to indicate separate origins (Fig. 9.12). The tongue of salamanders is closely associated with the rod-shaped elements of the larvae hyoid. But that of frogs is attached to the dorsal surface of the plate-like hyoid, and has no internal bony elements.

The frog tongue consists of two major muscles. The more superficial is the genioglossus medialis, whose anterior end is attached at the symphysis of the lower jaws. The more posterior and underlying hyoglossus is so arranged that their integrated movements flip the tongue out of the mouth with the initially dorsal surface becoming ventral in position. Flipping of the tongue is augmented by another muscle, the submentalis, unique to frogs, located close to the symphysis of the lower jaw. This muscle is attached to the small mentomeckelian bones that are movable relative to the dentary.

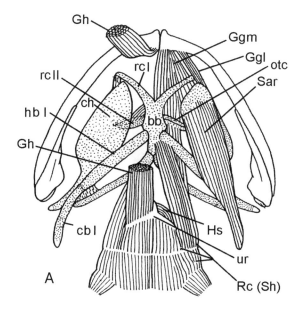

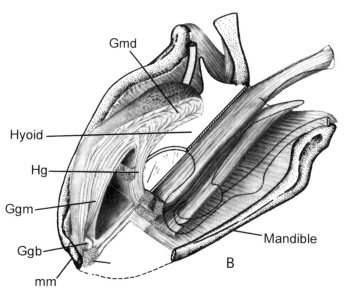

Figure 9.12. Tongue protruding apparatus of a primitive salamander (A) and a typical frog (B). The tongue protruding apparatus of salamanders involves the close integration of bony elements of the hyoid, but that of frogs is primarily muscular. From R. L. Carroll, 2007.

Their anterior surface is rotated ventrally so as to pull down on the anterior ventral extremity of the genioglossus (Gans and Gorniak, 1982).

The hyoglossus is relaxed as the tongue is protracted, but its contraction pulls the tongue and attached prey back into the mouth. As in salamanders, the tongue exudes a sticky substance that adheres to prey. The tongue of the very primitive living frog, *Ascaphus,* has the same basic structure and function as that of advanced bufonids and ranids, although it cannot be extended nearly as far (Peters and Nishikawa, 1999).

In addition to supporting the tongue, the hyoid of frogs is also associated with buccal respiration, during which it is moved dorsally and ventrally by muscles running between the lower jaws. This is anatomically and functionally different than in salamanders and caecilians, in which the movement of the hyoid apparatus is primarily in an anteroposterior direction.

LOCOMOTION. Almost every aspect of the postcranial skeleton of adult frogs is associated with their jumping (Figs. 9.7 and 9.13). The vertebral column is very short and functions as a rigid rod with no more than eight presacral vertebrae in living species; in some this number is reduced to five. In the adults the postsacral vertebrae fuse to form a rigid urostyle, but there is no flexible tail. No salamanders have lost their long, flexible tails, but this does occur in the very elongate caecilians. In contrast with all other amphibians, the rod-like ilium of frogs extends far forward from the acetabulum. The ischium is reduced to a small, semicircular element and the ventrally projecting pubis remains unossified. The femur is much elongated relative to that of primitive amphibians and the head is expanded and spherical to fit into the hemispherical acetabulum. The tibia and fibula are fused, and the tibiale and fibulare are elongated in the manner of the more proximal limb bones. The other tarsals are variably reduced or fused to one another, but there remain five digits with a phalangeal count generally of 2, 2, 3, 4, 3, or occasionally less.

The forelimb and girdle are also much modified in frogs, in a manner very distinct from that in salamanders. The dermal cleithrum and clavicle are retained, but the ventral portion of the girdle is much elaborated to form a flexible structure capable of absorbing the weight of the animal when it lands, like the shock absorbers of an automobile. The humerus is elongate, although not to the extent of the femur, and the ulna and radius are fused into a single unit. In contrast with the rear limb, the proximal carpals are not modified. All of these features of the modern anuran adult were in place by the Lower Jurassic.

All living frogs and their ancestors going back to the Lower Jurassic have a unique saltatory mechanism characterized by a vertical hinge joint between the sacral vertebra and the urostyle. The elongate, anteriorly directed ilium is securely attached to the urostyle via the cocygeo-illiacus muscles, and a rotational joint is present between the sacral diapophyses and the more ventrally situated iliac blade (Fig. 9.13). Extension of the trunk during jumping occurs through the contraction of the longissimus dorsi and coccygeo-sacralis muscles, accompanied by the thrust of the rear limbs (Jenkins and Shubin, 1998). Movement involves rotation at the ilio-sacral joints and extension at the sacro-urostylic joint. This system has no parallel in salamanders or the skeletons of any Paleozoic tetrapods. The rear limbs are always used in a symmetrical manner. Oddly, frogs retract their eyes when they jump, just as they do when swal-

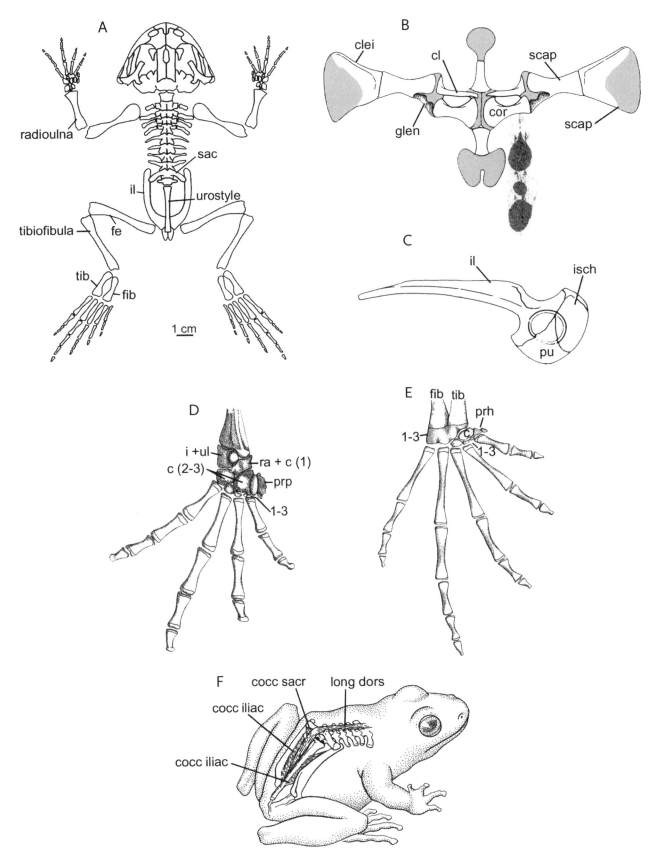

Figure 9.13. Skeleton of a frog and its mechanism for jumping. *A,* The Upper Jurassic frog *Notobatrachus,* which has already evolved the basic anatomy of modern frogs. From Sanchiz, 1998. *B* and *C,* Dorsal view of the pectoral girdle and lateral view of the left pelvis of a modern frog. From Duellman and Trueb, 1986. *D* and *E,* Manus and pes of the primitive living frog *Ascaphus truei.* Modified from Ritland, 1955. *F.* Cutaway view of the jumping apparatus of a frog prior to jumping. From Jenkins and Shubin, 1998.

lowing prey. It has been suggested that this reflects their most common direction of jumping—into the water to escape enemies—which does not require specific visual information.

LARVAE. The most conspicuous feature of adult frogs is their capacity for jumping, but even more spectacular changes have evolved in the nature of their larvae. With the exception of a few modern genera, all frogs have a highly specialized tadpole (Fig. 9.14). Although superficially resembling the early stages of salamander larvae, the tadpole is highly specialized in the late appearance of limbs and in having a mode of feeding that is exceptional among all vertebrates. In contrast with all other amphibian groups, nearly all tadpoles are herbivorous. This is associated with a unique jaw apparatus, a means of suspension feeding on tiny particles of food, and a complex pumping system.

A larval stage must have evolved during the transition from fish to land vertebrates, with the hatchlings initially lacking either paired fins or limbs. This was presumably followed by the progressive development of appendages capable of support and locomotion on land. Fossils of aquatic juveniles of labyrinthodonts are known from the end of the Lower Carboniferous and are very common in the Late Carboniferous and Early Permian of Europe. The larvae of branchiosaurids, described in Chapter 6, have many similarities with those of modern salamanders, as will be discussed in Chapter 11, and presumably fed on small animals that lived in large ponds and lakes (Boy and Sues, 2000). In contrast, modern frog larvae are most common in small, typically ephemeral bodies of water. This has the advantage of escape from predation by large fish or neotenic amphibians that depend on a permanently aquatic environment. However, such temporary ponds are likely to be deficient in animal prey that are obligatorily aquatic.

The ancestors of tadpoles made the great evolutionary step of achieving the ability to feed primarily on plant matter, a capacity not gained by any other amphibians, fossil or living. As primary producers, plants generate a much greater biomass than animals and so tadpoles may be much more numerous than salamander larvae in bodies of water where they live together. The invention of the tadpole larvae may also have contributed to the enormous diversity of frogs, compared with that of salamanders and caecilians.

On the other hand, feeding on plant material required a great many changes in the functional anatomy of the head and pharynx as well as the mode of digestion in the larvae. This was comparable to evolving an entirely new type of animal in which major modifications also occurred in the jaw musculature and its innervation, respiration, circulation of water, and entrapment of food particles. It also requires a drastic metamorphosis from the larvae to the adult that is nearly as great as that between a caterpillar and an adult butterfly. The evolution from a carnivorous to a herbivorous diet has occurred numerous times among different lineages of amniotes, including lizards, turtles, dinosaurs, and many distinct groups of birds and mammals, but without the drastic

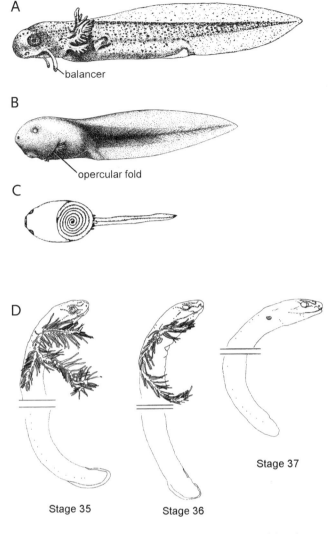

Figure 9.14. Larvae of modern amphibian orders. A, Stage 25 of the salamander Ambystoma maculatum. Modified from Harrison, 1969. Note conspicuous external gills and the balancers, extending from the back of the lower jaws. The balancers prevent early larvae from sinking into the sediment and help to maintain balance until the forelimbs develop. They are known in pond-dwelling hynobiids, salamandrids, and ambystomatids, but in no other modern amphibians. B, Stage 24 of the frog Rana pipiens. Modified from Shumway, 1940. The opercular fold is growing back over the external gills, which are completely covered in later larvae. C, Ventral view of the frog Scaphiopus multiplicatus, showing the highly coiled digestive tract where plant matter can be broken down by symbiotic microorganisms. Modified from Pfennig, 1992. D, The primitive caecilian Ichthyophis kohtaoensis. Two late embryonic stages showing external gills and a hatchling, in which the external gills have been lost. Modified from M. H. Wake and Dickie, 1998.

modifications of the body form that occurred in the ancestors of frogs.

We may discuss the changes in feeding mechanisms between primitive carnivorous larvae and herbivorous tadpoles in terms of the sequence of food processing. This involves its acquisition, breakdown into small particles, filtering, transport through the digestive system, and the process of digestion. In the fish ancestors of tetrapods, the larvae of early land vertebrates, and the probable ancestors of salamanders, prey was captured by gape-and-suck feeding. The jaws were

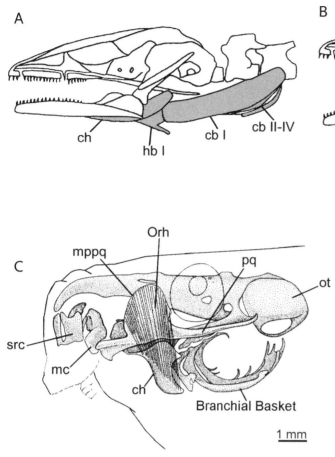

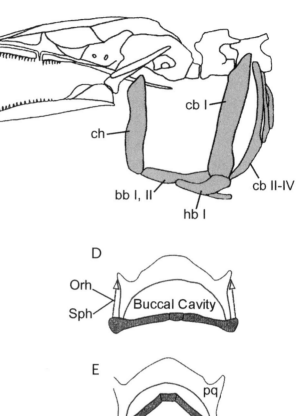

Figure 9.15. Comparison of the pumping apparatus of the larvae of modern salamanders and frogs. *A–B,* Closed and open positions of the hyoid apparatus during gape-and-suck feeding in salamanders. The ceratohyal and ceratobranchial change from an essentially horizontal to a vertical orientation, while the basibranchials remain horizontal but occupy a much lower position. These movements are associated with extensive expansion of the oropharyngeal chamber, which creates a vacuum and draws in water and food through the open mouth. *C,* Lateral view of the braincase and hyoid apparatus of the tadpole of the discoglossid frog *Alytes.* The orbitohyoideus muscle (homologous with the depressor mandibulae of the adult) and the subhyoideus pull up on the lateral lever arm of the ceratohyal, which lowers the medial portion of the bone and expands the buccal cavity. This pumping mechanism is at essentially right angles to that of salamanders, and does not involve the ventral and posterior elements of the branchial basket. *D* and *E,* Posterior views of the palatoquadrate and ceratohyals in tadpoles, showing the position of the muscles that expand and contract the buccal cavity. *A* and *B* from Deban and Wake, 2000; *C,* modified from Sanderson and Kupferberg, 1999; and *D* and *E* from Cannatella, 1999.

opened and the mouth and pharynx expanded by posteroventral movements of the hyoid apparatus (Fig. 9.15*A, B*). Large prey could be held by the jaws and the smaller food items trapped by the gill rakers as the water was expelled through the open gill slits.

In their ancestors, the gill chamber was covered by a bony operculum, but this was lost during the origin of tetrapods. External gills were present in larval temnospondyls and discosauriscids, indicating that there was not a fleshy covering of the gill chambers, as is the case for tadpoles. Larval temnospondyls and modern salamander larvae can respire continuously via the external gills without use of the hyoid pump, which operates only intermittently in relationship to feeding or taking in of atmospheric oxygen at the surface of the water.

In contrast, the ancestors of tadpoles must have modified all the structures of the mouth, pharynx, and digestive tract. While plant material is a great deal more plentiful than animal matter in aquatic environments, it is either thinly distributed in the water column, or represented by large, frequently attached structures that must be detached and broken down into small particles prior to digestion. The evolution of more effective means of filtering and trapping food particles that could not have been captured by the gill rakers of the more primitive larvae may have been an early stage in the origin of tadpole feeding. This could have occurred in larvae depending on filtering commonly occurring small plants such as algae and plant matter suspended in the water, prior to the evolution of the specialized jaws and pumping system of all modern frogs.

In fact, the structures used in the entrapment of small food particles in modern tadpoles may have been present much earlier in their evolution, since they have no bony elements and thus could not have been fossilized. In modern frogs, large particles are trapped by sieving on papillae in the buccal cavity and on ruffled epithelium within the branchial basket (Fig. 9.16). Smaller particles adhere to mucus secreted on the roof of the buccal cavity and from the dorsal and ventral velum, from which they are transported to the gut by cilia (Sanderson and Kupferberg, 1999). On the other hand, the denticles or denticular plates attached to the ceratobranchials in branchiosaurids and salamanders are completely absent in tadpoles.

Another change in the capacity for retention of small food particles can be seen in the reappearance of an operculum. The elaboration of external gills in Paleozoic branchiosaurids (Chapter 6) as well as larval salamanders precludes the presence of an extensive operculum, which would have made it more difficult for them to retain small prey or particles of organic matter suspended in the water. In order for the ancestors of tadpoles to make effective use of particular plant material, they would have had to re-evolve an operculum. This should not have been difficult, since the operculum begins to develop in salamander larvae, but rapidly regresses to allow the external gills to emerge.

Following the development of the operculum in tadpoles, the more distal, filamentous portion of the external gills degenerates, leaving a proximal portion referred to as the gill filters. These are associated with filter plates that arise from the ceratobranchials (Fig. 9.16B). The external gills of their ancestors have apparently been sacrificed in the formation of the operculum and the reorganization of the pharyngeal region. There is considerable debate as to how absence of the external gills has affected uptake of oxygen and the discharge of carbon dioxide in tadpoles. Ultsch et al. (1999, 196) state: "Probably the only universally important gas exchange organ among tadpoles is the integument." Functional lungs are also known in most tadpoles.

However, no matter how effective the filtering of small food particles is in association with periodic gape-and-suck feeding, this means of ingestion would only be sufficient for small, very young tadpoles. This is because very little of the available water is taken in during the short periods of pumping common to salamanders and assumed for known Paleozoic larvae. In order to gain sufficient nutrients, rapidly growing tadpoles had to develop a pumping system that operated continuously and so could process much greater quantities of suspended food matter.

This was achieved by the evolution of an entirely new pumping apparatus. To understand how this may have occurred, we need to go back to the fish ancestors of tetrapods. The hyoid apparatus and associated structures of the larvae of primitive living salamanders are sufficiently similar to those of the sarcopterygian ancestors of early land vertebrates and branchiosaurs for it to be used as a plausible model for the pattern in the antecedents of frogs. Beginning with *Eusthenopteron* (Fig.9.11A), the hyoid apparatus consisted of a median longitudinal series of basibranchials, from which extended a series of posterolaterally directed hyobranchials and ceratobranchials bearing denticulate bony plates. These elements served as a pump, powered by longitudinally arranged muscles that pulled the medial elements posteroventrally (Fig. 9.15A, B). This greatly expanded the volume of the oropharyngeal chamber, creating a vacuum that drew in water and suspended prey. When other muscles, running transversely between the lower jaws and below the throat, raised the central axis of the hyoid, the water was forced out through the gill slits between the ceratobranchials and across the gills that were attached at the end of the ceratobranchials. Moderate-sized prey was held by denticulate plates attached to the ceratobranchials, but smaller prey or particles of organic matter passed into the chamber beneath the bony operculum and were periodically swept away when the operculum opened. In modern salamanders and presumably their Paleozoic antecedents, this pumping system was used only intermittently, for feeding or taking up oxygen from the air.

The evolution of a new pumping system in ancestral frogs remains difficult to explain. Already in branchiosaurs and primitive living salamanders, some degree of structural separation had occurred between the anterior ceratohyals and the more posterior ceratobranchials. This is carried further in tadpoles, which fuse the intervening hyobranchials into a plate, attached to the branchial basket (Fig. 9.11B, D). Rather than having an integrated pump composed of the entire hyoid apparatus, the anterior and posterior portions function as two successive pumps. The nature of the branchial basket itself is significantly altered in tadpoles by the integration of the originally separate and articulating ceratobranchials into a flexible, basketlike structure, without the mechanical independence of its original elements. The branchial basket is contracted and expanded by muscles that attach to different areas of the basket, while the ceratohyals serve as the basis of an entirely new functional complex.

Early in the ancestry of frogs, the ceratohyals were reduced in anteroposterior extent, attached to the basibranchials, and formed an articulation with the palatoquadrate posterior to the area of articulation with Meckel's cartilage (Fig. 9.17B). Within the history of frogs, the ceratobranchials moved progressively more anterior in position relative to the back of the skull. The orientation of the articulating surface between the ceratohyals and the palatoquadrates is longitudinal, so that elevation and depression of the ceratohyals is at right angles to the longitudinal axis of the skull. This approaches a 90 degree angle with the action of the hyoid pump in the fish ancestors of tetrapods and Paleozoic labyrinthodonts (Fig. 9.15). Ventral excursion of the new hyoid pump, which expands the oropharyngeal chamber, is powered by a large muscle termed the orbitohyoideus, whose origin can be traced to the primarily jaw-opening muscle in adult frogs and salamanders, the depressor mandibulae. The more ventral interhyoideus muscle, running between the lower jaws, serves to reduce the volume of the mouth, as it does in salamanders (Fig. 9.15E).

The ceratohyal pump generates a continuous, posteriorly directed flow of water. Closure of the mouth and flaps on

A

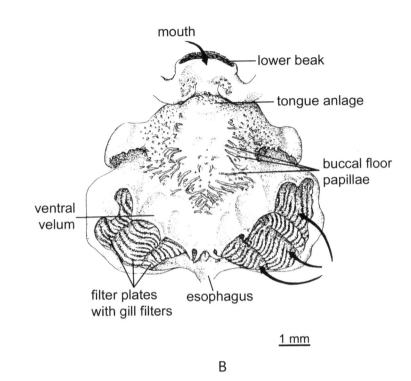

B

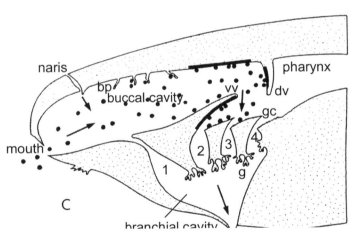

C

Figure 9.16. Comparison of the filtering apparatus of frogs and salamanders. *A,* Ventral view of the tissue between the lower jaws, the area of the cerato-branchials and attached gill rakers, and the external gills in the salamander *Ambystoma.* Food is trapped by the facing rows of gill rakers. Interdigitation of the gill rakers closes the external gill openings in order to maintain a vacuum in the oropharyngeal chamber. The gills are external, and serve to exchange respiratory gases with the water flowing across their extensive surface. *B,* Floor of the mouth of the discoglossid frog *Alytes,* showing the surfaces involved with trapping food particles, the position of the ventral velum, and the position and nature of the gill filters and filter plates that serve to trap food particles. The bones of the hyoid apparatus are not visible and gill rakers are totally absent. The entire filtering apparatus is modified from that of the larvae of salamanders and more primitive tetrapods. Modified from Wassersug and Rosenberg, 1979. *C,* Lateral view of the mouth and pharynx of a tadpole showing the relationships of three portions of the pumping apparatus. The buccal pump was shown in Figure 9.15. It can be separated from the pharyngeal pump by the ventral velum. The pharyngeal pump is driven by muscles attached to the branchial basket. It drives the water and suspended food particles through the gill slits to the opercular chamber. Muscles in the opercular chamber are occasionally activated to clear it of trapped particles. The arrows show the flow of water and the black dots the food particles, from the mouth into the buccal cavity, past the ventral velum, and past the gill filter plates. Large food particles are trapped on the buccal papillae and small particles adhere to mucus secreted on the roof of the buccal cavity and on the dorsal and ventral velum. Black dots indicate food particles. Arrows indicate direction of the flow of water. Modified from Sanderson and Kupferberg, 1999.

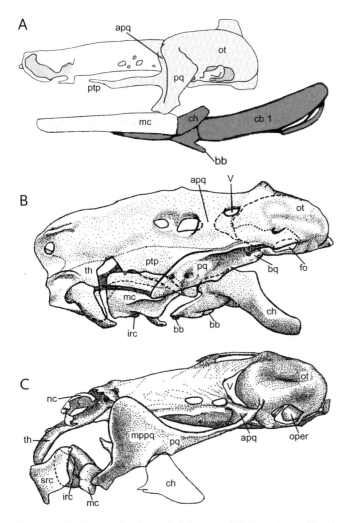

Figure 9.17. Braincase and endochondral elements of the lower jaw and hyoid apparatus in a larval salamander and tadpoles. *A,* The primitive living salamander *Ranodon,* which represents the pattern common to primitive bony fish and probably to Devonian tetrapods, based on the conservative nature of the hyoid and lower jaws in these animals. Modified from Rose, 2003. *B,* The primitive living frog *Ascaphus.* Modified from Pusey, 1943. *C,* The advanced frog *Rana.* Modified from Pusey, 1938. Note the progressive anterior movement of the palatoquadrate and its articulation with Meckel's cartilage, and of the ceratohyal relative to the otic capsule. The ceratohyal becomes connected with the palatoquadrate via a narrow articulating surface running parallel to the longitudinal axis of the skull. The ceratohyals hinge at the midline. Meckel's cartilage is greatly shortened, and is functionally replaced by the neomorphic superior and inferior rostral cartilages.

the nostrils prevents its anterior movement. Initially, posterior movement is limited by the ventral velum, which separates the mouth from the pharynx, but the velum drops as the buccal chamber is contracted (Sanderson and Kupferberg, 1999). More posteriorly, the branchial basket acts as a second pump, moving the water into the opercular chamber and digestive tract.

LARVAL JAWS. At some stage in the elaboration of the filtering apparatus and the initiation of a continuous rather than intermittent system of pumping, the lower jaws of the ancestral larvae, which had extended the entire length of the skull, were greatly shortened in relationship to the anterior extension of the palatoquadrate. The ancestral jaws were

functionally replaced in the evolving tadpole by two or more pairs of smaller elements, the suprarostral and infrarostral cartilages, both of which are covered by keratinized beaks (Fig. 9.17). These serve to detach and comminute plant material. The suprarostrals are thought to have evolved ontogenetically from the rostral portion of the sphenethmoid and the infrarostrals from the anterior portion of Meckel's cartilage, which is expressed in the adult frog as the mentomeckelian bones (Roček, 2003). The larval jaws are manipulated by the same group of muscles that close the jaws in adult frogs, but their orientation has shifted from primarily vertical to nearly horizontal (Fig. 9.18). In addition, nearly all modern frogs have several rows of keratinized denticles on the surface of the lips, above and below the mouth, for scraping plant material from the substrate.

Presumably, the processes of procuring and transporting particulate plant matter that evolved in the ancestors of tadpoles were accompanied by changes in their digestive capacities that enabled them to make use of this material. It has long been suspected that this depended, as in most herbivorous amniotes, on developing a symbiotic arrangement with microorganisms capable of breaking down cellulose. This was recently demonstrated by Pryor and Bjorndal (2005), who showed that 20% of the daily energy requirements of the common bullfrog larva consists of the fermentation products of complex structural carbohydrates supplied by symbiotic bacteria, protozoa, and nematodes that occupy a thick mucous coat lining the colon wall. One may assume that all aquatic animals, now and in the Paleozoic, naturally ingested a variety of small organisms capable of breaking down complex carbohydrates and so had the potential for evolving symbiotic relationships with them. However, among amphibians, only the frogs have been able to take advantage of these organisms and throughout their known history nearly all have retained herbivorous larvae.

METAMORPHOSIS. Surprisingly, all this evolutionary effort was expended on a larval stage that must later undergo metamorphosis, for the tadpole carries no sex organs that would enable it to reproduce. Unlike salamanders, no frog undergoes neoteny.

The most challenging feature of the frog tadpole is that it requires drastic changes to achieve the adult condition, far beyond those that occur in salamanders or caecilians. All of the processes of feeding, digestion, and locomotion change with the initiation of metamorphosis. Keith Alley (1990) has provided a particularly informative discussion regarding how changes have occurred in the muscles and associated nerves of the feeding apparatus. In relationship to the shift from the far anteriorly placed larval jaws to the posteriorly hinged adult jaws, the position, arrangement, and even number of jaw muscles change radically between larvae and adults. The jaw-closing muscles (the adductors) are repositioned and reoriented from the horizontal plane, ventral to the orbit, to a vertical plane, posterior to the orbit. The five larval adductors reestablish their identity in the adults, but the five jaw depres-

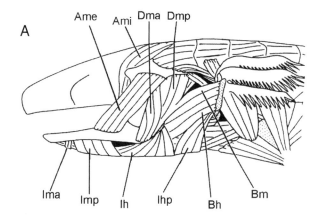

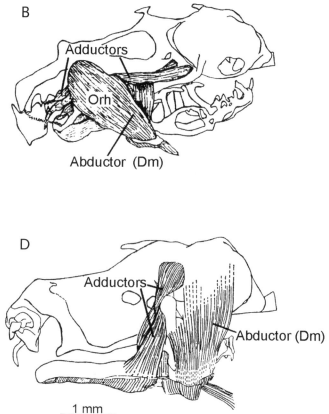

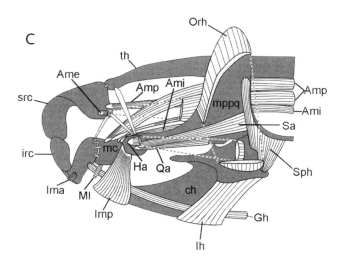

Figure 9.18. Evolution and ontogeny of jaw musculature. *A,* Diagram of the distribution and orientation of jaw muscles in the larvae of the salamandrid *Taricha granulosa,* a plausible model for primitive tetrapods. From Deban and Wake, 2000. *C,* The tadpole of *Rana catesbeiana.* From Cannatella, 1999. Note the clear difference in the orientation of the adductor mandibulae externus (Ame) and internus (Ami) and the position, origin, and insertion of the depressor mandibulae (Dma and Dmp) of the salamander and their ho-

mologue in frogs, the orbitohyoideus (Orh). *B* and *D,* Transformation of jaw muscles during metamorphosis in the frog *Rana.* Modified from De Jongh, 1968. The adductors reverse the evolutionary changes seen between the larvae of primitive amphibians and tadpoles to restore their primitive vertical orientation. The abductor loses its role in pumping and re-establishes its function as a jaw-opening muscle, the depressor mandibulae.

sors (abductors) coalesce into a common depressor mandibulae that is vertically oriented posterior to the tympanic ring (Fig. 9.18).

Even more impressive is that changes in the position and orientation of these muscles involve a complete dissolution of the larval muscle fibers, cell by cell, and their replacement by new muscle fibers. This process can be followed microscopically. In advanced tadpoles of the common frog *Rana pipiens* (stage XV), all the original muscle cells are intact, but within 7 to 10 days (stage XXIII) they are completely lost and replaced by a new generation of cells. This is accomplished by digestion of the original cells by invading phagocytes and replacement by new muscles arising from satellite cells adjacent to the larval myofibers. Both the death of the old cells and the elaboration of the new cells are influenced by an increase in thyroid hormone.

At the same time, there is a retrofitting of the neuromuscular circuits. In contrast with the muscle fibers, there is no loss of the original neurons, but a reattachment to the in-

vesting basal lamina of the larval myofibers that survived the loss of the muscle cells per se. The cellular geometry and synaptic input to these motor neurons are modified so that the much altered manner in which the adult muscles are used can be implemented. There is no evidence in the patterns of development of either salamanders or caecilians that presages the specializations of the anuran tadpole. We can only presume that phylogenetic divergence occurred long prior to the emergence of the earliest anurans.

In addition to changes in feeding, respiration, and digestion, the mode of locomotion is entirely altered during metamorphosis. Until the beginning of metamorphosis, neither paired fins nor limbs are evident. All locomotion depends on the highly flexible tail. With metamorphosis, the tail is rapidly reduced, long before effective fore- and hind limbs emerge, leaving the transforming frogs without the capacity for effective locomotion in either the water or on land. This results in very high mortality from predation; only a small percentage survive.

Essentially modern tadpoles were described by Chipman and Tchernov (2002) from the Lower Cretaceous, demonstrating the minimum time of their divergence from salamanders or caecilians, but suggesting a much earlier origin (Plate 12).

In common with primitive salamanders, nearly all frogs practice external fertilization. The only genus to have a copulatory organ is the primitive genus *Ascaphus*, which makes use of a tail-like structure to pass the sperm to the female. This is necessary because they breed in fast-moving streams.

Caecilians

Modern caecilians are very easy to differentiate from frogs and salamanders since they lack any trace of the appendicular skeleton, but have an extremely long vertebral column. The skulls of caecilians also differ in having an almost continuous covering of dermal bones, in contrast with the fenestrate condition of frogs and salamanders (Fig. 9.8). The orbital openings are small, and in some genera covered with bone. None are known to have image-forming eyes, for most spend much of their lives in burrows. One family, the Typhlonectidae, is aquatic.

The maxilla and palatine are fused to one another in the adults, and there are median rows of teeth on both the palate and the inner surface of the lower jaw. There are no appreciable interpterygoid vacuities, but the internal nares open medially to the area usually occupied by the maxilla. In relationship to their burrowing habits, the many bones that make up the posterior portion of the braincase in early tetrapods have fused into a single ossification termed the os basale, incorporating the supraoccipital, basioccipital, exoccipitals, opisthotic, prootic, and parasphenoid. The sphenethmoid is a massive element, whose lateral projections serve to buttress the anterior portion of the skull against the forces of compression resulting from burrowing through compacted sediments.

To compensate for the great reduction or complete loss of vision, a novel sensory organ, the tentacle, has evolved by making use of sensory tissue of the Jacobson's organ that occupies the area of the nasal capsule of other amphibians and muscles associated with the rudimentary eye. This enables caecilians to detect olfactory clues of prey living in the ground through which they burrow. The tentacle protrudes just in front of the orbital opening.

One of the most striking features of the caecilian skull is the nature of their jaw-closing musculature. Caecilians have a very narrow skull to enable them to push it through the substrate while burrowing. This greatly limits the size of the adductor chamber in the cheek region that contains the jaw-closing muscles in all tetrapods. Caecilians have gotten around this problem by making use of another muscle, the interhyoideus posterior, which serves to constrict the throat in other amphibians. In caecilians, this muscle is much enlarged posteriorly and inserts at the end of the very large retroarticular process at the back of the lower

jaw (Fig. 9.19). Because the lower jaw rotates around the articulating surface of the quadrate bone at the back of the skull, the retroarticular process serves as a lever to close the mouth.

Pulling the back of the jaw posteroventrally would tend to dislocate it from the skull, but in living caecilians this is avoided by the elaboration of another jaw-closing muscle, the pterygoideus, which runs from the posterodorsal surface of the palate posteroventrally and underneath the back of the lower jaw, just anterior to the jaw articulation. This serves as a sling to keep the articular surface of the lower jaw in contact with the quadrate. The configuration of the articulating surfaces of the quadrate and articular bones is also modified in modern caecilians to maintain close contact.

Not only is the jaw-closing musculature unique in caecilians, but so is the overall structure of the lower jaw (Fig. 9.19B–E). Frogs, salamanders, and caecilians have all simplified their jaws by reduction and/or fusion of the many elements present in primitive Paleozoic amphibians. This is taken to its extreme in caecilians, which have only two areas of ossification in the adults, an anterior pseudodentary, bearing the marginal dentition and a shorter medial row of teeth, and the pseudoangular, which forms the area of articulation with the quadrate, behind which is a very long retroarticular process. The pseudoangular also has a large internal process, beneath which runs one portion of the pterygoideus musculature. Meckel's cartilage ossifies anteriorly in the area occupied by the mentomeckelian bones in frogs and salamanders.

The hyoid apparatus of caecilians retains the complex of rod-shaped elements seen in salamanders and many Paleozoic tetrapods, but the hypobranchials are lost (Fig. 9.11). The specific function of the hyoid apparatus in respiration and feeding is less thoroughly studied than in frogs or salamanders. The tongue cannot be protruded, but is occupied by a mass of vascular tissue that presumably functions to hold the prey and facilitate swallowing. Other aspects of feeding are discussed by O'Reilly (2000).

Locomotion in caecilians involves only the axial skeleton and its associated musculature. The number of vertebrae in living species ranges from 86 to 285 (M. H. Wake, 2003). Locomotion in primitive caecilians was presumably sinusoidal, in common with other elongate vertebrates with much reduced or absent limbs that remain on the surface of the ground, with alternate contraction of the musculature on either side of the trunk progressing from fore to aft. For burrowing, modern caecilians make use of concertina locomotion, in which the musculature associated with the vertebrae and the body wall can act independently of one another (Summers and O'Reilly, 1997; O'Reilly et al., 2000). The vertebral column can be bent laterally while the body wall musculature holds the surface of the animal against the wall of the tunnel that it is making. When the vertebral column is straightened, the skull is forced forward to extend the tunnel, while the body wall maintains

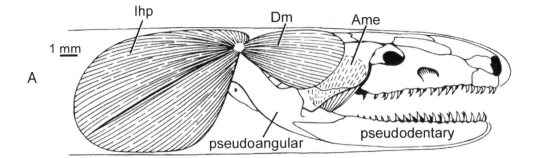

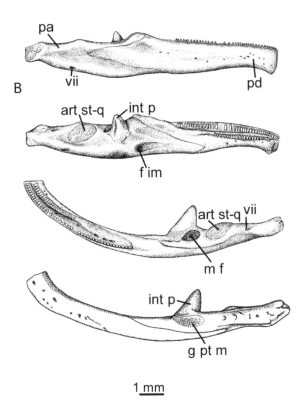

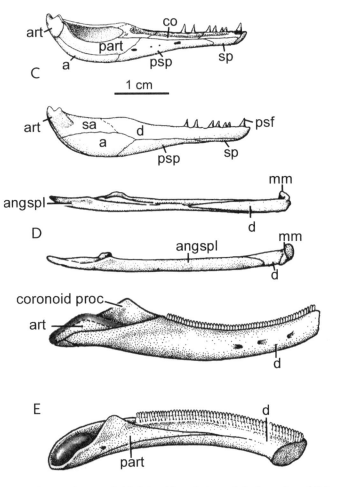

Figure 9.19. A, Primary muscles for opening and closing the jaws in caecilians, as exemplified by the primitive genus *Ichthyophis.* Modified from Nussbaum, 1983. *B,* Lateral, medial, dorsal, and ventral views of the lower jaw of the most primitive known caecilian, *Eocaecilia* from the Lower Jurassic. From Jenkins et al., 2007. *C,* Medial and lateral views of the lower jaw of *Balanerpeton,* representative of early tetrapods. *D,* Lower jaw of the hylid frog *Gastrotheca.* From Duellman and Trueb, 1986. *E,* Lower jaw of *Salamandra salamandra.* Modified from Francis, 1934.

the position of the more posterior trunk by hydrostatic pressure.

Modern caecilians have very large muscles that attach to the ventral surface of the back of the braincase to flex the skull ventrally. These are opposed by muscles running from the dorsal surface of the neck vertebrae to the top of the skull that bend the skull dorsally, thus forming a moving wedge to

displace the soil in front of the animal. Comparable ventral muscles are also present in salamanders. In common with other lissamphibians, caecilians practice cutaneous respiration, but unlike frogs and salamanders, primitive species retain small scales.

Caecilians are also distinguished from frogs and salamanders in their mode of reproduction. All species possess an

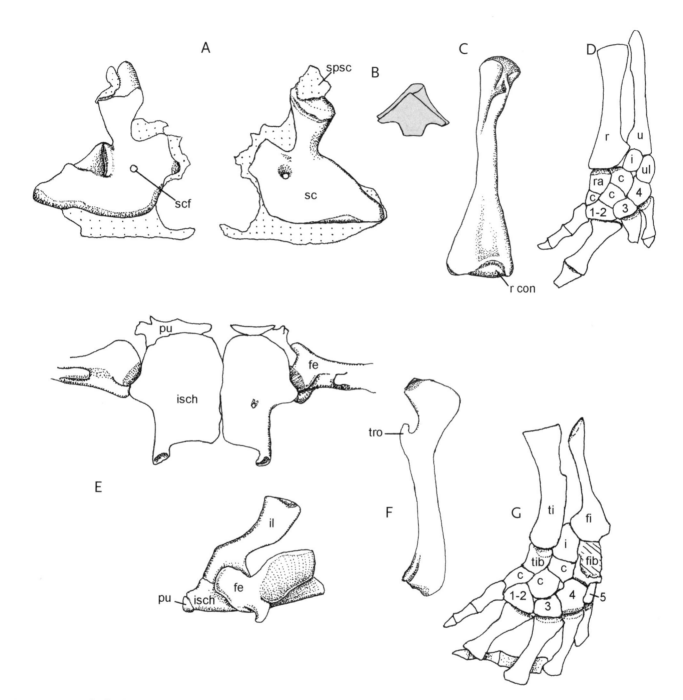

Figure 9.20. Appendicular skeleton of salamanders. All but *B* are based on a skeleton of *Hynobius nigrescens*, one of the most primitive salamanders (Harvard specimen, 22513). *A*, Lateral and medial views of the scapulocoracoid; the dorsal, anterior, and ventral surfaces are extended in cartilage. *B*, Cartilaginous sternum of *Salamandra salamandra*. Modified from Francis, 1934. *C*, Humerus. *D*, Lower forelimb. *E*, Ventral and left lateral views of the pelvis. *F*, Femur. *G*, Lower hind limb. Identification of carpals and tarsal is based on similarities with Paleozoic labyrinthodonts.

eversible copulatory organ enabling internal fertilization. The most primitive living caecilians lay their eggs on land, rather than in the water, and the parents guard the eggs until they hatch. The larvae then wriggle into the nearby water. More advanced caecilians give birth to live young (M. H. Wake, 1993). Like modern salamanders, external gills are present early in development, but by the time of hatching or live birth, the gills are lost. Hence it is very unlikely that external gills would be preserved in fossils.

Salamanders

Frogs and caecilians mark the extremes of locomotor adaptation and modes of reproduction in lissamphibians. Salamanders appear to retain more characters in common with ancestral tetrapods. Primitive genera retain four limbs. They have four toes in the hand and five in the rear (as do frogs), and a phalangeal count comparable to that of temnospondyl labyrinthodonts (Table 5.1). They retain the sinusoidal movements of the trunk common to primitive tetrapods, with the left and right limbs moving alternatively.

The skull is fenestrate, as in frogs, rather than having an extensive covering of dermal bones as seen in caecilians. However, details of cranial anatomy clearly distinguish all salamanders from anurans. No salamander has an otic notch in the squamosal, nor do the adults retain any trace of the impedance matching middle ear structure of frogs. In primitive salamanders, the squamosal has a hinge-like connection with the skull roof, so that the distal extremity and the attached quadrate can move laterally. This may facilitate gape-and-suck feeding in the larvae. Mediolateral movement of the cheek would be impossible in frogs, without the stapes puncturing the tympanum. Mobility of the cheek region is further facilitated by the lack of bony connection between the base of the squamosal and the maxilla and the presence of a synovial joint between the pterygoid and the base of the braincase. The palatal process of the pterygoid also fails to reach the maxilla. On the other hand, salamanders retain the lacrimal and prefrontal bones that are lost in frogs.

The number of presacral vertebrae range from 11 (just one more than in the most primitive frogs) to 61 (considerably less than in the shortest-living caecilians). All are unipartite structures, broadly resembling those of nectrideans. Primitively the number is about 20; there is never more than a single sacral. The most primitive centra are notochordal, but the more derived have evolved a ball-and-socket joint, with the concavity at the posterior end of the centrum (an opistocoelous joint), or rarely a procoelous joint.

Despite the similarity of the fore- and hind limbs to those of primitive tetrapods, the girdles are significantly altered, but in ways very distinct from those of frogs (Fig. 9.20). In contrast with frogs, all the dermal bones of the shoulder girdle—the cleithrum, clavicle, and interclavicle—are lost. The scapulocoracoid broadly resembles those of primitive

tetrapods, although much of the margin of these bones remains cartilaginous. A new endochondral bone, termed a sternum, has evolved at the midline but is obviously not homologous with a bone of that name in mammals, dinosaurs, or birds. The pelvis broadly resembles that of primitive tetrapods, but the pubis is much reduced and remains cartilaginous in all but the primitive hynobiids.

Salamanders are largely primitive in their mode of reproduction. None have copulatory organs. Fertilization is external in the primitive hynobiids, cryptobranchids, and perhaps sirenids, but all more advanced families package the sperm into a spermatophore that the female is induced to pick up with the lips of her cloaca and draw into her reproductive tract. Only one genus, *Salamandra,* is capable of giving birth to live young. Most salamanders lay their eggs in the water, as do frogs, but most members of the advanced family Plethodontidae lay their eggs on land.

Salamander larvae resemble those of Paleozoic temnospondyls and discosauriscids in having external gills, paired limbs, and a tail supported by vertebrae. Among the modern families, a biphasic lifestyle was almost certainly primitive, but more derived genera may be neotenic—retaining an aquatic way of life into adulthood, or hatching out on land, without a larval stage.

In contrast with the relatively primitive nature of the adult skeleton in salamanders, its sequence of chondrification and ossification in the larvae is unique compared with that of other living amphibians and, in fact, of tetrapods in general. These features will be discussed in Chapter 11, in reference to establishing their affinities among Paleozoic amphibians.

SEARCH FOR ANCESTRY

The anatomy and ways of life of each of the modern amphibian orders reveal a host of definitive characters that demonstrate highly divergent evolutionary pathways, but these features can be used to determine their possible relationships to particular groups of Triassic and Paleozoic tetrapods. These will be investigated in the following three chapters.

10

The Ancestry of Frogs

FULL UNDERSTANDING OF THE soft anatomy, physiology, behavior, and way of life of frogs, salamanders, and caecilians can only be gained on the basis of living organisms. Nevertheless the very similar skeletal remains going back 170 to nearly 200 million years provide the basis for attempting to link them with even older but much more primitive amphibians. Lower Jurassic frogs superficially appear much like their living descendants, differing in only a few, minor features.

JURASSIC FROGS

The oldest known frog, *Prosalirus bitis* (Fig. 9.7E), was discovered by a field party led by Farish Jenkins to the Early Jurassic Kayenta Formation of Arizona. It is known from four disarticulated specimens, representing most of the postcranial skeleton but only scattered elements of the skull (Shubin and Jenkins, 1995). The frontal and parietal have already fused to form the characteristic frontoparietal of frogs, but it has a posterolateral flange that would have overlapped the otic region and retains a parietal foramen as in more primitive tetrapods. The maxilla bears approximately 60 pedicellate teeth. The stapes has the configuration of modern frogs that have an impedance matching middle ear, including a process that would have articulated with the ventral rim of the fenestra ovalis (Bolt and Lombard, 1985). The presence of an angulosplenial bone indicates the modern configuration of the lower jaw.

Unfortunately, the exact number of presacral vertebrae cannot be determined, but individually they resemble those of archaic living frogs in being notochordal and amphicoelous. As in modern frogs, the atlas lacks transverse processes for the attachment of ribs (Fig. 10.1). The paired cotyles are separated by a small notochordal fossa that is retained in other early fossil frogs and primitive living genera. Four to five pairs of short ribs would have attached to successive vertebrae, one of which was fused. Some have uncinate processes. The more posterior presacral vertebrae

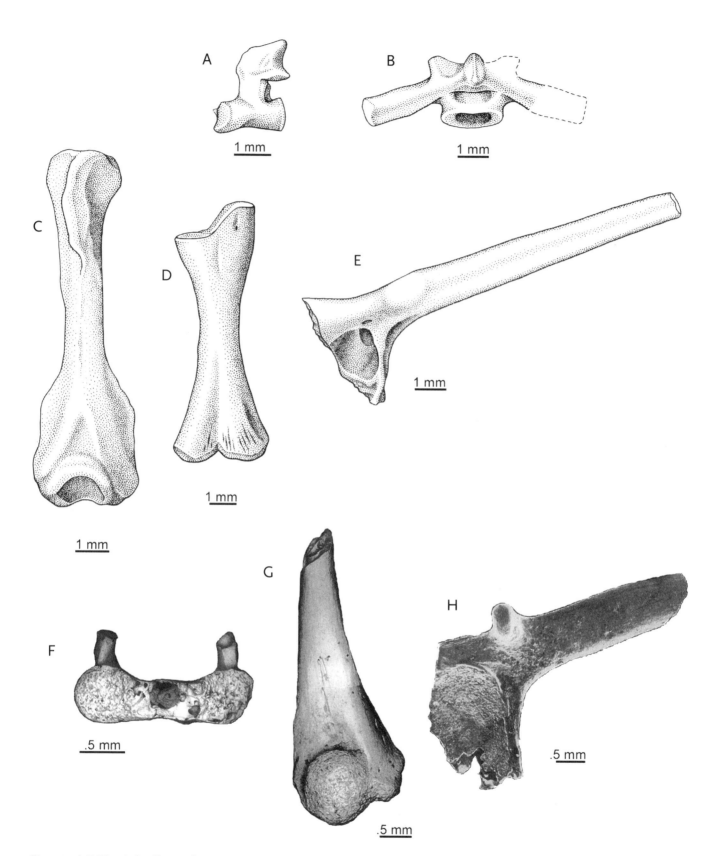

Figure 10.1. A–E, Disarticulated bones of *Prosalirus bitis,* the oldest frog, from the Lower Jurassic of Arizona. *A,* Atlas in left lateral view. Note the absence of transverse processes. *B,* Sacral vertebra in dorsal view. The parapophyses articulate with the dorsal surface of the iliac blades. Posterior zygopophyses are lost to facilitate hinging with urostyle. *C,* Ventral view of humerus. Note single area for articulation with the fused ulna-radius. *D,* Ulna-radius. *E,* Right ilium in lateral view. From Jenkins and Shubin, 1998. *F–H,* isolated elements of the Lower Triassic salientian, *Czatkobatrachus. F,* Atlas vertebra in anterior view showing double condyle. *G,* Distal end of humerus in ventral view. *H,* Right ilium in lateral view. From Evans and Borsuk-Białynicka, 1998.

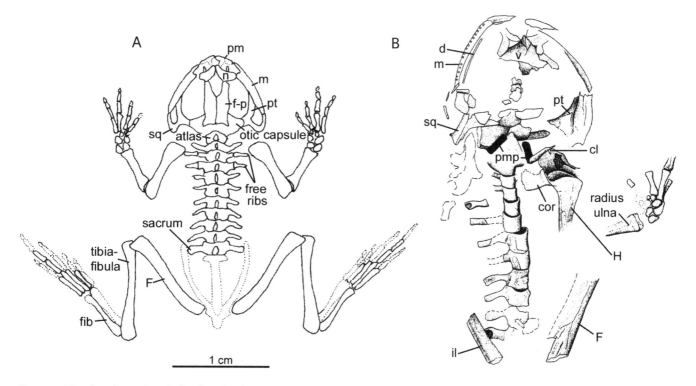

Figure 10.2. Vieraella, a Lower Jurassic frog from South America. *A,* Reconstruction of the skeleton in dorsal view showing fusion of ulna and radius and tibia and fibula and elongation of proximal tarsals. *B,* Ventral view of specimen. Posterior medial processes of hyoid drawn in black. *A,* modified from Estes and Reig, 1973; *B,* from Roček, 2000.

bear short transverse processes. The sacral vertebra has lost postzygopophyses, so as to facilitate a hinging articulation with the urostyle, but like the primitive living frogs *Ascaphus* and *Leiopelma,* lacks the condylar structure that evolved in more advanced anurans.

One of the most significant features of *Prosalirus* and all other frogs is that the transverse process (or parapophysis) of the sacral vertebra articulates with the dorsal edge of the ilium, rather than being firmly attached to its medial surface. As emphasized by Jenkins and Shubin (1998), the iliac blade serves as a fulcrum about which the vertebral column rotates during the initial stage of a jump, and transmits upwardly directed forces from the hind limbs through the ilium to the vertebral column. Behind the sacrum is a short, conical urostyle with which it presumably had a cartilaginous attachment. Advanced tadpoles described by Špinar (1972) document the development of the urostyle from the coalescence of caudal vertebrae, at least four of which remain distinct in the larvae. The urostyle of *Prosaliurus* would have equaled about the length of four trunk centra.

The well-preserved appendicular skeleton is striking in the number of advanced anuran features that it exhibits. As in all other frogs, the iliac blade extends far anteriorly from the acetabulum, in a manner unique among vertebrates. The pubis is presumably unossified, as in all other frogs. The rear limb is substantially longer than the forelimb. The tibia and fibula are fused to form the tibiofibula, which is longer than the femur. A single tarsal bone is known that might be either the tibiale or the fibulare, both of which are elongated

in frogs. Although deposited in strata dated as approximately 186 million years old, *Prosalirus* differs only slightly in the nature of the jumping apparatus from that of frogs living today.

The scapula and coracoid ossify separately instead of forming a unified scapulocoracoid as in many Paleozoic amphibians. The scapula probably supported a suprascapular cartilage, as in modern frogs, but the elongate form of the glenoid is more primitive. The coracoid is elongate and extends more or less ventrally. The radius and ulna are fused into a single radio-ulna, which articulates with a single condyle on the humerus.

The slightly younger *Vieraella,* from the Lower Jurassic of Argentina, is nearly complete, except for the absence of the urostyle and pelvis (Fig. 10.2). It has 10 notochordal and amphicoelous presacral vertebrae, the most of any frog. The tibia and fibula are fused and the tibiale and fibulare elongate. Neither of these genera can be placed in modern families, but they are unquestionably frogs. A third frog, *Notobatrachus,* is known from many specimens from the Upper Jurassic of South America (Fig. 9.13). It is placed in the living family Leiopelmatidae by Sanchiz (1998), and, like them, is primitive in the retention of free ribs on the second, third, and fourth vertebrae.

CRETACEOUS TADPOLES

Despite the ephemeral occurrence and inherent fragility of tadpoles, their fossil record is fairly extensive. Numerous species are known from various localities and horizons in the Tertiary, including extensive growth series. One of the richest localities is that of the Lower Miocene of northern Bohemia, described by Špinar (1972), from which have been found more than 1000 specimens, most of which belong to the Palaeobatrachidae, the only family of frogs to have become extinct.

A particularly informative assemblage has recently been described by Chipman and Tchernov (2002) from approximately 140 million-year-old Lower Cretaceous deposits in Israel. Nearly 300 specimens are known, ranging from advanced embryos 3.8 mm in total body length, with no bones yet ossified, to those near the end of metamorphosis that were approximately 25 mm in snout-vent length (Fig. 10.3 and Plate 12). All belong to a single species, *Shomronella jordanica,* a member of the superfamily Pipoidea, which are among the most primitive of living frogs but were the most common of anurans from the Mesozoic. Most living pipoids are aquatic as adults, but the absence of any fully metamorphosed specimens in the Israel locality, a small, seasonally fluctuating pond, suggests that they were terrestrial as adults.

Because of the highly specialized larval stage, which is totally without ossification, the sequence of bone development is entirely different from that of other tetrapods, as exemplified by the salamanders discussed in Chapter 11. Rather than early ossification of the tooth-bearing elements of the skull, the first bone to ossify in *Shomronella* tadpoles is the parasphenoid, followed by the frontoparietal and then the neural arches of the vertebrae and the exoccipitals. They are followed by the ilium, and, in proximal to distal sequence, the bones of the hind limbs. Finally, the bones of the pectoral girdle, the forelimb, and the urostyle ossify. There are 9 or occasionally 10 presacral vertebrae.

Pipoids in general ossify more fully before completion of metamorphosis than other modern frogs, indicating that their metamorphosis is more gradual. Otherwise, these tadpoles are already highly derived early in the history of frogs, with no intermediates known between them and those of Paleozoic amphibians. Tadpoles have recently been reported from the Middle Jurassic of Inner Mongolia by Yuan and his colleagues (2004). They were not illustrated or described in detail, but already resemble those of modern frogs. No fossils are known from the Paleozoic that approach the structure of the anuran tadpole.

TRIASSIC SALIENTIANS

Frogs are the only modern amphibian order with which Triassic fossils can be closely compared (Roček and Rage, 2000a). A single, nearly complete fossil of *Triadobatrachus,* preserved as impressions of both the dorsal and ventral surfaces, was described from Madagascar by Piveteau in 1937 (Figs. 10.4 and 10.5). It had been found in a chief's hut, but the nature of its preservation indicates that it was from the Lower Triassic Sakamena Formation, approximately 247 million years old, roughly 61 million years prior to the oldest known anuran, *Prosalirus*. It was immediately recognized as a possible relative of frogs on the basis of the very large skull, relatively short vertebral column (14 presacrals, including the atlas), and forwardly directed blade of the ilia. Jenkins and Shubin (1998) argued that the apparent shortening of the tail, with the rapid diminution of the size of the centra posteriorly, indicates an initial stage in recruitment of the tail musculature to translate force from the hind limb to the axial skeleton during a jump.

The frontoparietals have fused at the midline, but retain a pineal opening, as does *Prosalirus*. Posteriorly, they extend over most of the width of the otic capsules and were articulated with the squamosals, which would have precluded the extension of the adductor mandibulae posteriorly over the otic capsule, as is the case in modern frogs, but apparently not in *Prosalirus*. All the circumorbital bones common to Paleozoic amphibians have been lost, except perhaps the prefrontal, as well as those making up the back of the skull table. The squamosal forms a large otic notch that presumably supported a tympanum capable of responding to high-frequency airborne vibrations. The stapes is much reduced in size. The pterygoid is loosely attached to the parasphenoid and the palatal ramus extends to the maxilla, as in modern frogs but not salamanders. The narrow palatine bone is transversely oriented, as in modern frogs.

Unfortunately, the nature of preservation of this fossil does not allow determination of whether the teeth were pedicellate or not. There were apparently no teeth in the lower jaw, in common with nearly all modern frogs.

A particularly striking feature is the presence of paired posterior medial processes of the hyoid plate, in exactly their position in primitive frogs, as well as a *Y*-shaped parahyoid. This suggests that the larynx and tongue were supported in the same manner as living anurans (Fig. 9.11E). The presence of these bones may be interpreted as evidence that *Triadobatrachus* called in the manner of modern frogs. This is further indicated by the large size of the squamosal notch to support the large tympanum necessary for an impedance matching middle ear. Since these hyoid bones are unique to *Triadobatrachus* and anurans, their evolution might be linked with changes in the more posterior elements of the hyoid apparatus that are associated with the origin of the tadpole pumping system. Together with the reduction of the tail, it suggests that the tadpole may have evolved by this time as well.

The number of presacral vertebrae is only four greater than that of the primitive Mesozoic frog, *Vierella,* but ribs may have been present on all the trunk vertebrae. In contrast with all later frogs and their relatives, the atlas vertebra is described as bearing double-headed ribs. More posterior ribs are single headed. There are six distinct caudal vertebrae, in striking contrast with the unified urostyle of frogs.

As in frogs, but not salamanders, both the clavicles and

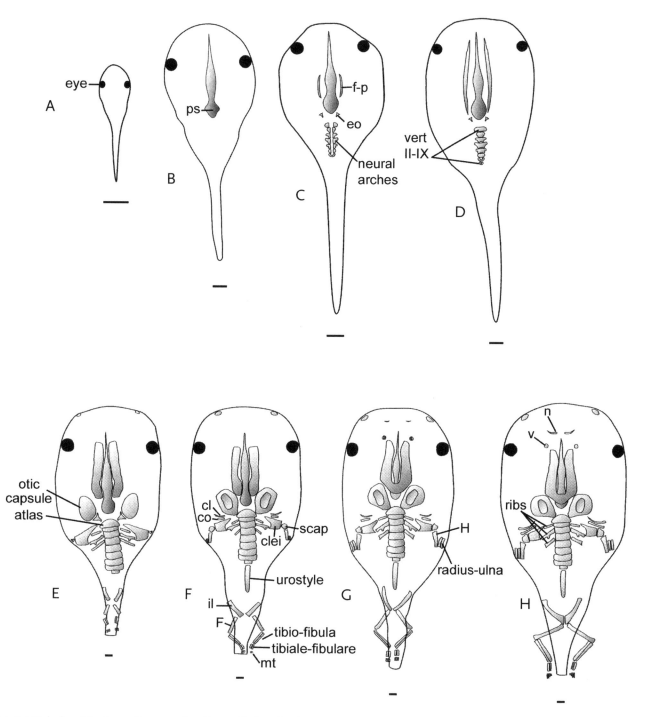

Figure 10.3. Tadpoles of the Lower Cretaceous frog *Shomronella* from Israel showing progressive ossification of the skeleton. Modified from Chipman and Tchernov, 2002. Scales are 1 mm in length. See also Plate 12.

cleithra are ossified. The scapula is large and well ossified and may have supported a suprascapula dorsally. The humerus is long, but massive. The ulna and radius are not fused, as they are in all frogs.

Although *Triadobatrachus* had an anteriorly elongated ilium, as in frogs, the transverse process of the sacral rib re-

tained a fixed attachment to its internal surface, as in other, more primitive terrestrial vertebrates. This suggests that *Triadobatrachus* represented an early phase in the evolution of the anuran jumping behavior. An even earlier stage in the evolution of jumping may be reflected in the habits of both frogs and salamanders to lunge forward when closely approaching

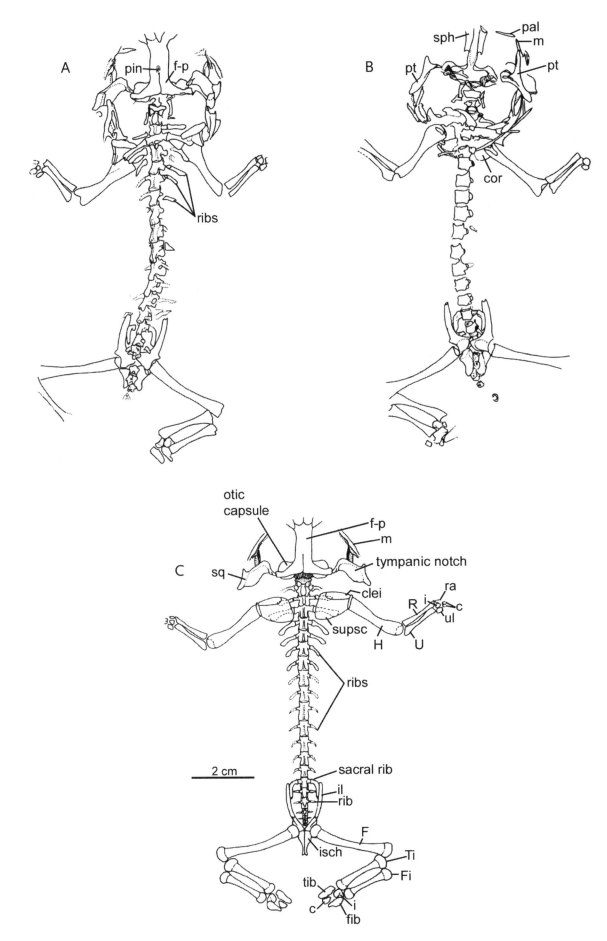

Figure 10.4. The Lower Triassic salientian *Triadobatrachus* from Madagascar. *A* and *B*, Dorsal and ventral views of skeleton, preserved as a natural mold in a sandstone nodule. *C*, Reconstruction of *Triadobatrachus* in dorsal view. Modified from Roček and Rage, 2000a.

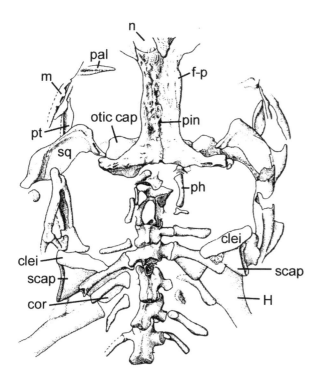

Figure 10.5. Dorsal and ventral views of the head and shoulder region of *Triadobatrachus.* The parahyal (the ossified portion of the hyoid plate) and posterior medial processes (which support the larynx) are retained in living anurans. From Roček and Rage, 2000a.

prey (D. B. Wake and Deban, 2000). Even without the rotational movement between the ilia and the sacral diapophyses, the posterior position of the acetabulum relative to the sacrum suggests some specialization involving forward movement of the trunk relative to the rear limbs.

A further change in the function of the appendicular skeleton is seen even later in the evolution of frogs. Among the adults of most living frogs, the rear limbs are used for aquatic locomotion in the same symmetrical manner as that of jumping. In contrast, Abourachid and Green (1999) have demonstrated that the most primitive frogs, *Leiopelma* and *Ascaphus,* always retain the primitive, asymmetrical movement of the rear limbs in swimming, in contrast with all other genera that have been studied. When swimming, their body swings from side to side, as is the case for tadpoles. It is also of interest that the posture of the rear limbs in *Triadobatrachus* closely resembles that of the oldest known frog tadpoles from the Lower Cretaceous (Plate 12).

Roček and Rage (2000a) argued that the sole specimen of *Triadobatrachus* belonged to a subadult individual based on the fact that the carpals and tarsals were ossified, but not the articulating surfaces at the ends of the limbs.

Although *Triadobatrachus* is distinguished from all modern families of frogs (the order Anura) by primitive characters, including the absence of a urostyle and the fixed attachment of the sacral rib to the ilium, the froglike features of the skull and the position of the acetabulum far behind the sacrum

justify its being placed in the more inclusive superorder, Salientia (Sanchiz, 1998).

While *Triadobatrachus massinoti* is the only froglike amphibian from the Triassic of the Southern Hemisphere, another species, *Czatkobatrachus polonicus,* is known from Poland. It is, however, represented by only isolated skeletal elements and so cannot be reconstructed. On the other hand, the individual bones are very similar to those of *Triadobatrachus,* although a few seem more similar to those of later frogs (Fig. 10.1F–H).

Hundreds of isolated bones of *Czatkobatrachus* were discovered in a karst deposit, that is, in fissure fillings of a tunnel or cave system dissolved from a limestone deposit. They were accompanied by bones of many small reptiles, including a small predatory archosaur, three or four genera of procolophonids, a prolacertiform, and one or two genera of lepidosaurs, which served to demonstrate their age as early Olenekian, the same age as *Triadobatrachus* (Borsuk-Białynicka et al., 2003).

Czatkobatrachus was considerably smaller than *Triadobatrachus,* with an estimated snout-vent length of approximately 5 cm versus approximately 10 cm, but appears more mature in the ossification of the articulating surfaces of the limb bones. The elbow joint is fully ossified, as is the radial condyle of the humerus. As in *Triadobatrachus,* the ulna and radius, and the tibia and fibula, remain separate, but the sacral ribs are fused to the centrum rather than being sutur-

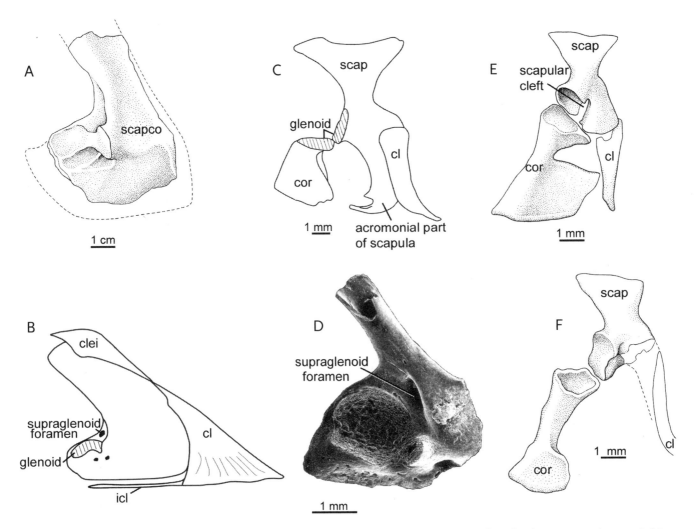

Figure 10.6. Shoulder girdles of temnospondyls and salientians. A, Scapulo-coracoid of the temnospondyl *Dissorophus*. Dermal bones are not drawn. B, Reconstruction of the shoulder girdle of the temnospondyl *Sclerocephalus*. From R. L. Carroll and Holmes, 2007. C, Shoulder girdle of *Triadobatrachus*, based on the reconstruction of Roček and Rage. D, Scapulocoracoid of *Czatkobatrachus*. E, Shoulder girdle of the modern frog *Leiopelma*. F, Reconstruction of the shoulder girdle of *Prosalirus*. A, E, and F from Jenkins and Shubin, 1998. B, C, and D from Borsuk-Białynicka and Evans, 2002.

ally attached. There is no trace of neurocentral sutures in the trunk vertebrae. The reduced transverse processes on the isolated vertebrae suggest reduction or loss of trunk ribs. The atlas vertebra, as in all frogs, lacks ribs.

The scapulocoracoid (Fig. 10.6) retains the primitive fusion of the scapula and coracoid common to well-ossified Paleozoic tetrapods, but shows initial elongation of the supraglenoid foramen into the distinctive cleft that occurs in some modern frogs (Borsuk-Białynicka and Evans, 2002). The glenoid is hemispherical in shape, allowing the head of the humerus to rotate freely, as in frogs, but in contrast with the stereotyped twisting motions resulting from the screw-shaped articulating surface common to Paleozoic tetrapods. Borsuk-Białynicka and Evans (2002) argued that the change in the configuration of the glenoid may reflect an evolutionary shift from the mainly locomotor role of the forelimbs in primitive tetrapods to encompass static functions, including support, balance, and eventually shock-absorption, as in modern anurans.

Considering the many features of *Triadobatrachus* and *Czatkobatrachus* that suggest the initiation of saltatory locomotion and even the emergence of the frog tadpole, it is surprising that fully anuran genera, with an essentially modern skeleton, are not known from the fossil record until approximately 61 million years later. This is almost as long a period of time as the entire history of Cenozoic mammals! It is particularly surprising in view of the much shorter gap in time (approximately 26 million years) between *Triadobatrachus* and its most plausible Permian antecedents. Considering the extremely high degree of constancy of the skeletal anatomy of frogs between the Lower Jurassic and the Recent, one may wonder how much earlier it might have been attained.

The fact that only two genera of salientians (each restricted to a single locality) are known over a period of approximately 87 million years between the mid-Permian and the Lower Jurassic documents how very incomplete our knowledge of their fossil record must be. However, this is a different type of problem than the long gap in the known fossil record of amphibians during the first 25 million years of the Carboniferous, when almost no tetrapods of any kind are known. In the case of the time span from the end of the Lower Permian until the beginning of the Jurassic, there is a very extensive fossil record of other tetrapods, including the ancestors of mammals and dinosaurs as well as the stereospondyls. What is missing during this period is specifically representatives of lineages leading to the modern amphibian groups. This suggests a particular bias against the existence and/or preservation of these small amphibians living at the water-land interface.

A great many small amphibians are known from the Late Carboniferous of North America and Europe because of their preservation in the very extensive coal swamp deposits. They were extensively collected and studied primarily because of the economic value of the coal. Even over that time period, however, the geographical area and number of highly productive localities were limited to a few square miles, as represented by the deposits at Jarrow, Mazon Creek, Linton, and Nýřany.

Putative sister-taxa of frogs, salamanders, and caecilians are known in the Lower Permian of North America and Europe from coal swamps and deltaic deposits, but by the mid-Permian most of the known horizons represent much drier conditions, where small amphibians were less likely to have lived. Following the mass extinction at the end of the Permian, there was a period lasting from 250 to 243 million years ago when no coal deposits are known anywhere in the world, and the level of coal production did not recover until 230 million years ago. Retallack et al. (1996) attributed the absence of coal deposits to the mass extinction of peat-forming plants at the end of the Permian, which were only replaced by newly evolved plants capable of growing under these conditions about 10 million years later. The restriction of environments favorable to animals living at the water-land interface, and the lack of geological interest in deposits of less commercial value both contribute to our incomplete knowledge of small amphibians from this period of time. This gap in knowledge results in our continuing difficulty of tracing the ancestry of the modern amphibian orders back into the Paleozoic.

PUTATIVE ANTECEDENTS OF SALENTIANS

The high degree of consistency in the overall anatomy of salentians (the anurans and triadobatrachids) demonstrates how closely they must be related to a common ancestor. This is supported by a long list of skeletal synapomorphies that unquestionably distinguishes them from salamanders and caecilians, as well as from most Paleozoic tetrapods.

Below the level of *Triadobatrachus,* no amphibians are known that show the conspicuous specializations associated with saltatory locomotion in salentians, nor their highly derived larvae. In the absence of such unique derived characters, we must look for other anatomical characteristics that might be indicative of sister-group relationships with more primitive tetrapods. The distinctive features of salentians are listed below. Autapomorphies (characters present only in salentians) are indicated by an asterisk. Other characters are putative synapomorphies of salentians and some more primitive tetrapods.

1. Large orbits, bordered by the frontal
2. Fusion of frontal and parietal*
3. Loss of circumorbital bones and those at the back of the skull table*
4. Large interpterygoid vacuities, extending to vomers
5. Absence of palatal fangs
6. Loss of ectopterygoid
7. Overlapping rather than solid sutural connection between pterygoid and parasphenoid
8. Pedicellate teeth
9. Large squamosal notch
10. Early ossification of quadrate
11. Slender stapedial stem
12. Footplate of stapes hinges with ventral margin of fenestra ovalis
13. Lower jaw consisting of dentary, angulosplenial (forming the articular surface of the mandible), Meckel's cartilage, and mentomeckelian bones*
14. Dentary missing teeth in almost all species*
15. Anterior portion of hyoid apparatus ossifies as median parahyoid and paired posterior medial processes of hyoid*
16. No hyoid bones integrated with the protrusable tongue muscles*
17. Bicondylar occipital-atlas articulation
18. Unipartite atlas, lacking ribs
19. No more than 10 presacral vertebrae*
20. Pleurocentra fused to neural arch in adults
21. Neural arches ossify prior to centra
22. Parapophysis of sacral vertebra forms an articulating surface with the dorsal surface of the iliac blade*
23. Sacral attachment well anterior to acetabulum*
24. Anterior caudal vertebrae fused into urostyle that articulates with the sacral vertebra*
25. Trunk ribs short
26. Loss of interclavicle*
27. Scapula and coracoid ossify separately*
28. Ilium elongate and directed anteriorly*
29. Hind limb substantially longer than forelimb*
30. Humerus elongate, with single condyle for articulation with ulna-radius*
31. Ulna and radius fused*
32. Femur elongate*
33. Tibia and fibula fused*
34. Tibiale and fibulare elongate*
35. Phalangeal formulas 2, 2, 3, 3 and 2, 2, 3, 4, 3 or close to that number
36. Loss of scales

Although the majority of these characters are unique to salentians, almost half were already expressed in more primitive tetrapods and must have evolved prior to the appearance of *Triadobatrachus*. In order to find plausible antecedents of salentians we must look among all the amphibians described in the preceding chapters and identify those that share a significant number of these characters with anurans and the Triassic salentians.

Some of the Paleozoic groups can be ruled out immediately. It is extremely unlikely that frogs would have evolved from elongate animals that had reduced or lost their limbs. These include most of the lineages grouped as lepospondyls—the aïstopods, adelogyrinids, *Acherontiscus*, lysorophids, and several families of microsaurs. More generally, lepospondyls appear as improbable sister-taxa since there is no evidence for the presence of external gills in the larvae, even in the smallest and most immature specimens, but they are present in early larvae of frogs. This is also the case for all temnospondyls for which larvae are known. The very small size of most lepospondyls, as both larvae and adults, may have enabled the young to survive on cutaneous respiration. On the other hand, even the smallest known lepospondyls, which are immature in other characters, have fully ossified, cylindrical caudal centra, while the larvae of all groups of frogs ossify their vertebrae very slowly, beginning with paired rudiments of the neural arches, not the centra. Of particular importance, all lepospondyls have lost the squamosal notch of early tetrapods and so are unlikely to have evolved an impedance matching middle ear with the characteristics of salentians.

Some of the labyrinthodont groups have also increased the number of presacral vertebrae and/or significantly reduced their limbs, including crassigyrinids, colosteids, and advanced embolomeres. Anthracosaurs are also distinguished by the presence of five, rather than four, digits in the manus, and a higher phalangeal count than any frogs, but these elements were reduced in many other groups as well.

This leaves only one group of Paleozoic amphibians as plausible sister-taxa of salentians, the temnospondyls. As a group, temnospondyls are the only assemblage that has a middle ear structure comparable to that of *Triadobatrachus* and anurans, with a large squamosal showing evidence of support for a tympanum and a lightly built stapes with a footplate that hinges on the ventral margin of the fenestra ovalis (Figs. 10.7 and 10.8). Seymouriamorphs may have had an impedance matching middle ear, but the shape of the stapes and its position relative to the parasphenoid is very different as is the tubular extension of the otic capsule leading to the fenestra ovalis (Fig. 6.2). Other synapomorphies of temnospondyls and anurans include the very large interpterygoid vacuities extending forward to the vomers, the relatively large size of the orbit, and a common phalangeal count of 2, 2, 3, 3 in the manus and 2, 2, 3, 4, 3 in the pes. Temnospondyls and frogs also retain the primitive sequence of vertebral ossification, beginning with paired neural arches and only later ossifying the central elements. In common with many smaller temnospondyls, adults of frogs are commonly terrestrial, especially when feeding, and lack lateral line canals. Among the temnospondyls, the dissorophoids and specifically the amphibamids, show the most synapomorphies.

Micropholis

Aside from the vast horde of aquatic stereospondyls that extended into the Cretaceous, a single more conservative temnospondyl, *Micropholis,* was a contemporary of *Triadobatrachus* and lived in southern Africa, which was then contiguous with Madagascar. *Micropholis* was described in Chapter 6 as an amphibamid dissorophoid (Fig. 6.16). It was relatively small, with large orbits, interpterygoid vacuities, and otic notches, but no other more specifically anuran features. However, another amphibamid, *Doleserpeton,* from the mid-Permian, some 26 million years earlier, shows many more features suggestive of frogs.

Doleserpeton

Doleserpeton annectans was discovered and initially described by John Bolt (1969) from a fissure filling of mid-Permian age (Leonardian) in a limestone quarry in Fort Sill, Oklahoma (Fig. 10.7). In common with the fissure fillings from which *Czatkobatrachus* was discovered, it was dated on the basis of the rich fauna of other amphibians and small reptiles with which it was found. It is classified as an amphibamid, but it is more advanced than other genera in its high degree of ossification at very small size and the nearly smooth surface of the bones forming the skull roof. Of much greater importance is the presence of pedicellate teeth. Very few crowns are in place, but many specimens show a uniform series of pedicels that terminate as smooth surfaces, as do those of true frogs.

In contrast with small specimens of other labyrinthodonts, the quadrate is highly ossified, with a clearly defined dorsal process resembling that of modern frogs, in which it forms the base of the area where the tympanic annulus develops (Bolt and Lombard, 1985). DeBeer (1985) illustrated the manner in which the tympanic annulus is attached to the squamosal in *Rana,* and a comparable area for its attachment can be seen in *Doleserpeton* (Sigurdsen, 2008). No specimens of *Doleserpeton* show palatal fangs, but rows of small denticles on the vomers resemble those in modern frogs. The absence of palatal fangs may be attributed to the small size of *Doleserpeton,* reflecting the paedomorphic genesis of this anuran character. The absence of an ectopterygoid in *Doleserpeton,* as in frogs, is associated with the very great lateral extent of the interpterygoid vacuities.

Adults of *Doleserpeton* resemble all lissamphibians in having a unipartite atlas vertebra with a bicondylar articulation with the occiput but lacking transverse processes. No vertebral column is preserved in its entirely. The best-articulated specimen shows 22 vertebrae extending from the atlas to the sacrum, but including a space that might have been occupied by one or more additional vertebrae.

Several examples of articulated trunk vertebrae illustrate a succession of changes during growth (Fig. 10.7D–F). The smallest are more advanced than typical temnospondyls in having crescentic rather than paired pleurocentra, but at this stage they are clearly separate from the neural arches and smaller than the crescentic intercentra. With increased size, and also visible as a succession of changes from posterior to anterior along the vertebral column of a single individual, the pleurocentra grow more rapidly than the intercentra, come in contract with the neural arches, and subsequently fuse with them to form highly notochordal holospondylus centra as in the most primitive living frogs. The ontogeny of *Doliserpeton* clearly reflects the phylogeny of anurans.

Doleserpeton expresses by far the most saliention synapomorphies of any Paleozoic tetrapod, but shows none of the characters associated with saltation. The limbs and girdles show little if any advance suggestive of the initiation of anuran locomotion, except, perhaps, the initiation of elongation of the tibiale and fibulare.

Amphibamus grandiceps

The earliest known amphibamid, *Amphibamus grandiceps*, from the lower Westphalian D of Mazon Creek, Illinois, is slightly smaller than *Doliserpeton*, but already highly ossified, except for the carpals and tarsals. It was long ago suggested as having froglike characters (Watson, 1940), but has few clear-cut synapomorphies other than the general form of the skull and the conspicuous otic notches. It lacks palatal fangs, but retains the ectopterygoid, and the frontal does not enter the margin of the orbit. It was reconstructed by J. T. Gregory (1950) as having 19 presacral vertebrae (Fig. 6.14).

Andrew Milner (1982) identified several larvae from Mazon Creek as belonging to *Amphibamus grandiceps*. They are broadly similar to those of micromelerpetontids in the presence of external gills, the early ossification of the circumorbital bones, and the formation of the neural arches prior to the centra. However, some preserve infillings of the stomach and intestine that are comparable to those of carnivorous rather than herbivorous tetrapods. None show specific derived characters in common with anuran larvae. The fleshy portion of the tail is long, but only the most anterior seven pairs of neural arches are ossified. They lack any evidence of the hyoid apparatus, which is preserved in other larvae from the same locality that Milner referred to as branchiosaurids. Neither type of larva shows evidence of the branchial denticles by which branchiosaurids are distinguished from micromelerpetontids. *Amphibamus grandiceps* is the oldest known dissorophoid, going back roughly 310 million years.

The Earliest Temnospondyls

Early amphibamids can be traced back to the earliest known temnospondyls, *Dendrerpeton* (Westphalian A) and *Balanerpeton* (late Viséan), discussed in Chapter 4, which exhibit the initial appearance of a conspicuous otic notch with a reduced stapes, large interpterygoid vacuities, and, in the smaller *Balanerpeton*, large orbits. Like amphibamids, they show no evidence of lateral line canals. *Balanerpeton* is known from one well-preserved juvenile specimen showing the early development of paired neural arches prior to ossification of the centra, in common with all temnospondyls and anurans. No earlier tetrapods show evidence of the emergence of any specifically anuran characters and no putative sister-taxa of temnospondyls are recognized among other Carboniferous tetrapod lineages.

Nested Synapomorphies

The possible sequence of evolution of anuran characters can be listed according to their appearance in the fossil record.

Primitive Characters Retained from Primitive Temnospondyls

Tabular not in contact with parietal, distinguishing them from anthracosaurs

Retention of paired pleurocentra

Ossification of neural arches prior to centra

Medial process of pterygoid overlapping parasphenoid, not suturally attached

Characters Derived at the Level of *Balanerpeton*

1. Conspicuous otic notch and reduced size of stapes
2. Large interpterygoid vacuities, extending to vomers
3. Large size of orbits, relative to skull length

Characters Derived at the Level of *Dissorophoids*

4. Loss of intertemporal bone
5. Relatively small size

Characters Derived at the Level of *Amphibamus grandiceps*

6. Palatal fangs replaced by rows of denticles
7. Further reduction of adult body size

Characters Derived at the Level of *Doleserpeton*

8. Orbits bordered by frontals and palatines
9. Pedicellate teeth
10. Early ossification of quadrate
11. Presence of dorsal process of quadrate
12. Double occipital condyle
13. Fusion of elements of atlas into a unified structure in adults
14. Holospondylous trunk vertebrae
15. Elongation of tibiale and fibulare

Characters Derived at the Level of *Triadobatrachus*

16. Loss of circumorbital bones and those at the back of the skull table
17. Fusion of frontal and parietal into paired frontoparietal
18. Neomorphic hyoid with parahyoid and posteromedial processes
19. Number of trunk vertebrae reduced to fourteen
20. Possible reduction of caudal vertebra to number close to that of larval frogs

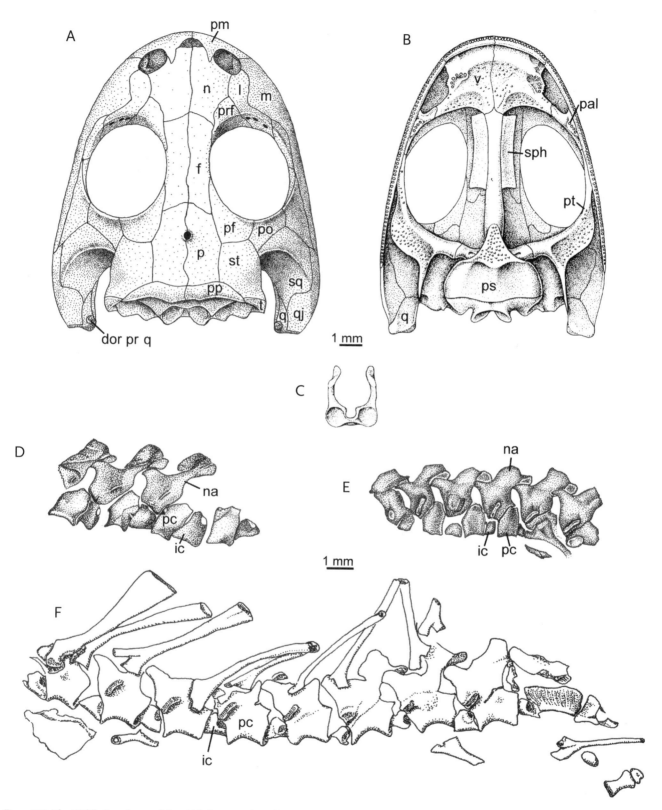

Figure 10.7. The Middle Permian amphibamid *Doliserpeton* from fissure fillings of Fort Sill, Oklahoma. *A* and *B,* Skull in dorsal and ventral views. *C,* Atlas in anterior view. Modified from Bolt, 1969. *D, E,* and *F,* sequence of ossification of vertebrae with increasing size.

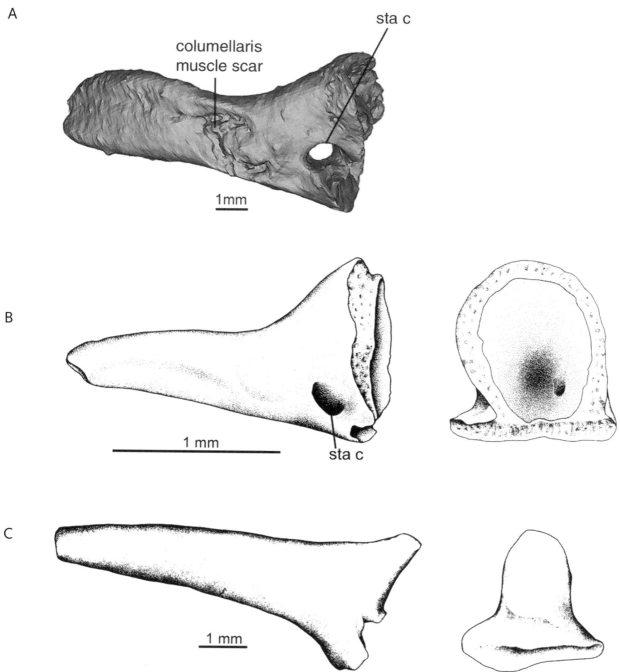

A

columellaris
muscle scar

sta c

1mm

B

sta c

C

1 mm

1 mm

Figure 10.8. Stapes of temnospondyls and a frog. A, Stapes of the early temnospondyl *Dendrerpeton*. From Robinson et al., 2005. B–C, Stapes of the amphibamid *Doliserpeton* and *Rana*, lengthwise and in medial views, showing the ventral edge that hinges on the margin of the fenestra ovalis. From Bolt and Lombard, 1985.

21. Anteriorly directed iliac blade
22. ? Initiation of tadpole characters

    Characters Derived at the Level of Lower Jurassic Frogs
23. Posterolateral extension of frontoparietal lost, allowing passage of adductor mandibulae posterior over otic capsule
24. Fusion of caudal vertebrae to form urostyle
25. Separate ossification of scapula and coracoid
26. Formation of rotating articulation between sacral rib and iliac blade
27. Further elongation of iliac blade
28. Fusion of ulna and radius and tibia and fibula
29. Elongation of all major limb bones

The recognition of a nested sequence of synapomorphies, beginning with *Balanerpeton* and running through the amphibamid dissorophoids and triadobatrachids, enables us to trace the evolution of anuran characters back to the base of known temnospondyls but, at present, no farther. Their possible affinities with salamanders and caecilians will be considered in the following chapters.

11

The Ancestry of Salamanders

RACING THE ANCESTRY OF SALAMANDERS is considerably
more difficult than that of frogs because of the much greater diversity of
the anatomy and ways of life of the living families (Plate 13). All frogs retain
very stereotyped body plans in relationship to the nature of their larvae and the saltatory locomotion of the adults, and nearly all are primarily terrestrial, following a
striking metamorphosis. In contrast, salamanders, as far back in the fossil record as
the living families can be traced, vary from a classically amphibious way of life with
aquatic larvae and primarily terrestrial adults to groups that are neotenic (maintaining larval characteristics throughout their lives). These different ways of life are accompanied by very distinct body forms. The terrestrial and metamorphosing genera
tend to have short bodies and stout limbs, while many of the obligatorily aquatic
groups have extremely elongate bodies, with the limbs highly reduced and even lost.
The greatest contrasts can be seen between sirenids, which are obligatorily aquatic
and retain only tiny forelimbs, to the many plethodontids, which lay their eggs on
land and may never go into standing water.

Such extreme body forms are known as far back as the Middle Jurassic (Fig. 11.1).
Their skeletal diversity suggests a considerably earlier time of divergence, but makes
it difficult to find a particular suite of skeletal characters that allow us to recognize
their possible antecedents in the Triassic and Paleozoic. Fortunately, there are numerous consistent features of the skulls and some aspects of the postcranial skeleton to support the monophyly of all the crown groups (Figs. 11.2–11.4) . Among
living salamanders, the family Hynobiidae is the most primitive, using Paleozoic
tetrapods as a general out-group. This is demonstrated by the following list of
characters:

Distinctive Features of Conservative Salamanders (Cryptobranchoids, Ambystomatids, and Salamandrids)

Skull

1. Clearly sequential pattern of ossifying dermal bones of the skull during ontogeny
2. Loss of most circumorbital bones of Paleozoic tetrapods, except for the prefrontal and lacrimal (retaining a lacrimal duct)
3. Loss of all primitive bones at the back of the skull table: tabular, supratemporal, intertemporal, and postparietal
4. Lack of squamosal notch
5. Absence of tympanic annulus and dorsal process of quadrate
6. Large orbits
7. Extensive interpterygoid vacuity
8. Stapes with short stem, initially with stapedial foramen
9. Circular opercular bone fitting into fenestra ovalis and attached to shoulder girdle by opercularis muscle
10. Absence of bony connection between maxilla and jaw suspension
11. Ventral end of squamosal moves laterally relative to braincase via a hinge-like articulation with the parietal and/or otic capsule
12. Absence of bony contact between pterygoid and maxilla
13. Medial process of pterygoid articulates with base of braincase rather than overlapping the parasphenoid as occurs in frogs
14. Pedicellate teeth
15. Bicondylar occipital condyle
16. Hyoid apparatus retains separate hypobranchials and ceratobranchials that articulate with one another
17. Ceratobranchials become ossified or calcified at the time of metamorphosis in neotenic species (for example, cryptobranchids)
18. Gill rakers not attached to bony plates, but arranged in six rows that are capable of interdigitation so as to preclude flow of water through gill slits during gape-and-suck feeding
19. Lower jaw consists of Meckel's cartilage, which ossifies anteriorly to produce the mentomecklian bone and posteriorly to form the articular bone, as well as the dentary, prearticular, angular, and coronoid

Postcranial Skeleton

20. Holospondylous, notochordal centra, fused to neural arches in adults
21. Single sacral vertebra
22. Retention of extensive tail in all species
23. Primitively, ossification of neural and haemal arches prior to centra
24. Short ribs on all trunk vertebrae except atlas
25. Single pair of sacral ribs
26. Loss of all dermal bones of shoulder girdle
27. Extensive elaboration of cartilage around endochondral shoulder girdle
28. Neomorphic sternum
29. Pubis greatly reduced in size and degree of ossification
30. Unique fusion of first and second distal carpals and tarsals to form basale commune
31. Phalangeal formula primitively 2, 2, 3, 2 and 2, 2, 3, 4, 3
32. Preaxial dominance in sequence of ossification of bones of appendages

In addition, numerous aspects of the soft anatomy, functional mechanics, and mode of reproduction, discussed in Chapter 9, demonstrate the strong probability that they shared a common ancestry, separate from that of frogs and caecilians.

Hynobiids are more primitive than all other living salamanders in that some (but not all species) ossify their neural arches before the centra (Boisvert, 2009). Hynobiids are also among the most primitive living salamanders in not practicing internal fertilization, but this trait is also shared with cryptobranchids, which are otherwise very different in body form and way of life. The common absence of internal fertilization has led to hynobiids and cryptobranchids being united in a single superfamily, Cryptobranchoidea. All other salamanders, with the possible exception of sirenids, in which their means of fertilization is not known with certainty, have internal fertilization. Salamanders lack a copulatory organ. Instead, the male produces a spermatophore that is picked up by the cloacal lips of the female. This capacity is approached in the hynobiid *Ranodon sibericus,* in which the female discharges her eggs on top of a spermatophore deposited by the male (Pough et al., 2004).

Hynobiids and cryptobranchids seem to be close to the base of the phylogeny of modern salamanders. Vertebrae attributed to the modern hynobiid *Ranodon* have been described from the Pliocene of Kazakhstan and from the Miocene and Pleistocene of Romania (A. R. Milner, 2000). Cryptobranchids are known as far back as the Middle Jurassic of China, as well as from the Oligocene and Miocene of Europe.

THE OLDEST KNOWN SALAMANDERS

The fossil record of the 10 currently recognized families of salamanders (Fig. 11.1) is limited almost entirely to the Cenozoic, going back only about 66 million years. Three of these families are known from scattered remains in the Late Cretaceous. In strong contrast with the scant evidence in the Cretaceous and for most of the Cenozoic, a wealth of fossils, representing at least four major lineages, is known from the mid-Jurassic of China. This does not indicate the actual rarity of salamanders in the intervening time period, but reflects unusual conditions of preservation.

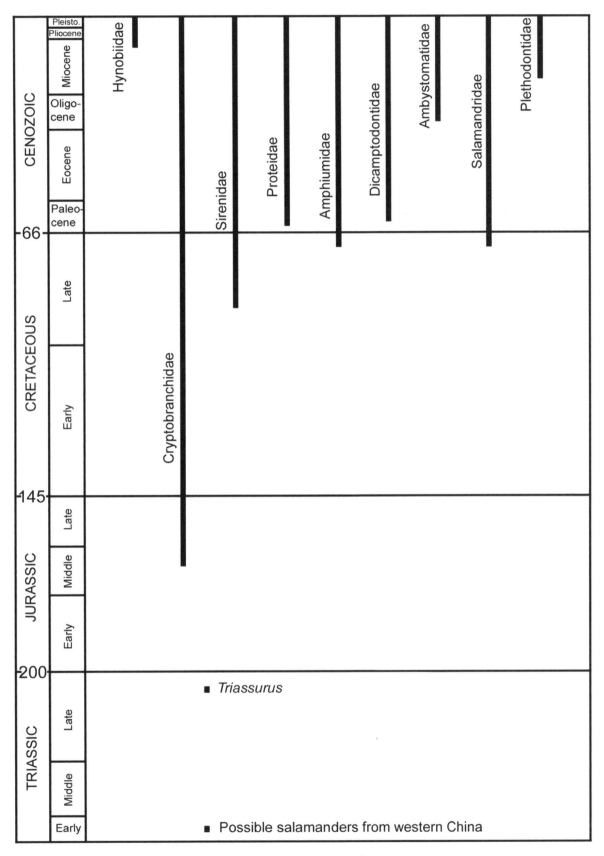

Figure 11.1. The temporal duration of the modern salamander families as indicated by the fossil record. The Rhyacotritonidae has no fossil record and there are very large gaps in our knowledge of the history of all other families. Based on data from A. R. Milner, 2000.

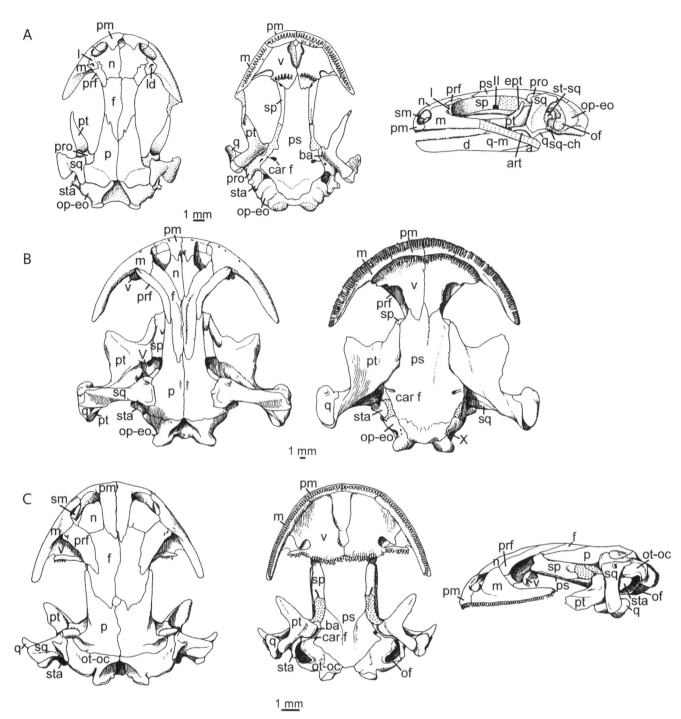

Figure 11.2. Skulls of modern salamander families. A, The hynobiid *Batra-chuperus sinensis* in dorsal and palatal views, and a lateral view of *Hynobius naevius*. B, Dorsal and palatal views of *Cryptobranchus alleghaniensis*. C, The ambystomatid *Ambystoma maculatum* in dorsal, palatal, and lateral views. From R. L. Carroll and Holmes, 1980.

Within the past 10 years, a wealth of fossils, termed the Jehol Biota, have been discovered in a band of sediments extending for hundreds of miles across northern China and Inner Mongolia (Norell and Ellison, 2005). These deposits reveal a host of primitive birds, feathered dinosaurs, ancestral mammals, and the oldest known flowering plants, but also a growing number of salamanders (Chang, 2003). The specimens include not only complete, articulated skeletons, but also clearly defined impressions of soft tissue representing the skin and external gills. The Jehol fossils resemble those of the Chengjiang fauna and the Burgess Shale in documenting a previously unknown biota.

The beds were deposited in a series of lakes that were subject to ash falls from volcanic eruptions that occurred over a span of several million years. Toxic gases overwhelmed all groups of organisms living in this area, from plants and insects to dinosaurs, but also killed the microorganisms that would normally have broken down their organic remains.

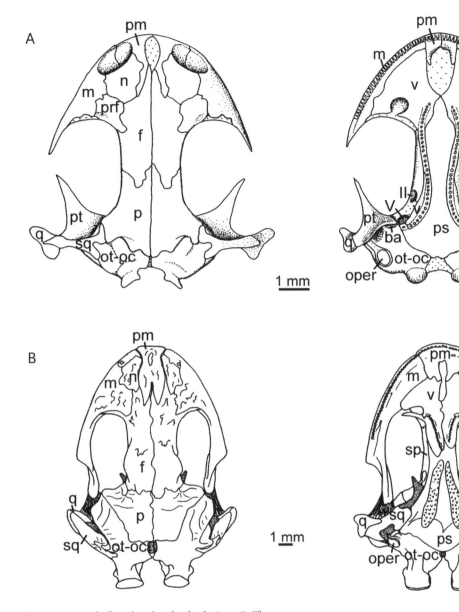

Figure 11.3. A, Salamandra atra, skull in dorsal and palatal views. *B*, The plethodontid *Phaeognathus hubrichti*, skull in dorsal and palatal views. From R. L. Carroll and Holmes, 1980.

This resulted in the most informative assemblage of fossil salamanders in their entire evolutionary history.

There has been some debate regarding the exact age of the fine lacustrine deposits in which the best-preserved salamanders have been found. As discussed by Liu and Liu (2005), the intercalated volcanic tuffs of the Tiaojishan Formation, in which these and other fossils in the vicinity of Daohugou have been found, were dated by radiometric means as approximately 165 million years old. This places them in the Middle Jurassic (Bathonian, in European terminology). In contrast, Wang (2004) had argued for an Upper Jurassic or Lower Cretaceous age of about 145 million years. However, even the 20 million years difference between these dates

is not very long in terms of the slow rate of salamander evolution.

Several highly distinct genera are currently recognized from the Daohugou beds. The closest in appearance to a particular family of modern salamanders is *Chunerpeton tianyiensis*, described by Gao and Shubin (2003) as a cryptobranchid, one of the most primitive of living families. It is now known from numerous exquisitely preserved specimens revealing not only the complete skeleton, but also the soft tissue covering the entire body (Figs. 11.5–11.7). *Chunerpeton* resembles modern cryptobranchids in being paedomorphic, that is, retaining larval characteristics in the largest known specimens that appear adult in the high degree of ossification of the der-

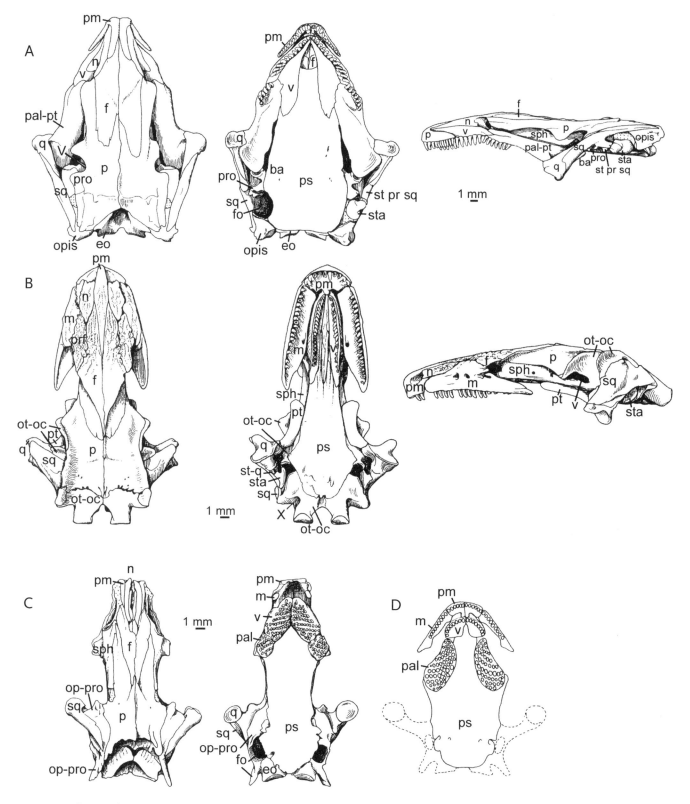

Figure 11.4. A, The proteid *Necturus,* skull in dorsal, palatal, and lateral views. *B, Amphiuma,* skull in dorsal, palatal, and lateral views. *C, Siren,* in dorsal, lateral, and palatal views. *D,* The Late Cretaceous sirenid *Habrosaurus.* From R. L. Carroll and Holmes, 1980.

mal bones of the skull. The largest fossils have a total body length approaching 20 cm.

The skulls are flattened, as in modern genera, and the row of palatal teeth on the vomers parallels those in the premax-

illa and maxilla. They retain a quadratojugal bone, and have a large fontanelle between the premaxillae but may have lost their nasals. However, *Chunerpeton* is conspicuously more primitive than the living cryptobranchids in retaining three pairs of fully developed external gills, whereas these are missing in adult specimens of *Cryptobranchus*.

Another important difference from modern cryptobranchids is the configuration of the hyoid apparatus. *Cryptobranchus* has elongate hypobranchials and ceratobranchials that are fused to one another to form a very long rod, but *Chunerpeton* has two pairs of short hypobranchials and a triradiate basibranchial, somewhat as in branchiosaurids and *Dvinosaurus* (Figs. 11.6C and 6.12F). Impressions of the body surface are smooth, in contrast with the strikingly wrinkled appearance in the living species. At least the anterior cervical ribs have two clearly separate areas of articulation with the vertebrae, a feature typical of most modern salamanders, whereas they are single headed in extant cryptobranchids. The fore- and hind limbs are well developed, but without ossification of the carpals or tarsals.

Another salamander from Daohugou, *Jeholotriton* (Fig. 11.7B), was identified as a hynobiid by Wang and Rose (2005), and like some living members of that family ossified the haemal and neural arches in the caudal region prior to the centra. However, the palatal dentition appears very much like that of modern sirenids, with tooth plates on the vomers and palatines. There is no elongation of the trunk or reduction of the limbs, but the retention of external gills in large specimens shows them to be paedomorphic, unlike any hynobiids. Its relationship remains enigmatic.

A highly distinct species, represented by a single immature specimen, appears to ossify at very small size and to have had relatively gracile limbs (Figs 11.7C and 11.8A). The trunk has approximately 16 vertebrae. The tail, ossified to the end, is as long as the rest of the body. An impression of the body surface is striking in demonstrating the presence of costal grooves, as in hynobiids, salamandrids, ambystomatids, and plethodontids. In these genera they are used to transport water from a wet substrate to the dorsal portion of the trunk by capillary action so as to dampen the skin for effective cutaneous respiration. The presence of costal grooves suggests that this animal, although immature, was already highly terrestrial, which may explain its rarity in lacustrine deposits. These derived features suggest that this animal may have belonged to the assemblage of advanced salamanders, the Salamandriformes. It somewhat resembles *Valdotriton* from the Early Cretaceous Las Hoyas locality in Spain, described by Evans and Milner (1996) (Fig. 11.8B). Unfortunately, the small Chinese specimen is too immature to show details of bone structure such as the presence or absence of spinal nerve foramina in the trunk or tail that are used to distinguish the more advanced families of salamanders.

An additional, clearly distinct salamander from the Daohugou beds, *Liaoxitriton daohugouensis*, was described by Wang in 2004 (Fig. 11.8D). It is smaller than either *Chunerpeton* or *Jeholotriton*, about 9 cm in snout-vent length, and yet is much more completely ossified, including several of the bones in both the carpus and tarsus. The position of a tarsal proximal to the first two metatarsals suggests its identity as a basale commune (a bone unique to salamanders), but the carpus shows two bones in this position. The orbitosphenoid is also ossified in *Liaoxitriton*, but a large fontanel remains between the vomers. In contrast with *Jeholotriton*, the caudal centra ossify prior to the neural and haemal arches.

Surprisingly, in a highly ossified, presumably terrestrial specimen, the configuration of the hyoid apparatus seems close to that of adult proteids and cryptobranchids, with a single pair of hypobranchials and ceratobranchials forming a continuous elongate structure. These were described as hypobranchial and ceratobranchial II by Wang, but with only a single pair preserved, they cannot be distinguished from hypobranchial I and ceratobranchial I, as they occur in modern paedomorphic salamanders (Duellman and Trueb, 1986, 302).

Another very striking individual specimen collected from the Daohugou beds is a beautifully preserved larva illustrated by Gao and Shubin (2003), without any obvious affinities with the previously described species represented primarily by adults (Fig. 11.9A). It broadly resembles the larvae of modern ambystomatids and other salamandriforms in its body form with a widely expanded caudal fin and fore- and hind limbs in the process of ossifying. It has three pairs of external gills and the area of the stomach is filled with conchostracans. The only primitive feature is the clear ossification of the neural and haemal arches of the tail before the caudal centra.

Continuing study of this material from the Jurassic of China will assist in establishing more specific affinities with later salamanders, and may help to determine the interrelationships and times of divergence of the modern families. On the other hand, no fossils older than the Middle Jurassic have yet been discovered that are plausibly related to specific living families or superfamilies.

PRIMITIVE CAUDATES

The specimens from the Jurassic and Lower Cretaceous document an early radiation of diverse body plans and ways of life among salamanders, ranging from obligatorily aquatic cryptobranchids to what were probably fully terrestrial forms. As in the case of frogs, in which members of living families and their immediate sister-taxa are included in the Anura but members of more primitive sister-taxa (the triadobatrachids) are grouped in a more inclusive assemblage, the Salientia, so the modern salamander families and their immediate sister-taxa are grouped as the Urodela, but more primitive taxa, thought to have diverged from a more inclusive assemblage, are united with them in the Caudata.

Several genera that may be closely related to the ancestry of salamanders, but are not thought to have close ties with any of the living families, are also known from the Middle Jurassic through Lower Cretaceous. Among the best known are *Karaurus*, from the Upper Jurassic of Kazakhstan, and *Kokartus*, from the Middle Jurassic of Kirghizstan

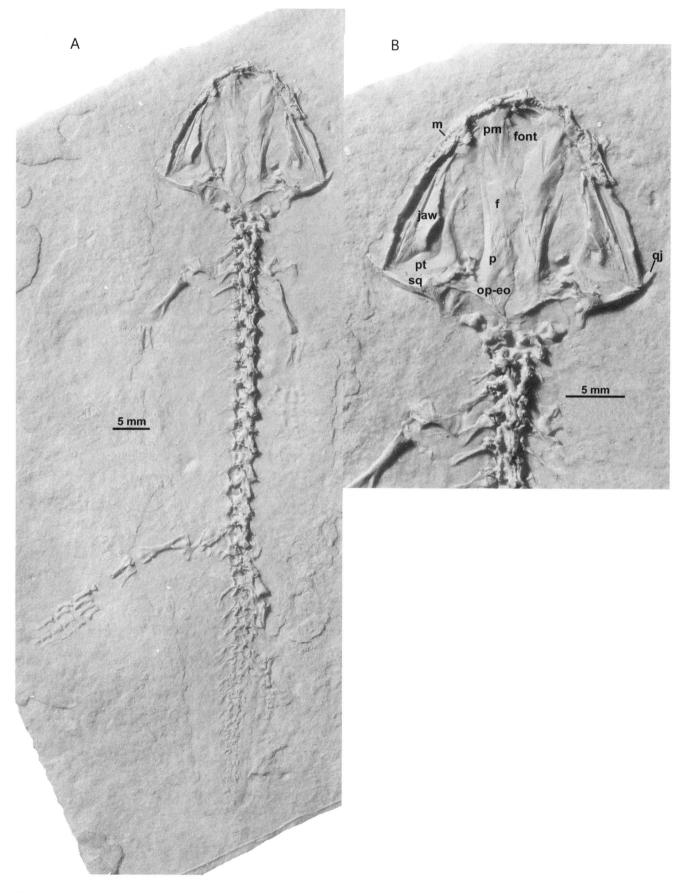

Figure 11.5. A, Dorsal view of entire skeleton of the paedomorphic salamander *Chunerpeton,* from the Middle Jurassic of Inner Mongolia. *B,* Skull in
dorsal view. Based on cast of specimen in the Inner Mongolia Museum.

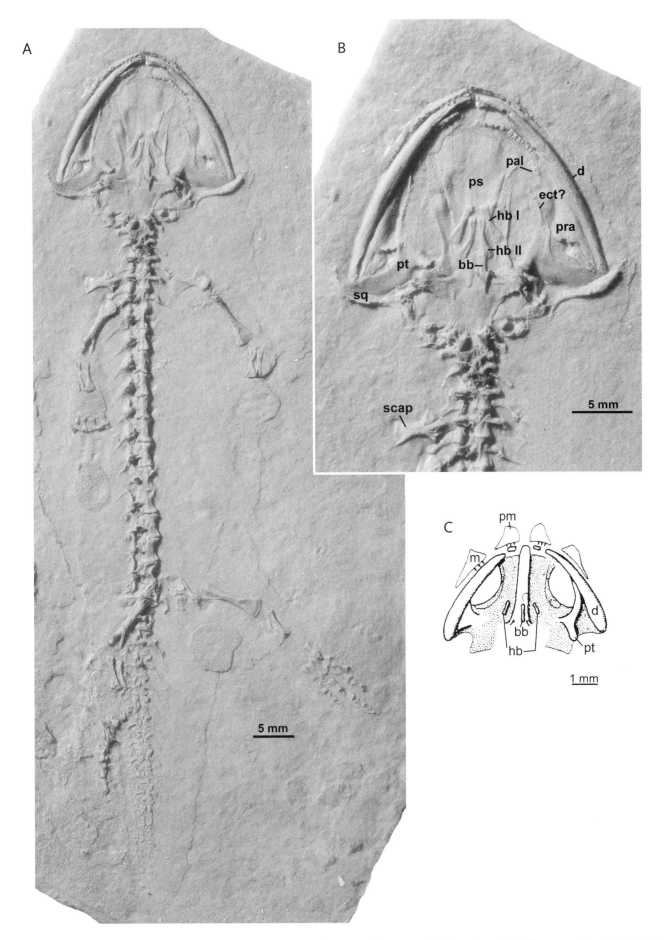

Figure 11.6. A and *B*, Ventral view of entire skeleton of *Chunerpeton* and larger-scale view of palate. Based on cast of specimen in the Inner Mongolia Museum. *C*, Palatal view of a possible branchiosaurid from the Westphalian D of Mazon Creek, Illinois. From A. R. Milner, 1982. Note similarity in the arrangement of the branchial elements.

Figure 11.7. Soft anatomy of salamanders from the Middle Jurassic Daohugou locality in Inner Mongolia. *A*, A cryptobranchid distinct from *Chunerpeton*. *B*, The hynobiid *Jeholotriton*. *C*, Unnamed genus of small, apparently terrestrial genus with costal grooves. Specimens in the Inner Mongolia Museum.

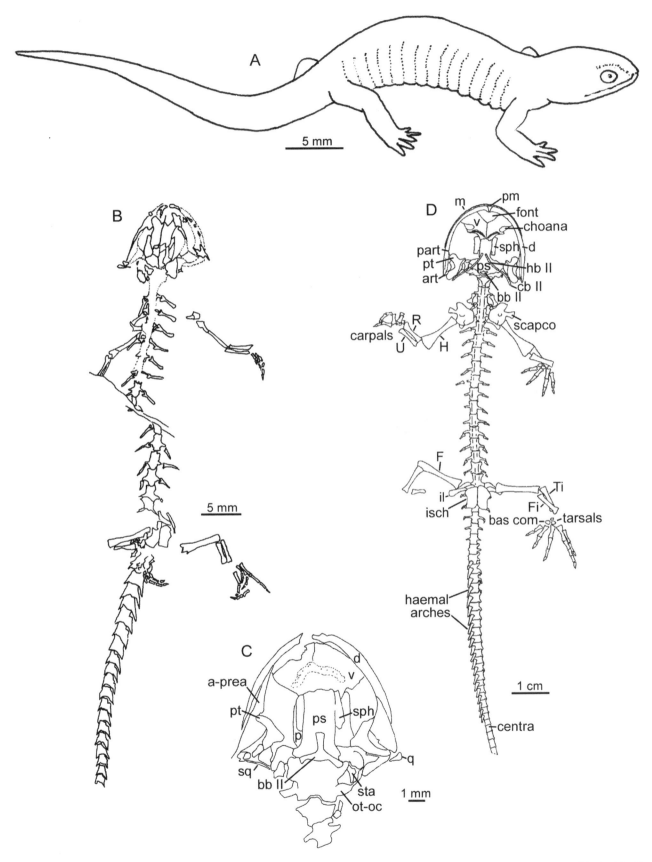

Figure 11.8. Middle Jurassic and Lower Cretaceous salamanders. *A*, Reconstruction of the body outline of the unnamed salamander from the Daohugou locality, Inner Mongolia. *B* and *C*, Skeleton of *Valdotriton* in dorsal view and detail of palate, from the Lower Cretaceous of Spain. After Evans and Milner, 1996. *D, Liaoxitriton daohugouensis,* Family *incertae sedis* from the Middle Jurassic of the Daohugou locality. From Wang, 2004.

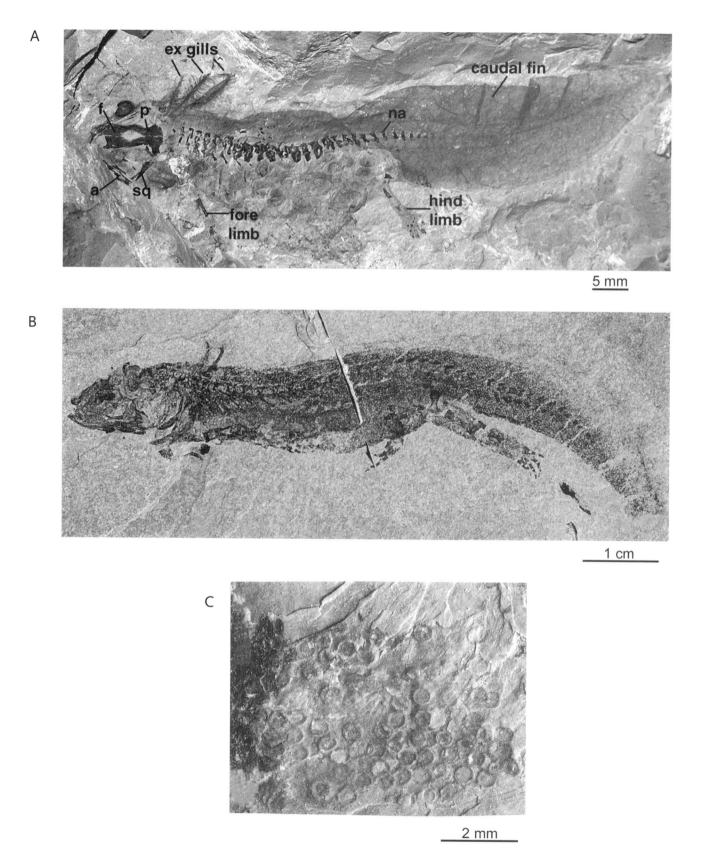

Figure 11.9. Fossils of eggs and larvae. *A,* Larva of an unnamed species of salamander from the Middle Jurassic of China. From Gao and Shubin, 2003. *B.* Larva of *Apateon pedestris* from the Lower Permian (specimen 44276 in the collection of the Royal Ontario Museum, Toronto). *C,* Fossil eggs from the upper part of the Lower Permian of Texas, putatively identified as belonging to antecedents of lissamphibians. From Mamay et al., 1998.

A

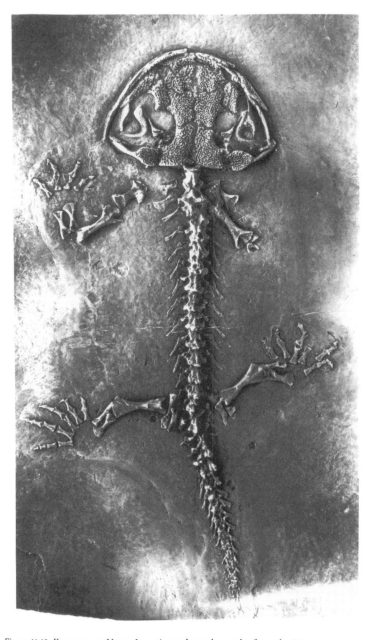

B

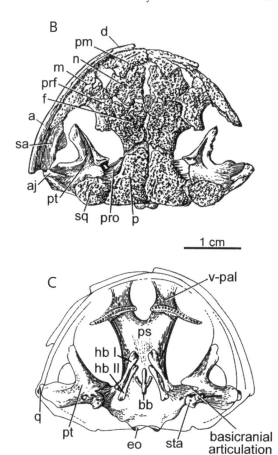

Figure 11.10. Karaurus, an Upper Jurassic caudate salamander from the Upper Jurassic of Russia. From Ivachnenko, 1978. Dorsal view of entire skeleton, and dorsal and palatal views of skull.

(Ivachnenko, 1978; A. R. Milner, 2000). They are placed in a distinct family, the Karauridae (Fig. 11.10). *Karaurus* appears primitive in the structure of the skull, with heavy sculpturing on the dermal bones. The sculpturing on the lateral margin of the parietal and adjacent squamosal precludes the passage of the superficial branch of the adductor mandibulae internus across the occiput, as occurs in modern salamanders. This supports the hypothesis of Holmes and Carroll (1980) that extension of the jaw muscles out of the adductor chamber happened separately in frogs and salamanders. *Karaurus* seems somewhat intermediate between aquatic and terrestrial genera in having stout limbs and a relatively short

trunk, but resembles the primitive Chinese cryptobranchids in the near identity of the hyoid elements and the common presence of a row of vomerine teeth parallel to the marginal dentition.

The most tantalizing fossil that might represent an earlier caudate is a single specimen from the Upper Triassic of Uzbekistan described by Ivachnenko (1978) as *Triassurus sixtelae.* It is a juvenile, in fact, probably a larval form, in which the neural arches of the vertebrae remain paired and no centra are yet ossified (Fig. 11.11). This condition is reminiscent of temnospondyls, larval frogs, and the most primitive hynobiid salamanders. The skull seems specifically salamander-like in

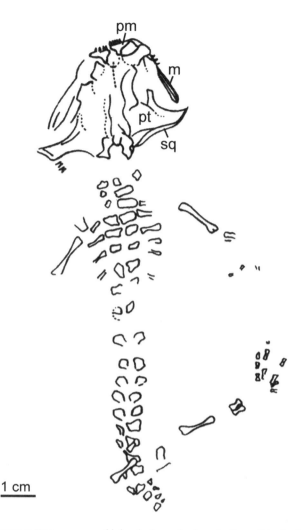

Figure 11.11. *Triassurus,* a possible basal salamander from the Upper Triassic of Uzbekistan. From Ivachnenko, 1978.

the long gap between the maxilla and the jaw suspension, and the lack of bony connection between the pterygoid and the maxilla. There are about 20 presacral vertebrae and limbs of modest proportions. Estes (1981) questioned its identity as a salamander, but noted similarities of the cheek with branchiosaurs. It remains the most similar to salamanders of any specimen from the Triassic.

Fossils from the Lower Triassic of western China were referred to by Gao and colleagues in 2004 as being either branchiosaurids or salamanders, but have not yet been described.

PALEOZOIC ANTECEDENTS OF SALAMANDERS
Evidence from Larvae

No amphibians are yet known from the Paleozoic that provide an obvious model for the ancestors of salamanders based on their adult morphology. Nearly all have more completely ossified skulls and lack the specializations of the ver-

tebrae and appendicular skeleton common to living urodeles. In the past, little attention was given to the larvae of Paleozoic amphibians as an additional basis of comparison. Laurin and Reisz (1997) stated that larval characters were excluded from their data matrix because they were not definitive characteristics of the taxa, but only an aspect of early development that was unavailable for study in most species.

It is true that larvae have only been described from a relatively small number of extinct vertebrates, but thousands of fossil larvae are known from the Upper Carboniferous and Lower Permian that reveal vital aspects of the evolutionary history of amphibians. In fact, it is the larvae that provide much of the data that enable us to trace the ancestry of salamanders into the Paleozoic.

Sequential Cranial Ossification

As we saw in Chapter 6, larvae are known for seymouriamorphs and a variety of temnospondyls, but the most informative are those of the dissorophoids, especially the families Branchiosauridae, Micromelerpetontidae, and Amphibamidae. Boy and Sues (2000) provided a very informative review of all branchiosaurs, but Rainer Schoch (1992) was the first to gain extensive evidence for a uniquely sequential pattern of ossification of the dermal bones of the skull that occurs in the branchiosaurid *Apateon*. He excavated and studied more than 600 specimens, all belonging to a single genus, from a single locality near Kusel in the Saar-Nahe Basin of southwestern Germany. This locality lies just above volcanic tuffs dated at 297 million years, slightly above the Carboniferous-Permian boundary.

These fossils were preserved in anoxic deposits at the bottom of a lake and show extraordinary anatomical details. Specimens can be assembled in long growth series, extending from recently hatched larvae, preserved as impressions of the skin surrounding the body but showing no ossification of the skull, to animals nearing metamorphosis. In nearly all branchiosaurs that had been described previously, the smallest preserved specimens had already ossified nearly all the dermal bones of the skull. Complete ossification of the skull at a very small size occurred at least as far back in tetrapod history as *Eusthenopteron* from the Upper Devonian (Fig. 11.12A). In marked contrast, the specimens from Kusel show a highly consistent sequence of ossification in skulls of increasing size, as seen in Plate 14 and Figure 6.20A. This begins with the tooth-bearing bones of the palate and upper and lower jaws, and continues with the squamosal, parietal, and frontal followed by the bones of the skull table and finally those surrounding the orbit (Table 11.1).

What we see in *Apateon* is a much extended period of larval growth, during which the first bones that ossify form a framework along the margins of the skull. This configuration provides a very flexible head, which would have allowed the oropharyngeal chamber to expand dorsoventrally and laterally during gape-and-suck feeding, as is presumably the case in modern salamanders.

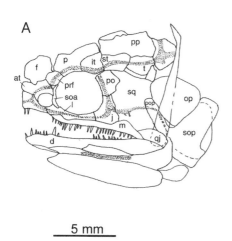

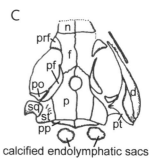

calcified endolymphatic sacs

1 mm

Figure 11.12. Early ossification of most of the dermal bones of the skull in the smallest known specimen of a choanate fish and early tetrapods. *A,* The choanate fish *Eusthenopteron.* From Schultze, 1984. *B,* The temnospondyl

Micromelerpeton. From Schoch and Carroll, 2003. *C, Amphibamus grandiceps* from the Westphalian D of Mazon Creek. From A. R. Milner, 1982.

Table 11.1. Sequence of Ossification of Dermal Skull Bones in the Branchiosaurid *Apateon caducus* and the Hynobiid *Ranodon sibiricus*

Apateon caducus	Ranodon sibiricus
Parasphenoid, pterygoid	Parasphenoid, pterygoid
Palatine	Question of homology of palatine
Premaxilla, vomer	Premaxilla, vomer
Maxilla	
Squamosal, parietal, frontal	Squamosal, parietal, frontal
Supratemporal, ectopterygoid	(Not ossified in salamanders)
Postparietal	? Posterior parietal
Quadratojugal	Quadratojugal in *Salamandrella*
Nasal	Nasal
	Maxilla
Prefrontal, lacrimal	Prefrontal, lacrimal
Postfrontal	(Not ossified in salamanders)
Tabular	(Not ossified in salamanders)
Postorbital	(Not ossified in salamanders)
Jugal	(Not ossified in salamanders)
Septomaxilla	Septomaxilla

Source: Schoch and Carroll, 2003.

Note: The sequence of ossification of bones common to *Apateon* and *Ranodon* is very similar, except for the much later ossification of the maxilla in *Ranodon.*

At present, no branchiosaurids other than those of the genus *Apateon* clearly show the protracted sequence of ossification of the dermal skull. This may be because of the rarity of this condition even within this single family, or a result of exceptional conditions of preservation at this particular locality. Very large numbers of *Micromelerpeton* are known from many localities, but none of them show sequential ossification, although very tiny specimens are known, in which the entire skull is ossified, as well as the dermal bones of the shoulder girdle, but no other bones in the body (Fig. 11.12*B*). There are also fairly well preserved branchiosaurids, assigned to the genus *Branchiosaurus,* known from Nýřany and a few at Mazon Creek, both of which are at least 10 million years older than the Kusel locality. Some of these specimens are as small as those at Kusel and still retain the calcium reserves in the otic capsules that are present in the most immature individuals, but already exhibit extensive ossification of the circumorbital bones (Fig. 11.12*C*). This can be interpreted as indicating that the capacity of branchiosaurs to ossify the dermal bones of the skull in a slow, sequential manner evolved only later, between the Westphalian D and near the end of the Carboniferous.

When I first examined the smaller specimens of *Apateon* from Kusel under the microscope, I immediately thought of the pattern of the dermal skull of adult salamanders. These branchiosaurid larvae had, through delayed ossification, evolved a cranial configuration closely resembling that of mature hynobiids, ambystomatids, and salamandrids, lacking most of the circumorbital bones and those at the back of the skull table (Schoch and Carroll, 2003). But with growth, they ossified all the bones common to adult dissorophoids, and so the mature specimens appeared as quite normal, old-fashioned labyrinthodonts. It may have taken their descendants more than 100 million years, until sometime in the Triassic, to evolve the means to truncate the ontogeny of the skull bones and achieve the anatomy of salamanders as adults. A further salamander-like feature of the skull is recognized in another branchiosaur, *Schonfielderpeton* from the Lower Permian, in which the cheek region is emarginated, even in adults, with a gap between the jugal and the maxilla.

What is equally striking is that some very large specimens of *Apateon* have pedicellate teeth (Fig. 11.13), although they are much stouter than the slender pedicels of modern sala-

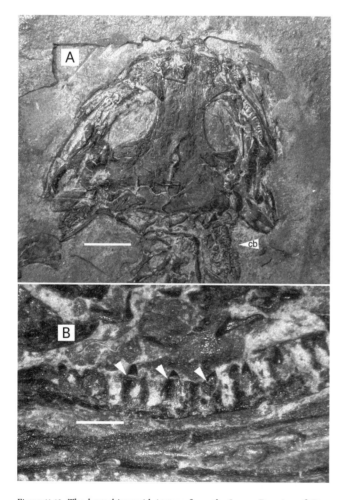

Figure 11.13. The branchiosaurid *Apateon* from the Lower Permian of Germany. *A,* Skull of a mature individual, showing the calcification of the ceratobranchials and the associated gill rakers. *B,* Detail of the lower jaw, showing the pedicellate nature of the teeth; arrows point to line of division between crown and pedicle. Upper scale bar 5 mm, lower scale bar 1 mm. From Schoch and Carroll, 2003.

manders. This specimen also exhibits calcified ceratobranchials. Both these and the appearance of pedicellate teeth are characters that appear in adult salamanders at the beginning of metamorphosis, but are not evident in the larvae.

Structure and Function of the Hyoid Apparatus

The occurrence of a uniquely salamander-like sequence of early cranial ossification and the presence of pedicellate teeth are among several lines of evidence that branchiosaurids are plausible antecedents of salamanders. Further support emerged from studies of a series of *Apateon* specimens that I had been sent by Hans-Dieter Sues, then working at the Royal Ontario Museum (R. L. Carroll et al., 2004). These showed the hyoid skeleton in superb detail. Figure 11.14 illustrates the nearly identical arrangement of the pharyngeal denticles relative to the ceratobranchials of branchiosaurids and primitive ambystomatid salamanders. In both groups, the individual dermal denticles (also called gill rakers) are attached to the ceratobranchials in such a way that they interlock across the gill slits like the teeth of a zipper. This enables the gill slits to be sealed during gape-and-suck feeding so that the vacuum in the oropharyngeal chamber is not lost, as described in modern salamanders by Lauder and Reilly (1988). They also provide the means for entrapping small prey.

This configuration of the dermal denticles is much altered from that in all other early labyrinthodonts in which the larvae are known. In these groups, as well as in the choanate fish *Eusthenopteron,* the dermal denticles were not separate, but attached to four rows of small dermal plates, each associated with a single ceratobranchial. In *Apateon* and other branchiosaurids, the dermal plates were lost and the dermal denticles arranged in six rows. This had already been achieved in branchiosaurids by sometime in the Upper Carboniferous.

Polarity of Ossification of the Limb Bones

Another striking feature of salamander development has been known since the early years of the 20th century as distinguishing them from all other terrestrial vertebrates. All living amniotes (reptiles, birds, and mammals) as well as frogs develop the bones that constitute the limbs in a posterior to anterior direction. This is termed posterior dominance. This is not apparent in the most proximal limb bones, the humerus and femur, but is for the lower limb: the ulna chondrifies and ossifies in advance of the radius, the fibula before the tibia, and the posterior before the anterior elements of the carpus and tarsus (Fig. 11.15) (Caldwell, 1994). The pattern of the metacarpals and metatarsals is somewhat more complex, but follows the same general direction. Ossification of the digits is in the order 4-(5 or 3)-2-1.

As mentioned in Chapter 9, Holmgren (1933) recognized that salamanders were exceptional among all vertebrates in developing the limb elements in the reverse order (anterior dominance). The order of ossification in the long bones is the opposite of that just listed, and that of the digits of the

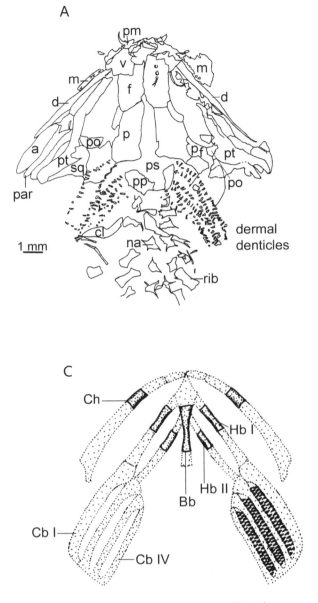

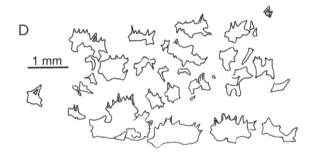

Figure 11.14. Pharyngeal denticles in salamanders and branchiosaurs. *A,* Palatal view of the branchiosaurid *Apateon.* The pharyngeal denticles, also termed gill rakers, are arranged in six rows so that the denticles interlock across the gill slits. *B,* The same pattern, seen in the modern salamander *Ambystoma tigrinum.* Note the external gills distal to the ceratobranchials.

C, Reconstruction of the arrangement of the hyoid apparatus and the gill rakers in branchiosaurids. *D,* The four rows of dental plates with attached denticles that occur in *Eusthenopteron* and larval labyrinthodonts other than branchiosaurids. From R. L. Carroll, 2007.

forelimb is the order 2-(1-3)-4 in the forelimb and 2-(1-3)-4-5 in the hind limb. He felt that this was such a fundamental difference that it suggested a separate origin of the tetrapod limb. Salamanders are also distinct in fusing proximal carpals and tarsals 1 and 2 into a single ossification, the basale commune.

The exquisitely preserved developmental sequences provided by the specimens of *Apateon* from Kusel cast additional light on this problem. Fröbisch et al. (2007) demonstrated that branchiosaurids showed essentially the same anterior

to posterior sequence of ossification as do modern salamanders (Fig. 11.16). The most parsimonious explanation for the phylogenetic distribution of this complex character is that it evolved only once, within the branchiosaurids, which otherwise share a large number of characters in common with salamanders. If branchiosaurids are not the sister-taxon of salamanders, the trait must have evolved twice. Or, if the trait evolved earlier in tetrapod evolution, it must have been lost in the ancestors of both amniotes and frogs, which would also have required a second or third evolutionary change.

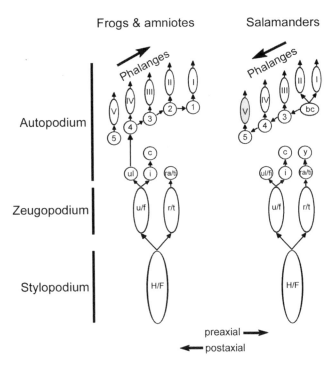

Figure 11.15. Axial dominance in the sequence of chondrification and ossification of the front and hind limbs of tetrapods. Frogs and amniotes develop their limb bones in a posterior to anterior sequence, ulna and fibular before radius and tibia, digit 4 before 3, 2, and 1. Salamanders are unique in developing their limb bones in an anterior to posterior sequence, as indicated by the large arrows. The letter y indicates an early center of tissue condensation of salamander limbs. From Fröbisch et al., 2007.

Overall Similarities between Branchiosaurids and Primitive Salamanders

These studies of branchiosaurids, specifically *Apateon*, reveal a number of unique derived characters in common with salamanders:

1. Sequential pattern of ossification of the dermal bones of skull
2. Stapes with short stem
3. Absence of dorsal process of quadrate
4. Pedicellate teeth develop at the time of metamorphosis
5. Ceratobranchials become calcified at maturity in paedomorphic species
6. Six rows of individual gill rakers capable of interdigitating across gill slits
7. Preaxial dominance of ossification of limb elements

In addition, primitive salamanders and branchiosaurs both retain numerous primitive characters of earlier temnospondyl amphibians:

1. Large orbits
2. Extensive interpterygoid vacuities
3. Hyoid apparatus retains separate hypobranchials and ceratobranchials
4. External gills in larvae and paedomorphic adults

5. Single sacral vertebra
6. Retention of extensive tail
7. Ossification of neural arches prior to centra
8. Short trunk ribs
9. Single pair of sacral ribs
10. Phalangeal formula close to 2, 2, 3, 2, and 2, 2, 3, 4, 3

You will note, of course, that the pattern of evolutionary change in the ancestry of salamanders converges back in time with that of frogs, as both approach the time of initial radiation of dissorophoids in the early Upper Carboniferous. This evidence strongly supports the sister-group affinities of frogs and salamanders within the modern fauna. Their putative antecedents, the amphibamids and branchiosaurids, were both present and already distinct at the time of deposition of the Mazon Creek fauna, but neither have yet been traced to earlier horizons. All the characters discussed in the previous chapter that argue against the origin of frogs from any group other than the temnospondyls also apply to salamanders.

A recent analysis of relationships among the Branchiosauridae by Schoch and A. R. Milner (2008) supports the monophyletic origin of this family, but suggests that they might be nested within the Amphibamidae. This would indicate that frogs and salamanders were essentially sister-taxa among the primitive dissorophoids. However, knowledge of the fossil record is not sufficient to determine the specific sequence of divergence of amphibamids and branchiosaurids relative to trematopids, zatracheids, or micromelerpetontids.

THE TRANSITION BETWEEN BRANCHIOSAURIDS AND SALAMANDERS

Although primitive salamanders share more unique derived features with branchiosaurids than with any other group of Paleozoic tetrapods, there are still a great number of differences between even the most primitive salamanders, fossil or living, and the most derived Lower Permian branchiosaurids. These include the loss of the following bones of the skull roof and palate: jugal, postorbital, postfrontal, supratemporal, tabular, postparietal, ectopterygoid, and palatine. The dorsal extremity of the squamosal comes to articulate with the otic capsule and the pterygoid forms a synovial joint with the base of the braincase rather than overlapping the parasphenoid. The cheek region is greatly modified with the loss of the squamosal notch, the dorsal process of the quadrate, the tympanum, and the middle ear cavity common to dissorophoids. This area became occupied by a much altered arrangement and larger mass of the adductor jaw muscles (R. L. Carroll and Holmes, 1980).

The magnitude of these changes is not surprising in view of the 120 million-year gap in the fossil record and presumably major changes in the ways of life between these two distinct periods in the evolutionary history of salamanders. For comparison, an even larger number of bones of the dermal skull roof were lost during the approximately 26 million

Figure 11.16. Sequence of limb bone ossification in the Upper Carboniferous branchiosaurid *Apateon.* Asterisk indicates the anterior side of the limb. Forelimbs are above, hind limbs below. Specimens increase in size from left to right. From Fröbisch et al., 2007.

years between the amphibamid *Doleserpeton* and the salentian *Triadobatrachus.*

Most, if not all, labyrinthodonts from the Paleozoic and Triassic almost certainly laid their eggs in the water, from which they hatched out as aquatic larvae (Fig. 11.9C). We now have examples of growth series for many of the families (Schoch, 2004; Witzmann, 2006; Witzmann and Pfretzschner, 2003). However, few show a clear metamorphosis be-

tween the larvae and adults; rather, they exhibit a more or less continuous sequence of change between the obligatorily aquatic larvae and facultatively terrestrial adults. In this they are clearly distinct from both frogs and most modern salamanders. Only one clear-cut example has been described by Schoch and Fröbisch (2006). Two specimens of the species *Apateon gracilis* are the only known branchiosaurids, out of thousands of known larvae, which show accelerated change in skeletal anatomy over a short period of time (as indicated by the limited change in body size) and achieved anatomical characters indicative of the terrestrial adults of other temnospondyl families. In addition to the expression of all the dermal bones of the skull, those of the braincase and also the quadrate are fully ossified, and the jaw articulation extended well behind the occiput, beneath a large otic notch. In these features they appear similar to the adults of amphibamids, in which the entire skeleton is highly ossified at a very small size.

All other species of *Apateon* and other branchiosaurids lack evidence of metamorphosis to fully terrestrial adults. They do achieve full ossification of the dermal bones of the skull and broadly resemble other adult temnospondyls, but retain larval features such as external gills and/or calcified ceratobranchials with associated gill rakers indicative of continued life in the water (Fig. 11.13). All of these animals may be considered neotenic, in reproducing while still retaining larval characteristics, or paedomorphic in the retention of larval anatomy and ways of life into adulthood.

The variable capacity for metamorphic change among branchiosaurids is a very important character in common with nearly all groups of living salamanders (Rose, 2003). What is surprising is that only a single branchiosaurid species is known to have had a rapid metamorphosis to an apparently fully terrestrial way of life. This may be attributed to the different environments from which their fossils have been recovered. The early history of branchiosaurids is known primarily from two localities in the Westphalian D—Nýřany in the Czech Republic and Mazon Creek in Illinois. No adults of these animals are known. Both localities have a large and diverse flora and fauna, and are assumed to represent low-altitude coal swamp environments within the Paleozoic tropics. Curiously, no branchiosaurs have ever been found at the Linton locality, although the remainder of the fauna is equally diverse and closely resembles that of Nýřany. Until recently, no branchiosaurs of any kind have been described from later horizons in North America. However, remains have been reported from the Virgilian (Late Pennsylvanian) of New Mexico (Werneburg and Lucas, 2007). The two metamorphosed specimens of *Apateon gracilis* came from a very diverse fauna in a Lower Permian locality near Dresden, Germany. Like Mazon Creek and Nýřany, the fauna near Dresden is thought to have lived in a lowland, equatorial environment.

In contrast, many of the Lower Permian branchiosaurids, collected from a series of basins that extended from France, across Germany, and into the Czech Republic, lived in a very different environment. They are known almost exclusively from offshore facies of relatively deep lakes. These had formed in intermontane basins of the Variscian mountains at elevations up to 2000 m or more above sea level (Boy and Sues, 2000). Even in the tropics, the air would have been cold at such an altitude and relatively little appropriate prey, such

as insects and other arthropods, would have been available on land.

Large, deep lakes, on the other hand, may have retained more amenable temperatures for small amphibians. However, no fossils of metamorphosed branchiosaurids are known from the lakes or in their immediately surrounding environment, which would be expected if they had bred in the area, even if they did not live there habitually. In contrast, adults are known of other amphibian groups of larger body size. This strongly suggests that these branchiosaurids did not metamorphose, but remained permanently neotenic. Nevertheless, these Lower Permian larvae do not yet show any overall differences compared with those found in Late Carboniferous localities with more equable environments.

No branchiosaurids are known with certainty after the Lower Permian. One can only imagine what might have happened if branchiosaurids had continued to live in large, cold lakes, or other water bodies in which the surrounding environment was unsuitable for their emergence as fully terrestrial adults. If branchiosaurids had been limited to strictly aquatic environments for millions of years, natural selection would have acted to maintain and elaborate systems associated with aquatic feeding, locomotion, respiration, and reproduction. These were already well established in branchiosaurids and continue to be expressed in the larvae of living salamanders. The gape-and-suck mode of feeding, common to larval labyrinthodonts and larval salamanders, would have been retained, in contrast with the evolution of an entirely different means of feeding in tadpoles. The calcification of the ceratobranchials common to living neotenic salamanders had already evolved in neotenic branchiosaurids (Fig 11.13), but did not occur in fully terrestrial specimens of adult *Apateon gracilis*.

We have seen that advanced branchiosaurids, specifically *Apateon pedestris* and *A. caducus*, retained a reduced complement of dermal skull bones comparable to that in modern salamanders until late in development, after which the circumorbital bones and those at the back of the skull table ossified. Only after the Lower Permian did the more immediate ancestors of salamanders evolve the capacity to halt development of these bones so that they never ossified. The ossification of two other bones, the quadratojugal and lacrimal, continued among primitive cryptobranchoid salamanders prior to their loss in more derived families.

Schmalhausen (1968) described elements of the middle ear as initiating development in early larval stages of a primitive hynobiid salamander, before being lost. This supports the hypothesis that the ancestors of salamanders possessed a middle ear, but if so, it was lost early in salamander evolution. As long as branchiosaurids habitually lived in the water, they would not have benefited from a system suited for reception of high-frequency airborne vibrations. Known branchiosaurids do have large otic notches formed by the squamosal, as in other dissorophoids, but they lack the well-ossified dorsal process of the quadrate that is associated with development of the tympanic annulus in modern frogs. In order to expose the tympanum to the air, neotenic branchiosaurids would have had to lift their heads out of the water. This would have exposed their external gills, which can only function when submerged. The two systems could work alternately, but not simultaneously. On the one hand, ancestral salamanders lost the use of high-frequency calls for species recognition, which may partially explain the much smaller number of extant species of salamanders than of frogs. But in the long run, they were able to use the area previously occupied by the very large middle ear cavity in primitive branchiosaurids and anurans for expansion of the area of the adductor jaw musculature.

In large, deep lakes, external fertilization (as in modern cryptobranchids and most hynobiids) may have been retained, as in nearly all frogs that breed in the water, even if they feed on land.

Unfortunately, the lack of fossil evidence of ancestral salamanders between the end of the Lower Permian and the Middle Jurassic makes it impossible to establish the timing and sequence of these changes, as well as the specific factors of selection that led to their expression. However, we can be certain that the potential for their development had been available long before the appearance of the diverse genera now known from the Middle Jurassic. This gap also makes it very difficult to establish the phylogenetic relationships among the early salamanders or their specific affinities with the modern families.

MODERN SALAMANDER FAMILIES

Many of the definitive skeletal characteristics of modern salamanders had evolved by the Middle Jurassic, but the paucity of fossils from the later Mesozoic makes it impossible to link them with any but the most primitive extant families. The gap in the fossil record of nearly 100 million years before the first known appearance of most of the modern families also makes it very difficult to establish their relationships with one another (Fig. 11.17). Eleven families are currently recognized (Pough et al., 2004). One, based on a single genus, *Rhyacotriton*, has no fossil record at all. All are briefly described to illustrate their distinctive features (Plate 13).

HYNOBIIDAE. We will begin with the most conservative of living salamanders, the hynobiids. They are known entirely from Asia, but broadly resemble the terrestrial salamandrids and ambystomatids from Europe and North America. Most members of these families have the distinctive costal grooves already seen in the specimen from the Middle Jurassic illustrated in Figures 11.7C and 11.8A. However, hynobiids are distinctly more primitive in having external fertilization. They breed in ponds, streams, and marshes. The hatchlings have balancers, as in the early larvae of ambystomatids. All have a complete metamorphosis to terrestrial adults with

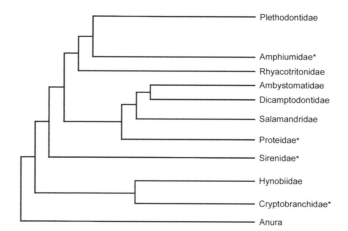

Figure 11.17. Salamander relationships based on Bayesian analysis of the combined molecular and morphological data, with paedomorphic taxa (*) coded as unknown for adult morphology. Modified from figure 8 in Wiens et al., 2005.

stout limbs and a short trunk. Hynobiids are relatively small, from 10 to 25 cm. The adults have short transverse rows of teeth on the vomers, somewhat resembling the tooth patches of *Doleserpeton* (Fig. 10.7B).

CRYPTOBRANCHIDAE. Cryptobranchids are also considered among the most primitive of salamanders because they retain external fertilization, but they differ greatly from hynobiids in their much greater size, up to nearly 2 m in length, and their obligatorily aquatic way of life. They have only an incomplete metamorphosis. They retain gill clefts, never develop eyelids, and lack the tongue pad necessary for terrestrial feeding. On the other hand, the adults of living species lose the external gills possessed by the larvae. Instead, they rely on large flaps of skin extending from the lateral trunk region to provide extra surface area for cutaneous respiration. Compared with other strictly aquatic families, their limbs are fairly well developed and retain the primitive number of toes.

The first known fossils of cryptobranchids were discovered in the 18th century from deposits in Switzerland that are now known to be Miocene in age. At the time, Scheuchzer (1726) thought them to be the remains of humans who had perished in the biblical flood and designated them as *Homo deluvii testis*—the man who had witnessed the flood. This species is now designated *Andrias scheuchzeri*.

SIRENIDAE. While cryptobranchoids maintain fairly stout fore- and hind limbs, the sirenids retain only the front limbs and have a long, eel-like body. The largest is nearly a meter in length. They also differ in retaining their external gills into the paedomorphic adults. They are grouped among the most primitive salamanders on the basis of the probable practice of external fertilization, but are otherwise highly derived. They have lost the premaxillae entirely and the maxilla is much reduced, to be replaced functionally by a keratinized beak. The primary dentition consists of large tooth plates on the vomers and palatines, similar to those of the Middle Jurassic *Jeholotriton* (Wang and Rose, 2005). The North American *Siren* is fully aquatic, but can aestivate during the dry season in a mucus cocoon in the mud beneath drying ponds. In contrast with other salamanders, sirenids lack pedicellate teeth.

PROTEIDAE. *Necturus,* more commonly termed the mud puppy, belongs to a third paedomorphic and obligatorily aquatic family, the Proteidae. It retains large external gills and a caudal fin. The premaxilla remains paired, but the maxilla is lost. There are only two pairs of larval gill slits, rather than the common larval number of three. *Necturus* lives in lakes, streams, and marshes, reaching a length of 45 cm. It maintains the normal four digits in the manus, but this is matched by the pes. The European *Proteus,* a blind cave salamander, has only three front digits and two on the hind limbs. Most individuals lack pigmentation.

AMPHIUMIDAE. The single genus *Amphiuma* is also paedomorphic, lacking eyelids and external gills. It has only one pair of gill slits and large pedicellate teeth. It reaches up to a meter in length, and the limbs are vestigial. The premaxilla is fused medially, the septomaxilla and lacrimal are lost, and the pterygoid is much reduced. The palatal teeth parallel those of the maxilla. They normally live in streams, rivers, and swamps, but can aestivate during dry spells.

The remaining salamander families are dominated by species that are relatively smaller, typically terrestrial as adults and with well-developed limbs.

AMBYSTOMATIDAE. The ambystomatids are limited to one genus, but it is widespread within North America and diverse at the species level. They average 10 to 30 cm in length. Like other advanced salamanders, they have a high number of intravertebral spinal nerves. Most are terrestrial, with short trunks and robust limbs, but one species, *Ambystoma mexicanum,* the Mexican axolotl, is obligatorily neotenic.

DICAMPTODONTIDAE. The dicamptodontids, generally placed close to the Ambystomatidae, are limited to one genus with four species. Most are terrestrial, but *D. copei* is permanently aquatic, living in cold streams and mountain lakes. Metamorphosis to fully terrestrial adults is facultative at the population and individual levels in other species. Both front and hind limbs are well developed at hatching, which would facilitate the locomotion of the larvae at the bottom of fast-flowing mountain streams, where they spend two to five years of their life.

RHYACOTRITON. This group consists of one genus, without a fossil record. The four species, all allopatric, are thought to have arisen from a very early radiation, based on genetic information. They are paedomorphic and distinctive in the loss of the operculum and opercularis muscle, as well as the great reduction of the lungs.

SALAMANDRIDAE. Salamandrids are common and diverse within both North America and Eurasia, constituting 15 genera and about 60 species. Most live in ponds and streams. The term newt is used for a number of genera within this family that have a distinct aquatic body form, other than the larvae, which is expressed during the breeding period or during some other portion of their adult life. Several other genera, including the common North American genus *Notophthalmus,* have a specialized terrestrial morph, the eft, within a life history that is primarily aquatic. *Salamandra* is unique among salamandrids in giving birth to live young. In some populations this is a facultative trait. Depending on the environment, the mother may give birth to a large number of advanced larvae or to a smaller number of terrestrial young.

PLETHODONTIDAE. Plethodontids are the most diverse, speciose, and derived of all salamander families, with 27 genera and more than 360 species. They are generally not paedomorphic, but all lack lungs. In some cases, this may be attributed to their living in fast-flowing streams, as is the case of lungless hynobiids and salamandrids. Many are very small and others slender and elongate, so that they can depend entirely on cutaneous respiration. Many tropical species are arboreal. They are also distinguished in the presence of a nasolabial groove, between the nostril and the margin of the lip, which is used in chemoreception. Plethodontids include the smallest of salamanders, as little as 3 cm in length.

EXTINCT FAMILIES OF CROWN-GROUP SALAMANDERS

Most of the salamanders known from the Late Cretaceous and Cenozoic can be placed in or close to the extant families, but there appear to be some groups that diverged at an early stage in urodele radiation, but failed to give rise to any living descendants. These have been placed in two families, the Batrachosauroididae and Scapherpetontidae (A. R. Milner, 2000). The batrachosauroidids are the better known, with skulls resembling those of proteids, accompanied by long bodies and short fore- and hind limbs. The scapherpetontids *Scapherpeton, Lisserpeton,* and *Piceoerpeton,* known from the Upper Cretaceous and Palaeocene, were large and superficially resembled cryptobranchids.

Primarily disarticulated material, of which the most informative elements are individual vertebrae, have been described from the Middle Jurassic of Kirtlington, England, and the Isle of Skye in Scotland (Evans and Waldman, 1996). These indicate the anatomical diversity and geographical distribution of primitive caudates, but none of these fossils help in establishing the specific origin or interrelationships of the living families.

ALBANERPETONTIDS

A further group that has been hypothesized as having affinities with salamanders are the albanerpetontids, known from the Mid-Jurassic into the Miocene (Fig. 11.18) (R. L. Carroll, 2007; McGowan, 2002). They appear superficially like small terrestrial salamanders, with a fairly long trunk (21 vertebrae in an articulated specimen) with a 15 mm long skull and well-developed limbs. However, details of the anatomy

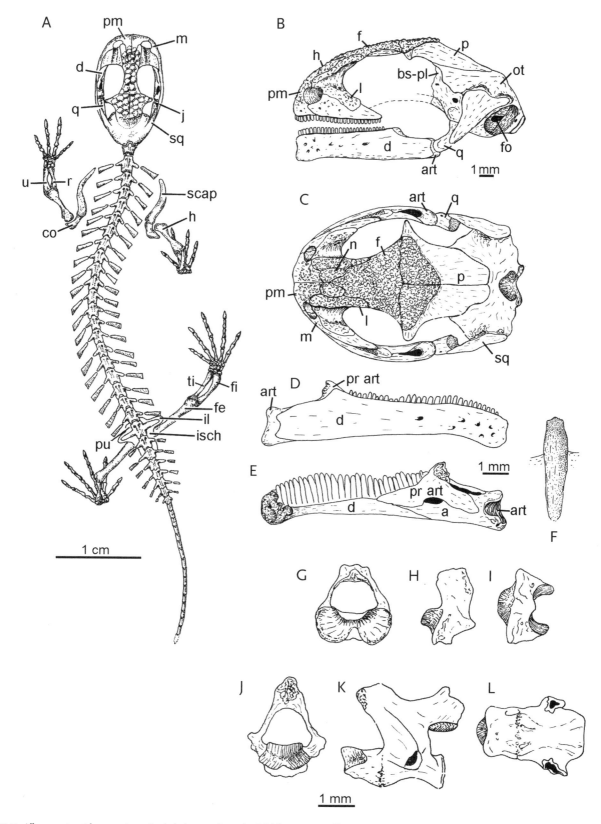

Figure 11.18. Albanerpetontids, an enigmatic clade known from the Middle Jurassic into the Miocene that has been suggested as being related to salamanders. *A,* Skeleton of *Celtedens megacephalus* from the Lower Cretaceous of Spain. Reproduced from McGowan and Evans, 1995. *B–L, Albanerpeton inexpectatum* from the upper Miocene of France. Modified from Estes and Hoffstetter, 1976. *B* and *C,* skull in lateral and dorsal views. *D* and *E,* lateral and medial views of lower jaw. *F,* single tooth (much enlarged). *G–I,* Atlas vertebra in anterior, lateral, and ventral views. *J–L,* Fused second and third vertebrae in anterior, lateral, and ventral views.

do not match those of any salamander. Throughout their long history, the skull is distinctive in the presence of medially fused frontal bones in sutural contact with paired parietals. Nearly all the bones making up the back of the braincase and otic capsules are fused to one another without trace of sutures. The jaw articulation is unique in being formed by a deep transverse groove at the posterior end of the articular into which fits the articulating surface of the quadrate. The symphysis of the lower jaws in all known species is formed by an asymmetrical peg-and-socket joint. A very large prearticular, in the position of a coronoid, has a conspicuous rugose area that presumably served for attachment of the adductor jaw musculature. None of these characteristics are present in any known salamanders, fossils or living. The teeth are not pedicellate.

The structure of the anterior cervical vertebrae is unique in the fusion of the centra of the atlas and axis, and the latter bone lacks a neural arch. The transverse processes of the trunk centra are short and lack the bicondylar articulation surfaces for the ribs common to most salamanders. Although the distal carpals and tarsals are ossified, the first and second have not fused to form the basale commune that is characteristic of urodeles, and the pubis is well ossified. The phalangeal formula is 2, 3, 3, 3, and 2, 3, 4/5, 3, unlike that of any salamanders.

The size and general appearance of albanerpetontids may have resembled those of contemporary terrestrial salamanders, and their way of life might have been similar, but there is no evidence of shared derived characters that would support common ancestry. On the other hand, the nature of the trunk vertebrae and the specific configuration of the articulation between the occiput and the atlas vertebra broadly resemble those of some microsaurs, but this is insufficient evidence for postulating a close relationship.

INTERRELATIONSHIPS OF THE MODERN SALAMANDER FAMILIES

All persons now studying the relationships of the living salamander families agree that they belong to a single, monophyletic assemblage. However, the interrelationships of these families among one another remain the subject of extensive debate. This is clearly shown by two recent articles in the journal *Systematic Biology* by Wiens et al. (2005) and Weisrock et al. (2005) that make use of various combinations of mitochondrial genomic sequences, nuclear genomic data, and morphological characters. There is strong support for the monophyly of all salamanders and for the early divergence of two lineages, one including hynobiids and cryptobranchids, and the other including all families practicing internal fertilization. There is also strong support for the monophyly of each of the currently accepted families. On the other hand,

there is not strong support for any branches that link two or more of the monophyletic families with one another, other than that between cryptobranchids and hynobiids.

Absence of evidence for such affinities can be attributed to the antiquity of the initial divergence of all of the salamander families, and the short duration of common ancestry of any pairs of related families. Short periods of common ancestry would presumably result in the accumulation of relatively few shared derived characters that were unique to particular pairs of taxa, compared with the much great number of characters that might have evolved separately in each of the divergent lineages. As summarized by Weisrock and his colleagues (2005, 775): "Statistical support for a mistaken topology is perhaps most likely to occur when resolving relatively short internal branches located deep in the phylogenetic history of a group." This appears to be a particularly serious problem in dealing with mtDNA and nuclear ribosomal RNA, which result in derived phylogenetic positions for cryptobranchids and hynobiids, although they remain sister-taxa of one another.

Wiens et al. (2005) dealt specifically with the problems of paedomorphic families, which bulk large in the evolution of salamanders. They showed that databases making extensive use of characters expressed in the adults of paedomorphic taxa result in strong support for a monophyletic assemblage, including groups otherwise placed in divergent families. For example, their Figure 2 illustrated a single monophyletic clade including members of the Amphiumidae, Sirenidae, Proteidae, Plethodontidae, Ambystomatidae, and Dicamptodontidae, to the exclusion of other taxa normally placed in the latter three families. They argued that traits associated with paedomorphism per se were likely to be highly convergent, and thus likely to result in misleading trees. They suggested that this problem could be avoided by coding paedomorphic taxa as unknown for adult morphology. This approach, combined with molecular data, is shown in Figure 11.17. What is clear from most of these analyses is that the currently recognized families probably differentiated well back in the Cretaceous, and that their original divergence may have occurred in the Jurassic, if not earlier.

Recent discoveries of a diversity of fossil salamanders from the Middle Jurassic of China further extend the time of divergence of highly distinct morphotypes. It is not yet possible to be certain of the specific affinities of these early genera with extant families, but major differences are already evident between paedomorphic aquatic forms and animals at least broadly resembling, in their small size and high degree of ossification, terrestrial salamandriforms. The Jurassic genera also exhibit characters that occur in a confusing mosaic of combinations in later taxa that make it difficult to determine their probable homology.

12

Eocaecilia and the Origin of Caecilians

WHILE WE ARE ALL VERY familiar with frogs and salamanders, few people in the Northern Hemisphere have ever even seen a caecilian (also termed apodans or Gymnophiona). Depending on their size, ranging from as small as 11 cm in length to over 80 cm, caecilians resemble angleworms or substantial snakes. Their resemblance to angleworms is accentuated by their superficially segmental appearance resulting from circular ridges, or annuli, which surround the body (Plate 16). All are confined to the wet tropics. They are cryptic in their habits, as burrowing or aquatic animals. The term caecilian means blind, in reference to the very small size of their eyes, which are covered with skin, or in some species, the bone of the skull, and are incapable of forming images. The reduced capacity for sight is compensated for by the evolution of alternative sensory structures, the paired tentacles, extending from in front of the eyes. These are chemosensory, coupled with the olfactory Jacobson's organ in the snout, which helps them detect prey or predators in the dark of the tunnels in which they live.

MODERN CAECILIANS

Relatively little research has been carried out on this group compared with the much more accessible frogs and salamanders. Among the most notable recent studies have been those of Marvalee Wake and her students and colleagues (M. H. Wake, 2003), Ronald Nussbaum (1998), Mark Wilkinson (Wilkinson and Nussbaum, 1997a, b), and James O'Reilly (2000).

As pointed out in Chapter 9, caecilians are unique among modern amphibians in totally lacking girdles and limbs. Not even rudiments have been recognized in early stages of development. The skeleton hence consists of only the skull and jaws, hyoid apparatus, and vertebral column with ribs. All known species are highly divergent from frogs and salamanders, united only by the presence of pedicellate teeth, a highly glandular skin, the capacity for cutaneous respiration, and a distinct larval stage. All living families have an eversible copulatory organ, the phallodeum, enabling internal fertilization. Unlike frogs and salamanders, primitive species retain small, fishlike scales hidden under the skin folds.

Cranial Anatomy

Skulls of representatives of the six modern families of caecilians are illustrated in Figure 12.1. All have much more complete ossification of the skull roof and palate than typical frogs and salamanders, with no more than very small openings for the eyes. They are highly consistent in the presence of a medial row of teeth on the palate and lower jaws, parallel to the marginal dentition. In contrast with frogs and salamanders, the endochondral braincase is ossified as a massive unitary structure, extending most of the length of the skull (Fig. 12.2). This must have served to consolidate the skull for burrowing. All families other than the Scolecomorphidae have a stapes, but none have an operculum or an opercularis muscle, since there is no shoulder girdle for its attachment (Fig. 12.3.) There is no trace of a froglike otic notch.

Although superficially similar in appearance, caecilian families can be recognized on the basis of progressive divergence and successive advances in their anatomy as well as a nested pattern of modification in particular nuclear and mitochondrial genes (San Mauro et al., 2005). Caecilian families were initially differentiated on the basis of external features of the body, aspects of reproduction, and ways of life, but they can also be distinguished by skeletal features that reflect the overall direction of evolutionary change and hence can be used to establish their probable affinities with Paleozoic amphibians. For example, the opening for the tentacle in the otherwise most primitive family, Rhinatrematidae, is immediately anterior to the eye, slightly farther forward in the Ichthyophidae, and progressively more anterior in caeciliids. In the highly derived typhlonectids it approaches the external naris. Rhinatrematids and ichthyophids are also recognized as being primitive in the terminal position of the mouth, whereas it is recessed on the underside of the snout in most other caecilians. Uraeotyphlids retain most of the primitive features of rhinatrematids and ichthyophids, but are considered intermediates, leading to the more derived families, because of the recess of the mouth behind the snout and the more anterior opening for the tentacle.

The caeciliids, with 120 living species, are the most common and widely distributed caecilians. They form the core of more derived members of the order, characterized by reduction in the number of skull bones and absence of a tail supported by normal caudal vertebrae behind the anus. It is assumed that the two more derived caecilian families (or subfamilies), the scolecomorphids and the typhlonectids, evolved from among primitive caeciliids, but their specific interrelationships remain unresolved. The scolecomorphids are uniquely derived in the loss of the stapes, present in all other families, and the complete covering of the eye by bone. The typhlonectids are all aquatic, with a laterally compressed body and a posteriorly expanded dorsal fin, but retain the capacity to burrow in the sediments beneath the water in which they live.

There has long been controversy over whether the presence of a clearly defined opening in the temporal region, between the parietal and frontal medially and the squamosal laterally, was a primitive or a derived condition for the caecilians. The discovery of a large opening in the otherwise primitive rhinatrematid, Epicrinops (Nussbaum, 1977), was suggested as indicating that this feature had been inherited from the condition seen in ancestral frogs and salamanders, in which the opening surrounding the eye was greatly expanded (Fig. 12.1A). On the other hand, a temporal opening is also present in one of the most derived families, the aquatic typhlonectids (12.1E). The ichthyophids and many caeciliids, in contrast, have only a very narrow gap between the parietal and the squamosal, and some retain the prefrontal and other circumorbital elements that have been lost in Epicrinops. It is this pattern that is seen in the oldest and most primitive of all caecilians, Eocaecilia, discussed later in this chapter.

Jaws and Hyoid

The evolution of jaws in the history of frogs and salamanders has been one of progressive loss of individual bones. This is carried to an extreme in caecilians, in which only two areas of ossification are evident in the adults: the anterior pseudodentary, bearing both the marginal dentition and a parallel medial row of teeth, and the posterior pseudoangular, which includes the area of jaw articulation (Figs. 9.19 and 12.4). Mentomeckelian elements, expressed in the adults of both frogs and salamanders, are recognizable early in development in caecilians but are indistinguishable from the pseudodentary in the adult. The posterior portion of Meckel's cartilage, the articular, is fully integrated with the pseudoangular, which also bears a long retroarticular process and a unique internal process that serves as an area of insertion for a portion of the pterygoideus muscle that helps to prevent the disarticulation of the lower jaw. The pseudoangular and pseudodentary are joined by a very long, diagonal suture (see discussion of jaw mechanics in Chapter 9).

No aspects of the skull or lower jaws could be considered as synapomorphies supporting sister-group relationships with either frogs or salamanders. This applies also to the hyoid apparatus. The overall configuration of the hyoid apparatus is very consistent among adult caecilians (D. B. Wake, 2003). The larval structure has only been described in Epicrionops and Ichthyophis. Both adults and larvae are clearly distinct from those of frogs or salamanders (Figs. 9.11 and 12.5). All the hyobranchial elements have been lost. In the larvae, there is a series of four medial basibranchials, lateral to which there are paired ceratohyals and four ceratobranchials, the last of which is much shortened. In mature specimens of all families, the ceratohyal is joined medially to the most anterior basibranchial, which in turn is fused to the first ceratobranchial. Ceratobranchials II and III are each joined at the midline. The remnant of ceratobranchial IV is fused with the end of III in most caecilians but in *Epicrionops* it is resorbed.

M. H. Wake (1989) described a pattern of hyoid development in *Epicrionops* that is unique among amphibians. In the larvae of this genus, all elements except the medial basibranchials become ossified early in ontogeny, but the bone is replaced by cartilage at the time of metamorphosis. This sequence contrasts with the replacement of cartilage by bone or retention of the cartilaginous condition in other animals retaining any of these elements in the adults. In all other caecilians, the hyoid apparatus remains cartilaginous throughout life. In contrast with frogs and salamanders, the tongue of *Epicrionops* is reduced in size and becomes more fixed in position at metamorphosis. However, as in frogs and salamanders, there is a rapid period of metamorphosis, in contrast with the slow, progressive changes seen between larval and adult temnospondyls.

Vertebrae and Ribs

As greatly elongate animals without limbs, the vertebral column of caecilians is the primary basis for locomotion. It is not surprising that the vertebrae are greatly different from those of frogs and salamanders. In modern caecilians, the structure of the articulating surfaces of the occiput and the atlas vertebra differ significantly, presumably in relationship to the use of the skull as a burrowing implement (Fig. 12.6). In both frogs and salamanders, the condylar articulating surfaces on the skull (and the associated atlas) are clearly separated from one another, and do not approach the ventral midline (Figs. 10.1*F* and 11.3). This is associated with simple vertical hinging of the skull relative to the trunk.

Frogs have an odontoid process (also termed the interglenoid tubercle) that separates the cotyles of the atlas and extends anteriorly to the base of the foramen magnum. In salamanders the atlas has four points of articulation with the occiput. Two large, cup-shaped atlantal cotyles articulate with the occipital condyles, and ventromedially, between the cotyles, the odontoid process has paired surfaces that articulate with the lateral walls of the foramen magnum. In living caecilians, by contrast, the occipital condyles are nearly confluent ventrally, and there is only a very narrow ridge at the midline of the atlas, separating the two cotyles (Fig. 12.6*A*).

Equally striking differences are evident in the trunk vertebrae and the nature of articulation of the ribs. Primitive living frogs and salamanders both have simple, cylindrical centra, but lack trunk intercentra. The trunk ribs in salamanders have the two heads, attached one above another (Fig. 12.6*F*), while in frogs, the ribs, where present, have a single head attached to the laterally projecting transverse process (Fig. 10.2). Modern caecilians have clearly separated rib heads, but their articulation with the centra differs from both frogs and salamanders. The tuberculum attaches to a short transverse process beneath the prezygapophysis, while the capitulum articulates with a unique process, the parapophysis, which extends from the ventral margin of the centrum to underlie

the more anterior centrum (Fig. 12.6*G*). This configuration had evolved by the Late Cretaceous (Evans et al., 1996), but was not present in the only known Early Cretaceous genus (Fig. 12.6*H*). Evans and Sigogneau-Russell (2001) argued that the presence of the large parapophyses of crown-group caecilians was related to the evolution of the unique pattern of their axial musculature. The position of the parapophysis and its articulation with the capitulum of the ribs support its homology with the intercentrum of primitive tetrapods.

FOSSIL RECORD OF CAECILIANS

The fossil record of caecilians is limited to five horizons, ranging in age from the Lower Jurassic to the Upper Paleocene. Different species may be represented at each of these localities, but only three are sufficiently well preserved to justify being named:

—*Apodops pricei:* A single posterior trunk vertebra from the Upper Paleocene of Brazil. Similar to the West African caeciliid *Geotrypetes* and the Central American *Dermophis* (Estes and M. Wake, 1972).
—Unnamed: A poorly preserved trunk vertebra from the Lower Paleocene of Bolivia, not sufficiently complete to compare with any living family (Rage, 1986).
—Unnamed: Four trunk vertebrae from the Upper Cretaceous of Sudan, broadly resembling those of modern caeciliids (Evans et al., 1996).
—*Rubricacaecilia monbaroni:* Lower jaw bones, numerous vertebrae and a possibly associated femur from the Lower Cretaceous of Morocco (Evans and Sigogneau-Russell, 2001). Definitely caecilian in the nature of the lower jaw, but primitive in having a nearly horizontal retroarticular process. The atlas is considerably more primitive than any modern caecilians in the retention of a prominent odontoid process and the trunk vertebrae lack the long, anteriorly directed parapophysis of all living caecilians (Fig. 12.6*B, H*).

By far the most important caecilian fossils are those of *Eocaecilia micropodia* from the Kayenta Formation, Lower Jurassic (Pleisbachian), of Arizona, collected by a field party led by Farish Jenkins (Jenkins et al., 2007). Forty specimens can be recognized from the same locality as the oldest known frog, *Prosalirus bitis,* discussed in Chapter 10. Nearly every bone in the body is known, including nine skulls (Figs. 12.2*D*, 12.4, and 12.6–12.8).

Eocaecilia is immediately recognizable as a caecilian by the overall configuration of the skull and lower jaws, but strikingly primitive in the configuration of the vertebrae and the presence of fore- and hind limbs. The skull has relatively small openings for the eyes, and there is a notch in the anterior margin of the eye socket comparable to that in *Epicrinops* for passage of the tentacle. The outline of the skull and its size resemble those of primitive living caecilians such as *Ichthyophis* (Fig. 12.10), but several additional bones can be recognized: a pair of postparietals behind the parietal, a small

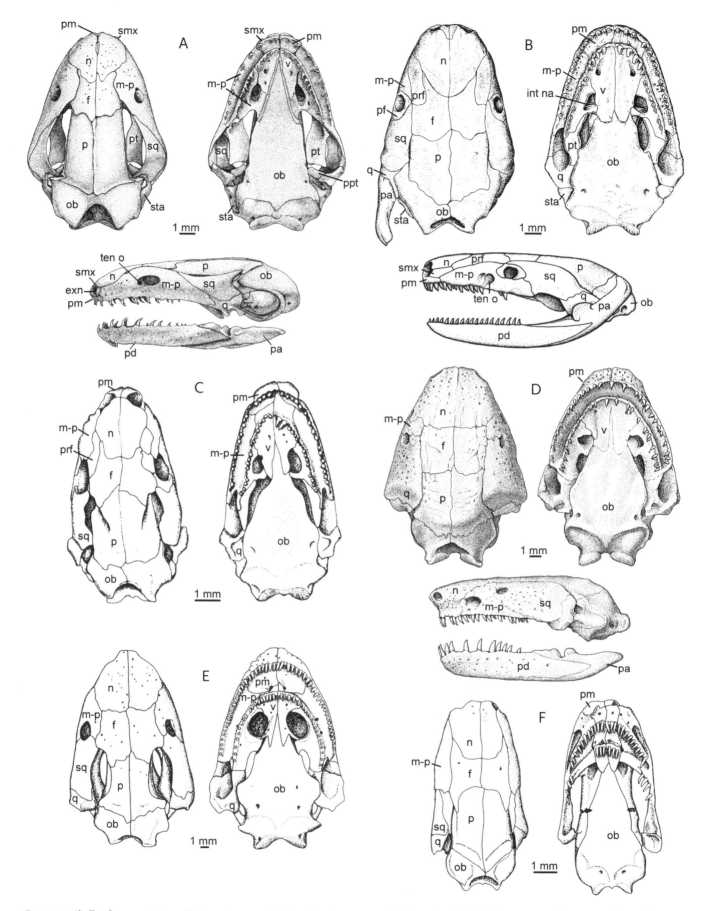

Figure 12.1. Skulls of representatives of the modern caecilian families. *A*, The rhinatrematid *Epicrionops* in dorsal, palatal, and lateral views. Modified from Nussbaum, 1977. *B*, The ichthyophid *Ichthyophis*, in three views. Modified from Jenkins et al., 2007. *C*, The uraeotyphlid *Uraeotyphlus*. *D*, The caeciliid *Dermophis*. Modified from M. H. Wake, 2003. *E*, The typhlonectid *Typhlonectes*. *F*, The scolecomorphid *Scolecomorphus*. *C*, *E*, and *F* from R. L. Carroll and Currie, 1975.

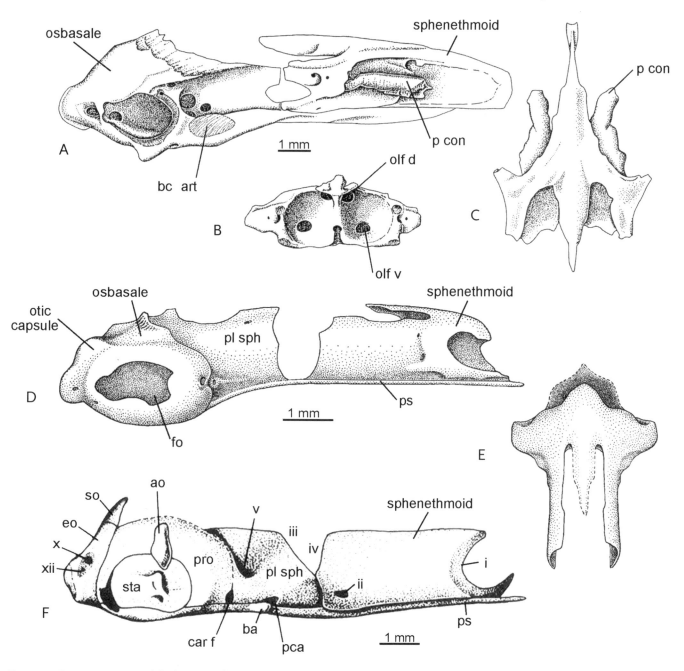

Figure 12.2. Comparative views of the braincase of the advanced modern caecilian *Dermophis* (A–C), the Lower Jurassic caecilian *Eocaecilia* (D and E), and the Lower Permian microsaur *Rhynchonkos* (F). A, D, F, lateral views. B, Anterior view of sphenethmoid. C, E, Dorsal views of sphenethmoid. From R. L. Carroll, 2007.

element between the parietal and the squamosal in the position of a tabular in Paleozoic tetrapods, a postfrontal, and a jugal.

The palate resembles that of all modern caecilians in having a row of teeth extending the length of the vomer and palatine, parallel to the marginal dentition. In all the articulated specimens, the lower jaw is so crushed into the skull that it is not possible to determine whether or not an ectopterygoid was present. It is reported in several modern caecilians (R. L. Carroll and Currie, 1975), but not in frogs or salamanders. The narial opening is separated from the premaxilla by the lateral contact between the palatine and vomer. The interpterygoid vacuities are more extensive than in modern caecilians, but do not lie beneath the orbits as in frogs and salamanders.

The jaw articulation is well anterior to the occipital condyles, but does not have the specialized structure of modern caecilians, in which the articulating surface of the quadrate fits into a deep groove in the pseudoangular. There was apparently not a well-defined area of articulation between the pterygoid and the base of the braincase, as there is in modern caecilians (Fig. 12.2A). The cultiform process of the parasphenoid is narrow, as in many Paleozoic amphibians, not greatly expanded as in all modern caecilians.

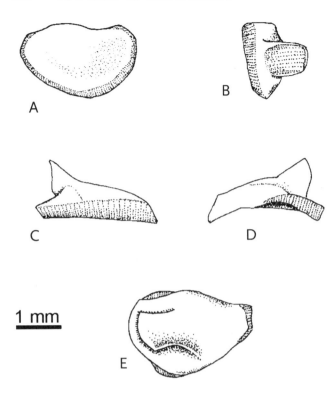

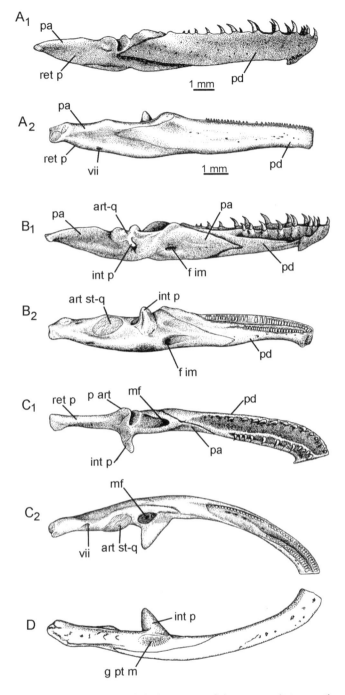

Figure 12.3. Stapes of the modern caecilian *Dermophis mexicanus* in *A,* Medial. *B,* Anterior. *C,* Ventral. *D,* Dorsal. *E,* Lateral views. Drawn from specimen described by M. H. Wake and Hanken (1982) in the collection of Marvalee H. Wake, Department of Integrative Biology, University of California, Berkeley.

The braincase broadly resembles that of extant caecilians in having two solidly ossified elements, the anterior sphenethmoid, which extends toward the lateral surface of the snout, and the os basale, which incorporates all the endochondral bones at the back of the skull, including the area of the pleurosphenoid, between the otic capsule and the area of the optic nerve. In these features, it is clearly distinct from that of frogs or salamanders. *Eocaecilia* is slightly more primitive than modern caecilians in that the parasphenoid is not fused to the base of the sphenethmoid.

In relationship to the use of the skull as an implement for burrowing, the posterior portion of the skull roof and the base of the parasphenoid in modern caecilians are both marked by crescentic recesses that indicate the anterior limits of attachment of axial musculature. The dorsal area of attachment can be seen in *Eocaecilia,* but it occupies the posterior portion of the postparietal, rather than the parietal. The parasphenoid does not exhibit a specific area for the attachment of the subvertebralis muscle.

One of the largest and most massive of the skull bones in *Eocaecilia* represents the fusion of the separate quadrate and stapes common to early tetrapods (Fig. 12.8). Its derivation from a stapes is demonstrated by its position lateral to the fenestra ovalis of the otic capsule and the presence of a stapedial foramen. However, there is not a clearly defined footplate that could have fit into the fenestra ovalis, as does the stapes in Paleozoic and modern amphibians. Its role as

Figure 12.4. Comparisons of the lower jaws of the primitive living caecilian *Epicrionops,* A$_1$-C$_1$, and the Lower Jurassic caecilian *Eocaecilia,* A$_2$-C$_2$ in lateral, medial, and dorsal views. *D,* ventral view of the lower jaw of *Eocaecilia,* showing a groove for the passage of the pterygoideus muscle (g pt m), which extends under the lower jaw to maintain its attachment with the stapes-quadrate against the force of the interhyoideus posterior. From R. L. Carroll, 2007.

a quadrate is indicated by the presence of an area of unfinished bone that served to articulate with the lower jaw. The stapes-quadrate also has a clearly defined surface for sutural attachment with the pterygoid.

The surface of this composite bone for articulating with the lower jaw resembles that of most primitive tetrapods, but not that of modern caecilians, where it is in the form of

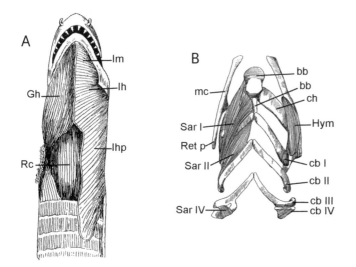

Figure 12.5. Throat and hyoid musculature of the caecilians *Caecilia* and *Ichthyophis*. Modified from Edgeworth, 1935.

a narrow, nearly transversely oriented blade that fits into a narrow grove in the articular. Retention of the contract between the stapes-quadrate and the lower jaw must have been maintained by the peculiar orientation and areas of insertion of the pterygoideus muscle, as described for the living species in Chapter 9. Since the quadrate and stapes are separate in modern caecilians, and the quadrate serves the usual role of linking the skull and lower jaw, it must be assumed that the condition seen in *Eocaecilia* is a unique derived character, precluding this particular genus from giving rise to later caecilians. On the other hand, most other aspects of the skeleton are nearly ideal for that role.

An additional oddity of the ear region of *Eocaecilia* is the close association in several specimens of an oval, concave ossicle in the area that would have overlain the fenestra ovalis (Fig. 12.8*D*). In the size and shape of its medial surface it resembles the footplate of the stapes in the modern caecilian *Dermophis*, but lacks the short posterolaterally facing stem seen in that genus (Fig. 12.3). As an element in addition to the stapes-quadrate in *Eocaecilia*, it is logical to consider this bone as an operculum, although neither an operculum nor an opercularis muscle is known in modern caecilians. It may, nevertheless have played a role in transmitting vibrations in *Eocaecilia*, where it was in a position to be attached to an opercularis muscle running to the shoulder girdle, but not in modern caecilians that lacked a scapula.

Nothing of the hyoid apparatus is known in *Eocaecilia*.

The lower jaw (Figure 12.4) has nearly all the major features of modern caecilians. There are only two areas of ossification, the pseudodentary and pseudoangular. The pseudodentary bears a medial row of teeth and the pseudoangular bears an internal process, a tiny Meckelian fenestra, and a very long retroarticular process. The internal process is much larger than in modern species, but already shows small anterior foramina that would have served for attachment of the fibers of the pterygoideus muscle, just as in modern caecil-

ians. Adjacent to the base of the internal process there is a broad but shallow groove in the pseudoangular that would have served for the passage of the remainder of the pterygoideus as it passed from the upper surface of the pterygoid to the lateral surface of the back of the lower jaw. This would have been vital in *Eocaecilia* to hold the lower jaw against the back of the skull in the absence of the complex, interlocking jaw articulation in modern caecilians. In contrast to the deep recess in the pseudoangular to receive the sharp ridge of the quadrate in modern caecilians, the articulating surfaces in *Eocaecilia* matched only a shallow concavity in the lower jaw with a slight convexity in the skull (Jenkins et al., 2007).

We saw that the Lower Cretaceous caecilian *Rubricacaecilia* retained primitive features of the vertebrae, including the presence of an odontoid process that fitted between the occipital condyles and the absence of parapophyses (Fig. 12.6*B*, *H*). It is not surprising that *Eocaecilia* also expressed these primitive character states, as well as the presence of intercentra throughout the trunk region. The discovery of a femur with the cranial remains of *Rubricacaecilia* was also matched by *Eocaecilia*, in which the natural association of this bone is strongly confirmed. Although no specimens of *Eocaecilia* are fully articulated, study of several overlapping sections of the vertebral column supports the number of presacral vertebrae as close to 49, compared with 75 to 78 in the most primitive living caecilian, *Epicrionops* (Taylor, 1968). What is very surprising is that two well-preserved vertebrae are definitely sacrals, as indicated by their relatively shorter length than those more anterior and posterior in position and their relatively larger transverse processes. The sacrum is immediately adjacent to the femora. Unfortunately, the pelvic girdle is too badly crushed for illustration or description.

Another specimen includes approximately 13 caudal vertebrae, of progressively smaller size. They are associated with several well-developed haemal arches, resembling those of Paleozoic labyrinthodonts, microsaurs, and amniotes. Distinct haemal arches, although without a proximal crossbar, can also be seen in larvae of *Epicrionops* (Fig. 12.6*J*). Progressive changes in the configuration of the haemal arches within the caudal region of this genus support the hypothesis that they are homologous with the parapophyses of the trunk, which have the same relative position as the intercentra in *Eocaecilia* and earlier tetrapods.

The forelimb and girdle are restored as having been in the position of the fifth and sixth vertebrae behind the skull (Fig. 12.9). There is no evidence of the dermal cleithrum, clavicle, or interclavicle, which may have been completely lost by this stage in caecilian evolution. The scapulocoracoid, however, is well ossified, although it is much reduced relative to the size of the skull and vertebrae, but in proportion to the limbs. The scapular blade is narrowly constricted at its midpoint, which is oval in cross section. The blade is expanded dorsally, and the distal extremity ends in an unfinished margin that was probably continued with a suprascapular cartilage. The glenoid articulation for the humerus faces posteroventrally

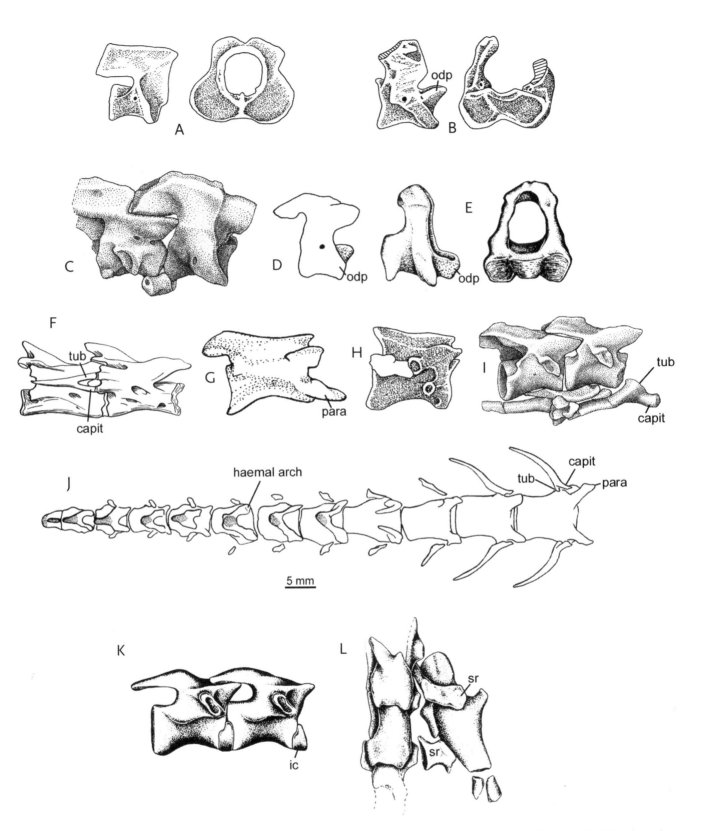

Figure 12.6. Vertebrae of caecilians and a microsaur. *A,* Right lateral and anterior views of the atlas of the modern genus *Ichthyophis. B,* Comparable views of the Lower Cretaceous genus *Rubricacaecilia. C,* Atlas and axis vertebrae of the Lower Jurassic *Eocaecilia,* in right lateral view. *D,* Sketch of the atlas of *Eocaecilia* in right lateral view. *E,* Atlas of the Lower Permian microsaur *Rhynchonkos,* in right lateral and anterior views. *F,* Trunk vertebrae of a modern salamander. *G,* 50th trunk vertebrae of the modern caecilian *Dermophis,* in right lateral view. *H,* trunk vertebra of *Rubricacaecilia. I,* Trunk vertebrae and adjacent ribs of the Lower Jurassic *Eocaecilia. J,* Caudal vertebrae of *Epicrionops* in ventral view. *K,* Trunk vertebrae of the Lower Permian microsaur *Rhynchonkos. L,* Dorsal view of sacral vertebrae, ribs, and adjacent femur of *Rhynchonkos. A, B, D,* and *H* from Evans and Sigogneau-Russell, 2001. *C* and *I* from Jenkins et al., 2007. *E, K,* and *L* from R. L. Carroll and Gaskill, 1978. *F,* modified from Romer and Parsons, 1977. *G, J,* from M. H. Wake, 2003.

and slightly laterally, with an overall spiral configuration common among Paleozoic tetrapods. A faint suture joining the scapula and coracoid crosses the glenoid. A circular fossa of unknown function occurs ventral to the anterior half of the glenoid. A similar depression occurs in a few microsaurs (R. L. Carroll and Gaskill, 1978; R. L. Carroll, 1991), but there is no evidence that they are homologous.

The humerus, a little more than 4 mm in length, broadly resembles those of small, but well-ossified Paleozoic tetrapods, with moderately expanded proximal and distal areas of expansion. The proximal area retains the primitive spiral configuration, but distally there is a comparatively large, hemispherical area of articulation for the radius. There is a distinct trochlea for articulation of the ulna, but as in small microsaurs, no entepicondylar foramen. The radius and shaft of the ulna are about one-half the length of the humerus. Disarticulated elements near the lower limbs have been recognized, but none can be specifically attributed to the manus.

A slender rod in the area of the sacrum, with a surface of articulation at one end, may be an ilium, but no other pelvic elements are sufficiently complete to be recognized. The femur has a bulbous oval head. The internal trochanter is conspicuous, as is that of *Rubricacaecilia* and living salamandrids. The tibia and fibula are approximately one-half the length of the femur. Articulated specimens allow the identification of three proximal tarsals: the tibiale, intermedium, and fibulare. Several specimens show elements of three digits, some of which have three phalanges.

No scales have been reported.

PUTATIVE ANTECEDENTS OF CAECILIANS

Using *Eocaecilia* as a model for the most primitive caecilians, the following skeletal characters distinguish basal members of this order from ancestral frogs and salamanders.

Skull

1. Skull solidly roofed, except for small openings for the eyes and nares
2. Line of loose attachment between the squamosal and the parietal
3. Retention of postparietals, a second bone in the temporal region, postfrontal, and jugal
4. Groove anterior to eye to accommodate the tentacle
5. Internal naris not bordered by premaxilla
6. Row of teeth on vomer parallel to those of premaxilla and maxilla
7. Interpterygoid vacuity not located beneath orbits
8. Fusion of all bones of the braincase into two large areas of ossification, an os basale posteriorly and the sphenethmoid anteriorly

Jaw

9. Adults express only two elements, the pseudodentary, bearing lateral and medial tooth rows, and the pseudoangular, bearing the jaw articulation, a large internal process, and a long retroarticular process
10. Very narrow mandibular foramen

Vertebrae and Ribs

11. At least 49 presacral vertebrae
12. Two pairs of sacral ribs
13. Crescentic intercentra throughout the trunk
14. Approximately 13 caudal vertebrae

Appendicular Skeleton

15. Extensive reduction in size and some loss of elements in girdles and limbs

About the only derived osteological feature that basal caecilians share with frogs and salamanders is pedicellate teeth, but that character is also seen in a putative ancestor of salamanders, *Apateon,* and the plausible sister-taxon of frogs, *Doleserpeton,* both of which are otherwise highly divergent from caecilians.

In common with Jurassic urodeles and anurans, *Eocaecilia* had already attained many of the skeletal features of its living descendants. The elongation of its vertebral column and the high degree of similarity of the cranial anatomy indicate that adaptation to a burrowing way of life was already well established by the Lower Jurassic. While basal frogs and salamanders resemble one another in having very open skulls, a relatively short vertebral column, and substantial limbs, basal caecilians were already highly divergent in these character complexes.

Although gaps of 150 to 180 million years separate stem-group anurans and urodeles from plausible Paleozoic antecedents, Permocarboniferous amphibamids and branchiosaurids document the sequential origin of many of the key characteristics of the extant taxa. The degree of expression of these characters converges toward a similar morphology within the Westphalian, indicating an ultimate common ancestry of salamanders and frogs from among the oldest known temnospondyl labyrinthodonts. Among the early temnospondyls, *Balanerpeton* was already derived in its small body size as an adult, high degree of ossification, and loss of lateral line canals on the skull, indicating a basically terrestrial way of life. Immature specimens of *Balanerpeton* suggest an aquatic larval stage, comparable to those of branchiosaurids and amphibamids, which probably had conspicuous external gills.

In contrast, no temnospondyls, as that group is currently recognized (R. L. Carroll and Holmes, 2007), shows any tendency toward extensive elongation of the trunk or limb reduction, and most retain a conservative pattern of cranial elements with relatively large orbits, very extensive interpterygoid vacuities, and large otic notches suggestive of impedance matching middle ears.

Double occipital condyles evolved separately among branchiosaurids and amphibamids, as did pedicellate teeth, in clades that otherwise show none of the unique derived characters of caecilians. In contrast with the antecedents

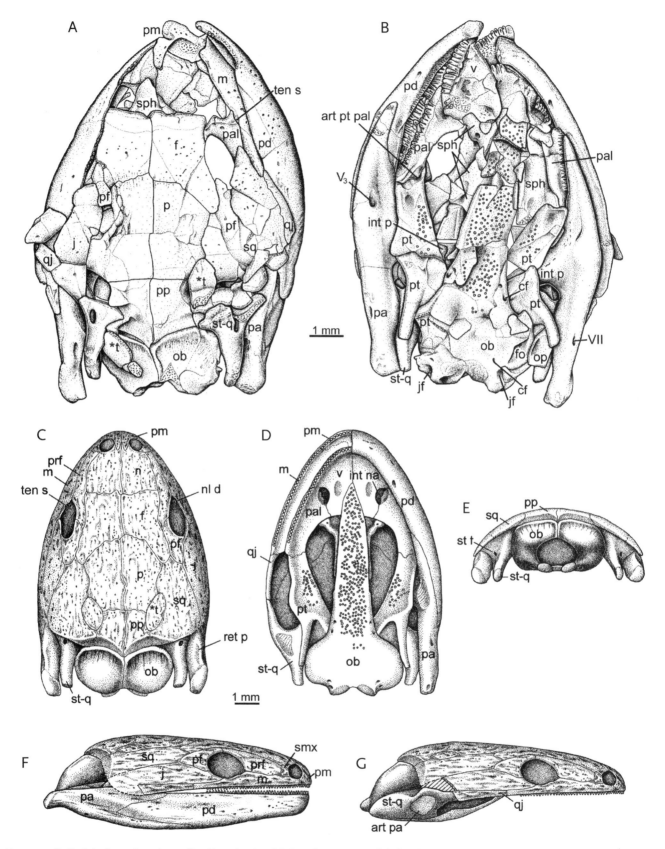

Figure 12.7. Skull of the Lower Jurassic caecilian *Eocaecilia*. *A* and *B*, Dorsal and ventral views of skull as preserved. *C–E*, Dorsal, palatal, and occipital views as reconstructed. *F*, Lateral view of skull with jaw in place. *G*, Lateral view of skull with jaw and posterior portion of squamosal removed to show stapes-quadrate and its surface for articulation with the lower jaw. From Jenkins et al., 2007. *t, bone questionably identified as a tabular.

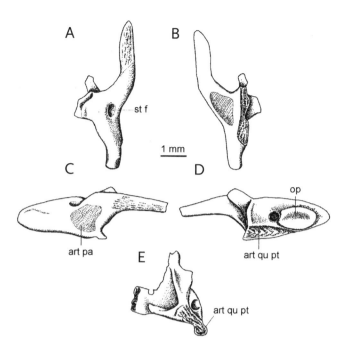

Figure 12.8. Reconstruction of the right stapes-quadrate of *Eocaecilia*. From Jenkins et al., 2007.

of salamanders, no larvae or juveniles are known from the Paleozoic whose general appearance suggests their ancestry to those of caecilians. Although extant caecilians have well-developed external gills in utero or within the egg, they are lost at the time of birth or hatching and would not likely be capable of fossilization as are those of branchiosaurs.

Looking back at all the families discussed in Chapters 4 and 6, the known Paleozoic fossils provide no support for a sister-group relationship between caecilians and any temnospondyls. The other major group of Paleozoic labyrinthodonts, the anthracosauroids, among which the discosauriscids also have gill-bearing larvae, also lack any obvious synapomorphies with caecilians (Špinar, 1972; Smithson, 2000; Laurin, 2000). This also holds for the smaller labyrinthodont families.

This leaves the other assemblage of Paleozoic tetrapods, the lepospondyls, for consideration. Lepospondyls share the following general similarities with caecilians:

1. Even the smallest known specimens of the six orders show no evidence of external gills
2. Ossification of cylindrical centra extending to the end of the tail, at a very early stage of development
3. Absence of an otic notch
4. Some members of all orders except the Nectridea evolved an elongate trunk region and some families show reduction or complete loss of limbs
5. All have cylindrical pleurocentra
6. All except microsaurs and *Acherontiscus* lack trunk intercentra

Aïstopods and adelogyrinids had lost all trace of limbs by their first appearance in the fossil record in the Lower

Carboniferous, making them improbable sister-taxa of a clade that still retained limbs in the Lower Jurassic. The only known specimen of *Acherontiscus* also lacks limbs, as well as being unique among lepospondyls in the presence of cylindrical intercentra and pleurocentra. Lysorophids retained limbs of about the same proportions as those in *Eocaecilia*, but they had already attained 69 trunk vertebrae in the Upper Carboniferous and 97 by the Lower Permian, whereas *Eocaecilia* had only approximately 49 in the Lower Jurassic. Lysorophids are otherwise highly improbable sister-taxa of caecilians in light of their great reduction in the number of skull bones in all known species (Fig. 4.55). Nectrideans are unique in fusing their haemal arches midway in the length of the caudal centra.

The only group of lepospondyls that includes genera that are sufficiently primitive to be plausible sister-taxa of caecilians are the microsaurs (Chapters 4 and 6). These are the most diverse assemblage of lepospondyls, with some 12 recognized families, 30 genera, and a fossil record extending from the upper part of the Lower Carboniferous through the Lower Permian. Most early taxa had a short vertebral column and fairly robust limbs, but more derived genera in the Lower Permian had more than 30 presacral vertebrae and several lineages reduced their limbs. Tiny juveniles from the Upper Carboniferous are typical of lepospondyls in ossifying the vertebral centra as complete cylinders at a very early stage.

Microsaurs have previously been suggested as possible antecedents of caecilians (J. T. Gregory et al., 1956), but the genus showing the most derived characters in common with caecilians is the Lower Permian genus *Rhynchonkos* (R. L. Carroll and Currie, 1975; R. L. Carroll, 2000). *Rhynchonkos* is known from numerous specimens from the Lower Permian of Oklahoma (Figs. 12.9 and 12.10). None are fully articulated, but there are approximately 37 presacral vertebrae. The vertebrae resemble those of *Eocaecilia* in having cylindrical pleurocentra fused to the neural arch and crescentic intercentra. The atlas has a large odontoid that articulated with the recessed basioccipital between double condyles.

The limbs of *Rhynchonkos* are abbreviated to about the same extent as in *Eocaecilia*, but the girdles retain the typical elements of Paleozoic tetrapods. The ilium has a long, narrow dorsal process supported by two pairs of sacral ribs, as in *Eocaecilia* (Fig. 12.6L). Both carpals and tarsals are well ossified and the rear limb appears to retain five digits, with the phalangeal count of the first three 2, 3, 3.

The skull of *Rhynchonkos* is broadly similar to that of *Eocaecilia* in its small size, relatively small orbits, and position of the jaw articulation well anterior to the occipital condyles (Fig. 12.10). The palate and lower jaw both resemble caecilians in having medial rows of teeth on the vomer, palatine, and ectopterygoid of the palate and on the medial surface of the lower jaw. The dermal bones of the skull correspond almost exactly with those of *Eocaecilia*, except for the retention of a large lacrimal and both a postorbital and postfrontal behind the eye where only a single bone is retained in this

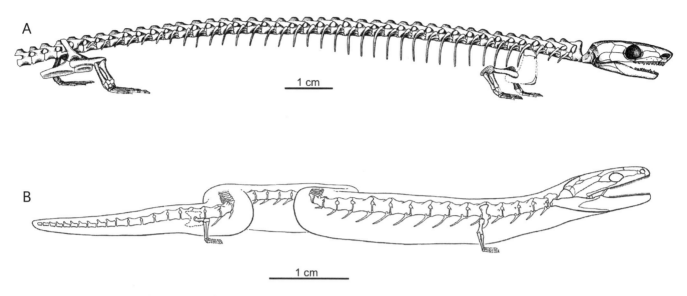

Figure 12.9. A, Reconstructions of the skeleton of the Lower Permian microsaur *Rhynchonkos,* and *B,* the Lower Jurassic caecilian *Eocaecilia.* From R. L. Carroll, 2007.

position in the Jurassic genus. *Rhynchonkos* does differ in having small bony plates that presumably reinforced the eyelid, but these do occur in one modern caecilian, *Atretochoana* (Chapter 13).

Of particular importance is the high degree of ossification of the endochondral braincase (Fig. 12.2). In contrast with most other Paleozoic amphibians, and certainly from ancestral salamanders and frogs, it extends continuously from the occiput through the sphenethmoid. The Xth nerve is wholly incorporated into the exoccipital, perhaps an early stage in the integration of all bones of the back of the braincase into a single os basale. Primitively, a separate supraoccipital is retained. The exoccipital, opisthotic, and prootic ossify separately, but the area surrounding the foramen for the Vth nerve is formed by an extensive pleurosphenoid that reaches anteriorly to link with the posterior margin of the sphenethmoid. In contrast with living caecilians, the parasphenoid is not fused with the other bones that make up the os basale, but as in *Eocaecilia,* passes freely beneath the sphenethmoid. The footplate of the stapes of *Rhynchokos* is very large, but there is a gap in ossification between the posteroventral margin of the footplate and the margin of the otic capsule in the position that is occupied by an operculum in frogs and salamanders. There is, however, no evidence of any ossification in this area. On the other hand, there is a small additional ossicle resting just above the stem of the stapes in this and several other microsaurs that has been called an accessory ear ossicle (R. L. Carroll and Gaskill, 1978). However, an ear ossicle in the position of that of *Eocaecilia,* medial to the interior surface of the footplate of the stapes, has not been observed in *Rhynchonkos.*

The lower jaw of *Rhynchonkos* retains most of the elements common to primitive Paleozoic tetrapods, except for the reduction in the number of coronoid bones from three to two. As in caecilians, there is a medial row of teeth on the lower jaw, but it is on one of the coronoids, rather than being attached to the pseudodentary. More important, *Rhynchonkos* shows the presence of a short, but clearly defined retroarticular process.

In summary, the following characters, derived above the level of basal tetrapods, are expressed in both *Rhynchonkos* and primitive caecilians. The characters indicated by an * are common to lepospondyls, but not to temnospondyls or anthracosauroids.

1. Orbital openings small relative to those of most comparable sized early tetrapods
2. Jaw articulation well anterior to occipital condyle
3. Absence of fangs on palatal bones*
4. Rows of teeth on vomer and palatine, parallel to those of premaxilla and maxilla
5. Loss of intertemporal and supratemporal bones common to early temnospondyls*
6. Ossification of pleurosphenoid
7. Double occipital condyle, not present in Carboniferous temnospondyls or anthracosaurs*
8. Cylindrical centra fused to neural arch early in development, in contrast with labyrinthodonts*
9. Greater elongation of vertebral column than any temnospondyls, anthracosauroids, nectrideans, or early amniotes*

Rhynkonkos exhibits more derived characters in common with caecilians, especially *Eocaecilia,* than any other Paleozoic genus that has been described. Many of the osteological differences recognized between these genera can be attributed to continuation of evolutionary trends that can be observed between more primitive microsaurs and *Rhynchonkos,* such as further elongation of the trunk, reduction of the limbs,

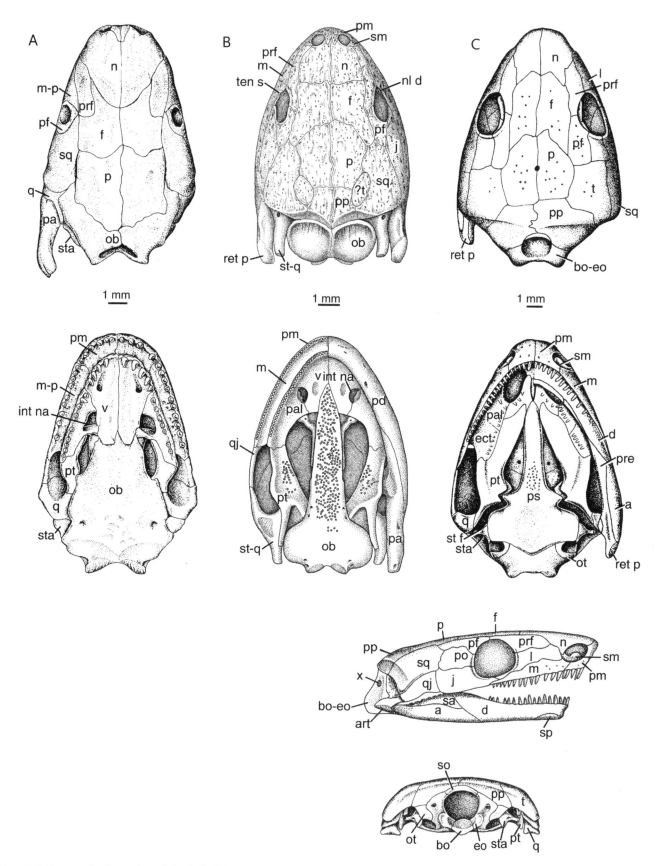

Figure 12.10. Comparative illustrations of the skull of the primitive modern caecilian *Ichthyophis* (A) in dorsal and palatal views, the Lower Jurassic caecilian *Eocaecilia* (B) in dorsal and palatal views, and the Lower Permian mic-rosaur *Rhynchonkos* (C) in dorsal, palatal, lateral, and occipital views. From R. L. Carroll, 2007.

NESTED SYNAPOMORPHIES LEADING TO CROWN GROUP CAECILIANS

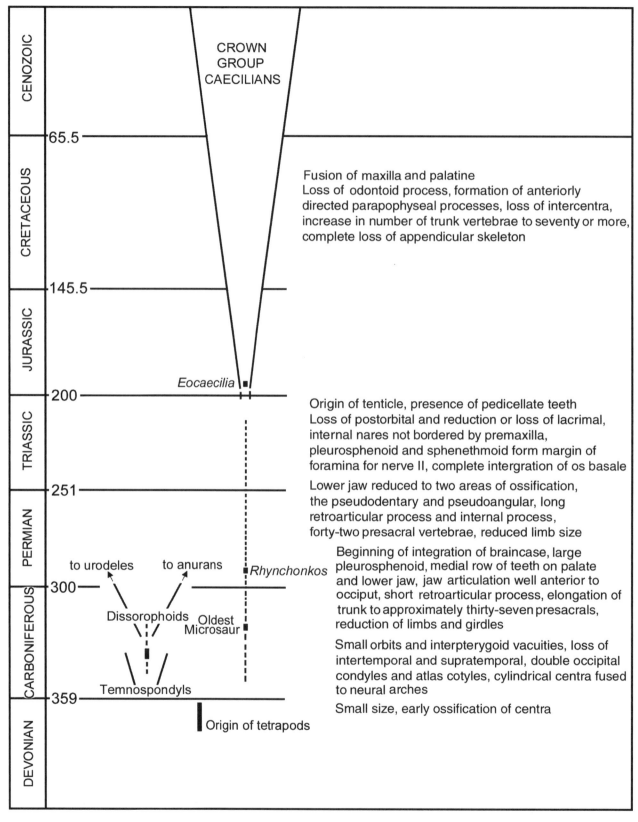

Figure 12.11. Nested synapomorphies leading from the oldest known microsaur to crown-group caecilians. If caecilians evolved from among microsaurs, their point of divergence preceded that between frogs and salamanders, assuming that the latter groups shared a common ancestry among early temnospondyls. The specific interrelationships among temnospondyls, microsaurs, and the antecedents of amniotes cannot yet be established because of our inadequate knowledge of the fossil record of uppermost Devonian and Lower Carboniferous tetrapods. From R. L. Carroll, 2007.

extension of the retroarticular process, and further consolidation of the braincase. Only the lower jaws have undergone a major reorganization.

Few, if any, skeletal changes would be considered reversals, and all may be attributed to continuing specialization for a burrowing way of life. There remains a long gap in time, and a substantial change in morphology, but *Rhynkonkos* provides the best-known example of a Paleozoic amphibian that could have been close to the ancestry of caecilians. However, it provides no support for close affinities with anurans or urodeles.

Rhynchonkos stovalli is the only known species within the Goniorhynchidae. This family resembles the much more diverse and widespread Gymnarthridae, which extends from the Westphalian A of the Upper Carboniferous into the Lower Permian of North America and Europe. Among the Lower Permian gymnarthrids, *Cardiocephalus peabodyi* is known to have had 37 presacral vertebrae and 2 sacrals, exactly as in *Rhynchonkos stovalli*. However, a well-articulated specimen of another genus of the same family from the Upper Carboniferous of the Czech Republic, *Sparodus* (Fig. 6.5A), has approximately 25, indicating the general time frame within which trunk elongation had occurred in a related clade of microsaurs.

An unnamed genus from the Namurian E2 (Serpukhovian) of Illinois demonstrates the earliest known occurrence of microsaurs (Figs. 4.1 and 4.30). Although none of the eight intertwined specimens are fully exposed or articulated, a minimum of 34 presacral vertebrae is well established, as is the presence of strap-shaped intercentra, similar to those of both *Rhynchonkos* and *Eocaecilia*. In contrast to any other lepospondyl, this species retains paired proatlas elements and clearly defined areas for their articulation with the atlas arch, as in the primitive tetrapod *Acanthostega*, but in no other microsaur. The centrum of the atlas is expanded laterally to form a bicotylar articulation with the occipital condyles, as in all other microsaurs. Not all of the limb elements are preserved, but the humerus and femur are approximately the length of three trunk vertebrae, as is also the case for *Rhynchonkos*. The humerus, however, is more primitive in retaining the entepicondylar foramen, as in other primitive microsaurs. This is the earliest known species that expresses some derived traits shared with the oldest known caecilian, *Eocaecilia*. The nested sequence of synapomorphies leading toward caecilians is shown in Figure 12.11.

This microsaur, described by Lombard and Bolt (1999), is about 5 million years younger than the horizon from which *Balanerepton* has been described. Hence, the age of the earliest currently known species that shows some derived characters in common with caecilians is approximately as great as the temnospondyls that exhibit characters in common with frogs and salamanders. This implies that the time of initial divergence of the clades leading to batrachians (frogs + salamanders) and caecilians was within the first 30 million years of the Carboniferous. This brings us back to a time when the fossil record is very poorly known and provides little evidence for the interrelationships among any of the major tetrapod clades.

As discussed in Chapter 5, no clear pattern of interrelationships among lepospondyls, or between lepospondyls and any other groups of Carboniferous tetrapods, has yet been convincingly documented. Among lepospondyls, early microsaurs may retain the largest number of labyrinthodont characteristics, for example, the presence of crescentic intercentra in the trunk region, intercentral haemal arches in the tail, and proatlantes linking the skull and atlas. They may also have lost the smallest number of dermal bones from the skull. Nevertheless, no sister-group relationship has been convincingly established between microsaurs and any other specific group of early tetrapods. There is strong support for the temnospondyl ancestry of both frogs and salamanders, but the broader affinities of temnospondyls are no closer to resolution than that of microsaurs. All we can be fairly confident of, from the standpoint of skeletal anatomy, is that frogs and salamanders have much closer affinities with one another than either has to caecilians. This is also supported by molecular studies, especially those of San Mauro and his colleagues (2005), who found that caecilians diverged from batrachians prior to the divergence of frogs and salamanders from one another, based on 44 amphibian nucleotide sequences of a portion of the RAG1 gene. The specific times of these divergences were further discussed by Lee and Anderson (2006) and are shown diagrammatically in Figure 12.11.

Unfortunately, there are as yet no means of establishing the patterns of relationship or relative times of divergence of microsaurs, temnospondyls, or amniotes either among one another, or with any of the other known lineages of Upper Devonian and Lower Carboniferous tetrapods. This precludes determination of whether batrachians (frogs + salamanders) or caecilians shared a more recent common ancestry with amniotes or each other.

13

The Success of Modern Amphibians

THE PREVIOUS THREE CHAPTERS have traced the ancestry of frogs, salamanders, and caecilians back toward their antecedents in the Paleozoic, using the most primitive members of the modern families as models. This chapter concentrates on their evolution during the Mesozoic and Cenozoic, to investigate the structural, behavioral, and geographical patterns of divergence that have led to the emergence of the modern fauna.

Frogs, salamanders, and caecilians may appear as minor players in today's world, but with 5948 currently recognized species they outnumber mammals, which have only about 4800 species, and approach reptiles with over 8000 species and birds with approximately 9000 (Frost et al., 2006; Pough et al., 2004).

ANURANS

Modern frogs show great diversity in their external appearance and behavior (Plate 11) and yet have retained a highly conservative skeletal anatomy that has remained almost unchanged for at least 200 million years (Handrigan and Wassersug, 2007). They have dominated the water-land interface for longer than any other group of amphibians. Their small body size and rapid rates of reproduction enable them to populate new areas quickly, while their low metabolic rate and capacity to adjust to wide ranges of temperature make it possible for them to adapt rapidly to environmental differences in time and space.

Evolutionary History
Anurans are the most diverse of living amphibian orders with 32 families, 372 genera, and nearly 6000 species (Frost et al., 2006). The entire fossil record of anurans, including extant species, has recently been discussed by Sanchiz (1998), Roček (2000), Roček and Rage (2000a, b), and Báez (2000). Sanchiz recognized 15 of the 32 modern families from the fossil record (Fig. 13.1). Only one family, the Palaeobatrachidae,

which was common in the Cenozoic, has become extinct. On the other hand, only about 80 modern genera are known from the fossil record, which includes roughly 600 species (Table 13.1).

The overall impression of the fossil record of anurans suggests a progressive increase in their geographical distribution, anatomical diversity, and numbers of genera and higher taxonomic ranks. Only two frogs are known from the Lower Jurassic, *Prosalirus bitis* from Arizona and *Vieraella herbstii* (Fig. 10.2) from Argentina. Borja Sanchiz (1998) placed both in the category of "Family incertae sedis," indicating that they could not be placed in any other recognized families on the basis of unique derived characters, but were distinguished only by primitive characters, including 10 presacral vertebrae that were notochordal and amphicoelous, retention of four to five free ribs, and absence of a condylar articulation between the sacral vertebra and the urostyle. Numerous specimens of a slightly more advanced genus, *Notobatrachus* (Fig. 9.13), with nine presacral vertebrae, are known from the Middle and Upper Jurassic of Argentina (Roček, 2000).

No comparable primitive frogs are known from the remainder of the Mesozoic or the Tertiary, but two genera are known from the modern fauna, *Ascaphus truei* from the northwestern United States and adjacent Canada, and *Leiopelma* from New Zealand, both with nine amphicoelous presacral vertebrae, freely articulating anterior trunk ribs, and primitive, asymmetrical movements of the rear limbs when swimming (Abourachid and Green, 1999). Judging from their primitive anatomy, they must have diverged from the very base of frog radiation. They have been specifically allied with the Lower Jurassic genera, and/or with one another as members of a single family designated either the Leiopelmatidae or the Ascaphidae. On the other hand, Green and Cannatella (1993) pointed out a host of anatomical differences between the living genera, and argued for their distinction as separate families. Sanchiz (1998) and Roček (2000) both place *Notobatrachus* in the Leiopelmatidae.

The next level of anuran evolution was achieved by members of the family Discoglossidae (including the Bombinatorinae), which first appeared in the Middle Jurassic of England. They were the most diverse of known frogs during the Upper Cretaceous, and are now represented by 16 species present in Europe, North Africa, the Middle East, and Asia. They are advanced in the reduction of the number of presacral vertebrae to eight, which are opisthocoelous (the centra are concave posteriorly), but they retain four pairs of free ribs. The specific interrelationships among the Ascaphidae, Leiopelmatidae, and Discoglossidae remain unresolved, but they have long been informally grouped as archaeobatrachians.

A further advance in anuran evolution is represented by the assemblage termed the Mesobatrachia (the middle level of anuran advancement), consisting of the living families Pipidae, Rhinophrynidae, Megophryidae, Pelodytidae, and Pelobatidae, and the one major group of frogs that are extinct, the primarily European Palaeobatrachidae, which are thought to have been largely aquatic as adults with habits similar to those of living pipids. The vertebral centra of palaeobatrachids are procoelous (concave anteriorly) rather than opistocoelous, as in other advanced anurans. The pipids and Rhinophrynidae are highly distinctive in having flattened bodies and lacking a movable tongue. All living species have webbed feet and are aquatic throughout their life. *Pipa* is distinctive in that the eggs are embedded in pockets in the skin of the back, from which they emerge as tiny frogs. Pipid tadpoles from the Lower Cretaceous were discussed in Chapter 10. Pelobatids, whose recent members are termed spadefoot toads, have the richest fossil record, going back to the Upper Jurassic.

All other frogs are grouped as the Neobatrachia, none of which are known prior to the Upper Cretaceous, but represent 96% of the living species. At least 20 families are recognized, but their specific interrelationships remain poorly resolved (Fig. 13.2; Frost et al., 2006). Sanchiz (1998) recognized only seven families from the fossil record, of which the Leptodactylidae alone is known from the Upper Cretaceous. The Bufonidae appeared in the Upper Paleocene, the Myobatrachidae in the Lower Eocene, and the Ranidae, Rhacophoridae, and Microhylidae in the Upper Eocene. The Hylidae were certainly present by the Lower Oligocene when they appeared in Saskatchewan. However, specimens from the Upper Paleocene of Brazil were reported by Estes and Reig in 1973, but never described.

Recent efforts to classify frogs have focused primarily on molecular analyses of the genome (Frost et al., 2006; Roelants and Bossuyt, 2005), but since these are restricted entirely to living species, there is no way in which they can indicate the absolute times of branching, nor the vast number of species and potentially higher taxa that emerged in the Mesozoic and early Cenozoic but subsequently became extinct without leaving any living descendants. Neither can they provide an actual history of the patterns and modes of change in structural features, which can only be gained from the recovery of fossil remains.

Geographical Distribution

As we have seen in the events leading to the end-Permian extinction (Chapter 8, Fig. 8.16), all the major land masses were connected to one another during the Triassic, forming a gigantic world continent. There were no major oceans separating any regions, making it possible for amphibians and other terrestrial vertebrates to move to the most distant areas of the earth, given a sufficient period of time and appropriate local environments. The configuration of the earth's surface at that time provided the best conditions for global disbursal of any period in the history of land vertebrates.

The immediate antecedents of anurans, *Triadobatrachus* and *Czatkobatrachus*, were already present in both the Northern and Southern Hemispheres by the Early Triassic, and Early Jurassic anurans were present in North and South

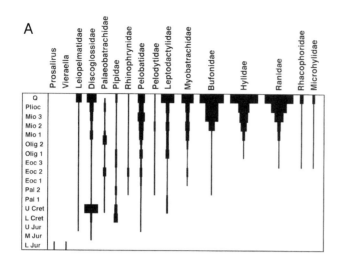

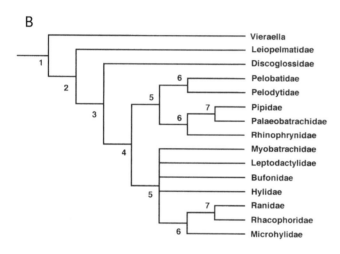

Figure 13.1. A, Temporal distribution of anuran families with a fossil record. *B*, Phylogenetic relationships among modern frog families that have a fossil record. From Sanchiz, 1998.

Table 13.1. Comparison of Numbers of Families, Genera, and Species of Frogs, Salamanders, and Caecilians Known from Fossils and from the Modern Fauna

	Families		Genera		Species	
	Modern	Fossil	Modern	Fossil	Modern	Fossil
Caecilians	6	3	33	5	173	5
Salamanders	10	9	62	40	548	40
Frogs	32	16	372	80	5227	~600

Sources: Numbers of living taxa from Frost et al, 2006. Number of fossil taxa from F. A. Jenkins et al., 2007; A. R. Milner, 2000; and Sanchiz, 1998.

America (Fig. 13.3). The presence of *Ascaphus* in North America may reflect its area of origin, but there is no record of when, or by what route, the ancestors of *Liopelma* reached New Zealand. The oldest known fossils of the Discoglossidae were collected from the Middle Jurassic of England. This family is also reported from North America in the Upper Jurassic and in China by the Lower Cretaceous (Chang, 2003). By the Upper Cretaceous they had spread to Europe, central Asia, Mongolia, and Uzbekistan. They are now restricted to western Europe, North Africa, and the Holy Land.

The oldest known mesobatrachians are represented by a pelobatid from the Upper Jurassic of North America. This family is now widely distributed in North America and Europe. Pipids are known in considerable numbers by the Lower Cretaceous in the Near East and in the Upper Cretaceous of Niger and Argentina. They are now restricted to northern South America and Africa south of the Sahara. Pelodytids were first known from the Upper Oligocene of North America, but are now limited to France, the Iberian Peninsula, and the land between the Black and Caspian Seas.

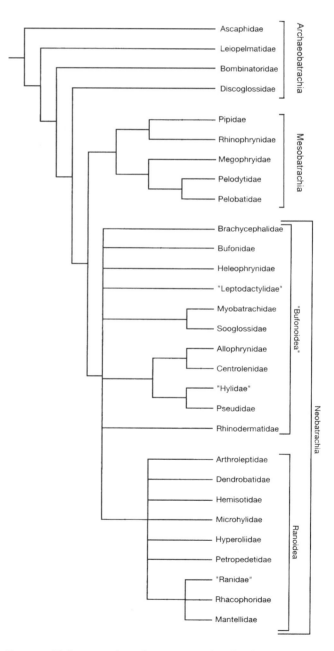

Figure 13.2. Phylogenetic relationships among modern frog families based on morphological and molecular characters. Based on Pough et al., 2004.

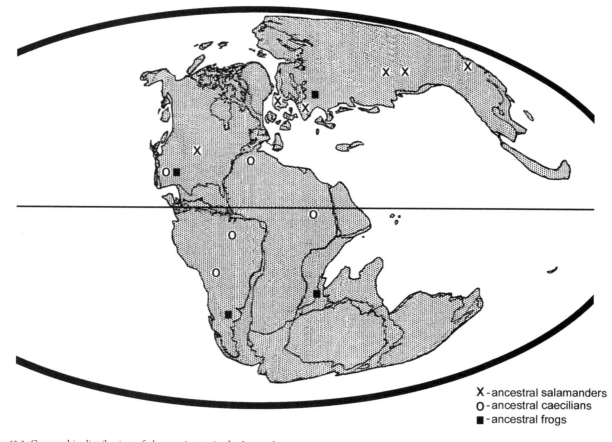

X - ancestral salamanders
O - ancestral caecilians
■ - ancestral frogs

Figure 13.3. Geographic distribution of the continents in the Lower Jurassic (Hettangian) showing the localities of fossils of the oldest known frogs, salamanders, and caecilians.

Neobatrachians were first recognized from the Upper Cretaceous of Argentina, represented by the family Leptodactylidae. This group remains dominant in South America and now inhabits Central America and the most southern United States as well as islands of the Caribbean but is not known to have spread to other land masses. As commonly defined, the Leptodactylidae is divided into 5 subfamilies, including 57 genera and 1243 species. According to Frost et al. (2006), this family is united by geography, but not by synapomorphies. He has recently placed it within a larger taxonomic grouping—the Leptodactyliformes—that he suggests as being a sister-group of the Hylidae.

Hylids (with 48 genera and 806 species) are not known with certainty until the lower Oligocene, when they appeared in southern Canada. By the uppermost Oligocene, they had reached Europe, where they are currently widespread. They also occur today throughout all but the southern extremities of South America, southeastern Asia, and northern Africa but without leaving a fossil record. They are also widely distributed in Australia, but with only a dubious record prior to the Quaternary.

The family Bufonidae is now essentially worldwide in distribution, except for Saharan Africa, New Guinea, and Australia. They are first known as fossils from the Upper Paleocene of Brazil and now include 35 genera with 485 spe-

cies. Fossils are also known from the Miocene of Columbia and Mio-Pliocene of Argentina, but they had reached North America by the Oligocene. Their fossil record extends from the Miocene to the Recent in Europe, Africa, and western Asia.

The Microhylidae includes 432 species grouped in 69 genera. Their current distribution extends from northern South America into southern North America, across southern Africa, eastern India, southeastern Asia, and Indonesia to northernmost Australia. Perhaps because of their small size and fragile skeletons, their only confirmed fossils are from the Miocene of Australia, consisting of one extinct genus with two species, and from the Pleistocene of North America and Japan. There are only unconfirmed reports of fossils from Europe.

The current distribution of the Rhacophoridae is limited to southern Africa, southeast Asia, and Indonesia, from which have been described 10 genera with 267 species. Their fossil record is limited to an unconfirmed report from the Upper Eocene of France and one member of a living genus from the Upper Pleistocene of Japan.

The Myobatrachidae includes 122 species allocated to 23 genera. Their current distribution includes Australia, New Guinea, and Tasmania. Two extinct species of living genera have been reported from Australia, as well as extinct forms

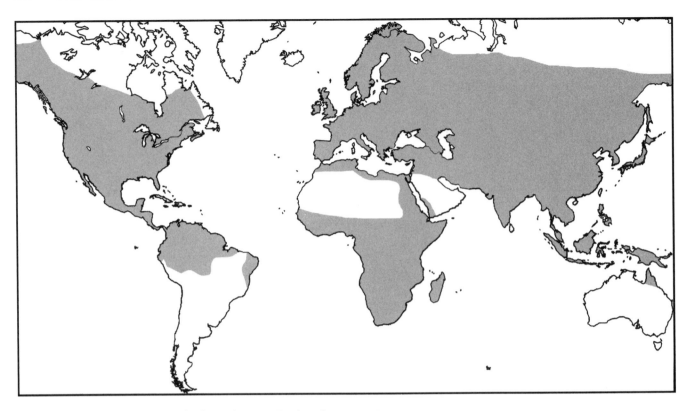

Figure 13.4. Modern distribution of the family Ranidae. From Pough et al., 2004.

from the Eocene, Oligocene, and Miocene. A fossil from the Eocene of India has been questionably attributed to this family.

The Ranidae, with 54 genera and 772 species, has the widest distribution of all frog families (Fig. 13.4). In common with bufonids, they are nearly worldwide in distribution, with the exception of southern South America, Saharan Africa, and Australia except for the far north. The earliest possible fossil record is from the Late Paleocene of France, but they were certainly present in Europe by the Late Eocene, where their remains are common throughout the later Cenozoic. The oldest North American reports are from the Miocene, followed by a major radiation in the Pliocene. Specimens from the Upper Pliocene and Pleistocene can all be placed in modern species. Ranids are also known from the Miocene of Turkey, Afghanistan, and China, the Middle Miocene of Morocco, the Oligocene of Kazakhstan, the Middle Miocene of Thailand, and the Pliocene of Azerbaydzhan.

The obvious incompleteness of the fossil record of frogs, as indicated by the absence of any evidence of the prior history of more than half of the 32 recognized living families, has led to the use of molecular data to reconstruct the pattern and timing of their evolutionary divergence (Roelants and Bossuyt, 2005), as well as those of salamanders and caecilians (Zhang et al., 2005; San Mauro et al., 2005; Roelants et al., 2007). These studies attempt to establish associations between major events of evolutionary divergence and large-scale tectonic movements of the continents. They conclude

that the initial phylogenetic radiation of all of these groups began early in the Mesozoic, when the continents were still grouped together. In nearly all cases, this is substantially earlier than the first appearance of the major families in the fossil record, but our knowledge of their fossils is so incomplete that this is hardly surprising. It is also possible that molecular divergence had occurred prior to the emergence of the skeletal traits by which we distinguish the individual taxa.

URODELES
Evolutionary History

Although typically amphibious in their life histories and sharing an ultimate common ancestry, the evolutionary history of salamanders has been very different from that of frogs. There remains a serious problem at the very base of salamander evolution as to whether members of the ancestral lineage underwent metamorphosis to terrestrial adults or were neotenic. Schoch and Fröbisch (2006) demonstrated that only one of the large number of branchiosaurid species is known to have undergone metamorphosis. It was also argued in Chapter 11 that specific attributes of the salamander skull, such as the loss of the middle ear and the slow sequential pattern of ossification of the skull, almost certainly evolved within an aquatic environment. Yet, both metamorphosing hynobiids and the neotenic cryptobranchids are included among the most primitive living salamanders, the Cryptobranchoidea, based on the absence of internal fertilization in both families. Cryptobranchids and hynobiids also have more chromosomes than other salamanders, resulting from the large number of microchromosomes. On the other hand, at least one of the salamanders from the Middle Jurassic Jehol fauna was clearly terrestrial as an adult (Figs. 11.7 and 11.8). This indicates that both metamorphosing and neotenic life histories and their consequences for adult anatomy had evolved by that time.

The relatively conservative anatomy of salamanders that undergo metamorphosis compared with the degree of difference between the highly derived anatomy of neotenic lineages also suggests that the anatomy of transforming adults represents a common ancestral pattern from which a series of unrelated neotenic groups diverged.

The amount of skeletal divergence among salamanders as a whole has been much greater than that of frogs. Frogs have retained a nearly constant morphotype from the Lower Jurassic to the present to the degree that it has been extremely difficult to differentiate the major living families on the basis of clearly defined anatomical synapomorphies. All the modern frog groups have evolved from a single bauplan, showing relatively little modification from the anatomy of the earliest known Lower Jurassic anurans. In contrast, salamanders as far back of the Middle Jurassic had already diverged into a number of highly distinctive body plans, associated with differing life history strategies (Plate 13; Figs. 13.5 and 13.6).

Specimens from the Middle Jurassic of Inner Mongolia (discussed in Chapter 11) illustrate at least three distinctive morphotypes: moderately large neotenic forms resembling *Cryptobranchus*, *Jeholotriton*, which retains primitive skeletal features of hynobiids but is also neotenic, and a gracile terrestrial genus that appears similar to the advanced salamandrids and ambystomatids. There is too great a gap in time and morphology to assign any of these early genera to specific Cenozoic and living families with assurance, but they demonstrate that for most of the history of salamanders they have been represented by lineages with highly distinct anatomical features and differing life histories.

Overall, salamanders appear to have a much more informative fossil record than anurans, with 9 of the currently accepted 10 families known from fossil remains, some extending back as far as the Jurassic (A. R. Milner, 2000). Of the 62 living genera accepted by Frost and his colleagues (2006), 40 are known from the fossil record, and/or have plausibly recognized immediate ancestors. However, the record at the species level is much less complete. Compared with the 548 living species, only about 40 are known from fossils (A. R. Milner, 2000).

Broadly, the anatomy of individual salamander families has remained relatively constant since the Late Cretaceous or Early Cenozoic, with the large neotenic genera being the most distinctive. The smaller, more terrestrial species are less likely to be preserved and are more difficult to differentiate at the level of species. This problem is certainly greatest among the Plethodontidae, in which a major radiation of the Bolito-

glossini (including more than one-half of living salamanders) occurred within northern South America after the formation of the Isthmus of Panama approximately 3 million years ago. Of the 15 or so genera, with approximately 222 species, only a few fossil vertebrae from the Upper Miocene of California have been attributed to this tribe.

Early Geographical Distribution

Although many amphibian fossils have been found in Late Permian and Triassic deposits of the Southern Hemisphere, none are suggestive of salamander ancestry. On the other hand, many putative antecedents have been found in the Upper Carboniferous and Permian of Europe and a few from a single locality in North America—Mazon Creek, Illinois. Subsequently, the record is restricted to Eurasia, with possible salamander remains from the Early Triassic of northern Gansu Province, northwestern China (Gao et al., 2004), and *Triassurus* (of uncertain affinities) from the Upper Triassic of southern Kyrgizstan, adjacent to western China. A primitive, but unquestioned salamander, *Kokartus*, appears in the Middle Jurassic of Kyrgizstan, followed by *Karaurus* from the Upper Jurassic of southern Kazakhstan (A. R. Milner, 2000). A much larger assemblage of beautifully preserved Middle and Upper Jurassic salamanders has recently been found in northern China and Inner Mongolia, suggesting that this region was central to the early radiation of the modern families (Figs. 11.5–11.9). In addition, several genera represented by fragmentary remains have been described from the Middle Jurassic of England and Scotland (Evans et al., 1988; Evans and Waldman, 1996). These Middle and Upper Jurassic salamanders form the basis for the radiation of the modern families, and demonstrate their initial restriction to the Northern Hemisphere (Wang and Gao, 2003).

Fossil Record of Modern Families

Although the life histories and adult anatomy of cryptobranchids and hynobiids are very distinct, they are united in the superfamily Cryptobranchoidea because of their common habit of external fertilization. Both also retain a separate angular bone, which is lost in most more advanced salamanders. Both of these families have Asian roots.

CRYPTOBRANCHIDAE. This family of large neotenic salamanders consists of only two living genera, *Andrias*, now limited to China and Japan, and *Cryptobranchus*, restricted to eastern North America. According to A. R. Milner (2000), the ancestors of the modern members of this family arose in eastern Asia by the Early Cretaceous. Two fossil genera of similar morphology, *Aviturus* and *Ulanurus*, are known from the Paleocene of Mongolia. Fossils designated *Zaissanurus*, but possibly representing a primitive species of *Andrias*, have been described from the Eocene-Oligocene of Mongolia and Russia, and *Andrias* itself is known in Europe from the Upper Oligocene, where it persisted into the Pliocene. The Asiatic lineage apparently migrated to North America via the Bering land bridge in the Upper Cretaceous or Paleocene. *Andrias*

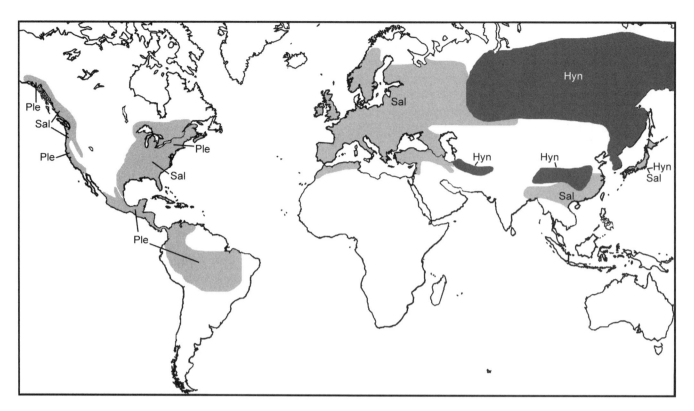

Figure 13.5. Modern distribution of salamander families. Plethodontids also occur in central Italy and Sardinia. Proteids occupy about the same area in eastern North America as the plethodontids, as well as a small area of north- eastern Italy and the adjacent Dalmatian coast. From Pough et al., 2004. Hyn, Hynobiidae. Ple, Plethodontidae. Sal, Salamandridae.

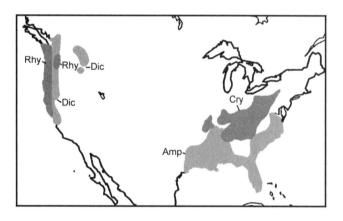

Figure 13.6. Modern distribution of salamander families. Cryptobranchidae also occupies central China and southern Japan. Range of sirenids almost identical with that of *Amphiuma* except for a small outlier in southern Michigan. From Pough et al., 2004. Amp, Amphiumidae. Dic, Dicamptodontidae. Rhy, Rhyacotritonidae. Cry, Cryptobranchidae.

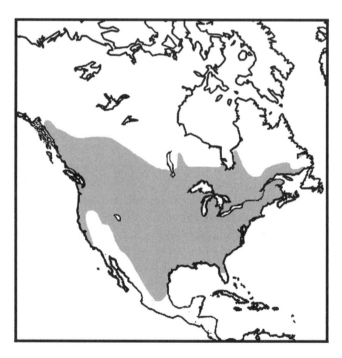

Figure 13.7. Modern distribution of the Ambystomatidae. This family does not occur outside of North America. From Pough et al., 2004.

maintained its primitive form into the Upper Miocene, followed by transformation into the modern *Cryptobranchus* by the Plio-Pleistocene.

HYNOBIIDAE. There are between seven and nine genera and approximately 50 species of small to medium-sized, metamorphosing hynobiids living primarily in China, Japan, and northeastern Russia but with some species extending into the Ural Mountains and Iran. The only reliably identified fossil representing this family is from the Upper Pliocene

of Kazakhstan, belonging to the modern species *Ranodon sibiricus.* Similar vertebrae have been collected from the uppermost Miocene and Lower Pleistocene of Romania, well beyond its current geographic range. Zhang and his colleagues

(2006) used clock-independent molecular dating to establish a putative time for the origination of the extant hynobiid species within the Middle Cretaceous. The morphology of the living species suggests that their common ancestor was a stream-adapted form.

SIRENIDAE. Sirenids are highly enigmatic. There is no evidence of structures associated with formation of spermatophores, nor behavioral evidence of internal fertilization. This suggests that they retained the primitive practice of external fertilization. They also lack pedicellate teeth, but this may be a result of their neoteny rather than suggestive of a major divergence from other salamanders. Modern species are known only from eastern North America. They are highly distinctive in having reduced forelimbs and no hind limbs at all. They have a massive anterior palatal dentition on the vomers and palatine bones, vaguely similar to that of *Jeholotriton* (Fig. 11.4). The maxilla and premaxilla are rudimentary and toothless. The trunk is very elongate, but only the most anterior vertebrae have rib-bearers. Oddly, sirenids have intravertebral spinal foramina throughout the vertebral column, which is a characteristic of the more advanced metamorphic families.

There are only two living genera, *Siren* and *Pseudobranchus*. They are limited to the southeastern Gulf Coast of the United States and northeastern Mexico, with a small outlier in southern Michigan. Fossils of both genera are known from the Eocene into the Pleistocene, from as far west as Wyoming. The more primitive *Habrosaurus* (Fig. 11.4D) is known from the Upper Cretaceous and Paleocene of North America.

Disarticulated remains resembling elements of sirenids have been discovered in the Upper Cretaceous of Bolivia and the Sudan and the Early Cretaceous of Niger. They lack pedicellate teeth, but are unique in having parasymphyseal teeth on the mandibles, as in labyrinthodonts. They would be the only record of fossil salamanders from the Southern Hemisphere (Evans et al., 1996). A. R. Milner (2000) argues that sirenids may have evolved and disbursed prior to the separation of Pangaea into northern Laurasia and southern Gondwana, but there is no fossil evidence.

NEOCAUDATA. All modern salamanders other than cryptobranchoids and sirenids are known to practice internal fertilization and may have evolved from a single ancestral lineage of facultatively terrestrial genera, such as the unnamed fossil from the Middle Jurassic illustrated in Figures 11.7 and 11.8. The subsequent geographic divergence of neocaudates from cryptobranchoids may be attributed to tectonic events in central Asia that led to the formation of the Turgai Sea, along what is now the eastern margin of the Ural Mountains, which formed a major north-south barrier between Asia and Euroamerica.

STEM NEOCAUDATES. The initial radiation of neocaudate lineages appears to have occurred primarily in Europe and North America. The earliest genera that have been recognized are *Iridotriton*, from the Upper Jurassic of Utah (Evans et al., 2005), and *Valdotriton*, from the Lower Cretaceous (Barremian) of Spain (Fig. 11.8). *Iridotriton* was a small but highly ossified and gracile animal. It has a single spinal nerve foramen in the tail (but none in the trunk), which may herald its further elaboration in more derived genera, and the parasphenoid lacks internal carotid foramina. On the other hand, the angular remains as an independent area of ossification in the lower jaw.

In *Valdotriton* the angular bone has become fused to the prearticular (Evans and Milner, 1996). The second basibranchial is in the shape of an inverted Y, as in modern metamorphosing adults, in contrast with the forwardly directed trident of the neotenic Middle Jurassic *Chunerpeton* and the Russia *Karaurus*. *Valdotriton* was small (40 mm snout-pelvis length) with well-developed limbs and girdles, but not yet fully ossified. The 16 presacral vertebrae are primitive in being notochordal and amphicoelous, but at least nine caudals have intravertebral foramina. The presacral vertebrae supported double-headed ribs in the manner of salamandrids and ambystomatids, but lack any intravertebral foramina except in the atlas.

Evans and Milner (1996) stated that there are no derived features to associate *Valdotriton* with any of the more advanced neocaudates, none of which are known until much later in the Cretaceous.

ADVANCED NEOCAUDATES. The seven modern families of neocaudates include a relatively conservative assemblage of primarily metamorphosing forms—the Salamandridae, Ambystomatidae, and Plethodontidae—and two whose adults are highly specialized for an aquatic way of life—the proteids and the amphiumids. The overall skeletal resemblance of the primitive members of the terrestrial families to Jurassic and Early Cretaceous stem neocaudates suggests that they shared a common ancestry, and that the neotenic lineages had evolved separately, at a later time. Unfortunately, there is no fossil evidence supporting the specific ancestry of either proteids or amphiumids. They tend to group together in phylogenetic analyses based on the common occurrence of adult characters that reflect aquatic adaptation. Wiens and his colleagues (2005) avoided this problem by coding their adult morphology, as well as that of other aquatic genera, as unknown, which resulted in a phylogeny that is more concordant with other morphological as well as molecular data (Fig. 11.17). Consequently, proteids appear as a sister-group of the Salamandridae and *Amphiuma* as a sister-group of plethodontids, despite their overall morphological disparity.

PROTEIDAE. This family consists of only two extant genera, *Necturus*, the mud puppy (with five species) from North America, and *Proteus*, a cave-dwelling salamander from northeastern Italy and the Dalmatian coast. Both are neotenous, with relatively short limbs, but not an especially long body. *Necturus* has only 18 presacral vertebrae. As in *Valdotriton*, the atlas is the only presacral vertebra to have intravertebral spinal foramina. Fossils assigned to *Necturus* have been reported from the Paleocene of Saskatchewan, and two genera resembling *Proteus* are known as early as the Miocene of Europe, the Caucasus, and Kazakhstan (A. R. Milner, 2000). There is no evidence of where their common ancestor may have lived.

SALAMANDRIDAE. The 18 living genera of salamandrids are widely distributed across Europe, Asia, and North America. They are distinguished anatomically by the presence of opisthocoelous vertebrae and intravertebral spinal foramina in all trunk vertebrae. Life histories are highly variable, but all stem from a basically terrestrial morphology. The oldest fossil salamandrids are identified on the basis of undescribed vertebrae from the uppermost Cretaceous of northern Spain.

A. R. Milner (2000) divides living salamandrids into four groups: basal salamandrids, the *Pleurodeles-Tylototriton* clade, the European newts, and the Asian and North American newts. Their fossil record is almost entirely limited to primitive members of modern genera or their immediate antecedents. Among the basal group, *Salamandra, Mertensiella, Chioglossa,* and *Salamandrina* are all primarily terrestrial and are known from fossils going back as far as the Eocene. All were restricted to Europe except *Mertensiella,* which occurs in Asia Minor. *Salamandra* remains widespread in Europe, but the rest occur as scattered relicts.

Pleurodeles, living in Spain, Portugal, Morocco, Algeria, and Tunisia, and *Tylototriton* and *Echinotriton,* living in southeast China, Okinawa, and the Ryukyu Islands, are characterized by a highly sculptured skull and orbitotemporal bar typical of more advanced salamandrids. Fossils are known from the Cenozoic of Europe, indicating subsequent migration to Asia. European newts, include *Triturus* and two other living genera, have fossil representatives going back to the Upper Paleocene of Europe and have now spread into Turkey, Iraq, and Iran.

Another clade of newts presumably arose in Europe but then disbursed to both Asia and North America. Fossils of *Procynops* (very similar to the living *Cynops*) are known from the Upper Miocene of Shantung Province in China. Two additional genera of this clade now range from India, Vietnam, and China to Japan. *Taricha,* now living on the Pacific coast of North America, and *Notophthalmus,* in the eastern states, are known from fossils from the Upper Oligocene of Oregon and the Miocene of Montana and Florida, suggesting that they entered North America via the Bering Strait prior to the Upper Oligocene.

DICAMPTODONTIDAE AND AMBYSTOMATIDAE. *Dicamptodon,* with four species now living in the Pacific Northwest, has alternatively been included with either the ambystomatids or the salamandrids. It is known from a Paleocene fossil collected in Alberta. It is neotenic in its habits, but not highly modified morphologically. It is primitive in lacking intravertebral spinal nerve foramina in the trunk, but has them in the tail.

The family Ambystomatidae has only a single genus but 30 species, all confined to North America but extending throughout much of the continent (Fig. 13.7). Fossils are known as early as the Lower Oligocene in Saskatchewan and great amounts of neotenic material have been collected from the Pliocene of Kansas and Texas.

RHYACOTRITONIDAE. The genus *Rhyacotriton* is known from four species, all living in northwestern North America. It is a primarily aquatic form, living in streams but resembling both ambystomatids and salamandrids, with which it has been alternatively classified. No fossils have been assigned to this genus.

AMPHIUMIDAE. This family is represented by only a single genus with three species, limited to southeastern North America. *Amphiuma* (referred as the Congo eel) is neotenous with a meter-long trunk and four greatly reduced limbs. It lacks external gills as an adult, but retains one pair of gill slits. As in *Siren,* ribs are missing from most of the trunk. Spinal foramina are limited to the mid-posterior caudal vertebrae. Fossils are known from the latest Cretaceous and Paleocene of Wyoming and Montana.

PLETHODONTIDAE. The Plethodontidae includes approximately 374 species, considerably more than all the other salamander families combined, grouped in 27 genera (Frost et al., 2006). All are lungless and fully metamorphosed as adults and possess a nasolabial groove that runs below the external naris to aid in chemoreception.

The Plethodontidae has been traditionally divided into two subfamilies, Desmognathinae, with only 2 genera and 20 species, and Plethodontinae, composed of 3 tribes—Plethodontini (3 genera, 62 species), Hemidactyliini (7 genera, 33 species), and Bolitoglossini (17 genera and 244 species) (Pough et al., 2004). Plethodontids are generally small and primarily terrestrial, although some are stream-dwellers. Most desmognathines and hemidactyliins have free-living larvae, but all plethodontines give birth to fully developed young.

The Plethodontidae is well established as an monophyletic assemblage on the basis of anatomical features, including the absence of lungs in all species, the presence of intravertebral spinal foramina throughout the column, and opisthocoelous vertebrae. Common ancestry is also confirmed by many molecular analyses (Chippindale et al., 2004; Macey, 2005; Mueller et al. 2004; Wiens et al., 2005). Absence of lungs is typically attributed to their ancestors having passed through a stage in which they inhabited cold, fast-flowing mountain streams in which buoyancy resulting from air-filled lungs would have interfered with feeding on the bottom. Cold, highly oxygenated water would have enabled them to make effective use of cutaneous respiration.

The small terrestrial and stream-dwelling Desmognathinae are restricted to three species in eastern North America, with a fossil record going back only to the Pleistocene. The Plethodontini, with only three genera, is common in both eastern and western North America; fossils are known from the Lower Miocene and Pleistocene. The Hemidactyliini is

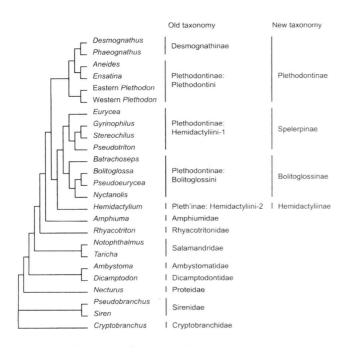

Figure 13.8. Relationships of Plethodontidae from Chippindale et al. (2004) based on parsimony analysis of 104 transformation series of morphology and 1493 informative sites of nuclear DNA (RAG-1) and mt DNA (cytochrome *c* and ND4a).

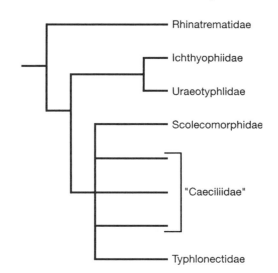

Figure 13.9. Phylogenetic relationships among caecilian families, based on morphology and molecular characters. From Pough et al., 2004.

known from eight genera from eastern North America to Texas, with two genera from the Pleistocene. The Bolitoglossini extends from western North America across Central America and into northern South America. The only pre-Pleistocene fossil is from the Upper Miocene of California. A possibly included genus, *Hydromantes/Speleomantes*, is known from France, Italy, and Sardinia and another has been reported from Korea (Min et al., 2005).

Recent molecular studies have led to differing views of relationships. Figure 13.8 from Chippindale et al. (2004) includes the Desmognathinae as a lineage within the Plethodontinae, some hemidactyliinis are placed in the newly designated subfamily Spelerpinae, and other members of the Hemidactyliini are designated as members of a subfamily Hemidactyliinae.

CAECILIANS

As was evident from Chapter 12, the fossil record of caecilians is far less informative than that of frogs or salamanders, consisting of only 5 species covering the 180 million years of their known existence. This compares with 173 caecilian species described from the modern fauna, none of which are recognized from fossil remains. We know that archaic caecilians were present as early as the Lower Jurassic in what is now North America. By the Upper Cretaceous they had evolved essentially modern vertebral characteristics, and their presence in the Sudan shows that they were widespread by that time. Isolated vertebrae, not distinguishable from modern genera of the family Caeciliidae, have been found in the Upper and Lower Paleocene in South America, but no fossils are known of the other five living families.

The presence of the significantly more primitive *Rubricacaecilia* in the Lower Cretaceous of Morocco could be interpreted as indicating that all the modern families had evolved subsequently from a significantly more advanced common ancestor. However, it is possible that *Rubricacaecilia* was a relict of an earlier radiation, and that more advanced caecilians are to be found elsewhere at this time. In the absence of fossil evidence, the phylogeny of caecilians is based entirely on living species (Fig. 13.9).

Today, caecilians are restricted primarily to the wet tropics of the Southern Hemisphere. They are widely distributed in northern South America and extend well into Central America. They are limited to damp coastal regions in Africa, but widely scattered in southeast Asia (Fig. 13.10). It is generally assumed that their primary disbursal occurred when the southern continents were still close to one another. At whatever time they disbursed, however, some presumably crossed over the northern continents to reach the Far East.

Among the more primitive families, the rhinotrematids are restricted to three areas in northern South America, but ichthyophids are scattered across southern India, Ceylon, southeast Asia, Indonesia, and the Philippines. Uraeotyphlids, however, occupy only a small area in southern India. The highly diverse caeciliids inhabit northern South America as well as Central America, the coastal region of the bight of Africa, small enclaves in East Africa, the Seychelles, and isolated areas of the Indian subcontinent. The highly derived scolecomorphids occupy three isolated regions in south-central Africa, while typhlonectids are known from large areas of northern and southeastern South America.

According to San Mauro and his colleagues (2005), the origin and early radiation of the caecilians preceded the breakup of Pangaea, which explains the discordance between their phylogenetic relationships and current pattern of dispersal. The earliest fossil caecilians are known from North America and northern Africa, well beyond the range of any of the

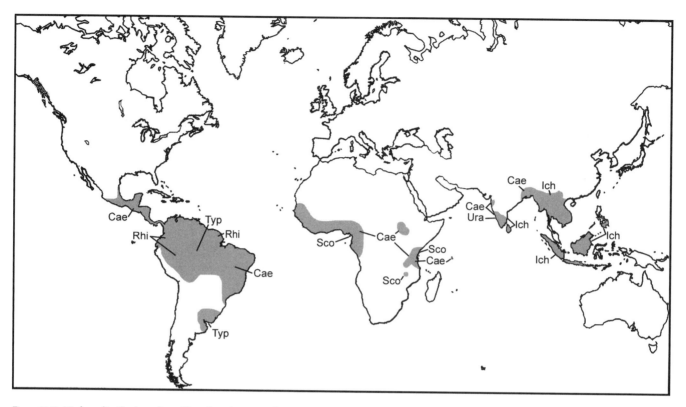

Figure 13.10. Modern distribution of caecilians, based on maps from Pough et al., 2004. Cae, Caeciliidae. Ich, Ichthyophiidae. Ura, Uraeotyphlidae. Rhi, Rhinatrematidae. Sco, Scolecormorphidae. Typ, Typhlonectidae.

modern families, but the most primitive living families—Rhinatrematidae, Ichthyophiidae, and Uraeotyphlidae—are separated on widely isolated continents. On the other hand, the much more advanced caeciliids occur across South America, Africa, and southeast Asia. Whatever their pattern of dispersal, the African *Rubricacaecilia* may have retained a primitive morphology long after the divergence of the more derived families.

THE "NEW" AMPHIBIANS

Although references to the soft anatomy, physiology, and behavior of modern amphibians have been made throughout the previous four chapters, these factors cannot be directly studied in fossils, making it difficult for us to determine what major biological advances have been made in the ancestry of modern frogs, salamanders, and caecilians and when these occurred. In contrast, living species demonstrate a number of significant changes in the modes of feeding, respiration, and reproduction that distinguish them clearly from our concepts of "typical" amphibians, but have been key to their success in the modern world.

Sexual Reproduction

Perhaps the most significant advances that have occurred in the biology of the modern amphibian orders were in various aspects of their reproduction. This has taken place sepa-

rately in multiple families of all three orders. Going back to our concept of amphibians discussed in the early chapters of this book, they have been distinguished by the retention of characters that were a heritage of their fish ancestors. Initially, they must have reproduced in the water, depending on external fertilization. Their eggs lacked extra-embryonic membranes, but had large enough supplies of yolk for the young to hatch out as free-living organisms. In order to transform from a fishlike hatchling into a terrestrial adult, they must have evolved a larval stage, although none is known from the fossil record until the mid-Carboniferous. Since the most primitive members of all three living amphibian orders have larval stages, they almost certainly reflect retention of a common ancestral condition. Loss of a larval stage has thus evolved separately in all three orders.

Knowledge of modern amphibians also indicates separate evolution of various means of internal fertilization, attainment of reproduction on land, and hatching of highly developed, terrestrial young. All modern caecilians, and some more derived genera of the other orders, lay their eggs on land. A smaller number retain their eggs within the reproductive tract and give birth to live young.

FROGS. Frogs are the most numerous, in terms of both individuals and species, of all living amphibian orders, and yet nearly all of the modern genera retain the primitive mode of external fertilization. A surprising exception is one of the most primitive species, *Ascaphus truei*, in which a highly vas-

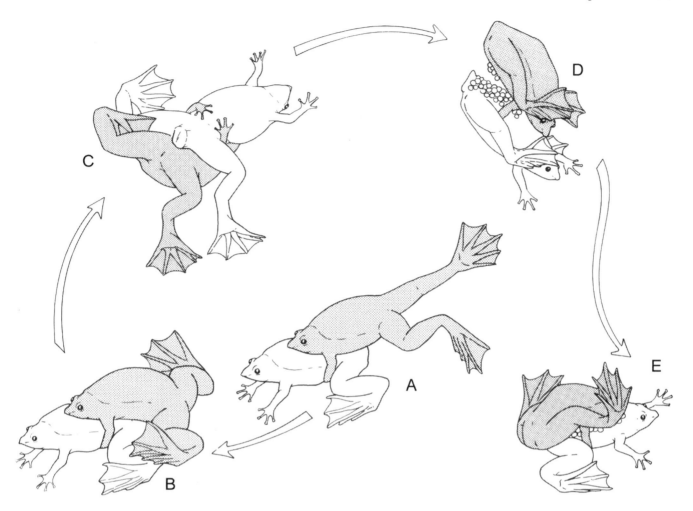

Figure 13.11. Mating maneuvers in *Pipa carvalhoi.* Male dark, female white. *A,* Midwater swimming with amplexus. *B,* Rest on bottom and push off. *C,* Ascent and turnover. *D,* Turnover and capture of eggs against belly of male.

E, Sink to bottom and placement of eggs on back of female. From Duellman and Trueb, 1986.

cularized extension of the male's cloaca is inserted into the female's cloaca with the help of muscles derived from the otherwise missing adult tail. The "tail" is supported by cartilaginous rods that are attached to the ventral part of the pelvic girdle. Internal fertilization in this species is necessitated by its living habitually in rapid currents that would disburse the sperm before the eggs could be effectively fertilized. The other exception is fertilization via cloacal apposition, which occurs in some species of *Eleutherodactylus* (Pough et al., 2004).

In order to assure that the eggs are effectively fertilized, nearly all frogs practice amplexus to ensure close contact between the sexes. Typically, the male grasps the female's trunk from behind; in primitive frogs, the grasp is immediately in front of the hind limbs, but in advanced frogs the attachment is farther forward, just behind the forelimbs. Except in a few species of fully terrestrial frogs, deposition of the eggs occurs in the water.

In most frogs, the eggs are not guarded or otherwise protected after fertilization, but there are interesting exceptions. In the mesobatrachian *Pipa* the fertilized eggs are pushed into

the tissue of the female's back by her mate where they sink into pockets in which they mature to small froglets before they swim away (Fig. 13.11). In other species the males or females carry strings of eggs wrapped around their legs or in pouches, make foam nests for their protection, or place them in holes in trees or aerial plants (Duellman and Trueb, 1986).

The leptodactylid *Eleutherodactylus* is exceptional in many ways. It is estimated to include about 700 species, most of which are ovoviviparous, with metamorphosis occurring within the egg. Two species, *E. coqui* and *E. jasperi,* both of which may be extinct, apparently had internal fertilization (via cloacal apposition) and are (were) viviparous. Another family, the Myobatrachidae, which is common in Australia, New Guinea, and Tasmania, is thought to have evolved from the Leptodactylidae or from bufonoids. Among the more than 100 species, two members of the genus *Rheobatrachus* are exceptional in having gastric brooding. After the eggs are laid and fertilized on land, they are swallowed by the female and development and metamorphosis occur in the stomach. Unfortunately, no members of these species have been seen in the wild since 1982 and they may be extinct.

SALAMANDERS. Fertilization is external in crypto-branchids, hynobiids, and probably sirenids. The eggs are typically deposited in the water and sperm is discharged over them. One hynobiid species, *Ranodon sibiricus,* approaches the more advanced salamanders in forming a spermatophore that is deposited in the water and the female places her eggs on top of it. However, the male and female are not in contact. All other salamander families practice internal fertilization. The male may deposit the spermatophore either on land or in the water, after which he entices the female to pick it up with the lips of her cloaca. The young of most salamander families hatch out in the water as free-swimming larvae.

As we have seen, salamander larvae exhibit the greatest degree of flexibility in their manner of maturation. Crypto-branchids, sirenids, proteiids, and amphiumids retain many larval characteristics as sexually mature adults and never leave the water. Hynobiids all metamorphose, but some remain primarily aquatic as adults. The salamandrid *Salaman-dra*'s choice of reproductive modes is highly flexible. Depending on the available food supplies, *S. salamandra* may give birth to a small number of fully metamorphosed offspring, whereas other populations deposit a larger number of eggs or advanced larvae that complete their development in the water over a period of up to two years (Pough et al., 2004). Among plethodontids, the Hemidactyliini all have an aquatic larval stage, whereas the Plethodontini and Bolitoglossini all lay their eggs on land where they develop directly, without metamorphosis.

CAECILIANS. In contrast with frogs and salamanders, all caecilians practice internal fertilization with an eversible copulatory organ referred to as a phallodeum, unique to this order. The primitive families Rhinotrematidae, Ichthyophii-dae, and Uraeotyphlidae all lay eggs on land or in soil cavi-ties, where they are protected by the mother. After hatching, the young make their way into the water (small streams or seepages) as larvae, with tiny gills and gill slits. They return to the land as adults, living in soil and leaf litter. This life history is retained in the most primitive of the Caeciliidae, *Praslinia,* but more advanced caeciliids hatch out on land as terrestrial juveniles, without an aquatic larval stage. Other genera, including *Schistometopum* (Plate 16) and *Dermophis,* give birth to live young. Scolecomorphids, a family with only two genera living in equatorial Africa, are also viviparous. All members of the family Typhlonectidae are viviparous as well. Their young are born aquatic, but quickly shed their gills and assume the anatomy of adults (Pough et al., 2004). According to M. H. Wake (1977), about 75% of all caecilian species bear live young.

Changes in Respiration and Feeding in Salamanders and Caecilians

Frogs, salamanders, and caecilians all have the capacity for cutaneous respiration, which is facilitated by their generally small body size and habitually moist and permeable skin. These factors limit their capacity for large body size in most

environments and their ability to adapt to arid environments. On the other hand, the presence of an alternative means of respiration has made it possible for some salamanders and one caecilian genus to lose the lungs, which has been asso-ciated with significant changes in the feeding apparatus, al-though in very different ways in the two groups.

As we saw in Chapters 9 and 10, the respiratory systems of the larvae of most groups of salamanders, going back to their Paleozoic ancestry among choanate fish and branchiosaurids, have remained very conservative. They are based on a gape-and-suck system, powered by the hyoid apparatus (Fig. 9.15). This has not changed significantly among neotenic salaman-ders, nor the larval stages of most transforming species. But its function has changed dramatically in salamanders that have reduced or lost their lungs. Lung reduction has occurred in several hynobiids, including *Ranodon,* the aquatic salaman-der *Rhyacotriton,* and the salamandrids *Chioglossa, Pachytri-ton,* and *Salamandrina.* Lung loss is complete in the hynobiid *Onychodactylus* and in all members of the Plethodontidae. Plethodontids are the most diverse and specious of all sala-manders, and the only family that has radiated extensively in the tropics. Reduction and loss of lungs in all these groups has been attributed to their adaptation, at some time in evo-lutionary history, to life in cold, fast-flowing streams that are highly oxygenated. The absence of lungs reduces their buoy-ancy, which allows them to remain at the bottom of streams for feeding (Bramble and Wake, 1985).

Although this way of life is assumed for the ancestors of plethodontids, most of the modern species in this family are now terrestrial or arboreal. In the larvae and adults of sala-manders, the hyoid apparatus is involved in feeding as well as respiration, with the anterior bones of the hyoid apparatus being thrust forward within the fleshy tongue. As long as the hyoid apparatus was also involved with pumping air into the lungs, it was not able to change its configuration significantly. However, with the loss of lungs in plethodontids, the hyoid apparatus was free to become specialized for prey capture. The capacity for extensive protrusion of the tongue was one of the factors that enabled the plethodontids to radiate ex-tensively in the Late Cenozoic, becoming the most specious of all salamander families, with 374 currently recognized spe-cies, almost 70% of the total.

TONGUE FEEDING IN ADVANCED SALAMAN-DERS. Primitive larval feeding in salamanders follows the gape-and-suck mechanism of their aquatic Paleozoic ante-cedents, based on retention of most of the hyoid apparatus of their fish ancestors. Only the dorsal, epibranchial elements have been lost. Primitive metamorphosing salamanders, such as ambystomatids, use the same mechanism for feeding in the water as adults, but not as efficiently, since the ingested water must be expelled through the mouth once the exter-nal gill openings are sealed (Lauder and Reilly, 1988). It is in-teresting to note that the modes of suction feeding used in permanently larval forms—*Necturus, Cryptobranchus, Siren,* and *Amphiuma*—were probably secondary, modified diver-

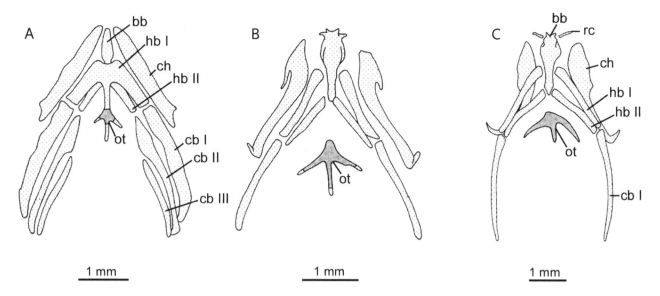

Figure 13.12. From left to right, three stages in the metamorphosis of the hyoid apparatus of the plethodontid *Eurycea.* From Rose, 2003.

gently from the pattern of Jurassic salamanders, some of which retained the features of branchiosaurids (Bramble and Wake, 1985).

Members of the families Hynobiidae, Ambystomatidae, Salamandridae, and Plethodontidae that metamorphose fully evolved protrusile tongues to feed on terrestrial prey. This is elaborated most effectively among the salamandrids and plethodontids, some of which can protrude the tongue nearly the length of their body. In those genera that retain an aquatic larval stage, such as ambystomatids, protrusion develops smoothly from suction feeding during metamorphosis, but without the capacity for great extension of the tongue. Gills are lost, gill slits closed, and the posterior elements of the branchial skeleton are missing. The ceratohyal becomes a large, blade-like structure, attached to the quadrate-stapes by a rod-like extension. At the midline there is a central basibranchial that bears one or two pairs of radii. The basibranchial supports the tongue pad. Two pairs of hyobranchials and a pair of certatobranchials are located posteriorly (Fig. 13.12).

The main muscle that propels the tongue is the subarcualis rectus I, which extends posteriorly from the ceratohyal and wraps around the hyobranchial anteriorly as well as the more posterior ceratobranchial (Fig. 13.13). When it contracts, the ceratohyal and basibranchial and attached tongue are thrust forward. This, however, is only possible in lungless salamanders, principally the plethodontids. In more primitive salamanders, these muscles are used in inflating the lungs, but once the lungs are lost, the ventral constrictor muscles of the throat serve to establish a solid platform that precludes posterior movement of the ceratobranchials. In addition, the genioglossus muscle that attaches to the front of the tongue in frogs and primitive salamanders is lost, so that there is no hindrance of anterior movement of the tongue.

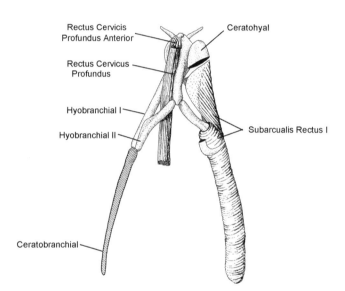

Figure 13.13. Dorsal view of the hyoid apparatus of the plethodontid *Eurycea longicauda,* which is moderately specialized for tongue projection. From Bramble and Wake, 1985.

Once prey is attached to the sticky anterior margin of the tongue, it is pulled back into the mouth by the rectus cervicis profundus. When it is at rest, this muscle is thrown into loops, like a coil of rope. When the tongue is protracted, it follows along until contact is made with the prey, and then is pulled back into the mouth, where the prey is lifted from the tongue by the palatal teeth. In advanced plethodontids, the ceratobranchial and rectus cervicis profundus extend midway in the length of the trunk.

The entire feeding mechanism is coordinated. The mandibular depressors, ventral constrictors, lingual protractors, and retractors all contract simultaneously, forming a completely predictable system that hits its target with amazing

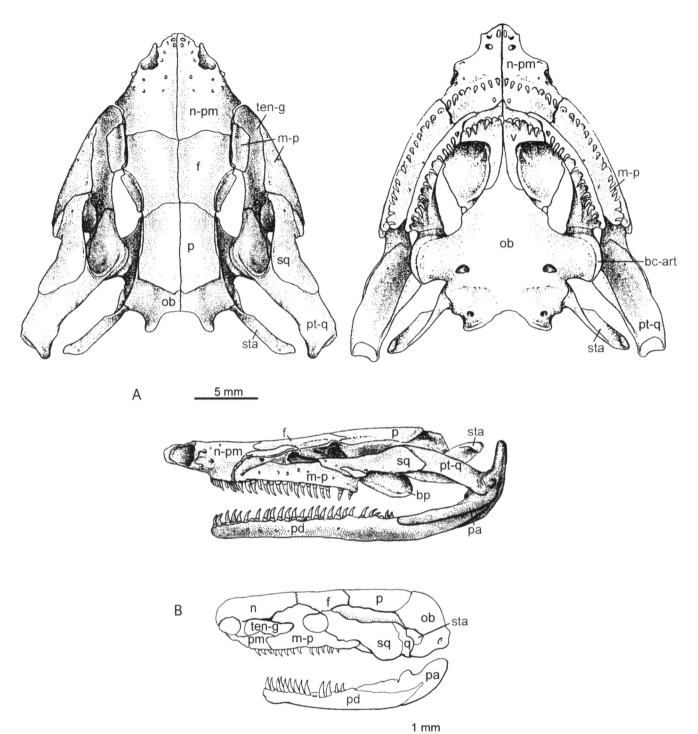

Figure 13.14. Typhlonectid caecilians. *A*, Dorsal, palatal, and lateral views of the largest known caecilian, *Atretochoana* from Brazil. From Wilkinson and Nussbaum, 1997a. *B*, Lateral view of the genus *Typhlonectes*. From M. H. Wake, 2003.

accuracy within 5 to 10 msec. It is a prepatterned system that is unmodulated by peripheral feedback, which is necessary because the mouth is opened so far that the eyes are facing dorsally rather than forward. This system is most fully perfected in the Bolitoglossini. This has certainly contributed to their extensive and rapid radiation during the 3 million years since they entered South America.

CAECILIANS. The general configuration of the skull is highly conservative among nearly all caecilians, back to *Eocaecilia* (Fig. 12.7). The one striking exception is *Atretochoana eiselti*, represented by two specimens. The first, originally described by Taylor (1968) from the Vienna Museum, was labeled as having come from South America. The second specimen, found in a collection at the University of Brasilia, has no precise locality data but suggests that both specimens had probably come from Brazil (Wilkinson and Nussbaum, 1997a, b). However, other typhlonectids, the family to which this genus belongs, have a wider distribution within South America (Fig. 13.10).

Comparison of the skull of *Atretochoana eiselti* (Fig. 13.14) with members of all groups of caecilians shows dramatic differences in their structure. Most striking is the great extension of the jaw suspension behind the occiput, in contrast with its position anterior to the occiput in all other caecilians, but loosely paralleling that of advanced snakes. The jaw articulation is also extended far more laterally than in other caecilians and the profile of the skull is much lower. In addition, the stapes has a very long stem that extends toward the extremity of the pterygoid-quadrate, and the bony margin of the orbit is lost, both anteriorly (for passage of the tentacle) and posteriorly, where the squamosal is widely separated from the parietal. According to Wilkinson and Nussbaum (1997a), these changes in the back of the skull are associated with major differences in the mode of respiration and feeding.

Atretochoana is unique among all adequately studied caecilians in the total absence of lungs, the closure of the internal nares in the palate, and the loss of pulmonary arteries and veins. Wilkinson and Nussbaum (1997a) argue that this condition, as in plethodontids, evolved in relationship to adaptation for life in cold, highly oxygenated, and rapidly flowing mountain streams, to reduce buoyancy. Because of the rich supply of oxygen in the water and the relatively low metabolic rate of caecilians in general, *Atretochoana* presumably relied entirely on cutaneous respiration for uptake of oxygen, although it is the largest caecilian known, with a total length of over 80 cm. This size distinguishes *Atretochoana* as the largest known terrestrial vertebrate lacking lungs.

The loss of lungs and reliance on the hyoid apparatus for respiration enabled *Atretochoana* to modify the cranial structure in ways unique among caecilians. This is reflected most strikingly in the far posterior position of the jaw articulation, supported by the extension of the quadrate far behind the occiput. This would have greatly increased the gape, presumably enabling *Atretochoana eiselti* to feed on much larger prey than any other caecilians, as also reflected in its total body size.

In addition, the cheek region, including the well-integrated maxillopalatine, squamosal, and fused pterygoid and quadrate, appears, from the absence of interdigitating sutures, to be freely movable on the central elements of the skull. This mobility is facilitated by articulation with the much enlarged basicranial articulation. Capacity for movement is also reflected in the gap in the skull table above the course of the tentacle and the loss of postorbital connection between the squamosal and the parietal. As a result, the area of the orbit is not surrounded by bone as it is in all other caecilians. The mobility of the cheek and the obliquely latero-ventral as well as posterior orientation of the adductor jaw muscles

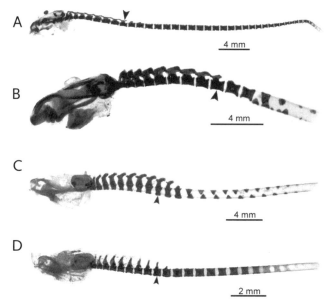

Figure 13.15. Four tadpoles of the Asian frog family Megophryidae showing ossification of the caudal vertebrae, unique to frogs. A, *Leptobrachella mjobergi*. B, *Leptolalax pelodytoides*. C, *Megophyrs lateralis*. D, *Ophryophryne microstoma*. From Handrigan and Wassersug, 2007.

led Wilkinson and Nussbaum (1997a) to suggest that kinesis of the cheek would allow the upper tooth rows to be everted outward.

Changes in the configuration of the stapes were accompanied by the elaboration of a novel branch of the pterygoideus muscle, the pars stapedialis, which runs between the stapes and the posterior margin of the maxillopalatine. According to Wilkinson and Nussbaum (1997a) contraction of this muscle would pull the cheek medially and posteriorly, complementing the action of the internal adductors. The ligamentous attachment common between the stapes and quadrate in other caecilians is lost. Presumably the presence of a large muscle running from the stapes to the fused pterygoid-quadrate affects hearing in some way, but this could only be tested in living specimens.

Overall, *Atretochoana* illustrates the greatest change in cranial anatomy since the ancestry of caecilians in the late Paleozoic, even overshadowing the differences between *Rhynchonkos* and *Eocaecilia* in general appearance. It can only be hoped that the discovery of two specimens from somewhere in Brazil reflects the continued existence of this highly divergent lineage of caecilians, rather than the last vestiges of an already extinct adaptive experiment.

Genome Size

Another feature of evolution, which is difficult to understand, is the tendency for salamanders to accumulate a very large amount of non-coding DNA in the chromosomes. According to T. R. Gregory (2005), the mean amount of DNA per nucleus in salamanders is nearly 10-fold higher than in frogs (36.7-millionth of a gram, compared with 4.6-millionth of a gram). The range in salamanders extends from approximately 14-millionth of a gram in *Desmognathus wrighti* to approximately 120-millionth of a gram in *Necturus*.

The amount of non-coding DNA is highly variable within and among vertebrate groups, but salamanders are exceptional in their very great amount. Variability in the quantity of non-coding DNA occurs in all organisms as the result of accumulation and duplication of transposable elements, or parasitic DNA, which has no apparent function in the operation of the cell, but is of little consequence in small amounts. However, in the quantity carried by most salamanders, its presence slows down cell division and increases the size of both the nucleus and the cell as a whole. Presumably, these factors serve as agents of selection to maintain relatively small amounts in most vertebrates, although roughly half of human DNA consists of such elements.

Why such large amounts are able to accumulate in salamanders has not yet been determined. It presumably accounts for the loss of the nucleus in most red blood cells, which occurs in both salamanders and mammals, where the small size of the cells results in a much higher surface to volume ratio and so more rapid transfer of oxygen. More striking is the fact that salamanders have only 5 to 10% the number of neurons in the brain as frogs of equivalent size (Roth et al., 1998). Urodeles compensate to some degree by having more dendrites per neuron. In addition, the paucity of neurons in especially small salamanders, such as bolitoglossines, may be compensated for by their being sit-and-wait predators, rather than actively searching for prey, and the possession of an effectively automated means of prey capture. One possible advantage of having a high and irregular amount of non-coding DNA is that only very closely related individuals could interbreed effectively, thus leading to the extremely rapid rate of speciation seen in bolitoglossines.

Tadpoles with Bony Tails

What is particularly striking about frogs, not only in contrast with salamanders and caecilians but also with most vertebrate groups (with bats as an informative exception), is the extreme degree of stereotypy of the general body form, as indicated by the skeleton. Specifically, all have 10 or fewer presacral vertebrae, a urostyle, greatly elongate rear limbs, and, as adults, no caudal vertebrae. All these characteristics are integral to their highly specialized saltatory locomotion.

The tadpoles of all frogs are also highly stereotypic in the globular head/trunk and highly flexible tail. In nearly all species it is without vertebral support. This increases the ca-

pacity for changing directions, which enables them to evade aquatic predators. Strong functional constraints on both aquatic and terrestrial locomotion are reflected in the highly stereotypic skeletal framework and hence overall form of the body. Although the superficial appearance of the body, patterns of behavior, and adaptation to specific habitats and ways of life vary greatly, these have limited expression in the skeleton and so make it difficult to recognize species-, genera-, or even family-level distinctions from the fossil record.

It is hence very exciting to discover that the tadpoles of one family of Asian frogs, the Megophryidae, related to pipids and pelobatids, include several genera that have caudal centra and neural arches. As described by Handrigan et al. (2007), the number of centra so far observed range from 5 to 6 in *Leptolalax* to about 30 in *Leptobrachella* (Fig. 13.15). Tadpoles of these and several other genera are typically found in fast-flowing streams, where their extended caudal skeleton anchors muscles that facilitate wiggling between rocks and even burrowing into the stream bed to avoid predation.

The specific nature of the centra differs from genus to genus. In *Leptobrachella* and *Ophryophryne,* the caudal centra ossify around the entire notochord, but in *Megophrys* and *Xenophrys* each centrum develops from dorsal and ventral pairs of ossification that expand to meet each other, broadly resembling the integration of paired pleurocentra and intercentra in primitive labyrinthodonts. These structures are not homologous in the phylogenetic sense of evolving from an immediate common ancestor, but presumably reflect the reactivation of a developmental potential latent in a long retained but suppressed ontogenetic pathway. In most anurans, the axial and appendicular elements of the skeleton of the larvae develop at about the same time, but in these megophryids the axial skeleton ossifies much earlier.

Handrigan and his colleagues (2007) attribute this evolutionary reversal to selection pressure associated with a riparian lifestyle. They argue that the capacity to redevelop caudal vertebrae was retained in mesobatrachians, but not among ranids and centrolenid genera that inhabit similar environments, because they have fewer developmental constraints than the later evolving Neobatrachian families.

The continuing structural, behavioral, and physiological advances and relatively recent taxonomic radiations among salamanders and frogs suggest a dynamic potential for future changes in ways of life and geographical disbursal, barring catastrophic changes to the environment of the earth.

14

The Future of Amphibians

THE FOSSIL RECORD OF AMPHIBIANS is rarely sufficiently complete to establish the specific time of either the origin or extinction of individual species. What we can see are relatively long periods of continuity among families and orders and the shorter duration or individual occurrences of species and genera. The long temporal gaps between members of particular families and orders and their sporadic geographical distribution provide some evidence of the degree of incompleteness of their fossil records.

The fossil record from the Carboniferous into the Lower Permian documents the great longevity of many families and orders that are represented by numerous genera and species, but does not indicate any major extinction events. In contrast, there was progressive reduction in the number of families throughout the later Permian. The loss of large, primarily terrestrial genera can be attributed to competition and predation by the expanding numbers, size, and diversity of amniote groups. The reduced fossil record of families common to coal swamp and other shallow water habitats may be explained by the increasing aridity of most localities yielding vertebrates.

Very few of the amphibian families living in the late Paleozoic are known to have persisted into the Triassic, following the most drastic period of extinction in the history of life. While a great number of stereospondyl lineages appeared at this time, the persistent rarity of smaller, semi-aquatic to terrestrial forms in the Triassic may be associated with the large-scale loss of the coal swamp flora that occurred at the end of the Permian (Retallack et al., 1996; Benton and Twitchett, 2003). However, at least three lineages, those leading to frogs, salamanders, and caecilians, must have survived this extinction event.

The immediate antecedents of frogs are represented by the Lower Triassic *Triadobatrachus* and *Czatkobatrachus*. Tentative remains of salamander antecedents have been described from the Triassic, but no trace of caecilians is known until the Lower Jurassic, contemporary with the oldest known anurans. All the modern amphibian orders are represented in the Middle Jurassic, suggesting survival without diminution through the end-Triassic extinction (Fraser et al., 2006).

The end-Cretaceous extinction terminated all remaining dinosaur lineages, as well as most groups of marine diapsids (other than crocodiles), although not turtles. But no amphibian families are known to have become extinct at this time. The fossil record of caecilians is too incompletely known to indicate whether the six living families diverged from one another before or after the end-Cretaceous extinction. One archaic family is known from the Lower Cretaceous, but its time of extinction is not known.

Only one family of frogs, the Palaeobatrachidae, is known to have become extinct during the Cenozoic, as well as two families of archaic salamanders, the Scapherpetontidae and Batrachosauroididae (A. R. Milner, 2000).

The fact that more than half of the living frog families have no fossil record indicates the degree of incompleteness of our knowledge throughout most of the Mesozoic and Cenozoic. On the other hand, the fossil record of frogs and salamanders is very well known during the Pleistocene, which was a time of great and repeated climatic change, especially well documented in the Northern Hemisphere. In fact, extensive collection of fossils from Pleistocene deposits fails to demonstrate significant extinction in North America, despite the violent climatic shifts caused by repeated advances and retreats of continental glaciations, with ice averaging from 2 to 3 km in thickness across most of Canada and spreading into the northern United States (Fig. 14.1).

The first continental glaciation began approximately 1.8 million years ago. For the first 900,000 years the intervals between periods of glaciation lasted for about 41,000 years, after which the interval increased to 100,000 years, resulting in a total of approximately 30 glacial episodes, during which time maximum glaciation would have wiped out nearly all the plants and animals living in the northern portion of North America. The last interglacial, extending from 130,000 to 116,000 years ago, was followed by a glaciation that reached its maximum extent about 17,000 years ago. Approximately 10,000 years ago there was a rapid retreat of the ice sheets, except for those covering Greenland and Antarctica as we see them today.

One would think that a catastrophe of such continental scale would have led to massive extinction. As pointed out by Holman (2000), there were serious extinctions among the mammals and birds. In continental North America, the British Isles, and the European continent, 8 families, 46 genera, and about 191 species of mammals became extinct during the Pleistocene. In the late Pleistocene alone, 2 families, 19 genera, and an undetermined number of species of North American birds were also lost. On the other hand, no genera or families of amphibians became extinct during that time. Only 2 species of the 44 frogs identified from the previous interglacial are currently considered to have become extinct, and there is some question about those. It would seem that

Figure 14.1. Shaded area indicates extent of ice sheet covering northern North America approximately 14,000 years ago. Modified from Kummel, 1961.

frogs and salamanders were at a considerable advantage over birds and mammals in surviving repeated destruction of the natural environment over millions of square miles.

This advantage can be attributed partially to their physiology. The low metabolic rates of frogs and salamanders would have enabled them to aestivate during long periods of cold. Some can freeze solid and thaw out in the spring with no harmful effects. They could not live for more than a few months without feeding and breeding, but they could survive intermittent cooling long enough to disperse, even if randomly, away from the advancing ice sheet. Amphibians also benefited from their large numbers, higher reproductive potential, wide ranges, and the ameliorating effect of moist environments, as well as wide ranges of temperature tolerance.

The rich fossil record and intense collecting by persons such as Alan Holman (2000) also provide evidence of the rapidity of reinvasion of northern habitats by amphibians at the end of the last period of continental glaciation. Frogs such as the freeze-tolerant *Rana sylvatica* may have reached southern Michigan by 13,000 years ago, when there were still vast ice sheets in the north. Less cold-tolerant species may not have advanced much farther until about 2500 years ago. Today, *Rana sylvatica* extends into the Arctic tundra. Other species are estimated as advancing at an average rate of about 0.1 km per year, or 100 km in 1000 years. But this is enough for them to escape the glacial front and to reclaim lost habitats sufficiently rapidly to avoid extinction.

However, the capacity of frogs and salamanders to have survived the dramatic and repeated changes in their natural environment during the Pleistocene may not provide a good model for evaluating their chances of survival in our modern world, dominated by human-induced changes. The most significant difference is the rapidity of change. Humans have done more to change the surface of the earth and its atmosphere in the last 2000 to 3000 years than occurred over the last 10,000 years, since the retreat of continental glaciation to the northern margins of the Northern Hemisphere. And the rate of change is continuing to increase.

Because the rate and scope of change today are beyond the level that can be studied from the fossil record, the remainder of this chapter will be devoted to a discussion of this subject by an expert on modern amphibians (especially frogs), David Green. He is professor of biology and director of Redpath Museum, McGill University, and was chairman of the Committee on the Study of Endangered Wildlife in Canada (COSEWIC), which was responsible for drafting recommendations regarding the protection of endangered species that were enacted into law by the Canadian Parliament.

THE STATE OF AMPHIBIANS TODAY

Since 1989, the shadow of amphibian population declines all over the world has sparked a surge of interest in amphibians. Indeed, there has been so much publicity about amphibian population declines over the last decade and a half that it seems like the most interesting thing about amphibians is that there are progressively fewer of them. In both the popular and scientific literature, amphibians' thin skins and biphasic life histories are routinely written off as liabilities dooming them to suffer from any and all possible sorts of environmental challenge. Yet amphibians have survived for hundreds of millions of years. If they are in trouble, then many other animals must be in trouble, too.

Fears about Amphibians

The amphibians of today are not just the relics of a distant and more glorious past. Today's frogs, salamanders, and caecilians are very different sorts of animals that are each marvelously successful in evolutionary and ecological terms. Although modern amphibians are rather conservative morphologically, they nevertheless outnumber mammals in terms of species, with more than 6000 so far described, which occupy an enormous range of terrestrial and freshwater habitats and far exceed all other terrestrial vertebrates in modes of reproduction and life history. Leaving larvae to fend for themselves in a pond until they metamorphose into adult form is only one of dozens of amphibian life-history strategies.

Amphibians, frogs especially, are also superlatively diverse and abundant. There are almost certainly more frogs, in terms of numbers of individuals, than any other terrestrial vertebrates. Adult amphibians and, even more spectacularly, larvae may exist in extremely high densities (Woolbright et al., 2006; Gehlbach and Kennedy, 1978; Deutschmann and Peterka,

1988). Tadpoles numbering in the hundreds of thousands are the dominant consumers in their ponds (Whiles et al., 2006). The ecological impact of amphibians is considerable.

During the 1980s, it came to the attention of numerous amphibian biologists that finding frogs in many places was progressively more difficult than it used to be. Around that time, I noticed that frogs in southern California and in the Sierra Nevada were proving to be scarcer than they had been before. Jars and jars of frogs preserved in museum collections attested to where they previously had been and to their former abundance. In 1983, the eminent amphibian biologist David Wake and I set out for Highland Lakes in Alpine County, California, to search for mountain yellow-legged frogs, *Rana muscosa*. Highland Lakes was a known locality for the species and, on a very fine day, we had every expectation that we would find them there. We carefully searched. We found not a one. The story of that frogless day, and other similar reports by other amphibian biologists, led by the end of the decade to a global realization that amphibian populations might be in serious decline (D. B. Wake, 1991, 2003). This generated a research program that captured headlines, mobilized research around the world, and spawned a phenomenal increase in our knowledge of amphibian populations (Blaustein and Wake, 1990; Lannoo, 2005).

"Canaries in the coal mine" became the popular phrase used to depict amphibian vulnerabilities to recent environmental change (Halliday, 2000; Storfer, 2003). It is not a bad analogy, considering the state of our world in many places, yet to be an effective sentinel, a coal-mine canary has to have at least some resilience to environmental problems and, most importantly, be noticeable. Coal mines once had canaries in them for a simple reason: as long as the birds sang the coal gas was not yet thick enough to kill the miners. Choruses of calling frogs are one of the most noticeable aspects of a healthy environment.

By 1990, reports of declining amphibian populations had come from all over the world and the stories had become news (Yoffe, 1992). Most serious were reports of species going extinct. The gastric-brooding frog, *Rheobatrachus silus*, of Queensland, Australia, and the golden toad, *Bufo periglenes*, from Monteverde, Costa Rica, both had apparently disappeared forever. *Rheobatrachus* was a spectacular animal; the females would swallow their fertilized eggs and incubate the larvae in their stomachs all the way through metamorphosis (Tyler, 1983). Then they would cough up the little froglets that would then swim away. During the development of the larvae, the females would not eat and would need to disable all digestion in the stomach. How they could do that will never be known, for they are all gone. When it was first discovered (Liem, 1973), *Rheobatrachus* was very common in southern Queensland streams yet, since 1981, not a single individual has been observed in the wild, despite extensive searches. Perhaps drought caused their decline, or overcollecting, or habitat degradation caused by the activities of loggers and gold-panners (Tyler and Davies, 1985).

Discovered in the mid-1960s (Savage, 1966), *Bufo periglenes* was spectacular in a wholly different way. Once a year, the gorgeous, yellow-orange males would assemble in great numbers at pools in the forest to compete for females, who were mottled white, black, and red (Jacobson and Vandenberg, 1991). The congregating toads became famous and a tourist attraction. Then, in 1987, the population crashed from tens of thousands to merely 29 individuals. The pools had dried up before their tadpoles had the chance to transform. By 1991, the toads were gone (Harding, 1993). Perhaps the climate had been affected by global warming or local rainfall patterns had been affected by deforestation. Perhaps they had succumbed to the fungal disease chytridiomycosis before anyone realized it (A. Pounds, 1996).

The fate of the Kihansi spray toad, *Nectophrynoides asperginis,* is a more recent tragedy. These toads lived only in a gorge bathed in the mists generated by the Kihansi Falls in the southern Udzungwa Mountains of Tanzania (Krajik, 2006). They were discovered in 1995, living in the path of a major hydroelectric project. The damming of the river eliminated the damp spray from the falls, irrevocably altering the toads' habitat (Quinn et al., 2005). Attempts were made to make mists artificially and a captive breeding program was begun in American zoos. By mid-2005, only a very few Kihansi spray toads were to be found in the Kihansi Gorge (Krajik, 2006). Today they exist only in the Toledo (Ohio) zoo.

In North America, massive declines in populations of the Western toad, *Bufo boreas,* have occurred throughout the Rocky Mountains and in California (Carey et al., 2005). The Oregon spotted frog, *Rana pretiosa,* is gone from most of its original range in the Puget Trough and Willamette Valley of the Pacific Coast of North America. The mountain yellow-legged frog, *Rana muscosa,* is virtually gone from the mountains ringing Los Angeles and from all but scattered localities in the Sierra Nevada (Bradford et al., 1993; Drost and Fellers, 1996). Many species of Harlequin frogs, genus *Atelopus,* have declined to near extinction in the Cordillera Central of Costa Rica and Panama (Ron et al., 2003). Northern leopard frogs, *Rana pipiens,* were once so abundant in southern Manitoba that they sustained a commercial harvest. Nearly 50,000 kg of frogs, equivalent to a million individuals, were collected each year during the early 1970s (Koonz, 1992) for biological supply houses. But by 1974, the harvest had declined to 5900 kg and then the population crashed catastrophically, ending collection. Since then, however, leopard frog abundance has rebounded to approach pre-1970s levels.

Decline and Reality

The appalling tales of amphibian extinctions generated genuine concern over the possibility of worldwide declines in their numbers. In 1990, the Declining Amphibian Populations Task Force (DAPTF) was formed under the auspices of the World Conservation Union (also referred to as IUCN) and working groups around the world were mobilized to study the phenomenon (Heyer and Murphy, 2005). At first there were skeptics, as the real picture is anything but clear. Declines were not everywhere nor among all species (Hairston and Wiley, 1993; Crump, 2005). For many species there was not enough data to detect trends at all (Pechmann et al., 1991). However, the DAPTF stimulated a tremendous outpouring of new research into amphibian population biology directed toward understanding whether amphibians were in decline and, if they were, what might be causing it.

Initially, most of these observations were little more than anecdotes. Some frogs appeared to be becoming rare, or at least more difficult to find. Yet rarity takes many forms (Gaston, 1997; Rabinowitz et al., 1986). Whenever a species consists of only a small number of individuals or is restricted to a limited geographic area, it can be considered rare. However, there is more than one way to look at population decline.

Looking only at the loss of individuals, Vial and Saylor (1993, 1) defined a decline as "a definite downward trend in numbers over a span of time appropriate to the species' life history, shown to be in excess of the normal fluctuations in population size." In some species that ordinarily have very stable population sizes, an accurate census may reveal a sustained loss in numbers of individuals. Losses of even a few whales or pandas, for instance, can be a significant population change for those species. A few amphibians, mainly those direct-developing species with very restricted ranges and small numbers of offspring, can be viewed in this manner.

But for most amphibians, this is not an effective measure. Because many amphibian populations fluctuate wildly, most aspects of their demographics are highly unpredictable. Such things as clutch size or age at maturity can vary greatly between individuals, depending on density, temperature, and chance events. Population-level characteristics, such as juvenile survival, overwintering survival, and predation rates, all vary from season to season and from year to year. These parameters are both stochastic and interdependent, the classic recipe for chaos and unpredictability. Fisheries scientists are more accustomed to this problem than most people who study terrestrial animals. It is impossible to predict how many fish there will be if all you really know is how many fish there were (Francis and Shotton, 1997). And most of the time nobody really knows how many fish there were either.

Thus the number of individual amphibians may be both difficult to estimate and highly subject to change. The hatching of any one bullfrog egg among the thousands in a single clutch, or the death of a single tadpole among tens of thousands in a pond, may have virtually no impact upon the bullfrog population as a whole. Populations of many amphibian species fluctuate by orders of magnitude, but these great swings in population size do not appear to be tightly coupled with their chances of extinction (Green, 2003). They are a natural part of their ecology. However, these species may become rare in ways other than in numbers of individuals. Thus, Green defined a population's decline on a landscape level as "the condition whereby the local loss of populations across the normal range of a species so exceeds the rate at which populations may be established, or re-established, that there is a definite downward trend in population number" (Green, 1997, 293). At the species level, we should think of the overall loss of populations, over their entire range, as the most serious form of decline.

The unstable nature of amphibian population size led many to suspect that the purported observations of amphibian decline were just the normal fluctuation of healthy populations (Pechmann and Wilbur, 1994). In most pond-breeding amphibians, successful reproduction in any one year is not guaranteed. Ponds can dry up before the larvae have transformed, disease may strike, or the weather may be too cold for the larvae to complete development. Consequently, the sizes of amphibian populations are highly unpredictable as they react to random events in the environment around them (Green, 2003). Long time-series are necessary to detect any trends in population size when fluctuations can be an order of magnitude or more (Blaustein et al., 1994), but not many populations of amphibians have been studied for more than 10 years in a row (Meyer et al., 1998).

Nevertheless, the weight of evidence is that amphibian populations are declining throughout the world (Houlahan et al., 2000; Storfer, 2003). This realization is based not on tracking single populations over many years, but by looking for trends among large numbers of populations simultaneously and by looking for geographic trends in the extent of species' ranges. Downturns in population size are both more numerous and of greater degree than increases; losses of populations outnumber the establishment of new populations. The recent Global Amphibian Assessment (Stuart et al., 2004) concluded that amphibians are highly threatened and are declining rapidly. The assessment puts 1856 species of the world's amphibians on their Red List and estimates that at least 2468 amphibian species are experiencing some form of population decrease. However, caution needs to be exercised in interpreting these results as the IUCN Red-List criteria allow species to be listed as threatened without specific evidence of decline. Nevertheless, 32.5% of amphibian species in the world are considered threatened at some level. The phenomenon of amphibian decline is real.

A Sea of Troubles

What is causing amphibians to decline, and why are some species more prone to decline than others (Biek et al., 2001; M. L. Crump, 2005)? Investigations into the plight of amphibian populations have been largely directed toward detecting decreases in numbers of individuals within populations and documenting possible proximate causes (Blaustein et al., 1994; Alford and Richards, 1999; Corn, 2000; Gardner, 2001; Kiesecker et al., 2001). Many possible causes have been touted, including emerging infectious diseases, increased ultraviolet radiation, chemical pollutants, introduced predators, habitat destruction, and climate change. Which particular factor is the culprit? Collins and Storfer (2003) considered that these threats assort into two classes of hypotheses, be-

ing either direct factors, Class I hypotheses, such as habitat change, or indirect factors, Class II hypotheses, including climate change, pollution, or infectious diseases. Nevertheless, each of these factors is as plausible as the next in many ways, and all are likely causes of decline to some degree. Equally true is that none is mutually exclusive. Indeed, none of them stands up to being the single underlying reason.

Among the most obvious initial candidates was acidifying precipitation. Increased acidity of ground and pond water exerts both lethal and sub-lethal effects on amphibian populations in a number of ways, including increased embryo and larval mortality, reduced egg and larval growth, reduced reproductive output, delayed hatching, and reduced adult body size (Freda and Dunson, 1986; Räsänen and Green, 2008). These factors may result in changes in the geographic distributions of species (Beebee et al., 1990) and altered predator-prey ratios (Waldman and Tocher, 1998; Alford and Richards, 1999). However, acidifying precipitation only affects regions with both high ion deposition from the atmosphere (generally downwind from industrialized areas) and low natural buffering capacity. Thus its effects on amphibians are also regional (Räsänen and Green, 2008). There is, in addition, evidence that amphibians can adapt to more acidic conditions (Räsänen et al., 2002; Glos et al., 2003).

No environment today is truly pristine but the occurrence of amphibian declines in numerous remote regions has led many to suggest that atmospheric pollutants may be a global cause of harm (Lips, 1999; Carey et al., 2001). Pesticides, endocrine disrupters, heavy metals, and acidifying precipitation were among the first culprits fingered in the search for the causes of amphibian decline (Sparling et al., 2001, Gardner, 2001).

A wide range of industrial and agricultural pollutants can be implicated in amphibian population declines (Russell et al., 1995; Sparling et al., 2001; Stallard, 2001). Nitrate fertilizers are known to affect larval development, feeding behavior, growth rates, and physical development at concentrations less than those officially recommended for agricultural application or for drinking water (Marco and Blaustein, 1999; Schuytema and Nebeker, 1999; Marco et al., 2001). Atrazine, endocrine disrupting chemicals, and polychlorinated biphenyls have also been implicated as agents of lethal and sublethal effects upon amphibians (Carey and Bryant, 1995; Bridges and Semlitsch, 2001). These chemical contaminants cannot be considered a universal cause of population declines and there are few data directly implicating contaminants in catastrophic population declines (Alford and Richards, 1999). It is nevertheless possible that environmental contamination may make an already bad situation worse (Carey et al., 2001; Stallard, 2001).

Another early candidate for a seemingly global cause of amphibian declines was increased ultraviolet (UV) radiation, specifically UV-B, presumed to be due to depletion of the atmosphere's ozone layer (Kerr and McElroy, 1993). This, however, is disputed by other scientists (Kiesecker and Blaustein,

1995; Nagl and Hofer, 1997; D. Crump et al., 1999; Adams et al., 2001; Macías et al., 2007). Licht (2003) concludes that the UV-B hypothesis for amphibian population declines is not supportable. UV-B appears to be, at most, simply another stressor among many in the environment.

On the other hand, the spread of exotic species can have profound effects on amphibian populations (Kats and Ferrer, 2003). In particular, exotic predators can be responsible for drastic population declines, particularly where the amphibians have no prior association with, or defense against, non-native predatory fish (Fellers and Drost, 1993; Fisher and Shaffer, 1996; Hecnar and M'Closkey, 1997). In the mountains of western North America, populations of frogs and salamanders have suffered dramatic reductions in distribution and abundance in naturally fishless lakes that had been artificially stocked with trout and other gamefish (Knapp et al., 2001; Orizaola and Braña, 2006). In California, red-legged frogs, *Rana aurora*, have been deeply affected by introduced mosquitofish, *Gambusia affinis*, which prey on their eggs and early larvae (Lawler et al., 1999) and Californian newts, *Taricha torosa*, have suffered from similar predation by both mosquitofish and the introduced crayfish, *Procambarus clarki* (Gamradt and Kats, 1996). Many species in Europe and Australia are also considered to be threatened by alien, predacious fish (Gillespie, 2001; Denoel et al., 2005). Removal of the predatory fish allows the amphibians to recover (Hoffman et al., 2004; Vredenburg, 2004).

An invasive plant, purple loosestrife, *Lythrum salicaria*, is detrimental to the survival of tadpoles of the American toad *Bufo americanus* (Brown et al., 2006), though gray tree-frog tadpoles, *Hyla versicolor*, are not similarly affected (Maerz et al., 2005). Ironically, some of the biggest threats to the well-being of amphibians are other amphibians. Invasive cane toads, *Bufo marinus*, have been implicated in the decline of native Australian frogs through both predation and competition (Crossland, 2000; Murray and Hose, 2005). The case against introduced bullfrogs, *Rana catesbeiana*, as predators upon other amphibians is equivocal. Certainly they have replaced red-legged frogs, *Rana aurora*, in many areas along the Pacific Coast of North America. Nevertheless, rather than preying upon or competitively excluding the red-legged frogs, they may simply be occupying habitats that are no longer usable by them or emptied of them through prior overharvest (Jennings and Hayes, 1985).

Emerging infectious diseases, especially chytridiomycosis caused by the ectopic fungus *Batrachochytrium dendrobatidis*, is one of the more recent menaces to amphibians to be identified (Longcore et al., 1999). Chytridiomycosis is particularly insidious and has been implicated in the tremendous die-offs and even exterminations of frogs in the highland of eastern Australia (Laurance, 1996) and lower Central America (Lips, 1999; Lips et al., 2006). Chytrids are, almost universally, harmless soil fungi and none were known to be pathogenic until the discovery of *B. dendrobatidis* infections in frogs (Lips, 1999; Berger et al., 1988; Weldon et al., 2004).

Some have considered chytridiomycosis to be *the* agent behind catastrophic amphibian declines worldwide (Skerrat et al., 2007) and certainly it has been widely found in association with lethal epidemics (Daszak et al., 1999; Lips, 1999; Laurance, 1996). Not everyone is convinced, however, and it appears to be ubiquitous, even in places where there is no history of amphibian die-offs (McCallum, 2004; Ouellet et al., 2005; Ron, 2005). There is support both for and against *B. dendrobatidis* as a novel pathogen (Morgan et al., 2007). Some species are resistant to the chytrid fungus (Garner et al., 2006). It does not appear to be wholly amphibian specific (Rowley and Alford, 2006) and, to that extent, understanding chytridiomycosis is a classic epidemiological question (Morse, 1995): What triggers an outbreak of disease? J. A. Pounds et al. (2006) correlated climate change with the spread of chytridiomycosis among frogs in Central America.

Amphibians, like all organisms, are particularly subject to changes in their external environment. The biphasic life histories and highly permeable skins of amphibians make them able to adapt to environmental changes in ways that differ from other animals. Simply because they tend to be small and rather sedentary ectothermic animals, it is predictable that they would be affected by significant changes in climate (J. A. Pounds and Crump, 1994; Lips et al., 2006; Kiesecker et al., 2001, Carey and Alexander, 2003; J. A. Pounds et al., 2006). Many amphibians are restricted to damp terrestrial habitats. Should the water regime in those habitats be disrupted by, for instance, global warming or changes in rainfall, amphibians should be expected to be affected. Global warming, with its predicted effects upon the variability and severity of weather patterns (N. E. Graham, 1995; Gitay et al., 2002) was implicated early as a contributing cause of regional amphibian declines (Corn and Fogelman, 1984; Berven, 1990; Stewart, 1995). The evidence, though, is equivocal as changes in climatic patterns cannot always be found to explain observed declines (Laurance, 1996; Alexander and Eischeid, 2001). Beebee's (1995) observation of three species of anurans tending to breed earlier and earlier in spring over a 10-year period in southern England could be seen in a number of other temperate-zone anuran populations (Blaustein et al., 2001). A compelling argument was made to explain the catastrophic decline of *Bufo periglenes* and Harlequin frogs, genus *Atelopus*, in the Monteverde cloud forests of Costa Rica in terms of climate change (J. A. Pounds and Crump, 1994). A. Pounds et al. (1999) suggest that climate-related changes are also correlated with declines of birds and reptiles.

There is not one magic bullet or smoking gun behind amphibian declines (Stuart, S. N., et al.). Concern over climate change as the primary threat to amphibians has lately resurfaced (Harvell et al., 2002; Carey and Alexander, 2003; J. A. Pounds et al., 2006), yet any of a number of mechanisms may explain the correlation of climatic changes with amphibian population declines, either singly or in combination (Boone and James, 2003; Epstein, 2001; Linder et al., 2003; Rohr et al., 2004). Pollutants, acidifying rain, climatic change, invasive species, and epidemic disease are all elements of habitat degradation and, ultimately, of habitat loss. Their effect is to increase the variance in population parameters and thereby increase extinction risk. These insults in combination naturally will work synergistically to make things even worse.

O BRAVE NEW WORLD?

In comparison with the factors just discussed, habitat destruction and fragmentation remain the most important and most often cited threats facing amphibians (Stuart et al., 2004). Habitat fragmentation is also widely invoked as a cause for decline of biodiversity for it isolates small populations from one another and inbreeding leads to loss of genetic variation within each remnant subpopulation (Harrison and Bruna, 1999; Fischer, 2000; A. G. Young and Clarke, 2000; Dudash and Fenster, 2000). Over the long term, this leaves them less able to adapt to further environmental change (Holsinger, 2000). But this is a long-term expectation. The eventual extinction of a remnant population requires not just its isolation but also further degradation of the habitat that remains. The more immediate effects of fragmentation may be profoundly different for species with differing dispersal abilities. Although toads, frogs, and salamanders have been known to disperse over distances exceeding 20 km in a year (M. A. Smith and Green, 2005), their abilities to disperse are modest compared to animals that fly or are borne on the wind (Hickling et al., 2006).

In the end, to comprehend amphibian population declines we need to understand the particular nature of their ecology to put their plight into context. It is humbling to recognize how little we really understood about amphibian population biology only 20 years ago. Understanding the causes of population declines in any group of organisms must take into account the biology of the organisms themselves. For amphibians, it is wrong-headed to think that all species behave the same way or confront similar environmental hazards. The extent of variation in the population dynamics of pond-breeding amphibians can be very different from that of terrestrial direct-developing amphibians, leading to different expectations of the probabilities of local extinctions and the necessity of dispersal (Green, 2003).

Despite all these problems, some amphibian species are on the increase, either benefiting from their abilities to contend with human-dominated environments such as cities or suburbs or invading environments far from their native ranges. The coqui (*Eleutherodactylus coqui*) has now reached such densities in Hawaii that it is considered a pest and source of noise pollution (Kraus et al., 1999; Kraus and Campbell, 2002) and *E. johnstonei*, which has colonized the whole of the Lesser Antilles of the Caribbean (Kaiser, 1997), are direct-developing species that are particularly adept at using what human habitations have to offer. The cane or marine toad (*Bufo marinus*), having been introduced throughout the tropical Pacific, has been expanding its range at a rapid rate across northern Australia from its original site of introduction near Brisbane. Its size, noxiousness, and tremendous reproductive potential in a land without any native predators able to cope with it have made its progress unstoppable (Crossland, 2000; Urban et al., 2007).

The amphibians today may not be any more vulnerable than other organisms. Many species of birds, especially insectivorous species, have been declining for decades (Rappole and McDonald, 1994; Julliard et al., 2004; Valiela and Martinetto, 2007), as have numerous species of butterflies and moths (M. S. Warren, 2001; Conrad et al. 2006) as well as reptiles (Gibbons et al., 2000). Population declines are occurring in every sort of organism, worldwide. Amphibians are not fragile victims susceptible to the slightest environmental upset. They are not inherently delicate creatures hampered by their water-absorbing skins and larvae. As any paleontologist will tell you, in the long run all species eventually go extinct. Amphibians as a whole have survived cataclysmic extinctions at the end of the Permian, the end of the Triassic, and the end of the Cretaceous and, in more recent times, a succession of continental glaciations. Amphibians are tough survivors. The world is changing at an astonishing pace, but they may survive the end of us too.

ABBREVIATIONS USED IN ILLUSTRATIONS

a: angular
a a, aa: atlas arch
a ic: atlas intercentrum
a pc: atlas pleurocentrum
abr, Abv: anterior branchial vessel
ac: anterior crest of humerus
ace: acetabulum
ad cr: adductor crest
ad r: adductor ridge
alp: anterolateral process of transverse process of vertebrae
Ame: adductor mandibulae externus
amf: anterior mandibular foramen
Ami: adductor mandibulae internus
Ami (pro): profundus head of the adductor mandibulae internus
Ami (pt): pterygoideus portion of adductor mandibulae internus
Ami (sup): superficial head of the adductor mandibulae internus
Amp: adductor mandibulae posterior
Amp (longus): adductor mandibulae posterior longus
an: anus
ang: angular
angspl: angulosplenial
ano: anocleithrum
ant n: anterior nostril
ao: accessory ear ossicle (unique to microsaurs)
aop: articular surface for opisthotic
ap: anterior process of ribs/anterior projection
apf: anterior palatine fenestra
apq: ascending process of quadrate
a-prea: fused angular-prearticular

apv: vomerine aperture
art: articular
art pa: surface for articulation with pseudoangular in caecilians
art pt pal: articulating surface between palatine and pterygoid in caecilians
art-q: area of articulation between quadrate and articular
art qu pt: articulating surface with quadrate ramus of pterygoid
art st-q: articulating surface for fused stapes and quadrate
art uln: articulation for ulna and radius
ast: astragalus
at: anterior tectal
at fl t: attachment for flexor tendon
atp: atriopore
ax a: axis arch
ax ic: axis intercentrum
ax pc: axis pleurocentrum
ba: branchial arches and basicranial articulation
bas com: basal commune
bb: basibranchial
bb1, bb2: basibranchial 1 and 2
bc art: basicranial articulation
bd: branchial denticles
bh: branchiohyoideus
bm: branchiomandibularis
bo: basioccipital
bp: basicranial process
bpp, bpt, bptp: basipterygoid process or surface for articulation
bq: basal process of quadrate
br: brain

bs: basisphenoid/body scales
bsp: branchiostegal plate
bs-pl: fused basi- and pleuro-sphenoid
buc: buccal cavity
c: foramen
cal: calcaneum
can: channel for nerves or blood vessels
cap: capitulum for articulation of radius
capit: capitulum of rib
car, car g: groove for carotid artery
car f: carotid foramen
cb i, ii, iii, iv: ceratobranchial 1–4
ces: cartilage deposit in inner ear
cf: carotid foramen
ch: choana/ceratohyal
cl: clavicle
clei: cleithrum
co: coronoid
coc: caudal rhombencephalic neural crest cells
cocc iliac: coccygeo iliacus
cocc sacr: coccygeo sacralis
cor: coracoid
cp: caudal projection
d: dentary
da: dorsal aorta
df: dorsal fin
dh: surface for articulation with dorsal process of hyomandibular
diap: diapophysis of vertebrae
Dm, dm: depressor mandibulae
Dma, dma: depressor mandibulae (anterior)
Dmp, dmp: depressor mandibulae (posterior)
do: dorsal osteoderms
do pr: dorsal process of ilium
dor pr q: dorsal process of quadrate
dpc: dorsopectoral crest
drs: dorsal median ridge scale
ds: denticular structure/dorsal shield
dv: dorsal velum
eb: epibranchial
ect: ectopterygoid/ectepicondyle
ect f, ectf: ectepicondylar foramen
ectr: ectepicondylar ridge
eg: endostyle glands
en: external naris
ent: entopterygoid/entepicondyl
entep, entf, ent f: entepicondylar foramen
eo: exoccipital
eo-op: exocciptal opisthotic
epi: epipterygoid
es: endostyle
esl: lateral extrascapular
esm: median extrascapular
esp: esophagus
ex: exoccipital
exn: external naris
F: femur

f: frontal
fe: femur
f im: intramandibular foramen
Fi: fibula
fib: fibulare
fm: foramen magnum
fo: fenestra ovalis
Fon, font: fontanelle
fp, f-p: frontoparietal
g: gonad/gular
g pt m: groove for the pterygoideus muscle to pass beneath the back of the lower jaw
gast: ventral gastralia
gc: gill filter plates
Ggb, ggb: genioglossus basilis
Ggl, ggl: genioglossus (lateral division)
Ggm, ggm: genioglossus (medial division)
Gh, gh: geniohyoideus
glen: glenoid
Gmd, gmd: genioglossus medialis distalis
gr ao: groove for aortic artery
gu: gular plate
H, h: humerus
ha: hyoangularis
hb i, ii: hypobranchial 1, 2
hbp: hypobranchial plate
hem: haemal arch
Hg, hg: hyoglossus
hh: hypohyal
Hs, hs: hebosteoypiloideus
ht: heart
hy: hyomandibular
hyb: hypobranchials
Hym, hym: hyomandibularis
i: intermedium
ic: intercentra/intercentrum
icf: internal carotid foramen
icl: interclavicle
ih: interhyoideus
Ihp, ihp: interhyoideus posterior
il: ilium
il pr: iliac process
im: intermandibularis
imo: intermandibular ossicles
in, ina: internarial bone
incl: interclavicle
ineps: interepipodial space
inf: internarial foramen
inp: internarial pit
int: intermedium/internal trochanter
int fossa: intertrochanteric fossa
int na: internal nostril
int p: internal process of lower jaw in caecilians
ipmf: interpremaxillary fenestra
iptv: interpterygoid vacuity
irc: infrarostal cartilage
isch: ischium

it: internal trochanter/intertemporal/intestine
it f: intertrochanteric fossa
j: jugal
j for, jf, jfor: jugular foramen
l, la: lacrimal
lc: laryngeal cartilage
ld: lacrimal duct
ldp: lateral dorsal process
le: lateral eye
lee: lateral exposure of ectopterygoid
lens: placode that will form the lens of the eye
lep: lateral exposure of palatine bone
lmf: lateral mandibular foramen
lof: lateral otic fissure
long dors: longissimus dorsi
lr: lateral rostral
m: maxilla
m f, mf: meckelian fenestra
m-pal, m pal: fused maxilla and palatine
MAMI (pro): adductor mandibulae internus (profundus)
MAMI (sup): adductor mandibulae internus (superficialis)
MAMP: adductor mandibulae posterior
mc: Meckel's cartilage/medial centrale/metacarpal
mds: median dorsal sensor line
meck f: meckelian fenestra
mes: mesencephalon
mfn: medial foramina
mfos: mandibular fossa
mg: midgut
mm: myomere and mentomeckelian bone
mnc: mesencephalic neural crest cells
mo: mouth opening
m-p: fused maxilla and palatine
mppq: muscular process of palatoquadrate
mrc: medial rostral cartilage
ms: myosepta
mt: metatarsal
n: nasal
na: neural arch
nb: nasal bone
nc: nasal capsule
Nc: neural cord
nl d, nld: naso-lacrimal duct
notoc: notochord
n-pm: fused nasal and premaxilla
nt: notochord
o: opercular bone
oa: occipital arch
ob f: obturator foramen
ob: os basale, bone resulting from fusion of exoccipitals basioccipital, supraoccipital, opisthotic, and prootic in caecilians
oc art: occipital articulation
occa: groove for occipital artery
odp: odontoid process
of: opercular fenestra
olf d: dorsal branch of olfactory tract

olf v: ventral branch of olfactory tract
olfa: olfactory placode
on: opening for notochord
op: accessory ear ossicle
oper: operculum
opis: opisthotic area of otic capsule
or: orbit
Orh, orh: orbitohyoideus
ot: otic capsule
otc: otoglossal cartilage
ot n, otn: otic notch
ot-oc: otic occipital
ov: optic vesicle
p: parietal
p con: processus conchoides
pa: pseudoangular
paf: parietal foramen
pal: palatine
par: prearticular
para: parapophysis
parp: paroccipital process for connection of the otic capsule with the tabular
part: prearticular
pc: pleurocentrum
pca: palatine canal
pd: pseudodentary
pdp: posterior distal process of ulna
pdph: posterodistal process of humerus
pe: posterior element
pf: postfrontal
pfor: pineal foramen
ph: pharyngeal cavity and parahyoid
pi: paired pineal opening
pin: pineal opening
pis: pisiform
pl sph: pleurosphenoid
pm: premaxilla
pmf: posterior mandibular foramen
pmp: posteromedial process
po: postorbital
pop: preopercular
pos pr, pos pro: posterior process of ilium
post n: posterior nostril
pot: posttemporal
pp: postparietal
ppc: palpebral cup
ppl: postparietal lappet
ppr: posterior process of ribs
ppt: posterior pterygoid
pq: palatoquadrate
pra, pr art: prearticular
prf: prefrontal
prh: prehallux
pro: prosencephalon of brain/prootic/position of otic capsule
pro a, proa: proatlas
prp: prepollex

ps: parasphenoid
psf: parasymphyseal fang
psp: postsplenial
pspp: postspiracular plate
psyf: parasymphyseal fang
psyp: parasymphyseal plate
pt: pterygoid
pt f, ptf: posttemporal fossa
ptp: posterior spine of transverse process/pterygoid process of palatoquadrate
ptq: pterygoquadrate
pu: pubis
pub pr: pubic process
q: quadrate
qa: quadratoangularis
qj: quadratojugal
R: radius
r: rib/rostral bone
r con: radial condyle
ra: radiale
rad: radial condyle
rap: retroarticular process
Rc, rc: radial cartilage
Rc: rectus cervicis
Ret p, ret p: retroarticular process
rhom: rhombencephalon
roc: rostral rhombencephalic neural crest cells
rs: shaft of rib
rt ar p: retroarticular process
sa: surangular
Sa: suspensorioangularis
sac: sacral vertebra
Sar, sar: subarcualis rectus muscle
sc: scapulocoracoid
scap: scapula
scapco, scap co: scapulocoracoid
scf: scapulocoracoid foramen
scl: sclerotic plates supporting eye
sclei: supracleithrum
scr: suprarostral cartilage
sgl f: supraglenoid foramen
Sh: sternohyoideus
sl: sublingual rod
slg: sensory line groove
sm: submentalis/septomaxilla
smx: septomaxilla
sn, sn c, snc: supraneural canal
snf: foramen for spinal nerve
so: supraoccipital
soa: accessory ossicle
so-op: fused supraoccipatal and opisthotic
sop: subopercular
sp: splenial
sph: sphenethmoid
Sph: suspensoriohyoideus
spnf: foramen for spinal nerve

spp: postsplenial
spsc: suprascapular
sq: squamosal
sq art: articulation between squamosal and skull roof
sq-ch: ligament connecting squamosal and ceratohyal
sqn: squamosal notch
sr: sacral rib
src: superior rostral cartilage
st: supratemporal
st pr sq, st sq: stapedial process of squamosal
sta: stapes
sta at: foramen for stapedial artery
sta f: stapedial foramen
st-q: fused stapes and quadrate in caecilians
sub: submandibulobranchiostegal
subm: submandibular
suo: subopercular
sup: supinator process
sup sc, supsc: suprascapula
syt: synotic tectum
t: tabular/tentacle-like structure
te: tectal
tel: telencephalon
tem os: temporal ossicles
ten o: tentacular opening
ten s: sulcus for tentacle in caecilians
ten-g: tentacular groove
th: trabecular horn
Ti: tibia
tr fl pt: tranverse flange of pterygoid
tro: trochanter
t-sq: tabular-squamosal
tub: tuberculum of rib
tz: transition zone between dorsal shield and body scales
U: ulna
uh: urohyal
ul, ula: ulnare
uno: unossified area of otic capsule
ur: urohyal
v: vomer
V$_1$: ophthalmic ramus of trigeminal nerve
V$_2$: maxillary ramus of trigeminal nerve
V$_3$: mandibular ramus of trigeminal nerve
v cr s: ventral cranial suture
va: ventral aorta
vest-co: vestibulo-cochlear placode
vf: ventral fin
vg: venus groove
vh: surface for articulation with ventral process of hyomandibular
vp: foramen for profundus branch of nerve V
vv: ventral velum

Roman numerals I to XII foramina for cranial nerves
ii, iv for foramina in scapulocoricoid

GLOSSARY

acritarch. Primitive, unicellular eukaryote.

akinetic. Absence of movement between skull bones.

amphicoelous. Description of vertebra with concavities on both anterior and posterior ends.

apical ectodermal ridge (AER). Line of tissue that develops at the tip of the limb bud. It is essential to limb growth and differentiation.

apomorphy. Derived character.

appendicular skeleton. Girdles and limbs.

archaebacteria. Kingdom of prokaryotes, capable of living in environments with extremely high temperatures, salt content, or other extreme conditions.

autapomorphy. Derived character that is unique to a particular taxon.

autopod. Distal part of the limb.

axial skeleton. Skull and vertebral column.

bilaterian. Multicellular animal with bilateral symmetry.

bioturbation. Mixing of sediments by burrowing organisms.

bone morphogenetic proteins (BMPs). Specific bone- and cartilage-inducing molecules.

character state. One of two or more degrees of expression of a character that can be used to support relationships.

character. In phylogenetic systematics, an attribute of an organism that occurs in two or more character states, and can be used to support specific relationships.

chondrification. Formation of a bone by cartilage.

clade. Group that can trace its origin to a single ancestral species.

cladogram. A branching diagram representing the relationships between characters or character states from which phylogenetic inferences can be made.

continental drift. Movement of continents associated with plate tectonics.

crown group. All the species that share a common ancestry with the living members of a monophyletic group.

dating, absolute. Dating of geological strata by radiometric means.

dating, relative. Dating of geological strata on the basis of their position relative to other strata that are dated by radiometric or other means.

dermal bone. Bone that forms superficially and without a cartilaginous precursor.

digital arch. Extension of the axis of development within the limb bud into the autopodium of tetrapod limbs.

endochondral bone. Bone that forms deep in the body and is preformed in cartilage.

eubacteria. Prokaryotes lacking the extreme characteristics of archaebacteria.

eukaryotes. Organisms with a nuclear membrane and linear chromosomes.

fenestra ovalis. Opening in the otic capsule into which fits the footplate of the stapes.

gastralia. Narrow, elongate scales protecting the abdominal surface of primitive amphibians and reptiles.

ghost lineage. Early portion of a lineage that is not known from the fossil record but is predicted to have existed on the basis of the earlier appearance of a sister-group.

Gondwana. A supercontinent made up of South America, Africa, Antarctica, and Australia.

half-life. The period of time required for a radioactive element to decay to one-half of its original number of atoms.

holospondylus vertebrae. Those with a cylindrical configuration.

homeobox genes. Genes containing a homeobox sequence.

homeobox. Highly conserved region of a gene that codes for a particular sequence of amino acids termed the homeodomain.

homeodomain. Sequence of amino acids in a protein that serves as a sequence-specific DNA-binding site for genes that control development.

homeotic genes. Genes that control the position in the embryo where a structure develops.

homologous. Term applied to structures occurring in two or more groups that evolved from a common ancestor that had that structure, although it may have different functions in the derived groups.

Hox **cluster.** Arrangement of *Hox* genes in a continuous linear sequence on a single chromosome.

Hox **genes.** Genes that make up the homeobox cluster.

introns. A non-coding region within RNA that facilitates the crossing over between two adjacent coding regions.

Laurasia. The continents of the Northern Hemisphere when North America, Europe, and Asia formed a single land mass.

lepidotrichia. Narrow elongate scales covering the fins of bony fish.

mantle plume. Flow of magma from the base of the mantle to the earth's surface without respect for the distribution of the crustal plates.

metazoan. Multicellular animal.

monophyletic group. Originally coined in reference to a group with a single common ancestor; redefined by Hennig as a single species and all of its descendants.

neoteny. The reproductive system matures while other aspects of the body retain a larval level of development.

notochord. Longitudinal rod that precedes the vertebral column both evolutionarily and developmentally as a means of supporting the trunk of chordates.

Ossification. Formation of bone.

Pangaea. Name given to the combined continents of the Northern and Southern Hemispheres when they were united during the Late Permian and Triassic.

paraphyletic group. Species and some, but not all, of its descendants.

PAUP. Phylogenetic Analysis Using Parsimony. A means of hypothesizing relationships on the assumption that evolutionary change typically occurs in a parsimonious manner.

phalangeal formula. Number of bones in each of the fingers and toes.

phylogenetic systematics. Methodology established by Hennig for determining relationships and classifying organisms that is based on evolutionary affinities.

plate tectonics. Movement, separation, and coalescence of lithosphere plates as a result of convection currents in the earth's mantle.

plesiomorphy. A primitive character or character state.

polyphyletic group. An assemblage of taxa with two or more putative ancestors.

polytomy. The apparent origin of three or more taxa from a single immediate common ancestor.

prokaryote. Cell lacking a nuclear membrane and having circular rather than linear chromosomes.

protist. Unicellular eukaryote, with or without chloroplasts.

protozoan. Single celled eukaryote that lacks chloroplasts.

sister-group. One of a pair of taxa that share an immediate common ancestry.

stem group. Members of a monophyletic group that do not share an immediate common ancestry with any of the living clades within that group.

subduction. Movement of one tectonic plate under another.

synapomorphy. Shared derived character that unites sister-groups.

synovial joint. Freely movable joint between bones, typically with a liquid-filled cavity, surrounded by a synovial membrane.

taxa. Plural of taxon.

taxon. A group of organisms belonging to a particular taxonomic rank, for example, a species, genus, or family.

tetrapod. Four-limbed terrestrial vertebrate.

unconformity. Boundary between rock units that do not have parallel bedding planes, indicating a period of folding and erosion between successive beds.

unguals. Terminal phalanges.

zone of polarizing activity (ZPA). Group of cells on the posterior margin of the limb bud that determines the anterior-posterior axis of the limb.

REFERENCES

Abourachid, A., and D. M. Green. 1999. Origins of the Frog-kick? Alternate-leg swimming in primitive frogs, families Leiopelmatidae and Ascaphidae. Journal of Herpetology 33: 657–663.

Adams, M. J., D. E. Schindler, and R. B. Bury. 2001. Association of amphibians with attenuation of ultraviolet-b radiation in montane ponds. Oecologia 128: 519–525.

Ahlberg, P. E. 1998. Postcranial stem tetrapod remains from the Devonian of Scat Craig, Morayshire, Scotland. Zoological Journal of the Linnean Society 122: 99–141.

Ahlberg, P. E., and J. A. Clack. 1998. Lower jaws, lower tetrapods—a review based on the Devonian genus *Acanthostega*. Transactions of the Royal Society of Edinburgh, Earth Sciences 89: 11–46.

———. 2006. A firm step from water to land. Nature 440: 747–749.

Ahlberg, P. E., J. A. Clack, and H. Blom. 2005. The axial skeleton of the Devonian tetrapod *Ichthyostega*. Nature 437: 137–140.

Ahlberg, P. E., J. A. Clack, and E. Lukševičs. 1996. Rapid braincase evolution between *Panderichthys* and the earliest tetrapods. Nature 381: 61–64.

Ahlberg, P. E., J. A. Clack, E. Lukševičs, H. Blom, and I. Zupins. 2008. *Ventaststega curonica* and the origin of tetrapod morphology. Nature 453: 1199–1204.

Ahlberg, P. E., and Z. Johanson. 1998. Osteolepiforms and the ancestry of tetrapods. Nature 395: 792–794.

Ahlberg, P. E., E. Lukševičs, and O. Lebedev. 1994. The first tetrapod finds from the Devonian (Upper Famennian) of Latvia. Philosophical Transactions of the Royal Society of London B 343: 303–328.

Albert, V. A. (ed.). 2005. Parsimony, Phylogeny, and Genomics. Oxford University Press, Oxford.

Alexander, M. A., and J. K. Eischeid. 2001. Climate variability in regions of amphibian declines. Conservation Biology 15: 930–942.

Alford, R. A., and S. J. Richards. 1999. Global amphibian declines: A problem in applied ecology. Annual Review of Ecology and Systematics 30: 133–165.

Alley, K. E. 1990. Retrofitting larval neuromuscular circuits in the metamorphosing frog. Journal of Neurobiology 21: 1092–1107.

Anderson, J. S. 2002. Revision of the aïstopod genus *Phlegethontia* (Tetrapoda: Lepospondyli). Journal of Paleontology 76: 1029–1046.

———. 2003a. A new Aïstopoda (Tetrapoda: Lepospondyli) from Mazon Creek, Illinois. Journal of Vertebrate Paleontology 23: 79–88.

———. 2003b. Cranial anatomy of *Coloraderpeton brilli,* postcranial anatomy of *Oestocephalus amphiuminus,* and reconsideration of Ophiderpetontidae (Tetrapoda: Lepospondyli: Aïstopoda). Journal of Vertebrate Paleontology 23: 532–543.

Anderson, J. S., R. L. Carroll, and T. B. Rowe. 2003. New information on *Lethiscus stocki* (Tetrapoda: Lepospondyli: Aïstopoda) from high-resolution computed tomography and a phylogenetic analysis of Aïstopoda. Canadian Journal of Earth Science 40: 1071–1083.

Andrews, S. M., and R. L. Carroll. 1991. The order Adelospondyli: Carboniferous lepospondyl amphibians. Transactions of the Royal Society of Edinburgh 82: 239–275.

Andrews, S. M., and T. S. Westoll. 1970. The postcranial skeleton of *Eusthenopteron foordi* Whiteaves. Transactions of the Royal Society of Edinburgh 68: 207–329.

Archibald, J. D. 2005. Eutheria (Placental Mammals). Encyclopedia of Life Sciences; pp. 1–4. John Wiley & Sons, Ltd. www.els.net.

Avery, R. A. 1982. Field studies of body temperatures and thermoregulation; pp. 93–166 in C. Gans (ed.), Biology of the Reptilia, vol. 12. Academic Press, New York.

Báez, A. M. 2000. Tertiary anurans from South America; pp. 1388–1401 in H. Heatwole and R. L. Carroll (eds.), Amphibian Biology, vol. 4, Palaeontology. Surrey Beatty & Sons, Chipping Norton, Australia.

Baird, D. 1952. Revision of the Pennsylvanian and Permian footprints *Limnopus, Allopus,* and *Baropus.* Journal of Paleontology 26: 832–840.

———. 1962. A rhachitomous amphibian, *Spathicephalus,* from the Mississippian of Nova Scotia. Breviora 157: 1–9.

Baker, J., M. Bizzarro, N. Witting, J. Connelly, and H. Haaack. 2005. Early planetesimal melting from an age of 4.5662 Gyr for differentiated meteorites. Nature 436: 1127–1131.

Barghoorn, E. S., and S. M. Tyler. 1965. Microfossils from the Gunflint chert. Science 147: 563–577.

Bartholomew, G. A., and R. C. Lasiewski. 1965. Heating and cooling rates, heart rate, and simulated diving in the Galápagos marine iguana. Comparative Biochemistry and Physiology 16: 573–582.

Beaumont, E. H. 1977. Cranial morphology of the Loxommatidae (Amphibia: Labyrinthodonta). Philosophical Transactions of the Royal Society of London B 280: 29–101.

Beaumont, E. H., and T. R. Smithson. 1998. Zoological Journal of the Linnean Society 122: 187–209.

Beebee, T. J. C. 1995. Amphibian breeding and climate. Nature 374: 219–220.

Beebee, T. J. C., R. J. Flower, A. C. Stevenson, S. T. Patrick, P. G. Appleby, C. Fletcher, C. Marsh, J. Atkanski, B. Rippey, and R. W. Battarbee. 1990. Decline of natterjack toad *Bufo calamita* in Britain: Paleoecological, documentary and experimental evidence for breeding site acidification. Biological Conservation 53: 1–20.

Behrensmeyer, A. K, J. D. Damuth, W. A. DiMichele, R. Potts, H.-D. Sues, and S. L. Wing. 1992. Terrestrial Ecosystems through Time. University of Chicago Press, Chicago and London.

Benton, M. J. 2003. When Life Nearly Died. Thames & Hudson, London.

Benton, M. J. , V. P. Tverdokhlebov, and M. V. Surkov. 2004. Ecosystem remodelling among vertebrates at the Permian-Triassic boundary in Russia. Nature 432: 97–100.

Benton, M. J., and R. J. Twitchett. 2003. How to kill (almost) all life: The end-Permian extinction event. Trends in Ecology & Evolution 18: 358–365.

Berger, L., R. Speare, P. Daszak, D. E. Green, A. A. Cunningham, C. L. Goggin, R. Slocombe, M. A. Ragan, A. D. Hyatt, K. R. McDonald, H. B. Hines, K. R. Lips, G. Marantelli, and H. Parkes. 1988. Chytridiomycosis causes amphibian mortality associated with population declines in the rain forests of Australia and Central America. Proceedings of the National Academy of Sciences USA 95: 9031–9036.

Berman, D. S., and A. C. Henrici. 2003. Homology of the astragalus and structure and function of the tarsus of Diadectidae. Journal of Vertebrate Paleontology 20: 253–268.

Berman, D. S., A. C. Henrici, S. S. Sumida, and T. Martens. 2000. Redescription of *Seymouria sanjuanensis* (Seymouriamorpha) from the Lower Permian of Germany based on complete, mature specimens with a discussion of Paleoecology of the Bromacker locality assemblage. Journal of Vertebrate Paleontology 20: 253–268.

Berman, D. S., R. A. Kissel, A. C. Henrici, S. S. Sumida, and T. Martens. 2004. A new diadectid (Diadectomorpha), *Orobates pabsti* from the Early Permian of Central Germany. Annals of the Carnegie Museum 35: 1–36.

Berman, D. S., R. R. Reisz, and D. A. Eberth. 1985. *Ecolsonia cutlerensis,* an Early Permian dissorophid amphibian from the Cutler Formation of north-central New Mexico. Circular 191 News Mexico Bureau of Mines & Mineral Resources. iii–31.

———. 1987. A new genus and species of trematopid amphibian from the Late Pennsylvanian of North-Central New Mexico. Journal of Vertebrate Paleontology 7: 252–269.

Berman, D. S., R. R. Reisz, and M. A. Fracasso. 1981. Skull of the Lower Permian dissorophid amphibian *Platyhystrix rugosus.* Annals of Carnegie Museum 17: 391–416.

Berven, K. A. 1990. Factors affecting population fluctuations in larval and adult stages of the wood frog (*Rana sylvatica*). Ecology 71: 1599–1608.

Bharathan, G., B.-J. Janssen, E. A. Kellogg, and N. Sinha. 1997. Did homeodomain proteins duplicate before the origin of angiosperms, fungi, and metazoa? Proceedings of the National Academy of Science USA 94: 13749–13753.

Biek, R., W. Funk, B. A. Maxell, and L. S. Mills. 2001. What is missing in amphibian decline research: Insights from ecological sensitivity analysis? Conservation Biology 16: 728–734.

Blanco, M. J., B. Y. Misof, and G. Wagner. 1998. Heterochronic differences of *Hox-11* expression in *Xenopus* fore- and hind limb development: Evidence for lower limb identity of the anuran ankle bones. Development, Genes, and Evolution 208: 175–187.

Blaustein, A. R., L. K. Belden, D. H. Olson, D. M. Green, T. L. Root, and J. M. Kiesecker. 2001. Amphibian breeding and climate change. Conservation Biology 15: 1804–1809.

Blaustein, A. R., and D. B. Wake. 1990. Declining amphibian populations—a global phenomenon. Trends in Ecology & Evolution 5: 203–204.

Blaustein, A. R., D. B. Wake, and W. P. Sousa. 1994. Amphibian declines: Judging stability, persistence and susceptibility of populations to local and global extinctions. Conservation Biology 8: 60–71.

Boisvert, C. A. 2005. The pelvic fin and girdle of *Panderichthys* and the origin of tetrapod locomotion. Nature 438: 1145–1147.

———. 2009. Vertebral development of modern salamanders provides insights into a unique event of their evolutionary history. Journal of Experimental Zoology (Mol. Dev. Evol.) 312B: 1–29.

Boisvert, C. A., E. Mark-Kurik, and P. E. Ahlberg. 2008. The pectoral fin of *Panderichthys* and the origin of digits. Nature 456: 636–638.

Bolt, J. R. 1969. Lissamphibian origins: Possible protolissamphibians from the Lower Permian of Oklahoma. Science 166: 888–891.

Bolt, J. R., and R. E. Lombard. 1985. Evolution of the amphibian tympanic ear and the origin of frogs. Biological Journal of the Linnean Society 24: 83–99.

———. 2000. Palaeobiology of *Whatcheeria deltae,* a primitive Mississippian tetrapod; pp. 1044–1052 *in* H. Heatwole and R. L. Carroll (eds.), Amphibian Biology, vol. 4. Surrey Beatty & Sons, Chipping Norton, Australia.

Boone, M. D., and S. M. James. 2003. Interactions of an insecticide, herbicide, and natural stressors in amphibian community mesocosms. Ecological Applications 13: 829–841.

Borsuk-Białynicka, M., and S. E. Evans. 2002. The scapulocoracoid of an Early Triassic stem-frog from Poland. Acta Palaeontologica Polonica 47: 79–96.

Borsuk-Białynicka, M., T. Maryańska, and M. A. Shishkin. 2003. New data on the age of the bone breccia from the locality Czatkowice 1 (Cracow Upland, Poland). Acta Palaeontologica Polonica 48: 153–155.

Bossy, K. A., and A. C. Milner. 1998. Order Nectridea; pp. 73–131 *in* R. L. Carroll, K. A. Bossy, A. C. Milner, S. M. Andrews, and

C. F. Wellstead, Lepospondyli Part 1, Encyclopedia of Paleo-herpetology. Verlag Dr. Friedrich Pfeil, München.

Boy, J. A. 1972. Die Branchiosaurier (Amphibia) des saarpfälzischen Rotliegenden (Perm, SW-Deutschland). Abhandlungen Hessisches Landesant fuer Bodenforschung 65: 1–137.

———. 1974. Die Larven der rhachitomen Amphibien (Amphibia: Temnospondyli; Karbon-Trias). Palaeontologische Zeitschrift 62: 107–132.

———. 1988. Uber einige Vertreter der Eryopoidea (Amphibia: Temnospondyli) aus dem europaischen Rotliegend (Hochstes Karbon-Perm). 1. Sclerocephalus. Palaontologische Zeitschrift 62: 107–132.

———. 1989. Uber einige Vertreter der Eryopoidea (Amphibia: Temnospondyli) aus dem europaischen Rotliegend (Hochstes Oberkarbon-Perm). 2. Acanthostomatops. Palaontologische Zeitschrift 63: 133–151.

———. 1995. Uber die Micromelerpetontidae (Amphibia: Temnospondyli). 1. Morphologie und Palaookologie des Micromelerpeton credneri (Unter-Perm; Sudwest-Deutschland). Palaontologische Zeitschrift 69: 429–457.

Boy, J. A., and K. Bandel. 1973. Bruktererpeton fiebigi n. gen. n. sp. (Amphibia: Gephyrostegida) Der erste tetrapode aus dem Rheinisch-Westfälischen Karbon (Namur B; W.-Deutschland). Palaeontolographica 145: 39–77.

Boy, J. A., and H.-D. Sues. 2000. Branchiosaurs: Larvae, metamorphosis and heterochrony in temnospondyls and seymouriamorphs; pp. 1150–1197 in H. Heatwole and R. L. Carroll (eds.), Amphibian Biology, vol. 4, Palaeontology. Surrey Beatty & Sons, Chipping Norton, Australia.

Boyd, M. J. 1982. Morphology and relationships of the Upper Carboniferous aïstopod amphibian Ophiderpeton nanum. Palaeontology 25: 209–214.

———. 1984. The Upper Carboniferous assemblage from Newsham, Northumberland. Palaeontology 27: 367–392.

Bradford, D. F., F. Tabatabai, and D. M. Graber. 1993. Isolation of remaining populations of the native frog, Rana muscosa, by introduced fishes in Sequoia and Kings Canyon National Parks, California. Conservation Biology 7: 882–888.

Bramble, D. M., and D. B. Wake. 1985. Feeding mechanisms of lower tetrapods; pp. 230–261 in M. Hildebrand, D. M. Bramble, K. F. Liem, and D. B. Wake (eds.), Functional Vertebrate Morphology. Harvard University Press, Cambridge, Massachusetts.

Brasier, M. D., O. R. Green, A. P. Jephcoat, A. K. Kleppe, M. J. Van Kranendonk, J. F. Lindsay, A. Steele, and N. V. Grassineau. 2002. Questing the evidence for Earth's oldest fossils. Nature 416: 76–81.

Brazeau, M., and P. E. Ahlberg. 2006. Tetrapod-like middle ear architecture in a Devonian fish. Nature 439: 318–321.

Bridges, C. M., and R. D. Semlitsch. 2001. Genetic variation in insecticide tolerance in a population of southern leopard frogs (Rana sphenocephala): Implications for amphibian conservation. Copeia 2001: 7–13.

Brocks, J. J., G. A. Logan, R. Buick, and R. E. Summons. 1999. Archean molecular fossils and the early rise of eukaryotes. Science 285: 1033–1036.

Broili, F., and J. Schröder. 1937. Beobachtungen an Wirbeltieren der Karrooformation. XXV. Über Micropholis Huxley. Sitz-Ber. Akad. Wiss. München 1938: 19–38.

Brown, C. J., B. Blossey, J. C. Maerz, and S. J. Joule. 2006. Invasive plant and experimental venue affect tadpole performance. Biological Invasions 8: 327–338.

Butterfield, N. J. 2000. Bangiomorpha pubescens n. gen., n. sp.: Implications for the evolution of sex, multicellularity, and the Mesoproterozoic/Newproterozoic radiation of eukaryotes. Palaeobiology 26: 386–404.

Bystrow, A. P. 1938. Dvinosaurus als neotenische form des Stegocephalen. Acta Zoologica 19: 209–295.

———. 1944. Kotlassia prima Amalitzky. Bulletin of the Geological Society of America 55: 379–416.

Caldwell, M. 1994. Developmental constraints and limb evolution in Permian and extant lepidosauromorph diapsids. Journal of Vertebrate Paleontology 14: 459–471.

Campbell, K. S. W., and M. W. Bell. 1977. A primitive amphibian from the Late Devonian of New South Wales. Alcheringa 1: 369–381.

Cannatella, D. 1999. Architecture: Cranial and axial musculoskeleton; pp. 52–91 in R. W. McDiarmid and R. Altig (eds.), Tadpoles: The Biology of Anuran Larvae. University of Chicago Press, Chicago and London.

Carey, C., and M. A. Alexander. 2003. Climate change and amphibian declines: Is there a link? Diversity and Distributions 9: 111–121.

Carey, C., and C. J. Bryant. 1995. Possible interrelations among environmental toxicants, amphibian development, and declines of amphibian populations. Environmental Health Perspectives 103: 3–17.

Carey, C., P. S. Corn, M. S. Jones, L. J. Livo, E. Muths, and C. W. Loeffler. 2005. Factors limiting the recovery of Boreal toads (Bufo b. boreas); pp. 222–240 in M. J. Lannoo (ed.), Amphibian Declines: The Conservation Status of United States Species. University of California Press, Berkeley.

Carey, C., R. W. Heyer, J. Wilkinson, R. A. Alford, J. W. Arntzen, T. Halliday, L. Hungerford, K. R. Lips, E. M. Middleton, S. A. Orchard, and A. S. Rand. 2001. Amphibian declines and environmental change: Use of remote sensing data to identify environmental correlates. Conservation Biology 15: 903–913.

Carroll, R. L. 1963. A microsaur from the Pennsylvanian of Joggins, Nova Scotia. Natural History Papers, National Museum of Canada no. 22: 1–13.

———. 1964a. The relationships of the rhachitomous amphibians Parioxys. American Museum Novitates 2167: 1–11.

———. 1964b. Early evolution of the dissorophid amphibians. Bulletin of the Museum of Comparative Zoology, Harvard 131: 163–250.

———. 1964c. The earliest reptiles. Journal of the Linnean Society (Zoology) 45: 61–83.

———. 1966. Microsaurs from the Westphalian B of Joggins, Nova Scotia. Proceedings of the Linnean Society of London 177: 63–97.

———. 1968. The postcranial skeleton of the Permian microsaur Pantylus. Canadian Journal of Zoology 46: 1175–1192.

———. 1969a. Problems of the origin of reptiles. Biological Reviews 44: 393–432.

———. 1969b. A middle Pennsylvanian captorhinomorph, and the interrelationships of primitive reptiles. Journal of Paleontology 43: 151–170.

———. 1969c. Origin of reptiles; pp. 1–43 in C. Gans (ed.), Biology of the Reptilia, vol. 1. Academic Press, London.

———. 1969d. A new family of Carboniferous amphibians. Palaeontology 12: 537–548.

———. 1970a. The ancestry of reptiles. Philosophical Transactions of the Royal Society of London B 257: 267–308.

———. 1970b. Quantitative aspects of the amphibian-reptilian transition. Forma et Functio 3: 165–178.

———. 1970c. The earliest known reptiles. Yale Scientific Magazine, 16–23.

———. 1980. The hyomandibular as a supporting element in the skull of primitive tetrapods; pp. 293–317 in A. L. Panchen (ed.), The Terrestrial Environment and the Origin of Land Vertebrates. Academic Press, London.

———. 1988a. Vertebrate Paleontology and Evolution. W. H. Freeman and Company, New York.

———. 1988b. An articulated gymnarthrid microsaur (Amphibia) from the Upper Carboniferous of Czechoslovakia. Acta Zoologica Cracoviensia 31: 441–450.

———. 1989. A juvenile adelogyrinid (Amphibia: Lepospondyli) from the Namurian of Scotland. Journal of Vertebrate Paleontology 9: 191–195.

———. 1991. Batropetes from the Lower Permian of Europe—a microsaur, not a reptile. Journal of Vertebrate Paleontology 11: 229–242.

———. 1995. Problems of the phylogenetic analysis of Paleozoic choanates. Bulletin du Muséum National d'Histoire Naturelle. 4e sér 17 Section C no. 1–4: 389–445.

———. 1997. Patterns and Processes of Vertebrate Evolution. Cambridge University Press, Cambridge.

———. 1998a. Cranial anatomy of ophiderpetontid aïstopods: Palaeozoic limbless amphibians. Zoological Journal of the Linnean Society 122: 143–166.

———. 1998b. Order Microsauria Dawson 1893; pp. 1–72 in P. Wellnhofer (ed.), Encyclopedia of Paleoherpetology, Part 1: Lepospondyli. Verlag Dr. Friedrich Pfeil, München.

———. 1998c. Order Aïstopoda Miall 1875; pp. 163–182 in P. Wellnhofer (ed.), Encyclopedia of Paleoherpetology, Part 1: Lepospondyli. Verlag Dr. Friedrich Pfeil, München.

———. 1999. Homology among divergent Paleozoic tetrapod clades; pp. 47–64 in B. K. Hall (ed.), Homology. Novartis Foundation Symposium. John Wiley & Sons, Chichester.

———. 2000. Eocaecilia and the Origin of Caecilians; pp. 1402–1411 in H. Heatwole and R. L. Carroll (eds.), Amphibian Biology, vol. 4, Paleontology. Surrey Beatty & Sons, Chipping Norton, Australia.

———. 2001a. Chinese salamanders tell tales. Nature 410: 534–536.

———. 2001b. The origin and early radiation of terrestrial vertebrates. Journal of Paleontology 75: 1202–1213.

———. 2007. The Palaeozoic ancestry of salamanders, frogs, and caecilians. Zoological Journal of the Linnean Society 150 (suppl. 1): 1–142.

Carroll, R. L., and D. Baird. 1968. The Carboniferous amphibian Tuditanus [Eosauravus] and the distinction between microsaurs and reptiles. American Museum Novitates 2337: 1–50.

———. 1972. Carboniferous stem-reptiles of the Family Romeriidae. Bulletin of the Museum of Comparative Zoology, Harvard 143: 321–364.

Carroll, R. L., C. Boisvert, J. Bolt, D. M. Green, N. Philip, C. Rolian, R. Schoch, and A. Tarenko. 2004. Changing patterns of ontogeny from osteolepiform fish through Permian tetrapods as a guide to the early evolution of land vertebrates; pp. 321–343 in G. Arratia, M. V. H. Wilson, and R. Cloutier (eds.), Recent Advances in the Origin and Early Radiation of Vertebrates. Verlag Dr. Friedrich Pfeil, München.

Carroll, R. L., K. A. Bossy, A. C. Milner, S. M. Andrews, and C. F. Wellstead. 1998. Lepospondyli; pp. xii + 216 in P. Wellnhofer (ed.), Encyclopedia of Paleoherpetology. Verlag Dr. Friedrich Pfeil, München.

Carroll, R. L., P. Bybee, and W. D. Tidwell. 1991. The oldest microsaur (Amphibia). Journal of Paleontology 65: 314–322.

Carroll, R. L., and J. Chorn. 1995. Vertebral development in the oldest microsaur and the problem of "Lepospondyl" relationships. Journal of Vertebrate Paleontology 15: 37–56.

Carroll, R. L., and P. J. Currie. 1975. Microsaurs as possible apodan ancestors. Zoological Journal of the Linnean Society 57: 229–247.

Carroll, R. L., and P. Gaskill. 1978. The order Microsaurs. Memoir of the American Philosophical Society 126: 1–211.

Carroll, R. L., and R. Holmes. 1980. The skull and jaw musculature as guides to the ancestry of salamanders. Zoological Journal of the Linnean Society 68: 1–40.

———. 2007. Evolution of the appendicular skeleton of amphibians; pp. 185–224 in B. K. Hall (ed.), Fins into Limbs. University of Chicago Press, Chicago and London.

Carroll, R. L, J. Irwin, and D. Green. 2005. Thermal physiology and the origin of terrestriality. Zoological Journal of the Linnean Society 143: 345–385.

Carroll, R. L., A. Kuntz, and K. Albright. 1999. Vertebral development and amphibian evolution. Evolution & Development 1: 36–48.

Carroll, S. B. 1995. Homeotic genes and evolution of arthropods and chordates. Nature 376: 479–485.

———. 2005. Endless Forms Most Beautiful. W. W. Norton & Co., New York and London.

Carroll, S. B., J. K. Grenier, and S. D. Weatherbee. 2001. From DNA to Diversity. Blackwell Science, Malden, Massachusetts.

Case, E. C. 1935. Description of a collection of associated skeletons of Trimerorhachis. Contributions from the Museum of Paleontology, University of Michigan 4: 227–274.

Chang, M. M. (ed.). 2003. The Jehol Biota. Shanghai Scientific & Technical Publishers, Shanghai.

———. 2004. Synapomorphies and scenarios—more characters of Youngolepis betraying its affinities to the Dipnoi; pp. 665–686 in G. Arratia, M. V. H. Wilson, and R. Cloutier (eds.), Recent Advances in the Origin and Early Radiation of Vertebrates. Verlag Dr. Friedrich Pfeil, München.

Chase, J. A. 1965. Neldasaurus wrightae, a new rhachitomous labyrinthodont from the Texas Lower Permian. Bulletin Museum of Comparative Zoology (Harvard) 133: 153–225.

Chen, J.-Y., D.-Y. Huang, and C.-W. Li. 1999. An early Cambrian craniate-like chordate. Nature 402: 518–522.

Chipman, A. D., and E. Tchernov. 2002. Ancient ontogenies: Larval development of the lower Cretaceous anuran Shomronella jordanica (Amphibia: Pipoidea). Evolution & Development 4: 86–95.

Chippindale, P. T., R. M. Bonett, A. S. Baldwin, and J. J. Wiens. 2004. Phylogenetic evidence for a major reversal of life-history evolution in plethodontid salamanders. Evolution 58: 2809–2822.

Clack, J. A. 1987. Pholiderpeton scutigerum Huxley, an amphibian from the Yorkshire coal measures. Philosophical Transactions of the Royal Society of London B 318: 1–107.

———. 1994. Silvanerpeton miripedes, a new anthracosauroid from the Viséan of East Kirkton, West Lothian, Scotland. Transactions of the Royal Society of Edinburgh: Earth Sciences 84: 369–376.

———. 1996. The palate of Crassigyrinus scoticus, a primitive tetrapod from the Lower Carboniferous of Scotland; pp. 55–64 in A. R. Milner (ed.), Studies on Carboniferous and Permian Vertebrates, Special Papers in Palaeontology 52.

———. 1998a. The Scottish Carboniferous tetrapod *Crassigyrinus scoticus* (Lydekker)—cranial anatomy and relationships. Transactions of the Royal Society of Edinburgh, Earth Sciences 84: 369–376.

———. 1998b. The neurocranium of *Acanthostega gunnari* Jarvik and the evolution of the otic region in tetrapods. Zoological Journal of the Linnean Society 122: 61–97.

———. 2000. The origin of tetrapods; pp. 979–1029 *in* H. Heatwole and R. L. Carroll (eds.), Amphibian Biology, vol. 4. Surrey Beatty & Sons, Chipping Norton, Australia.

———. 2001. *Eucritta melanolimnetes* from the Early Carboniferous of Scotland, a stem tetrapod showing a mosaic of characteristics. Transactions of the Royal Society of Edinburgh, Earth Sciences 92: 72–95.

———. 2002a. Gaining Ground: The Origin and Evolution of Tetrapods. Indiana University Press, Bloomington and Indianapolis.

———. 2002b. An early tetrapod from "Romer's Gap." Nature 418: 72–76.

———. 2002c. The dermal skull roof of *Acanthostega,* an early tetrapod from the Late Devonian. Transactions of the Royal Society of Edinburgh, Earth Sciences 93: 17–33.

———. 2003. A new baphetid (stem tetrapods) from the Upper Carboniferous of Tyne and Wear, U. K., and the evolution of the tetrapod occiput. Canadian Journal of Earth Sciences 40: 483–498.

———. 2005a. Getting a leg up on land. Scientific American 293: 100–107.

———. 2005b. The emergence of early tetrapods. Palaeogeography, Palaeoclimatology, Palaeoecology 232: 167–189.

Clack, J. A., and P. E. Ahlberg. 1998. A reinterpretation of the braincase of the Devonian tetrapod *Ichthyostega stensioei.* Journal of Vertebrate Paleontology 18 (suppl. 34A).

Clack, J. A., H. Blom, and P. E. Ahlberg. 2003. New insights into the postcranial skeleton of *Ichthyostega.* Journal of Vertebrate Paleontology 23: ABSTRACTS 41A.

Clack, J. A., and R. L. Carroll. 2000; pp. 1030–1043 *in* H. Heatwole and R. L. Carroll (eds.), Amphibian Biology, vol. 4. Surrey Beatty & Sons, Chipping Norton, Australia.

Clack, J. A., and M. I. Coates. 1995. *Acanthostega*—A primitive aquatic tetrapod?: pp. 359–372 *in* M. Arsenault, H. Lelièvre, and P. Janvier (eds.), Studies on Early Vertebrates. Bulletin du Muséum National d'Historie Naturelle, Paris 17.

Clack, J. A., and S. M. Finney. 2005. *Pederpes finneyae,* an articulated tetrapod from the Tournaisian of western Scotland. Journal of Systematic Palaeontology 2: 311–346.

Clack, J. A., and R. Holmes. 1988. The braincase of the anthracosaur *Archeria crassidisca* with comments on the interrelationships of primitive tetrapods. Palaeontology 31: 85–107.

Clark, J., and R. L. Carroll. 1973. Romeriid reptiles from the Lower Permian. Museum of Comparative Zoology Bulletin 144: 353–406.

Cloutier, R., and G. Arratia. 2004. Early diversification of actinopterygians; pp. 217–270 *in* G. Arratia, M. V. H. Wilson, and R. Cloutier (eds.), Recent Advances in the Origin and Early Radiation of Vertebrates. Verlag Dr. Friedrich Pfeil, München.

Coates, M. I. 1996. The Devonian tetrapod *Acanthostega gunnari* Jarvik: Postcranial anatomy, basal tetrapod relationships and patterns of skeletal evolution. Transactions of the Royal Society of Edinburgh, Earth Sciences 87: 363–421.

Coates, M. I., and J. A. Clack. 1990. Polydactyly in the earliest known tetrapod limbs. Nature 347: 66–69.

———. 1995. Romer's gap: Tetrapod origins and terrestriality; pp. 373–388 *in* M. Arsenault, H. Lelièvre, and P. Janvier (eds.), Studies on Early Vertebrates. Bulletin du Muséum National d'Historie Naturelle, Paris 17.

Coates, M. I., J. E. Jeffery, and M. Ruta. 2002. Fins to limbs: What the fossils say. Evolution & Development 4: 390–401.

Cogger, H. G., and R. G. Zweifel (eds.). 1998. Encyclopedia of Reptiles and Amphibians. Academic Press, San Diego.

Cohen, B. A., T. D. Swindle, and D. A. King. 2000. Support for the lunar cataclysm hypothesis from lunar meteorite impact melt ages. Science 290: 1754–1756.

Cohn, M. J., and C. Tickle. 1999. Developmental basis of limblessness and axial patterning in snakes. Nature 399: 474–479.

Collins, J. P., and A. Storfer. 2003. Global amphibian declines: Sorting the hypotheses. Diversity & Distributions 9: 89–98.

Conrad, K. F., M. S. Warren, R. Fox, M. S. Parsons, and I. P. Woiwod. 2006. Rapid declines of common, widespread British moths provide evidence of an insect biodiversity crisis. Biological Conservation 132: 279–291.

Conway, M. S. 1998. The Crucible of Creation. Oxford University Press, Oxford.

Cook, C. E., E. Jiménez, M. Akam, and E. Soló. 2004. The Hox gene complement of acoel flatworms, a basal bilaterian clade. Evolution & Development 6: 154–163.

Corn, P. S. 2000. Amphibian declines: Review of some current hypotheses; pp. 663–696 *in* D. W. Sparling, G. Linder, and C. A. Bishop (eds.), Ecotoxicology of Amphibians and Reptiles. SETAC Press, Pensacola, Florida.

Corn, P. S., and J. C. Fogelman. 1984. Extinction of montane populations of the northern leopard frog (*Rana pipiens*) in Colorado. Journal of Herpetology 18: 147–152.

Cote, S. R., R. L. Carroll, R. Cloutier, and L. Bar-Sagi. 2002. Vertebral development in the Devonian sarcopterygian fish *Eusthenopteron foordi* and the polarity of vertebral evolution in non-amniote tetrapods. Journal of Vertebrate Paleontology 22: 487–502.

Crossland, M. R. 2000. Direct and indirect effects of the introduced toad *Bufo marinus* (Anura: Bufonidae) on populations of native anuran larvae in Australia. Ecography 23: 283–290.

Crump, D., M. Berril, D. Coulson, D. Lean, L. McGillivray, and A. Smith. 1999. Sensitivity of amphibian embryos, tadpoles, and larvae to enhanced UV-B radiation in natural pond conditions. Canadian Journal of Zoology 77: 1956–1966.

Crump, M. L. 2005. Why are some species in decline but others not?; pp. 7–9 *in* M. J. Lannoo (ed.), Amphibian Declines: The Conservation Status of United States Species. University of California Press, Berkeley.

Currie, P. J. 1977. A new haptodontine sphenacodont (Reptilia: Pelycosauria) from the Upper Pennsylvanian of North America. Journal of Paleontology 51: 927–942.

Daeschler, E. 2005. Biogeography of the Middle and Late Devonian (late Givetian-Frasnian) ichthyofauna from the Okse Bay Group, Nunavut Territory, Canada. Journal of Vertebrate Paleontology 25 (suppl. 49A).

Daeschler, E. B., N. H. Shubin, and F. A. Jenkins Jr. 2006. A Devonian tetrapod-like fish and the evolution of the tetrapod body plan. Nature 440: 757–763.

Daeschler, E. B., N. H. Shubin, K. S. Thomson, and W. W. Amaral. 1994. A Devonian tetrapod from North America. Science 265: 639–642.

Daly, E. 1994. The Amphibamidae (Amphibia: Temnospondyli), with a description of a new genus from the Upper Pennsylvanian of Kansas. University of Kansas Museum of Natural History Miscellaneous Publication no. 85: 1–59.

Damiani, R., C. A. Sidor, J. S. Steyer, R. M. H. Smith, H. C. E. Larsson, A. Maga, and O. Ide. 2006. The vertebrate fauna of the Upper Permian of Niger. V. The primitive temnospondyl *Saharastega moradiensis*. Journal of Vertebrate Paleontology 26: 559–572.

Daszak, P., L. Berger, A. A. Cunningham, A. D. Hyatt, D. E. Green, and R. Speare. 1999. Emerging infectious diseases and amphibian population declines. Emerging Infectious Diseases 5: 735–748.

Davidson, E. 2001. Genomic Regulatory Systems. Academic Press, San Diego.

Davis, A. P., and M. R. Capecchi. 1996. A mutational analysis of the 5′ HoxD genes: Dissection of genetic interactions during limb development in the mouse. Development 122: 1175–1185.

Davydov, V., B. R. Wardlaw, and F. M. Gradstein. 2004. The Carboniferous period; pp. 222–248 in F. Gradstein, J. Ogg, and A. Smith (eds.), A Geological Time Scale 2004. Cambridge University Press, Cambridge.

Dawson, J. W. 1891. On the mode of occurrence of remains of land animals in erect trees at the South Joggins, Nova Scotia (abstract). Transactions of the Royal Society of Canada 19: 4–5.

Dawson, J. W., and R. Owen. 1862. Description of specimens of fossil Reptilia discovered in the Coal-Measures of the South Joggins, Nova Scotia. Quarterly Journal of the Geological Society 18: 237–244 + plates IX and X.

Deban, S. M., and D. B. Wake. 2000. Aquatic feeding in salamanders; pp. 65–94 in K. Schwenk (ed.), Feeding: Form, Function, and Evolution in Tetrapod Vertebrates. Academic Press, San Diego.

deBeer, G. R. 1985. The Development of the Vertebrate Skull. University of Chicago Press, Chicago and London.

De Jongh, H. J. 1968. Functional morphology of the jaw apparatus of larval and metamorphosing *Rana temporaria* L. Netherlands Journal of Zoology 18: 1–108.

DeMar, R. E. 1966. The phylogenetic and functional implications of the armor of the Dissorophidae. Fieldiana Geology 16: 55–88.

Denoel, M., G. Dzukic, and M. L. Kalezic, 2005. Effects of widespread fish introductions on paedomorphic newts in Europe. Conservation Biology 19: 162–170.

Deutschmann, M. R., and J. J. Peterka. 1988. Secondary production of tiger salamanders in three North Dakota lakes. Canadian Journal of Fisheries and Aquatic Sciences 45: 691–697.

Dias-Da-Silva, S., C. Marsicano, and C. Leandro. 2006. Rhytidosteid temnospondyls in Gondwana: A new taxon from the Lower Triassic of Brazil. Palaeontology 49: 381–390.

Dilkes, D. W. 1990. A new trematopsid amphibian (Temnospondyli: Dissorophoidea) from the Lower Permian of Texas. Journal of Vertebrate Paleontology 10: 222–243.

Dilkes, D. W., and R. R. Reisz. 1987. *Trematops milleri* Williston, 1909 identified as a junior synonym of *Acheloma cumminsi* Cope, 1882, with a revision of the genus. American Museum Novitates 2902: 1–12.

DiMichele, W. A., and R. W. Hook. 1992. Paleozoic terrestrial ecosystems; pp. 205–325 in A. K. Behrensmeyer, J. D. Damuth, W. A. DiMichele, R. Potts, H-D. Sues, and S. L. Wing (eds.), Terrestrial Ecosystems through Time. University of Chicago Press, Chicago and London.

Doolittle, R. F. 1995. The origins and evolution of eukaryotic proteins. Philosophical Society of London B 349: 235–240.

Drossopoulu, G., K. E. Lewis, J. J. Sanz-Ezquerro, N. Nikbakht, A. P. McMahon, and C. Hofmann. 2000. A model for antero-posterior patterning of the vertebrate limb based on sequential long- and short-range Shh signalling and Bmp signalling. Development 127: 1337–1348.

Drost, C. A., and G. M. Fellers. 1996. Collapse of a regional frog fauna in the Yosemite area of the California Sierra Nevada, USA. Conservation Biology 10: 414–425.

Dudash, M. R., and C. B. Fenster. 2000. Inbreeding and outbreeding depression in fragmented populations; pp. 35–53 in A. G. Young and G. M. Clarke (eds.), Genetics, Demography and Viability of Fragmented Populations. Cambridge University Press, Cambridge.

Duellman, W. E., and L. Trueb. 1986. Biology of Amphibians. McGraw-Hill Book Company, New York.

Dutuit, J.-M. 1988. *Diplocaulus minimus* n.sp. (Amphibia: Nectridea) Lepospondyl of the Argana Formation (Moroccan Occidental Atlas). Compte rendu de l'Academie des Sciences, Paris Serie II 307: 851–854.

Dzialowski, E. M., and M. P. O'Connor. 1999. Utility of blood flow to the appendages in physiological control of heat exchange in reptiles. Journal of Thermal Biology 24: 21–32.

Eaton, T. H., 1959. The ancestry of the modern amphibia: A review of the evidence. University of Kansas Publications. Museum of Natural History 12: 155–180.

———. 1973. A Pennsylvanian trematopsid amphibian from Kansas. Occasional Papers of the Museum of Natural History, University of Kansas 14: 1–8.

Edgeworth, F. H. 1935. The Cranial Muscles of the Vertebrates. Cambridge University Press, Cambridge.

Ellis, R. S. 2005. The infrared dawn of starlight. Nature 438: 39.

Engel, M. S., and D. A. Grimaldi. 2004. New light shed on the oldest insect. Nature 427: 627–630.

Epstein, P. R. 2001. Climate change and emerging infectious diseases. Microbes and Infection 3: 747–754.

Erwin, D. H. 1993. The Great Paleozoic Crisis: Life and Death in the Permian. Columbia University Press, New York.

———. 2006. Extinction: How Life on Earth Nearly Ended 250 Million Years Ago. Princeton University Press, Princeton, New Jersey.

Estes, R. 1981. Gymnophiona, Caudata; pp. 1–115 in R. Wellnhofer (ed.), Handbuch der Paläoherpetologie, vol. 2. Gustav Fischer Verlag, Stuttgart.

Estes, R., and R. Hoffstetter. 1976. Les Urodèles du Miocène de La Grive-Saint-Alban (Isère, France). Bulletin due Muséum National D'histoire Naturelle 3ᵉ série (no. 398): 297–343.

Estes, R., and O. A. Reig. 1973. The early fossil record of frogs: A review of the evidence; pp. 11–63 in J. L. Vial (ed.), Evolutionary Biology of the Anurans. University of Missouri Press, Columbia.

Estes, R., and M. Wake. 1972. The first fossil record of caecilian amphibians. Nature 239: 228–231.

Evans, S. E., and M. Borsuk-Białynicka. 1998. A stem-group frog from the Early Triassic of Poland. Acta Palaeontologica Polonica 43: 573–580.

Evans, S. E., C. Lally, D. C. Chure, A. Elder, and J. A. Maisano. 2005. A Late Jurassic salamander (Amphibia: Caudata) from the Morrison Formation of North America. Zoological Journal of the Linnean Society 143: 599–616.

Evans, S. E., and A. R. Milner. 1996. A metamorphosed salamander from the Lower Cretaceous of Spain. Philosophical Transactions of the Royal Society of London B 351: 627–646.

Evans, S. E., A. R. Milner, and F. Mussett. 1988. The earliest known salamanders (Amphibia: Caudata): A record from the Middle Jurassic of England. Geobios 21: 539–552.

Evans, S. E., A. R. Milner, and C. Werner. 1996. New salamander and caecilian material from the Late Cretaceous of the Sudan. Palaeontology 39: 77–95.

Evans, S. E., and D. Sigogneau-Russell. 2001. A stem-group caecilian (Lissamphibia: Gymnophiona) from the Lower Cretaceous of North America. Palaeontology 44: 259–273.

Evans, S. E., and M. Waldman. 1996. Small reptiles and amphibians from the Middle Jurassic of Skye Scotland; pp. 219–226 in M. Morale (ed.), The Continental Jurassic. Museum of Northern Arizona Bulletin 60, Flagstaff, Arizona.

Feder M. E., and W. W. Burggren. 1992. Environmental Physiology of the Amphibians. University of Chicago Press, Chicago.

Fedonkin, M. A., J. G. Gohling, K. Grey, G. M. Norbonne, and P. Vickers-Rich. 2007. The Rise of Animals. Johns Hopkins University Press, Baltimore.

Fellers, G. M., and C. A. Drost. 1993. Disappearance of the Cascades frog: Rana cascadae at the southern end of its range, California, USA. Biological Conservation 65: 177–181.

Felsenstein, J. 1978. Cases in which parsimony or compatibility methods will be positively misleading. Systematic Zoology 27: 401–410.

Fischer, M. 2000. Species loss after habitat fragmentation. Trends in Ecology and Evolution 15: 396.

Fisher, R. N., and H. B. Shaffer. 1996. The decline of amphibians in California's Great Central Valley. Conservation Biology 10: 1387–1397.

Francis, E. T. B. 1934. The Anatomy of the Salamander. Clarendon Press, Oxford.

Francis, R., and R. Shotton. 1997. Risk in fisheries management: A review. Canadian Journal of Fisheries and Aquatic Science 54: 1699–1715.

Fraser, N. C., J. O. Farlow, and D. Henderson. 2006. Dawn of the Dinosaurs: Life in the Triassic. Indiana University Press, Bloomington and Indianapolis.

Freda, J., and W. A. Dunson. 1986. Effects of low pH and other chemical variables on the local distribution of amphibians. Copeia 1986: 454–466.

Fritsch, A. 1876. Über die Fauna des Gaskohle des Pilsner und Rakonitzer Beckens. Sitzungsberichte der K. Böhmischen Gesellschaft der Wissenschaften Prag 1875: 70–79.

———. 1883. Fauna der Gaskohle des Pilsner und der Kalksteine der Permformation Böhmens. Vol. 1. Prague.

Fritzsch, B., and M. H. Wake. 1988. The inner ear of gymnophione amphibians and its nerve supply: A comparative study of regressive events in a complex sensory system (Amphibia, Gymnophiona). Zoomorphology 108: 201–217.

Fröbisch, N. B. 2008. Ontogeny and phylogeny of small dissorophoid amphibians. Unpublished PhD thesis, McGill University, 1–357.

Fröbisch, N. B., R. L. Carroll, and R. R. Schoch. 2007. Limb ossification in the Paleozoic branchiosaurid Apateon (Temnospondyli) and the early evolution of preaxial dominance in tetrapod limb development. Evolution & Development 9: 69–75.

Frost, D. R., and 18 co-authors. 2006. The amphibian tree of life. Bulletin of the American Museum of Natural History 297: 1–371.

Gagnier, P-Y. 1991. Ordovician vertebrates from Bolivia; in R. Suarez-Soruco (ed.), Fosile y Facies de Bolivia, vol. I, Vertebrados. Revista Técnica de YPFB. 12(3–4): 371–379.

Gamradt, S. S., and L. B. Kats. 1996. Effect of introduced crayfish and mosquitofish on California newts. Conservation Biology 10: 1155–1162.

Gans, C. 1975. Tetrapod limblessness: Evolution and functional corollaries. American Zoologist 15: 455–467.

Gans, C., and G. C. Gorniak. 1982. How does the toad flip its tongue? Test of two hypotheses. Science 216: 1335–1337.

Gans, C., and F. H. Pough (eds.). 1982. Biology of the Reptilia. Vol. 12, Physiology. Academic Press, San Diego.

Gao, K.-Q., R. Fox, D. Li, and J. Zhang. 2004. A new vertebrate fauna from the Early Triassic of northern Gansu Province, China. Journal of Vertebrate Paleontology 24 Abs.: 62A.

Gao, K-Q., and N. H. Shubin. 2003. Earliest known crown-group salamanders. Nature 422: 424–428.

Gardiner, B. G. 1984. The relationships of the palaeoniscoid fishes, a review based on new specimens of Mimia and Moythomasia from the Upper Devonian of Western Australia. Bulletin of the British Museum (Natural History) Geology 37: 173–428.

Gardner, T., 2001. Declining amphibian populations: A global phenomenon in conservation biology. Animal Biodiversity and Conservation 242: 1–20.

Garner, T. W. J., M. W. Perkins, R. Govindaragulu, D. Seglie, S. Walker, A. A. Cunningham, and M. C. Fisher. 2006. The emerging amphibian pathogen Batrachochytrium dendrobatidis globally infects introduced populations of the North American bull frog, Rana catesbeiana. Biology Letters 2: 455–459.

Gaston, K. J. 1997. What is rarity?; pp 30–47 in W. E. Kunin and K. J. Gaston (eds.), The Biology of Rarity. Chapman & Hall, London.

Gates, D. M. 1980. Biophysical Ecology. Springer-Verlag, New York.

Gatten, R. E., Jr., K. Miller, and R. J. Full. 1992. Energetics at rest and during locomotion; pp. 314–377 in M. E. Feder and W. W. Burggren (eds.), Environmental Physiology of the Amphibians. University of Chicago Press, Chicago and London.

Gehlbach, F. R., and S. E. Kennedy. 1978. Population ecology of a highly productive aquatic salamander (Siren intermedia). Southwestern Naturalist 23: 423–430.

Gehring, W. J. 1985. Homeotic genes, the homeobox and the genetic control of development. Cold Spring Harbor Symposium on Quantitative Biology 50: 243–251.

———. 1998. Master Control Genes in Development and Evolution: The Homeobox Story. Yale University Press, New Haven.

Geoffroy Saint-Hilaire, E. 1830. Principes de Philosophie Zoologique, discuté en Mars 1830, au Sein de l'Académie Royale des Sciences. Pichon et Didier, Paris.

Gesteland, R. F., T. R. Cech, and J. F. Atkins (eds.). 1999. The RNA World. Cold Spring Harbor Laboratory Press, Cold Spring Harbor.

Gibbons, J.W., D. E. Scott, T. J. Ryan, K. A. Buhlmann, T. D. Tuberville, B. S. Metts, J. L. Greene, T. Mills, Y. Leiden, S. Poppy, and C. T. Winne. 2000. The global decline of reptiles, déjà vu amphibians. BioScience 50: 653–666.

Gilbert, S. F. 1988. Developmental Biology. 2d ed. Sinauer Associates, Sunderland, Massachusetts.

Gillespie, G. R. 2001. The role of introduced trout in the decline of the spotted tree frog (Litoria spenceri) in south-eastern Australia. Biological Conservation 100: 187–198.

Gitay, H., A. Suárez, R. T. Watson, and D. J. Dokken (eds.). 2002. Climate Change and Biodiversity. Intergovernmental Panel on Climate Change Technical Paper V: 1–85. IPCC, Geneva, Switzerland.

Glaessner, M. F. 1983. The Dawn of Animal Life: A Biohistorical Study. Cambridge University Press, Cambridge.

Glos, J., T. U. Grafe, M. O. Rodel, and K. E. Linsenmair. 2003. Geographic variation in pH tolerance of two populations of the common frog, Rana temporaria. Copeia 2003: 650–656.

Godfrey, S. J. 1988. Isolated tetrapod remains from the Carboniferous of West Virginia. Kirtlandia 43: 27–36.

———. 1989. The postcranial skeletal anatomy of the Carboniferous tetrapod *Greererpeton burkemorani* Romer, 1969. Philosophical Transactions of the Royal Society of London 323: 75–153.

Godfrey, S. J., and R. Holmes. 1995. The Pennsylvanian temnospondyl *Cochleosaurus florensis* Rieppel, from the lycopsid stump fauna at Florence, Nova Scotia. Breviora 500: 1–25.

Gould, S. J. 1989. Wonderful Life: The Burgess Shale and the Nature of History. Norton, New York.

Gradstein, F., J. Ogg, and A. Smith (eds.). 2004. A Geological Time Scale 2004. Cambridge University Press, Cambridge.

Graham, J. B. 1997. Air-Breathing Fishes: Evolution, Diversity and Adaptation. Academic Press, San Diego.

Graham, N. E. 1995. Simulation of recent global temperature trends. Science 267: 666–671.

Graur, D., and W.-H. Li. 2000. Fundamentals of Molecular Evolution. 2d ed. Sinauer Associates, Inc., Sunderland, Massachusetts.

Gray, J., and W. Shear. 1992. Early life on land. American Scientist 80: 444–456.

Green, D. M. 1997. Perspectives on amphibian population declines: Defining the problem and searching for answers. Herpetological Conservation 1: 291–308.

———. 2003. The ecology of extinction: Population fluctuation and decline in amphibians. Biological Conservation 111: 331–343.

Green, D. M., and D. C. Cannatella. 1993. Phylogenetic significance of the amphicoelous frogs, Ascaphidae and Leiopelmatidae. Ethology, Ecology & Evolution 5: 233–245.

Greer, A. E. 1991. Limb reduction in squamates: Identification of the lineages and discussion of the trends. Journal of Herpetology 25: 166–173.

Gregory, J. T. 1950. Tetrapods of the Pennsylvanian nodules from Mazon Creek, Illinois. American Journal of Science 248: 833–873.

Gregory, J. T., F. E. Peabody, and L. I. Price. 1956. Revision of the Gymnarthridae American Permian microsaurs. Peabody Museum of Natural History, Yale University 10: 1–77.

Gregory, T. R. 2005. Genome size evolution in animals; pp. 3–87 *in* T. R. Gregory (ed.), The Evolution of the Genome. Elsevier, San Diego.

Gregory, W. K. 1951. Evolution Emerging. Macmillan, New York.

Gubin, Y. M. 1980. New Permian dissorophids from the Ural forelands. Paleontology Journal 14: 88–96.

———. 1991. Permian Archegosauroid Amphibians of the USSR. Nauka, Moscow.

Haas, A., S. Hertwig, and I. Das. 2006. Extreme tadpoles: The morphology of the fossorial megophryid larva, *Leptobrachella myobergi*. Zoology 109: 26–42.

Hairston, N. G., Sr., and R. H. Wiley. 1993. No decline in salamander (Amphibia: Caudata) populations: A twenty-year study in the southern Appalachians. Brimleyana 18: 59–64.

Halder, G., P. Callaerts, and W. J. Gehring. 1995. Induction of ectopic eyes by targeted expression of the *eyeless* gene in *Drosophila*. Science 267: 1788–1792.

Hall, B. K. 1999. The Neural Crest in Development and Evolution. Springer, New York.

———. 2005. Consideration of the neural crest and its skeletal derivatives in the context of novelty/innovation. Journal of Experimental Zoology 304B: 548–557.

——— (ed.). 2007. Fins into Limbs. University of Chicago Press, Chicago.

Hallam, A., and P. B. Wignall. 1997. Mass Extinctions and Their Aftermath. Oxford University Press, Oxford.

Halliday, T. 2000. Do frogs make good canaries? Biologist (London) 47: 143–146.

Handrigan, G. R., A. Haas, and R. J. Wassersug. 2007. Bony-tailed tadpoles: The development of supernumerary caudal vertebrae in larval megophryids (Anura). Evolution & Development 9: 190–202.

Handrigan, G. R., and R. J. Wassersug. 2007. The anuran *Bauplan:* A review of the adaptive, developmental, and genetic underpinnings of frog and tadpole morphology. Biological Reviews of the Cambridge Philosophical Society 82: 1–25.

Hanken, J. 1984. Miniaturization and its effect on cranial morphology in plethodontid salamanders, genus *Thorius* (amphibia: Plethodontidae): I: Osteological variation. Biological Journal of the Linnean Society 23: 55–75.

———. 1993. Adaptation of bone growth to miniaturization of body size; pp. 79–104 *in* B. K. Hall (ed.), Bone, vol. 7, Bone Growth. CRC Press, Boca Raton, Florida.

Harding, K. 1993. Conservation and the case of the golden toad. British Herpetological Bulletin, 44: 31–34.

Harrison, R. G. 1969. Organization and Development of the Embryo. Yale University Press, New Haven.

Harrison, S., and E. Bruna. 1999. Habitat fragmentation and large-scale conservation: What do we know for sure? Ecography 22: 225–232.

Hartman, H., and A. Fedorov. 2002. The origin of the eukaryotic cell: A genomic investigation. Proceedings of the National Academy of Science 99: 1420–1425.

Hartmann, C., and C. J. Tabin. 2004. Wnt-14 plays a pivotal role in inducing synovial joint formation in the developing appendicular skeleton. Cell 104: 341–351.

Harvell, C. D., C. E. Mitchell, J. R. Ward, S. Altizer, A. P. Dobson, R. S. Ostfeld, and M. D. Samuel. 2002. Climate warming and disease risks for terrestrial and marine biota. Science 296: 2158–2162.

Heaton, M. J. 1979. Cranial anatomy of primitive captorhinid reptiles from the late Pennsylvanian and early Permian of Oklahoma and Texas. University of Oklahoma Geological Survey Bulletin 127: 1–81.

Heatwole, H., and G. T. Barthalmus (eds.). 1994. Amphibian Biology. Vol. 1, The Integument. Surrey Beatty & Sons, Chipping Norton, Australia.

Heatwole, H., and R. L. Carroll (eds.). 2000. Amphibian Biology. Vol. 4, Palaeontology. Surrey Beatty & Sons, Chipping Norton, Australia.

Heatwole, H., and M. Davies. (eds.). 2003. Amphibian Biology. Vol. 5, Osteology. Surrey Beatty & Sons, Chipping Norton, Australia.

Heatwole, H., and M. Dawley (eds.). 1998. Amphibian Biology. Vol. 3, Sensory Perception. Surrey Beatty & Sons, Chipping Norton, Australia.

Heatwole, H., and B. K. Sullivan (eds.). 1995. Amphibian Biology. Vol. 2, Social Behaviour. Surrey Beatty & Sons, Chipping Norton, Australia.

Hecnar, J. S., and R. T. M'Closkey. 1997. The effects of predatory fish on amphibian species richness and distribution. Biological Conservation 79: 123–131.

Hellrung, H. 2003. *Gerrothorax pustuloglomeratus,* a temnospondyl (Amphibia) with a bony branchial chamber from the Lower Keuper of Kupferzell (South Germany). Stuttgarter Beiträge zur Naturkunde Serie B: 1–130.

Hennig, W. 1966. Phylogenetic Systematics. University of Illinois Press, Urbana, Chicago, and London.

Hess, H. H. 1962. History of ocean basins; pp. 599–620 in A. E. J. Engel, H. L. James, and B. F. Leonard (eds.), Petrologic Studies—A Volume in Honor of A. F. Buddington. Geological Society of America, New York.

Heyer, W. R., and J. B. Murphy. 2005. Declining Amphibian Populations Task Force; pp. 17–20 in M. J. Lannoo (ed.), Amphibian Declines: The Conservation Status of United States Species. University of California Press, Berkeley.

Hickling R., D. B. Roy, J. K. Hill, R. Fox, and C. D. Thomas. 2006. The distributions of a wide range of taxonomic groups are expanding polewards. Global Change Biology 12: 450–455.

Hitchcock, E. C. 1995. A functional interpretation of the anteriormost vertebrae and skull of Eusthenopteron. Bulletin du Muséum National d'Historie Naturelle, Paris 17: 269–285.

Hoffman, R. L., G. L. Larson, and B. Samora. 2004. Responses of Ambystoma gracile to the removal of introduced nonnative fish from a mountain lake. Journal of Herpetology 38: 578–585.

Holman, J. A. 2000. Pleistocene Amphibia: Evolutionary stasis, range adjustments, and re-colonization patterns; pp. 1445–1458 in H. Heatwole and R. L. Carroll (eds.), Amphibian Biology, vol. 4, Palaeontology. Surrey Beatty & Sons, Chipping Norton, Australia.

Holmes, R. 1977. The osteology and musculature of the pectoral limb of small captorhinids. Journal of Morphology 152: 101–140.

———. 1984. The Carboniferous amphibian Proterogyrinus scheelei Romer, and the early evolution of tetrapods. Philosophical Transactions of the Royal Society of London 306: 431–527.

———. 1989. The skull and axial skeleton of the Lower Permian anthracosauroid amphibian Archeria crassidisca Cope. Palaeontographica 207: 161–206.

———. 2000. Palaeozoic temnospondyls; pp. 1181–1120 in H. Heatwole and R. L. Carroll (eds.), Amphibian Biology, vol. 4, Paleontology. Surrey Beatty & Sons, Chipping Norton, Australia.

———. 2003. The hind limb of Captorhinus aguti and the step cycle of basal amniotes. Canadian Journal of Earth Sciences 40: 515–526.

Holmes, R., and R. Carroll. 1977. A temnospondyl amphibian from the Mississippian of Scotland. Bulletin of the Museum of Comparative Zoology 147: 489–511.

———. 1980. The skull and jaw musculature as guides to the ancestry of salamanders. Zoological Journal of the Linnean Society 68: 1–40.

Holmes, R., R. Carroll, and R. R. Reisz. 1998. The first articulated skeleton of Dendrerpeton acadianum (Temnospondyli, Dendrerpetontidae) from the Lower Pennsylvanian locality of Joggins, Nova Scotia, and a review of its relationships. Journal of Paleontology 18: 64–79.

Holmgren, N. 1949. On the tetrapod limb problem—again. Acta Zoologica (Stockholm) 30: 485–508.

———. 1939. Contribution to the question of the origin of the tetrapod limb. Acta Zoologica (Stockholm) 20: 89–124.

Holsinger, K. E. 2000. Demography and extinction in small populations; pp. 55–74 in A. G. Young and G. M. Clarke (eds.), Genetics, Demography and Viability of Fragmented Populations. Cambridge University Press, Cambridge.

Hook, R. W. 1983. Colosteus scutellatus (Newberry), a primitive temnospondyl amphibian from the Middle Pennsylvanian of Linton, Ohio. American Museum of Natural History Novitates 2770: 1–41.

Hou, X-G., R. J. Aldridge, J. Bergström, D. J. Siveter, D. J. Siveter, and F. Xiang-Hong. 2004. The Cambrian Fossil of Chengjiang, China. Blackwell Publishing, Malden, Massachusetts.

Houlahan, J. E., C. S. Findlay, B. R. Schmidt, A. H. Meyer, and S. L. Kuzmin. 2000. Quantitative evidence for global amphibian population declines. Nature 404: 752–755.

Ivachnenko, M. F. 1978. Urodelans from the Triassic and Jurassic of Soviet Central Asia. Paleontologicheskiy Zhurnal 1978: 84–89 [in Russian].

Jacobson, S. K., and J. J. Vandenberg. 1991. Reproductive ecology of the endangered golden toad (Bufo periglenes). Journal of Herpetology 25: 321–327.

Jarvik, E. 1942. On the structure of the snout of crossopterygians and lower gnathostomes in general. Zoologische Bidrag Uppsala 21: 235–575.

———. 1952. On the fish-like tail in the ichthyostegid stegocephalians with descriptions of a new stegocephalian and a new crossopterygian from the Upper Devonian of East Greenland. Meddr Grønland 114: 1–90.

———. 1954. On the visceral skeleton in Eusthenopteron with a discussion of the parasphenoid and palatoquadrate in fishes. K. Svenska Vetenskapsakad. Handl., 5, no. 1: 1–104.

———. 1960. Théories de l'Évolution des Vertébrés, Reconsidérés a la Lumière des Récentes Découvertes sur les Vertébrés Inférieures. Masson, Paris.

———. 1980. Basic Structure and Evolution of Vertebrates. Vol. 1. Academic Press, London and New York.

———. 1996. The Devonian tetrapod Ichthyostega. Fossils and Strata 40: 1–213.

Javaux, E., A. H. Knoll, and M. R. Walter. 2001. Ecological and morphological complexity in early eukaryotic ecosystems. Nature 412: 66–69.

Jenkins, F. A., and N. H. Shubin. 1998. Prosalirus bitis and the anuran caudopelvic mechanisms. Journal of Vertebrate Palaeontology 18: 495–510.

Jenkins, F. A., D. Walsh, and R. L. Carroll. 2007. Anatomy of Eocaecilia micropodia, a limbed gymnophionan of the Early Jurassic. Bulletin of the Museum of Comparative Zoology 158: 1–81.

Jennings, M. R., and M. P. Hayes. 1985. Pre-1900 overharvest of California red-legged frogs (Rana aurora draytonii): The inducement for bullfrog (Rana catesbeiana) introduction. Herpetologica 41: 94–103.

Julliard, R., F. Jiguet, and D. Couvet. 2004. Common birds facing global changes: What makes a species at risk? Global Change Biology 10: 148–154.

Kaiser, H. 1997. Origins and introductions of the Caribbean frog, Eleutherodactylus johnstonei (Leptodactylidae): Management and conservation concerns. Biodiversity and Conservation 6: 1391–1407.

Kashlinsky, A., R. G. Arendt, J. Mather, and S. H. Moseley. 2005. Tracing the first stars with fluctuations of the cosmic infrared background. Nature 438: 45–50.

Kats, L. B., and R. P. Ferrer. 2003. Alien predators and amphibian declines: Review of two decades of science and the transition to conservation. Diversity & Distributions 9: 99–110.

Kerr, J. B., and C. T. McElroy. 1993. Evidence for the large upward trends of UV-B radiation linked to ozone depletion. Science 262: 1032–1034.

Kiesecker, J. M., and A. R. Blaustein. 1995. Synergism between UV-B radiation and a pathogen magnifies amphibian embryo mortality in nature. Proceedings of the National Academy of Sciences of the USA 92: 11049–11052.

Kiesecker, J. M., A. R. Blaustein, and L. K. Belden. 2001. Complex causes of amphibian population declines. Nature 410: 681–684.

Klembara, J., and I. Bartík. 2000. The postcranial skeleton of *Discosauriscus* Kuhn, a seymouriamorph tetrapod from the Lower Permian of the Boskovice Furrow (Czech Republic). Transactions of the Royal Society of Edinburgh, Earth Sciences 90: 287–316.

Klembara, J., and M. Ruta. 2004a. The seymouriamorph tetrapod *Utegenia shpinari* from the ?Upper Carboniferous-Lower Permian of Kazakhstan. Part I: Cranial anatomy and ontogeny. Transactions of the Royal Society of Edinburgh, Earth Sciences 94: 45–74.

———. 2004b. The seymouriamorph tetrapod *Utegenia shpinari* from the ?Upper Carboniferous-Lower Permian of Kazakhstan. Part II: Postcranial anatomy and relationships. Transactions of the Royal Society of Edinburgh, Earth Sciences 94: 45–74.

———. 2005a. The seymouriamorph tetrapod *Ariekanerpeton sigalova*, from the Lower Permian of Tadzhikistan. Part II: Postcranial anatomy and relationships. Transactions of the Royal Society of Edinburgh, Earth Sciences 96: 71–93.

———. 2005b. The seymouriamorph tetrapod *Ariekanerpeton sigalovi* from the Lower Permian of Tadzhikistan. Part I: Cranial anatomy and ontogeny. Transactions of the Royal Society of Edinburgh, Earth Sciences 96: 43–70.

Knapp, R. A., K. R. Matthews, and O. Sarnelle. 2001. Resistance and resilience of alpine lake fauna to fish introductions. Ecological Monographs 71: 401–421.

Knoll, A. H. 1995. Proterozoic and Early Cambrian protists: Evidence for accelerating evolutionary tempo; pp. 63–83 *in* W. M. Fitch and F. J. Ayala (eds.), Tempo and Mode in Evolution. National Academy Press, Washington, D.C.

———. 2003. Life on a Young Planet. Princeton University Press, Princeton, New Jersey.

Koonz, W. 1992. Amphibians in Manitoba; pp. 19–20 *in* C. A. Bishop and K. E. Pettit (eds.), Declines in Canadian Amphibian Populations: Designing a National Monitoring Strategy. Canadian Wildlife Service, Occasional Paper no. 76.

Krajik, K. 2006. The lost world of the Kihansi toad. Science 311: 1230–1232.

Kraus, F., and E. W. Campbell. 2002. Human-mediated escalation of a formerly eradicable problem: The invasion of Caribbean frogs in the Hawaiian Islands. Biological Invasions 4: 327–332.

Kraus, F., E. W. Campbell, A. Allison, and T. Pratt. 1999. *Eleutherodactylus* frog introductions to Hawaii. Herpetological Reviews 30: 21–25.

Kummel, B. 1961. History of the Earth. W. H. Freeman and Company, San Francisco and London.

Langston, W., Jr. 1953. Permian amphibians from New Mexico. University of California Publications in Geological Sciences 29: 349–416.

Lannoo, M. J. (ed.). 2005. Amphibian Declines: The Conservation Status of United States Species. University of California Press, Berkeley.

Lauder, G. V., and M. S. Reilly. 1988. Functional design of the feeding mechanism in salamanders: Causal bases of ontogenetic changes in function. Journal of Experimental Biology 134: 219–233.

Lauder, G. V., and H. B. Schaffer. 1985. Functional morphology of the feeding mechanism in aquatic ambystomatid salamanders. Journal of Morphology 185: 297–326.

Laurance, W. F. 1996. Epidemic disease and the catastrophic decline of Australian rain forest frogs. Conservation Biology 10: 406–413.

Laurin, M. 2000. Seymouriamorphs; pp. 1064–1080 *in* H. Heatwole and R. L. Carroll (eds.), Amphibian Biology, vol. 4, Palaeontology. Surrey Beatty & Sons, Chipping Norton, Australia.

Laurin, M., and R. R. Reisz. 1997. A new perspective on tetrapod phylogeny; pp. 9–59 *in* K. L. M. Martin and S. S. Sumida (eds.), Amniote Origins. Academic Press, San Diego.

Lawler, S. P., D. Dritz, T. Strange, and M. Holyoak. 1999. Effects of introduced mosquitofish and bullfrogs on the threatened California Red-legged frog. Conservation Biology 13: 613–622.

Layne, J. R., and R. E. Lee. 1989. Seasonal variation in freeze tolerance and ice content of the tree frog *Hyla versicolor*. Journal of Experimental Zoology 249: 133–137.

Lebedev, O. A., and J. A. Clack. 1993. Upper Devonian tetrapods from Andreyevka, Tula Region, Russia. Palaeontology 36: 721–734.

Lebedev, O. A., and M. I. Coates. 1995. The postcranial skeleton of the Devonian tetrapod *Tulerpeton curtum* Lebedev. Zoological Journal of the Linnean Society 114: 307–348.

Lee, M. S. Y., and J. S. Anderson. 2006. Molecular clocks and the origin(s) of modern amphibians. Molecular Phylogenetics and Evolution 40: 635–639.

Levine, R. P., J. A. Monroy, and E. L. Brainerd. 2004. Contribution of eye retraction to swallowing performance in the northern leopard frog, *Rana pipiens*. Journal of Experimental Biology 207: 1361–1368.

Lewis, G. E., and P. P. Vaughn. 1965. Early Permian Vertebrates from the Cutler Formation of the Placerville Area Colorado. Geological Survey Professional Paper 503-C: C1–C50.

Licht, L. E. 2003. Shedding light on ultraviolet radiation and amphibian embryos. BioScience 53: 551–561.

Liem, D. S. 1973. A new genus of a frog of the family Leptodactylidae from SE Queensland, Australia. Memoirs of the Queensland Museum 16: 459–470.

Lillywhite, H. B., and F. A. Maderson. 1982. Skin structure and permeability; pp. 397–442 *in* C. Gans and F. H. Pough (eds.), Biology of the Reptilia vol. 12. Academic Press, New York.

Linder, G., S. K. Krest, and D. W. Sparling (eds.). 2003. Amphibian Decline: An Integrated Analysis of Multiple Stressor Effects. SETAC Press, Pensacola, Florida.

Lips, K. R. 1999. Mass mortality and anuran declines at an upland site in western Panama. Conservation Biology 13: 117–125.

Lips, K. R., F. Brem, R. Brenes, J. D. Reeve, R. A. Alford, J. Voyles, C. Carey, L. Livo, A. P. Pessier, and J. P. Collins. 2006. Emerging infectious disease and the loss of biodiversity in a Neotropical amphibian community. Proceedings of the National Academy Science USA 103: 3165–3170.

Liu, Y., and Y. Liu. 2005. Comments on "$^{40}Ar/^{39}Ar$ dating of ignimbrite from Inner Mongolia, northeastern China, indicates a post-Middle Jurassic age for the overlying Daohugou Bed" by H. Y. He et al. Geophysical Research Letters 32: L12314 1–3.

Long, J. 1995. The Rise of Fishes. Johns Hopkins University Press, Baltimore.

Longcore, J. E., A. P. Pessier, and D. K. Nichols. 1999. *Batrachochytrium dendrobatidis* gen. et sp. nov., a chytrid pathogenic to amphibians. Mycologia 91: 219–227.

Lombard, R. E., and J. R. Bolt. 1988. Evolution of the stapes in Paleozoic tetrapods; pp. 37–67 *in* B. Fritzsch, M. J. Ryan, W. Wilczynski, T. E. Hetherington, and W. Walkowiak (eds.), The

Evolution of the Amphibian Auditory System. John Wiley & Sons, New York.

———. 1995. A new primitive tetrapod *Whatcheeria deltae* from the Lower Carboniferous of Iowa. Palaeontology 38: 471–494.

———. 1999. A microsaur from the Mississippian of Illinois and a standard format for morphological characters. Journal of Paleontology 73: 908–923.

Lynch, M. 1999. The age and relationships of the major animal phyla. Evolution 53: 319–325.

Macey, J. R. 2005. Plethodontid salamander mitochondrial genomics: A parsimony evaluation of character conflict and implications for historical biogeography. Cladistics 21: 194–202.

Macías, G., A. Marco, and A. R. Blaustein. 2007. Combined exposure to ambient UVB radiation and nitrite negatively affects survival of amphibian early life stages. Science of the Total Environment 385: 55–65.

MacLeod, N. 2006. Extinction: How life on Earth nearly ended 250 million years ago. Palaeontology Association Newsletter no. 62: 126–131.

Marco, A., and A. R. Blaustein. 1999. The effects of nitrite on behavior and metamorphosis in Cascades frogs (*Rana cascadae*). Environmental Toxicology and Chemistry 18: 946–949.

Marco, A., D. Cash, L. K. Belden, and A. R. Blaustein. 2001. Sensitivity to urea fertilization in three amphibian species. Archives of Environmental Contamination and Toxicology 40: 406–409.

Maerz, J. C., C. J. Brown, C. T. Chapin, and B. Blossey. 2005. Can secondary plant compounds of an invasive plant affect larval amphibians? Functional Ecology 19: 970–975.

Mamay, S. H., R. W. Hook, and N. Hotton III. 1998. Amphibian eggs from the Lower Permian of North-Central Texas. Journal of Vertebrate Paleontology 18: 80–84.

Margulis, L. 1981. Symbiosis in Cell Evolution. W. H. Freeman, San Francisco.

Martin, R. A. 2002. A possible role for reproductive behavior in the origin of tetrapod feet. Paludicola 4: 6–9.

Martin, W., D. V. Grazhdankin, S. A. Bowring, D. A. D. Evans, and M. A. Fodonkin. 2000. Age of Neoproterozoic bilaterian body and trace fossils, White Sea, Russia: Implications for metazoan evolution. Science 288: 841–845.

Matthew, G. F. 1905. New species and a new genus of batrachian footprints of the Carboniferous system of eastern Canada. Transactions of the Royal Society of Canada 10: 77–101.

McCallum, H. 2004. Inconclusiveness of chytridiomycosis as the agent in widespread frog declines. Conservation Biology 19: 1421–1430.

McGowan, G. 2002. Albanerpetontid amphibians from the Lower Cretaceous of Spain and Italy: A description and reconsideration of their systematics. Zoological Journal of the Linnean Society 135: 1–32.

McGowan, G., and S. E. Evans. 1995. Albanerpetontid amphibians from the Cretaceous of Spain. Nature 373: 143–145.

Meier, S., and D. S. Packard Jr. 1984. Morphogenesis of the cranial segments and distribution of neural crest in the embryo of the snapping turtle. Chelydra Serpentina. Developmental Biology 102: 309–323.

Mencken, H. L. 1920. Prejudices. Second series. Knopf, New York.

Menneken, M., A. A. Nemchin, T. Geisler, R. T. Pidgeon, and S. A. Wilde. 2007. Hadean diamonds in zircon from Jack Hills, Western Australia. Nature 448: 917–920.

Mereschkowsky, C. 1905. Uber Nature und Ursprung der Chromatophoren im Pflanzenreiche. Biologische Centrablatt 25: 593–604.

Meyer, A. H., B. R. Schimdt, and K. Grossenbacher. 1998. Analysis of three amphibian populations with quarter-century long time series. Proceedings of the Royal Society London Series B, 265: 523–528.

Miall, L. C. 1875. Report of the committee on the structure and classification of the labyrinthodonts. Report of the British Association for the Advancement of Science 1874: 149–192.

Miles, R. S., and T. S. Westoll. 1968. The placoderm fish *Coccosteus cuspidatus* Miller ex Agassiz from the Middle Old Red Sandstone of Scotland. Part I. Descriptive morphology. Transactions of the Royal Society of Edinburgh 67: 373–476.

Miller, R. F., R. Cloutier, and S. Turner. 2003. The oldest articulated chondrichthyan from the Early Devonian period. Nature 425: 501–504.

Miller, S. L. 1953. A production of amino acids under possible primitive Earth conditions. Science 117: 528–529.

———. 1992. The prebiotic synthesis of organic compounds as a step toward the origin of life; pp. 1–28 *in* J. W. Schopf (ed.), Major Events in the History of Life. Jones and Bartlett, Boston.

Milner, A. C. 1994. The aïstopod amphibian from the Viséan of East Kirkton, West Lothian, Scotland. Transactions of the Royal Society of Edinburgh, Earth Sciences 84: 363–368.

Milner, A. C., and W. Lindsay. 1998. Postcranial remains of *Baphetes* and their bearing on the relationships of the Baphetidae (= Loxommatidae). Zoological Journal of the Linnean Society 122: 211–235.

Milner, A. R. 1980a. The temnospondyl amphibian *Dendrerpeton* from the Upper Carboniferous of Ireland. Palaeontology 23: 125–141.

———. 1980b. The tetrapod assemblage from Nýřany, Czechoslovakia; pp. 439–496 *in* A. L. Panchen (ed.), The Terrestrial Environment and the Origin of Land Vertebrates. Academic Press, London.

———. 1982. Small temnospondyl amphibians from the Middle Pennsylvanian of Illinois. Paleontology 25: 635–664.

———. 1986. Dissorophoid amphibians from the Upper Carboniferous of Nýřany; pp. 671–674 *in* Z. Roček (ed.), Studies in Herpetology. Charles University, Prague.

———. 1988. The relationships and origin of living amphibians; pp. 59–102 *in* M. Benton (ed.), The Phylogeny and Classification of the Tetrapods, vol. 1, Systematics Association. Clarendon Press, Oxford.

———. 1990. The radiations of temnospondyl amphibians; pp. 321–349 *in* P. D. Taylor and G. P. Larwood (eds.), Major Evolutionary Radiations. Clarendon Press, Oxford.

———. 1993. The Paleozoic relatives of lissamphibians. Herpetological Monograph no. 6: 8–27.

———. 1996. Early amphibian globetrotters? Nature 381: 741–742.

———. 2000. Mesozoic and Tertiary Caudata and Albanerpetontidae; pp. 1412–1444 *in* H. Heatwole and R. L. Carroll (eds.), Amphibian Biology, vol. 4. Surrey Beatty & Sons, Chipping Norton, Australia.

Milner, A. R., and S. E. K. Sequeira. 1994. The temnospondyl amphibians from the Viséan of East Kirkton, West Lothian, Scotland. Transactions of the Royal Society of Edinburgh, Earth Sciences 84: 331–361.

———. 1998. A cochleosaurid temnospondyl amphibian from the Middle Pennsylvanian of Linton, Ohio. U.S.A. Zoological Journal of the Linnean Society 122: 261–290.

———. 2004. *Slaugenhopia texensis* (Amphibia: Temnospondyli) from the Permian of Texas is a primitive tupilakosaurid. Journal of Vertebrate Paleontology 24: 320–325.

Min, M.-S., S.-Y. Yang, R. M. Bonett, D. R. Vieites, R. A. Brandon, and D. B. Wake. 2005. Discovery of the first Asian plethodontid salamander. Nature (London) 435: 87–90.

Minelli, A. 1993. Biological Systematics. Chapman & Hall, London.

Minguillón, C., J. Gardenyes, E. Serra, L. F. C. Castro, A. Hill-Force, P. W. H. Holland, C. T. Amemiya, and J. Garcia-Fernández. 2005. No more than 14: The end of the amphioxus Hox cluster. International Journal of Biological Sciences 1: 19–23.

Modesta, S. P., R. J. Damiani, J. Neveling, and A. M. Yates. 2003. A new Triassic owenettid pararreptile and the mother of mass extinctions. Journal of Vertebrate Paleontology 23: 715–719.

Modesta, S. P., H.-D. Sues, and R. Damiani. 2001. A new Triassic procolophonoid reptile and its implications for procolophonoid survivorship during the Permo-Triassic extinction event. Proceedings of the Royal Society of London, Series B: 2047–2052.

Morgan, J. A. T., V. T. Vredenburg, L. J. Rachowicz, R. A. Knapp, M. J. Stice, T. Tunstall, R. E. Bingham, J. M. Parker, J. E. Longcore, C. Moritz, C. J. Briggs, and J. W. Taylor. 2007. Population genetics of the frog-killing fungus *Batrachochytrium dendrobatidis*. Proceedings of the National Academy of Sciences of the USA 104: 13845–13850.

Morse, S. S. 1995. Factors in the emergence of infectious diseases. Emerging Infectious Diseases 1: 7–15.

Mossman, D. J., and W. A. S. Sarjeant. 1980. How we found Canada's oldest known footprints. Canadian Geographic 100: 50–53.

Moulton, J. M. 1974. A description of the vertebral column of *Eryops* based on the notes and drawings of A. S. Romer. Breviora. Museum of Comparative Zoology, Harvard 428: 1–44.

Moy-Thomas, J. A., and R. S. Miles. 1971. Paleozoic Fishes. 2d ed. Chapman & Hall, London.

Mueller, R. L., J. R. Macey, M. Jaekel, D. B. Wake, and J. L. Boore. 2004. Morphological homoplasy, life history evolution, and historical biogeography of plethodontid salamanders inferred from complete mitochondrial genomes. Proceedings of the National Academy of Sciences of the USA 101: 13820–13825.

Murray, B. R., and G. C. Hose. 2005. Life-history and ecological correlates of decline and extinction in the endemic Australian frog fauna. Austral Ecology 30: 564–571.

Mus, M. M., and J. Bergström. 2005. The morphology of hyolithids and its functional implications. Palaeontology 48: 1139–1167.

Nagl, A. M., and R. Hofer. 1997. Effects of ultraviolet radiation on early larval stages of the Alpine newt, *Triturus alpestris*, under natural and laboratory conditions. Oecologia 110: 514–519.

Nelson, C. E., B. A. Moran, A. C. Burke, E. Laufer, E. DiMambro, L. C. Murtaugh, E. Gonzales, L. Tessarollo, L. F. Parada, and C. Tabin. 1996. Analysis of *Hox* gene expression in the chick limb bud. Development 122: 1449–1466.

Norell, M. M., and M. Ellison. 2005. Unearthing the Dragon. Pearson Education, Inc., New York.

Novikov, I. V., M. A. Shishkin, and V. K. Golubev. 2000. Permian and Triassic anthracosaurs from eastern Europe; pp. 60–70 in M. J. Benton, M. A. Shishkin, and D. M. Unwin (eds.), The Age of Dinosaurs in Russia and Mongolia. Cambridge University Press, Cambridge.

Nussbaum, R. A. 1977. Rhinatrematidae: A new family of caecilians (Amphibia: Gymnophiona). Occasional Papers of the Museum of Zoology, University of Michigan 682: 1–30.

———. 1983. The evolution of a unique dual jaw-closing mechanism in caecilians (Amphibia: Gymnophiona) and its bear-

ing on caecilian ancestry. Journal of Zoology, London 199: 545–554.

———. 1998. Caecilians; pp. 52–59 in H. G. Cogger and R. G. Zweifel (eds.), Encyclopedia of Reptiles and Amphibians. Academic Press, San Diego.

Olson, E. C. 1961. Jaw mechanisms: Rhipidistians, amphibians, reptiles. American Zoologist 1: 205–215.

O'Reilly, J. C. 2000. Feeding in caecilians; pp. 149–166 in K. Schwank (ed.), Feeding. Academic Press, San Diego.

O'Reilly, J. C., A. P. Summers, and D. A. Ritter. 2000. The evolution of the functional role of trunk muscles during locomotion in adult amphibians. American Zoologist 40: 123–135.

Orizaola, G., and F. Braña. 2006. Effect of salmonid introduction and other environmental characteristics on amphibian distribution and abundance in mountain lakes of northern Spain. Animal Conservation 9: 171–178.

Ouellet, M., I. Mikaelian, B. D. Pauli, J. Rodrigue, and D. M. Green. 2005. Historical evidence of widespread chytrid infection in North American amphibian populations. Conservation Biology 19: 1431–1440.

Panchen, A. L. 1959. A new armoured amphibian from the Upper Permian of East Africa. Philosophical Transactions of the Royal Society of London B 242: 207–281.

———. 1966. The axial skeleton of the labyrinthodont *Eogyrinus attheyi*. Journal of Zoology, London 150: 199–222.

———. 1970. Anthracosauria; pp. 1–84 in O. Kuhn (ed.), Encyclopedia of Paleoherpetology. Gustav Fischer Verlag, Stuttgart.

———. 1972. The skull and skeleton of *Eogyrinus attheyi* Watson (Amphibia: Labyrinthodontia). Philosophical Transactions of the Royal Society of London 263: 279–326.

———. 1973. On *Crassigyrinus scoticus* Watson, a primitive amphibian from the Lower Carboniferous of Scotland. Palaeontology 16: 179–193.

———. 1975. A new genus and species of anthracosaur amphibian from the Lower Carboniferous of Scotland and the status of *Pholidogaster pisciformis* Huxley. Philosophical Transactions of the Royal Society of London B 269: 581–640.

———. 1977. On *Anthracosaurus russelli* Huxley (Amphibia: Labyrinthodontia) and the family Anthracosauridae. Philosophical Transactions of the Royal Society of London B 279: 447–512.

———. 1985. On the amphibian *Crassigyrinus scoticus* Watson from the Carboniferous of Scotland. Philosophical Transactions of the Royal Society of London B 309: 505–568.

Panchen, A. L., and T. R. Smithson. 1990. The pelvic girdle and hind limb of *Crassigyrinus scoticus* (Lydekker) from the Scottish Carboniferous and the origin of the tetrapod pelvic skeleton. Transactions of the Royal Society of Edinburgh, Earth Sciences 81: 31–44.

Panopoulou, G., and A. J. Poustka. 2005. Timing the mechanisms of ancient genome duplication—the adventure of a hypothesis. Trends in Genetics 21: 559–567.

Parsons, T. S., and E. E. Williams. 1962. The teeth of Amphibia and their relation to amphibian phylogeny. Journal of Morphology 110: 375–389.

———. 1963. The relationships of the modern Amphibia: A re-examination. Quarterly Review of Biology 38: 26–53.

Paton, R. L., T. R. Smithson, and J. A. Clack. 1999. An amniote-like skeleton from the Early Carboniferous of Scotland. Nature 398: 508–513.

Patten, B. M. 1958. Foundations of Embryology. McGraw-Hill, New York.

Pawley, K., and A. Warren. 2005. A terrestrial stereospondyl from the Lower Triassic of South Africa: The postcranial skeleton of *Lydekkerina huxleyi* (Amphibia: Temnospondyli). Palaeontology 48: 281–298.

Pechmann, J. H. K., D. E. Scott, R. D. Semlitsch, J. P. Caldwell, L. J. Vitt, and J. W. Gibbons. 1991. Declining amphibian populations: The problem of separating human impacts from natural fluctuations. Science 253: 892–895.

Pechmann, J. H. K., and H. Wilbur. 1994. Putting declining amphibian populations in perspective: Natural fluctuations and human impacts. Herpetologica 50: 65–84.

Peters, S. E., and K. C. Nishikawa. 1999. Comparison of isometric contractile properties of the tongue muscles in three species of frogs, *Litoria caerulea, Dyscophus guinetti,* and *Bufo marinus.* Journal of Morphology 242: 107–124.

Pfennig, D. W. 1992. Polyphenism in spadefoot toad tadpoles as a locally adjusted evolutionarily stable strategy. Evolution 46: 1408–1420.

Pianka, E. R., and L. J. Vitt. 2003. Lizards: Windows to the Evolution of Diversity. University of California Press, Berkeley.

Piveteau, J. 1937. Paléontologie de Madagascar. XXIII. Un Amphibien du Trias Inférieur-essai sur l'origine et l'évolution des amphibiens anoures. Annales de Paléontologie 26: 135–177.

Porter, S. M., and A. H. Knoll. 2000. Testate amoebae in the Neoproterozoic Era: Evidence from vase-shaped microfossils in the Chuar Group, Grand Canyon. Paleobiology 26: 360–385.

Pough, F. H., R. M. Andrews, J. E. Cadle, M. L. Crump, A. H. Savitzky, and K. D. Wells. 2004. Herpetology. Prentice Hall, Upper Saddle River, New Jersey.

Pounds, A. 1996. Conservation of the golden toad: A brief history. British Herpetological Bulletin 55: 5–7.

Pounds, A., M. Fogden, and J. Campbell. 1999. Biological response to climate change on a tropical mountain. Nature 398: 611–614.

Pounds, J. A., M. R. Bustamante, L. A. Coloma, J. A. Consuegra, M. P. L. Fogden, P. N. Foster, E. La Marca, K. L. Masters, A. Merino-Viteri, R. Puschendorf, S. R. Ron, G. A. Sánchez-Azofeifa, C. J. Still, and B. Young. 2006. Widespread amphibian extinctions from epidemic disease driven by global warming. Nature 439: 161–167.

Pounds, J. A., and M. L. Crump. 1994. Amphibian declines and climate disturbance: The case of the golden toad and the harlequin frog. Conservation Biology 8: 72–85.

Pryor, G. S., and K. A. Bjorndal. 2005. Symbiotic fermentation, digesta passage, and gastrointestinal morphology in bullfrog tadpoles (*Rana catesbeiana*). Physiological and Biochemical Zoology 78: 201–215.

Pusey, H. K. 1938. Structural changes in the anuran mandibular arch during metamorphosis, with reference to *Rana temporaria.* Quarterly Review of Microscopic Sciences 84: 479–552.

———. 1943. On the head of the liopelmid frog, *Ascaphus truei.* The chondrocranium, jaws, arches, and muscles of a partly-grown larva. Quarterly Review of Microscopic Sciences 84: 105–185.

Quinn, C. H., H. J. Ndangalasi, J. Gerstle, and J. C. Lovett. 2005. Effect of the Lower Kihansi Hydropower Project and post-project mitigation measures on wetland vegetation in Kihansi Gorge, Tanzania. Biodiversity and Conservation 14: 297–308.

Rabinowitz, D., S. Cairns, and T. Dillon. 1986. Seven forms of rarity and their frequency in the flora of the British Isles; pp. 182–204 *in* M. E. Soulé (ed.), Conservation Biology: The Science of Scarcity and Diversity. Sinauer Associates, Sunderland, Massachusetts.

Raff, R. 1996. The Shape of Life. University of Chicago Press, Chicago.

Rage, J.-C. 1986. Le plus ancien Amphibien apode (Gymnophiona) fossile. Remarques sur la répartition et l'historire paleobiographie des Gymnophioines. Comptes Rendus de l'Académie des Sciences Paris 302: 1033–1036.

Rappole, J. H., and M. V. McDonald. 1994. Cause and effect in population declines of migratory birds. Auk 111: 652–660.

Räsänen, K., and D. M. Green. 2008. Acidification and its effects on amphibian populations; *in* H. Heatwole (ed.), Amphibian Biology, vol. 8, Conservation and Ecology. Surrey Beatty & Sons, Chipping Norton, Australia (in press).

Räsänen, K., A. Laurila, and J. Merilä. 2002. Carry-over effects of embryonic acid conditions on development and growth of *Rana temporaria* tadpoles. Freshwater Biology 47: 19–30.

Reig, O. A. 1964. El Problema del Origen monofilético o polifilético de los anfibios, con consideraciones sobre las relaciones entre anoros, urodelos y ápodos. Ameghiniana 3: 191–211.

Reisz, R. R. 1972. Pelycosaurian reptiles from the Middle Pennsylvanian of North America. Bulletin of the Museum of Comparative Zoology 144: 27–62.

———. 1975. Pennsylvanian pelycosaurs from Linton, Ohio and Nýřany, Czechoslovakia. Journal of Paleontology 49: 522–527.

———. 1977. *Petrolacosaurus kansensis* Lane, the oldest known diapsid reptile. Science 196: 1091–1093.

———. 1981. A diapsid reptile from the Pennsylvanian of Kansas. University of Kansas Museum of Natural History, Special Publication 7: 1–74.

———. 2007. The cranial anatomy of basal diadectomorphs and the origin of amniotes; pp. 228–252 *in* J. S. Anderson and H.-D. Sues (eds.), Major Transitions in Vertebrate Evolution. Indiana University Press, Bloomington and Indianapolis.

Reisz, R. R., and D. Baird. 1983. Captorhinomorph "stem" reptiles from the Pennsylvanian coal-swamp deposit of Linton, Ohio. Annals of Carnegie Museum 52: 393–411.

Reisz, R. R., and D. S. Berman. 1986. *Ianthasaurus hardestii* n. sp., a primitive edaphosaur (Reptilia, Pelycosauria) from the Upper Pennsylvanian Rock Lake Shale near Garnett, Kansas. Canadian Journal of Earth Sciences 23: 77–91.

Repetski, J. E. 1978. A fish from the Upper Cambrian of North America. Science 200: 529–531.

Retallack, G. J., J. J. Veevers, and R. Morante. 1996. Global coal gap between Permian-Triassic extinction and Middle Triassic recovery of peat-forming plants. Geological Society of America Bulletin 108: 195–207.

Richards, T. A., and T. Cavalier-Smith. 2005. Myosin domain evolution and the primary divergence of eukaryotes. Nature 436: 113–118.

Ritland, R. M. 1955. Studies on the post-cranial morphology of *Ascaphus truei.* I. Skeleton and spinal nerves. Journal of Morphology 97: 119–177.

Robinson, J., P. E. Ahlberg, and G. Koentges. 2005. The braincase and middle ear region of *Dendrerpeton acadianum* (Tetrapoda: Temnospondyli). Zoological Journal of the Linnean Society 143: 577–597.

Roček, Z. 2000. Mesozoic anurans; pp. 1295–1331 *in* H. Heatwole and R. L. Carroll (eds.), Amphibian Biology, vol. 4. Surrey Beatty & Sons, Chipping Norton, Australia.

———. 2003. Larval development and evolutionary origin of the anuran skull; pp. 1877–1995 *in* H. Heatwole and M. Davies (eds.), Amphibian Biology, vol. 5, Osteology. Surrey Beatty & Sons, Chipping Norton, Australia.

Roček, Z., and J.-C. Rage. 2000a. Proanuran stages (*Triadobatrachus, Czatkobatrachus*); pp. 1283–1294 *in* H. Heatwole and R. L. Carroll (eds.), Amphibian Biology, vol. 4. Surrey Beatty & Sons, Chipping Norton, Australia.

———. 2000b. Tertiary Anura of Europe, Africa, Asia, North America, and Australia; pp. 1332–1387 *in* H. Heatwole and R. L. Carroll (eds.), Amphibian Biology, vol. 4. Surrey Beatty & Sons, Chipping Norton, Australia.

Roelants, K., and F. Bossuyt. 2005. Archaeobatrachian paraphyly and Pangaean diversification of crown-group frogs. Systematic Biology 54: 111–126.

Roelants, K., D. J. Gower, M. Wilkinson, S. P. Loader, S. D. Biju, K. Guillaume, L. Moriau, and F. Bossuyt. 2007. Global patterns of diversification in the history of modern amphibians. Proceedings of the National Academy of Sciences 104: 887–892.

Rogers, J. J. W., and M. Santosh. 2004. Continents and Supercontinents. Oxford University Press, Oxford.

Rohr, J. R., A. A. Elskus, B. S. Shepherd, P. H. Crowley, T. M. McCarthy, J. H. Niedzwiecki, T. Sager, A. Sih, and B. D. Palmer. 2004. Multiple stressors and salamanders: Effects of an herbicide, food limitation, and hydroperiod. Ecological Applications 14: 1028–1040.

Rolfe, W. D. I., E. N. K. Clarkson, and A. L. Panchen (eds.). 1994. Volcanism and Early Terrestrial Biotas. Royal Society of Edinburgh, Edinburgh.

Romer, A. S. 1930. The Pennsylvanian tetrapod of Linton, Ohio. Bulletin of the American Museum of Natural History 59: 77–147.

———. 1939. Notes on branchiosaurs. American Journal of Science 237: 748–761.

———. 1956. Osteology of the Reptiles. University of Chicago Press, Chicago.

———. 1957. The appendicular skeleton of the Permian embolomerous amphibian *Archeria*. Contributions of the Museum of Paleontology, University of Michigan 13: 103–159.

———. 1958. Tetrapod limbs and early tetrapod life. Evolution 12: 365–369.

———. 1963. The larger embolomerous amphibians of the American Carboniferous. Bulletin of the Museum of Comparative Zoology, Harvard 128: 415–454.

———. 1969. The cranial anatomy of the Permian amphibian *Pantylus*. Breviora 312: 1–37.

Romer, A. S., and T. S. Parsons. 1977. The Vertebrate Body. W. B. Saunders & Co., Philadelphia.

Romer, A. S., and L. L. Price. 1940. Review of the Pelycosauria. Geological Society of America Special Papers 28: 1–538.

Ron, S. R. 2005. Predicting the distribution of the amphibian pathogen *Batrachochytrium dendrobatidis* in the New World. Biotropica 37: 209–221.

Ron, S. R., W. A. Duellman, L. A. Coloma, and M. R. Bustamante. 2003. Population decline of the jambato toad *Atelopus ignescens* (Anura: Bufonidae) in the Andes of Ecuador. Journal of Herpetology 37: 116–126.

Rose, C. S. 2003. The developmental morphology of salamander skulls; pp. 1684–1781 *in* H. Heatwole and M. Davies (eds.), Amphibian Biology, vol. 5, Osteology. Surrey Beatty & Sons, Chipping Norton, Australia.

Rose, K., and D. Archibald (eds.). 2005. The Rise of Placental Mammals. Johns Hopkins University Press, Baltimore.

Rosing, M. T. 1999. C-13-depleted carbon microparticles in >3700-Ma sea-floor sedimentary rocks from west Greenland. Science 283: 674–676.

Roth, G., U. Dicke, and W. Wiggers. 1998. Vision; pp. 783–877 *in* H. Heatwole and E. M. Dawley (eds.), Amphibian Biology, vol. 3, Sensory Perception. Surrey Beatty & Sons, Chipping Norton, Australia.

Rowley, J. J. L., and R. A. Alford. 2006. The amphibian chytrid *Batrachochytrium dendrobatidis* occurs in freshwater shrimps in rainforest streams in northern Queensland, Australia. Ecohealth 3: 49–52.

Ruddle, R. H., J. L. Bartels, K. L. Bentley, C. Kappen, M. T. Murtha, and J. W. Pendleton. 1994. Evolution of *Hox* genes. Annual Review of Genetics 28: 423–442.

Russell, R.W., S. J. Hecnar, and, G. D. Haffner. 1995. Organochlorine pesticide residues in Southern Ontario spring peepers. Environmental Toxicology and Chemistry 14: 815–817.

Ruta, M., and J. A. Clack. 2006. A review of *Silvanerpeton miripedes*, a stem amniote from the Lower Carboniferous of East Kirkton, West Lothian, Scotland. Transactions of the Royal Society of Edinburgh, Earth Sciences 97: 31–63.

Ruta, M., M. I. Coates, and D. L. J. Quicke. 2003. Early tetrapod relationships revisited. Biological Reviews 78: 251–345.

Ruta, M., A. R. Milner, and M. I. Coates. 2002. The tetrapod *Caerorhachis bairdi* Holmes and Carroll from the Lower Carboniferous of Scotland. Transactions of the Royal Society of Edinburgh, Earth Sciences 92: 229–261.

San Mauro, D., V. Miguel, M. Alcobendas, R. Zardoya, and A. Meyer. 2005. Initial diversification of living amphibians predated the breakup of Pangaea. American Naturalist 65: 590–599.

Sanchiz, B. 1998. Salientia; pp. 1–275 *in* R. Wellnhofer (ed.), Handbuch der Paläoherpetologie, vol. 4. Verlag Dr. Friedrich Pfeil, München.

Sanderson, S. L., and J. J. Kupferberg. 1999. Development and evolution of aquatic larval feeding mechanisms; pp. 301–377 *in* B. K. Hall and M. H. Wake (eds.), Larval Forms. Academic Press, San Diego.

Sanz-Ezquerro, J. J., and C. Tickle. 2003. Fgf signaling controls the number of phalanges and tip formation in developing digits. Current Biology 13: 1830–1836.

Sarjeant, W. A. S. 1988. Fossil vertebrate footprints. Geology Today 4: 125–130 (and cover photograph).

Sarjeant, W. A. S., and J. D. Mossman 1987. Vertebrate footprints from the Carboniferous sediments of Nova Scotia: A historical review and description of newly discovered forms. Palaeogeography, Palaeoclimatology, Palaeoecology 23: 279–306.

Savage, J. M. 1966. An extraordinary new toad from Costa Rica. Revista de Biología Tropical 14: 153–167.

Säve-Söderbergh, G. 1932. Preliminary note on Devonian stegocephalians from East Greenland. Meddelelser on Grønland 98: 1–211.

Scheuchzer, J. J. 1726. Homo diluvii testis et theoskopos. Zurich.

Schmalhausen, I. I. 1968. The Origin of Terrestrial Vertebrates. Academic Press, New York.

Schmidt-Nielsen, K. 1989. Scaling: Why Is Animal Size So Important? Cambridge University Press, Cambridge.

Schoch, R. R. 1992. Comparative ontogeny of early Permian branchiosaurid amphibians from southwestern Germany: Developmental stages. Palaeontolographica A 222: 43–83.

———. 2004. Skeleton formation in the Branchiosauridae: A case study in comparing ontogenetic trajectories. Journal of Vertebrate Paleontology 24: 309–319.

Schoch, R. R., and R. L. Carroll. 2003. Ontogenetic evidence for the Paleozoic ancestry of salamanders. Evolution & Development 5: 314–324.

Schoch, R. R., and N. B. Fröbisch. 2006. Metamorphosis and neoteny: Alternative pathways in an extinct amphibian clade. Evolution 60: 1467–1475.

Schoch, R. R., and A. R. Milner. 2000. Stereospondyli; in P. Wellnhofer (ed.), Encyclopedia of Paleoherpetology. Verlag Dr. Friedrich Pfeil, München.

———. 2008. The intrarelationships and evolutionary history of the temnospondyl family Branchiosauridae; in Systematic Palaeontology (in press).

Schoch, R. R., and B. S. Rubidge. 2005. The amphibamid Micropholis from the Lystrosaurus assemblage zone of South Africa. Journal of Vertebrate Paleontology 25: 502–522.

Schopf, J. W. 1999. Cradle of Life. Princeton University Press, Princeton, New Jersey.

Schultze, H.-P. 1984. Juvenile specimens of Eusthenopteron foordi Whiteaves, 1881 (Osteolepiform rhipidistian, Pisces) from the Late Devonian of Miguasha, Quebec, Canada. Journal of Vertebrate Paleontology 4: 1–16.

———. 1996. The elpistostegid fish Elpistostega, the closest the Miguasha fauna comes to a tetrapod; pp. 316–327 in H.-P. Schultze and R. Cloutier (eds.), Devonian Fishes and Plants of Miguasha, Quebec, Canada. Verlag Dr. Friedrich Pfeil, München.

Schultze, H.-P., and R. Cloutier (eds.). 1996. Devonian Fishes and Plants of Miguasha, Quebec, Canada. Verlag Dr. Friedrich Pfeil, München.

Schuytema, G. S., and A. V. Nebeker. 1999. Comparative toxicity of ammonium and nitrate compounds to Pacific treefrog and African clawed frog tadpoles. Environmental Toxicology and Chemistry 18: 2251–2257.

Seilacher, A., L. A. Buatois, and M. G. Mángano. 2005. Trace fossils in the Ediacaran-Cambrian transition: Behavioral diversification, ecological turnover and environmental shift. Palaeogeography, Palaeoclimatology, Palaeoecology 227: 323–356.

Selden, P., and J. Nudds. 2004. Evolution of Fossil Ecosystems. University of Chicago Press, Chicago.

Sequeira, S. E. K. 1996. A cochleosaurid amphibian from the Upper Carboniferous of Ireland. Special Papers in Palaeontology 52: 65–80.

———. 1998. The cranial morphology and taxonomy of the saurerpetontid Isodectes obtusus comb. Nov. (Amphibia: Temnospondyli) from the Lower Permian of Texas. Zoological Journal of the Linnean Society 122: 237–259.

Sequeira, S. E. K., and A. R. Milner. 1993. The temnospondyl amphibian Capetus from the Czech Republic. Palaeontology 36: 657–680.

Shishkin, M. A. 1961. New data on Tupilakosaurus. Doklady Akademii Nauk SSSR 136: 938–941.

Shishkin, M. A., I. V. Novikov, and Y. M. Gubin. 2000. Permian and Triassic temnospondyls from Russia; pp. 33–59 in M. J. Benton, M. A. Shiskin, D. M. Unwin, and E. N. Kurochkin (eds.), The Age of Dinosaurs in Russia and Mongolia. Cambridge University Press, Cambridge.

Shu, D.-G., H.-L. Luo, S. Conway Morris, X.-L. Zhang, S.-X. Hu, L. Chen, J. Han, M. Zhus, Y. Li, and L.-Z. Chen. 1999. Lower Cambrian vertebrates from south China. Nature 402: 42–46.

Shubin, N. H., and P. Alberch. 1986. A morphogenetic approach to the origin and basic organization of the tetrapod limb. Evolutionary Biology 20: 319–387.

Shubin, N. H., E. B. Daeschler, and M. I. Coates. 2004. The early evolution of the tetrapod humerus. Science 304: 90–93.

Shubin, N. H., E. B. Daeschler, and F. A. Jenkins Jr. 2006. The pectoral fin of Tiktaalik roseae and the origin of the tetrapod limb. Nature 440: 764–771.

Shubin, N. H., and F. A. Jenkins, 1995. An Early Jurassic jumping frog. Nature 377: 49–52.

Shumway, W. 1940. Stages in the normal development of Rana pipiens. I. External form. Anatomical Record 78: 139–147.

Sidor, C. A., R. F. O'Keefe, R. Damiani, J. S. Steyer, R. M. H. Smith, H. C. E. Larsson, P. C. Sereno, O. Ide, and A. Maga. 2005. Permian tetrapods from the Sahara show climate-controlled endemism in Pangaea. Nature 434: 886–889.

Sigurdsen, T. 2008. The otic region of Doleserpeton (Temnospondyli) and its implication for the evolutionary origin of frogs. Zoological Journal of the Linnean Society 154: 738–751.

Skerrat, L. F., L. Berger, R. Speare, S. Cashins, K. R. McDonald, A. D. Phillott, H. B. Hines, and N. Kenyon. 2007. Spread of chytridiomycosis has caused the rapid global decline and extinction of frogs. Eco-Health 4: 125–134.

Smith, A. B., and K. J. Patterson. 2002. Dating the time of origin of major clades: Molecular clocks and the fossil record. Annual Reviews of Earth & Planetary Sciences 30: 65–88.

Smith, M. A., and D. M. Green. 2005. Are all amphibian populations metapopulations? Dispersal and the metapopulation paradigm in amphibian ecology. Ecography 28: 110–128.

Smithson, T. R. 1982. The cranial morphology of Greererpeton burkemorani Romer (Amphibia: Temnospondyli). Zoological Journal of the Linnean Society 76: 29–90.

———. 1985a. The morphology and relationships of the Carboniferous amphibian Eoherpeton watsoni Panchen. Zoological Journal of the Linnean Society 85: 317–410.

———. 1985b. Scottish Carboniferous amphibian localities. Scottish Journal of Geology 21: 123–142.

———. 1986. A new anthracosaur amphibian from the Carboniferous of Scotland. Palaeontology 29: 603–628.

———. 1994. Eldeceeon rolfei, a new reptiliomorph from the Viséan of East Kirkton, West Lothian, Scotland. Transactions of the Royal Society of Edinburgh, Earth Sciences 84: 377–382.

———. 2000. Anthracosaurs; pp. 1053–1063 in H. Heatwole and R. L. Carroll (eds.), Amphibian Biology, vol. 4, Palaeontology. Surrey Beatty & Sons, Chipping Norton, Australia.

Smithson, T. R., R. L. Carroll, A. L. Panchen, and S. M. Andrews. 1994. Westlothiana lizziae from the Viséan of East Kirkton, West Lothian, Scotland, and the amniote stem. Transactions of the Royal Society of Edinburgh, Earth Sciences 84: 383–412.

Sober, E. 1994. Let's razor Ockham's razor; pp. 136–157 in E. Sober (ed.), From a Biological Point of View. Cambridge University Press, Cambridge.

Sordino, P., F. van der Hoeven, and D. Duboule. 1995. Hox gene expression in teleost fins and the origin of vertebrate digits. Nature 375: 678–681.

Sparling, D. W., G. M. Fellers, and L. L. McConnell. 2001. Pesticides and amphibian population declines in California, USA. Environmental Toxicology and Chemistry 20: 1591–1595.

Špinar, Z. V. 1972. Tertiary Frogs from Central Europe. Academia Publishing House, Prague.

Spotila, J. R., M. P. O'Connor, and G. S. Bakken. 1992. Biophysics of heat and mass transfer; pp. 59–80 in M. E. Feder and W. W. Burggren (eds.), Environmental Physiology of the Amphibians. University of Chicago Press, Chicago.

Stallard, R. F. 2001. Possible environmental factors underlying amphibian decline in Eastern Puerto Rico: Analysis of U.S. government data archives. Conservation Biology 15: 943–953.

Stark, P. D., R. S. Ellis, J. Richard, J-P. Kneib, G. P. Smith, and M. R. Santos. 2007. A Keck survey for gravitationally lensed Lyα emitters in the redshift range 8.5 < z < 10.4: New constraints on the contribution of low-luminosity sources to cosmic reionization. The Astrophysical Journal 663: 10–28.

Steen, M. C. 1930. The British Museum Collection of Amphibia from the Middle Coal Measures of Linton, Ohio. Proceedings of the Zoological Society of London 45: 849–891 + 6 plates.

———. 1937. On Acanthostoma vorax Credner. Proceedings of the Zoological Society of London B: 491–500.

———. 1938. On the fossil amphibia from the Gas Coal of Nýřany, and other deposits in Czechoslovakia. Proceedings of the Zoological Society of London B 108: 205–283.

Stewart, M. M., 1995. Climate driven population fluctuations in rain forest frogs. Journal of Herpetology 29: 437–446.

Steyer, J. S., and R. Damiani. 2005. A giant brachyopoid temnospondyl from the Upper Triassic or Lower Jurassic of Lesotho. Bulletin de la Société Géologique de France 176: 243–348.

Steyer, J. S., R. Damiani, C. A. Sidor, F. R. O'Keefe, H. C. E. Larsson, A. Maga, and O. Ide. 2006. The vertebrate fauna of the Upper Permian of Niger. IV. Nigerpeton ricqlesi (Temnospondyli: Cochleosauridae), and the edopoid colonization of Gondwana. Journal of Vertebrate Paleontology 26: 18–28.

Storfer, A. 2003. Amphibian declines: Future directions. Diversity & Distributions 9: 151–163.

Strickberger, M. W. 2000. Evolution. 3d ed. Jones & Bartlett, Sudbury, Massachusetts.

Stuart, S., J. S. Chanson, N. A. Cox, B. E. Young, A. S. L. Rodrigues, D. L. Fishman, and R. W. Waller. 2004. Status and trends of amphibian declines and extinctions worldwide. Science 306: 1783–1786.

Stuart, S. N., M. Hoffman, J. S. Chanson, N. A. Cox, R. J. Berridge, P. Ramani, and B. E. Young. 2008. Threatened Amphibians of the World. Lynx Edicions, Barcelona, Spain.

Summers, A. P., and J. C. O'Reilly. 1997. A comparative study of locomotion in the caecilians Dermophis mexicanus and Typhlonectes natans (Amphibia: Gymnophiona). Zoological Journal of the Linnean Society 121: 65–76.

Swofford, D. L. 2001. PAUP*: Phylogenetic analysis using parsimony (*and other methods), version 4.0b10. Sinsauer Associates, Sunderland, Massachusetts.

Szarski, H. 1962. The origin of the Amphibia. Quarterly Review of Biology 37: 189–241.

———. 1968. The origin of vertebrate foetal membranes. Evolution 22: 211–214.

Taylor, E. H. 1968. The Caecilians of the World. University of Kansas Press, Lawrence.

Thomas, R. D. K., and E. C. Olson (eds.). 1980. A Cold Look at the Warm-Blooded Dinosaurs. AAAS Selected Symposium, 28. Westview Press, Boulder, Colorado.

Thomson, K. S. 1967. Mechanisms of intracranial kinetics in fossil rhipidistian fishes (Crossopterygii) and their relatives. Journal of the Linnean Society (Zoology) 46: 223–253.

Thomson, K. S., and K. H. Bossy. 1970. Adaptive trends and relationships in early Amphibia. forma et functio 3: 7–31.

Tice, M. M., and D. R. Lowe. 2004. Photosynthetic microbial mats in the 3,416-Myr-old ocean. Nature 431: 549–552.

Turner, S. 2004. Early vertebrates: Analysis from microfossil evidence; pp. 65–94 in G. Arratia, M. V. H. Wilson, and R. Cloutier (eds.), Recent Advances in the Origin and Early Radiation of Vertebrates. Verlag Dr. Friedrich Pfeil, München.

Turner, S., and R. F. Miller. 2005. New ideas about old sharks. American Scientist 93: 244–252.

Tyler, M. J. 1983. The Gastric Brooding Frog. Croom Helm, London.

Tyler, M. J., and M. Davies. 1985. The gastric brooding frog; pp. 469–470 in G. Grigg, R. Shine, and H. Ehmann (eds.), Biology of Australasian Frogs and Reptiles. Royal Zoological Society of New South Wales, Sydney.

Ultsch, G. R., D. F. Bradford, and J. Freda. 1999. Physiology: Coping with the environment; pp. 189–214 in R. W. McDiarmid and R. Altig (eds.), Tadpoles: The Biology of Anuran Larvae. University of Chicago Press, Chicago.

Urban, M. C., B. Philips, D. K. Skelly, and R. Shine. 2007. The cane toad's (Chaunus [Bufo] marinus) increasing ability to invade Australia is revealed by a dynamically updated range model. Proceedings of the Royal Society of London B 274: 1413–1419.

Valentine, J. W. 2004. On the Origin of Phyla. University of Chicago Press, Chicago.

Valiela, I., and P. Martinetto. 2007. Changes in bird abundance in eastern North America: Urban sprawl and global footprint? Bioscience 57: 360–370.

van Andel, T. H. 1994. New Views on an Old Planet. 2d ed. Cambridge University Press, Cambridge.

Vaughn, P. P. 1969. Further evidence of close relationship of the trematopsid amphibians with a description of a new genus and new species. Bulletin of the Southern California Academy of Sciences 68: 121–130.

Vial, J. L., and L. Saylor. 1993. The status of amphibian populations: A compilation and analysis. Declining Amphibian Populations Task Force Working Document no. 1. IUCN, Milton Keynes, United Kingdom.

Villeneuve, M. 2004. Radiogenic isotope geochronology; pp. 87–95 in F. Gradstein, J. Ogg, and A. Smith (eds.), A Geological Time Scale 2004. Cambridge University Press, Cambridge.

Vorobyeva, E. I. 1995. The shoulder girdle of Panderichthys rhombolepis (Gross) (Crossopterygii), Upper Devonian, Latvia. Geobios 19: 285–288.

Vorobyeva, E. I., and A. Kuznetsov. 1992. The locomotor apparatus of Panderichthys rhombolepis (Gross), a supplement to the problem of fish-tetrapod transition; pp. 131–140 in M. Mark-Kurik (ed.), Fossil Fishes as Living Animals. Tallin: Academy of Sciences of Estonia, Institute of Geology.

Vorobyeva, E. I., and H.-P. Schultze. 1991. Description and systematics of panderichthyid fishes with comments on their relationship to tetrapods; pp. 68–109 in H.-P. Schultze and L. Trueb (eds.), Origins of the Higher Groups of Tetrapods. Cornell University Press, Ithaca.

Vredenburg, V. T. 2004. Reversing introduced species effects: Experimental removal of introduced fish leads to rapid recovery of declining frog. Proceedings of the National Academy of Sciences 101: 7646–7650.

Wagner, G., and C.-H. Chiu. 2001. The tetrapod limb: A hypothesis on its origin. Journal of Experimental Zoology 291: 226–240.

Wagner, G., and H. Larsson. 2007. Fins and limbs in the study of evolutionary novelties; pp. 49–61 in B. K. Hall (ed.), Fins into Limbs. University of Chicago Press, Chicago.

Wake, D. B. 1991. Declining Amphibian Populations. Science (Washington, D.C.) 253: 860.

———. 2003. Foreword; pp. vii–x in R. D. Semlitsch (ed.), Amphibian Conservation. Smithsonian Institution, Washington, D.C.

Wake, D. B., and S. M. Deban. 2000. Terrestrial feeding in salamanders; pp. 95–116 in K. Schwenk (ed.), Feeding: Form, Function, and Evolution in Tetrapod Vertebrates. Academic Press, San Diego.

Wake, M. H. 1977. The reproductive biology of caecilians: An evolutionary perspective; pp. 73–102 *in* S. Guttman and D. Taylor (eds.), Reproductive Biology of the Amphibia. Plenum Press, New York.

———. 1989. Metamorphosis of the hyobranchial apparatus in *Epicrionops* (Amphibia: Gymnophiona: Rhinatrematidae): Replacement of bone by cartilage. Annales des Sciences Naturelles Zoologie 10: 171–182.

———. 1993. The evolution of oviductal gestation in amphibians. Journal of Experimental Zoology 266: 394–413.

———. 2003. The osteology of caecilians; pp. 1809–1876 *in* H. Heatwole and M. Davies (eds.), Amphibian Biology, vol. 5, Osteology. Surrey Beatty & Sons, Chipping Norton, Australia.

Wake, M. H., and R. Dickie. 1998. Oviduct structure and function and reproductive modes in amphibians. Journal of Experimental Zoology 282: 477–506.

Wake, M. H., and J. Hanken. 1982. The development of the skull of *Dermophis mexicanus* (Amphibia: gymnophiona), with comments on skull kinesis and amphibian relationships. Journal of Herpetology 19: 68–77.

Waldman, B., and M. Tocher. 1998. Behavioral ecology, genetic diversity and declining amphibian populations; pp. 394–443 *in* T. M. Caro (ed.), Behavioral Ecology and Conservation Biology. Oxford University Press, Oxford.

Wang, Y. 2000. A new salamander (Amphibia: Caudata) from the Early Cretaceous Jehol Biota. Vertebrata PalAsiata 38: 100–103.

———. 2004. A new Mesozoic caudate (*Liaoxitriton daohugouensis* sp. nov.) from Inner Mongolia, China. Chinese Science Bulletin 49: 858–860.

Wang, Y., and K.-q. Gao. 2003. Amphibians; pp. 77–85 *in* M.-m. Chang (ed.), The Jehol Biota. Shanghai Scientific & Technical Publishers, Shanghai.

Wang, Y., and C. S. Rose. 2005. *Jeholotriton paradoxus* (Amphibia: Caudata) from the Lower Cretaceous of southeastern Inner Mongolia, China. Journal of Vertebrate Paleontology 25: 523–532.

Warren, A. 2000. Secondarily aquatic temnospondyls of the Upper Permian and Mesozoic; pp. 1121–1149 *in* H. Heatwole and R. L. Carroll (eds.), Amphibian Biology, vol. 4. Surrey Beatty & Sons. Chipping Norton, Australia.

———. 2007 New data of *Ossinodus pueri*, a stem tetrapod from the Early Carboniferous of Australia. Journal of Vertebrate Paleontology 27: 850–862.

Warren, A., and M. N. Hutchinson. 1983. The last labyrinthodont? A new brachyopoid (Amphibia, Temnospondyli) from the early Jurassic Evergreen formation of Queensland, Australia. Philosophical Transactions of the Royal Society London B 303: 1–62.

Warren, A., T. H. Rich, and P. Vickers-Rich. 1997. The last last labyrinthodonts? Palaeontographica Abteilung A 247: 1–24.

Warren, A., and S. Turner. 2004. The first stem tetrapod from the Lower Carboniferous of Gondwana. Palaeontology 47: 151–184.

Warren, M. S. 2001. Rapid responses of British butterflies to opposing forces of climate and habitat change. Nature 414: 65.

Wassersug, R. J., and K. Rosenberg. 1979. Surface anatomy of branchial food traps of tadpoles: A comparative study. Journal of Morphology 159: 393–426.

Watson, D. M. S. 1940. The origin of frogs. Transactions of the Royal Society of Edinburgh 60: 195–231.

Watson, J. D., and F. H. C. Crick 1953. Molecular structure of nucleic acids. Nature 171: 737.

Weisrock, D. W., L. J. Harmon, and A. Larson. 2005. Resolving deep phylogenetic relationships in salamanders: Analyses of mitochondrial and nuclear genomic data. Systematic Biology 54: 758–777.

Welch, J. J., E. Fontanillas, and L. Bromham. 2005. Molecular dates for the "Cambrian Explosion": The influence of prior assumptions. Systematic Biology 54: 672–678.

Weldon, C., L. H. du Preez, A. D. Hyatt, R. Muller, and R. Speare. 2004. Origin of the amphibian chytrid fungus. Emerging Infectious Diseases 10: 2100–2105.

Wellstead, C. F. 1982. A Lower Carboniferous aïstopod amphibian from Scotland. Palaeontology 25: 193–208.

———. 1991. Taxonomic revision of the Lysorophia, Permo-Carboniferous lepospondyl amphibians. Bulletin of the American Museum of Natural History 209:1–90.

———. 1998. Order Lysorophia; pp. 133–148 *in* P. Wellnhofer (ed.), Encyclopedia of Paleoherpetology, vol. 1. Verlag Dr. Friedrich Pfeil, München.

Werneburg, R. 1998. Ein larvaler *Acanthostomatops* (Zatrachydidae, Amphibia) aus der Neiderhäslich-Formation (Unter-Perm) des Döhlen-Beckens. Veröff. Museum für Naturkunde Chemnitz 27: 49–52.

Werneburg, R., and S. Lucas. 2007. "Milnererpeton" from the Late Pennsylvanian of New Mexico is the first truly "European branchiosaurid" from North America. Journal of Vertebrate Paleontology Program and Abstracts 27: 164A.

Werneburg, R., and J. W. Schneider. 2006. Amphibian biostratigraphy of the European Permo-Carboniferous; pp. 201–215 *in* S. G. Lucas, G. Cassinis, and J. W. Schneider (eds.), Non-Marine Permian Biostratigraphy and Biochronology. Geological Society, London, Special Publications.

Whiles, M. R., K. R. Lips, C. M. Pringle, S. S. Kilham, R. J. Bixby, R. Brenes, S. Connelly, J. Checo Colon-Gaud, M. Hunte-Brown, C. Montgomery, and S. Peterson. 2006. The effects of amphibian population declines on the structure and function of Neotropical stream ecosystems. Frontiers in Ecology and Environment 4: 27–34.

White, M. E. 1990. The Flowering of Gondwana. Princeton University Press, Princeton, New Jersey.

Whiteman, H. H. 1994. Evolution of facultative paedomorphosis in salamanders. Quarterly Review of Biology 69: 205–221.

Wiens, J. J., R. M. Bonett, and P. T. Chippindale. 2005. Ontogeny discombobulates phylogeny: Paedomorphosis and higher-level salamander relationships. Systematic Biology 54: 91–110.

Wilkinson, M., and R. A. Nussbaum. 1997a. Comparative morphology and evolution of the lungless caecilian *Atretochoana eiselti* (Taylor) (Amphibia: Gymnophiona: Typhlonectidae). Biological Journal of the Linnean Society 62: 39–109.

———. 1997b. Evolutionary relationships of the lungless caecilian *Atretochoana eiselti* (Taylor) (Amphibia: Gymnophiona: Typhlonectidae). Biological Journal of the Linnean Society 62: 319–109.

Williston, S. W. 1910. *Cacops, Desmatospondylus:* New genera of Permian vertebrates. Bulletin of the Geological Society of America 21: 249–284.

Wilson, H. M., and L. I. Anderson. 2004. Morphology and taxonomy of Paleozoic millipedes (Diplopoda: Chilognatha: Archipolypoda) from Scotland. Journal of Paleontology 78: 169–184.

Wilson, T. J. 1965. A new class of faults and their bearing on continental drift. Nature 207: 343–347.

Witzmann, F. 2006. Developmental patterns and ossification sequence in the Permo-Carboniferous temnospondyl *Archegosaurus decheni* (Saar-Nahe basin, Germany). Journal of Vertebrate Paleontology 26: 7–17.

———. 2007. The evolution of the scalation pattern in temnospondyl amphibians. Zoological Journal of the Linnean Society 150: 815–834.

Witzmann, F., and H.-U. Pfretzschner. 2003. Larval ontogeny of *Micromelerpeton credneri* (Temnospondyli, Dissorophoidea). Journal of Vertebrate Paleontology 23: 750–768.

Woolbright, L. L., A. H. Hara, C. M. Jacobsen, W. J. Mautz, and F. L. Benevides Jr. 2006. Population densities of the coquí, *Eleutherdactylus coqui* (Anura: Leptodactylidae) in newly invaded Hawaii and in native Puerto Rico. Journal of Herpetology 40: 122–126.

Wray, G. A., J. S. Levinton, and L. H. Shapiro. 1996. Molecular evidence for deep Precambrian divergences among metazoan phyla. Science 274: 568–573.

Wright, E. P., and T. H. Huxley. 1866. On a collection of fossils from the Jarrow Colliery, Kilkenny. Geological Magazine 3: 165–171.

Xiao, S., and A. H. Knoll. 2000. Phosphatized animal embryos from the Neoproterozoic Doushantuo Formation at Weng'an, Guizhou, South China. Journal of Paleontology 74: 767–788.

Yates, A. M., and A. Warren. 2000. The phylogeny of the "higher" temnospondyls (Vertebrata: Choanata) and its implications for the monophyly and origins of the Stereospondyli. Zoological Journal of the Linnean Society 128: 77–121.

Yee, A. R., and J. W. Kronstad. 1998. Dual sets of chimeric alleles identify specificity sequences for the *bE* and the *bW* mating and pathogenicity genes of *Ustilago maydis*. American Society for Microbiology 18: 221–232.

Yoffe, E. 1992. Silence of the frogs. New York Times Magazine, December 13, 1992: 36–38, 64, 66, 76.

Young, A. G., and G. M. Clarke (eds.). 2000. Genetics, Demography, and Viability of Fragmented Populations. Cambridge University Press, Cambridge.

Young, G. C., V. N. Karatajute-Talimaa, and M. M. Smith. 1996. A possible Late Cambrian vertebrate from Australia. Nature 383: 810–812.

Yuan, C., H. Zhang, L. Ming, and J. Xinxin. 2004. Discovery of a Middle Jurassic fossil tadpole from Daohugou region, Ningcheng, Inner Mongolia, China. Acta Geologica Sinica 78: 145–149.

Zangerl, R. 1981. Chondrichthyes I: Paleozoic Elasmobranchii. in H-P Schultze and O Kuhn (eds.), Handbook of Paleoichthyology, vol. 3B. GV Verlag, Berlin.

Zhang, P., Y.-Q. Chen, H. Zhou, Y.-F. Liu, X.-L. Wang, T. J. Paperfuss, D. B. Wake, and L.-H. Qu. 2006. Phylogeny, evolution, and biogeography of Asiatic salamanders (Hynobiidae). Proceedings of the National Academy of Science 103: 7360–7635.

Zhang, P., H. Z. Zhou, Y.-Q. Chen, Y.- F. Liu, and L.-H. Qu. 2005. Mitogenomic perspectives on the origin and phylogeny of living amphibians. Systematic Biology 54: 391–400.

Zhu, M., and P. E. Ahlberg. 2004. The origin of the internal nostril of tetrapods. Nature 432: 94–97.

Zhu, M., P. E. Ahlberg, W. Zhao, and L. Jia. 2002. First Devonian tetrapod from Asia. Nature 420: 760–761.

Zhu, M., and X. Yu. 2004. Lower jaw character transitions among major sarcopterygian groups—a survey based on new materials from Yunnan, China; pp. 271–286 *in* G. Arratia, M. V. H. Wilson, and R. Cloutier (eds.), Recent Advances in the Origin and Early Radiation of Vertebrates. Verlag Dr. Friedrich Pfeil, München.

Zittel, K. A. von. 1890. Handbüch der Palaeontologie. III. Band Vertebrata (Pisces, Amphibia, Reptilia, Aves). München.

Zubay, G. 2000. Origins of Life on the Earth and in the Cosmos. 2d ed. Academic Press, San Diego.

INDEX